Fermentation
and
Biochemical Engineering

Volume I

Contents at a Glance

Fermentation
and
Biochemical Engineering

Volume I

K M Richard

S R Durbin

CBS

CBS Publishers & Distributors Pvt Ltd

New Delhi • Bengaluru • Chennai • Kochi • Kolkata • Mumbai
Bhopal • Bhubaneswar • Hyderabad • Jharkhand • Nagpur • Patna • Pune • Uttarakhand • Dhaka (Bangladesh)

Fermentation
and
Biochemical Engineering

Volume I

ISBN: 978-93-89185-91-1

Copyright © Authors and Publisher

First Edition: 2020

Published by Satish Kumar Jain and produced by Varun Jain for

CBS Publishers & Distributors Pvt Ltd

4819/XI Prahlad Street, 24 Ansari Road, Daryaganj, New Delhi 110 002, India.

Ph: 23289259, 23266861, 23266867   Fax: 011-23243014   Website: www.cbspd.com

e-mail: delhi@cbspd.com; cbspubs@airtelmail.in.

Corporate Office: 204 FIE, Industrial Area, Patparganj, Delhi 110 092

Ph: 4934 4934   Fax: 4934 4935   e-mail: publishing@cbspd.com; publicity@cbspd.com

Branches

- **Bengaluru:** Seema House 2975, 17th Cross, K.R. Road,
  Banasankari 2nd Stage, Bengaluru 560 070, Karnataka
  Ph: +91-80-26771678/79   Fax: +91-80-26771680   e-mail: bangalore@cbspd.com
- **Chennai:** 7, Subbaraya Street, Shenoy Nagar, Chennai 600 030, Tamil Nadu
  Ph: +91-44-26680620, 26681266   Fax: +91-44-42032115   e-mail: chennai@cbspd.com
- **Kochi:** 42/1325, 1326, Power House Road, Opp KSEB Power House,
  Ernakulam 682 018, Kochi, Kerala
  Ph: +91-484-4059061-65   Fax: +91-484-4059065   e-mail: kochi@cbspd.com
- **Kolkata:** 6/B, Ground Floor, Rameswar Shaw Road, Kolkata-700 014, West Bengal
  Ph: +91-33-22891126, 22891127, 22891128   e-mail: kolkata@cbspd.com
- **Mumbai:** 83-C, Dr E Moses Road, Worli, Mumbai-400018, Maharashtra
  Ph: +91-22-24902340/41   Fax: +91-22-24902342   e-mail: mumbai@cbspd.com

Representatives

• Bhopal	0-8319310552	• Bhubaneswar	0-9911037372	• Hyderabad	0-9885175004
• Jharkhand	0-9811541605	• Nagpur	0-9421945513	• Patna	0-9334159340
• Pune	0-9623451994	• Uttarakhand	0-9716462459	• Dhaka (Bangladesh)	01912-003485

Printed at: Glorious Printers, Daryaganj, Delhi, India

Preface

In broader terms, biotechnology is defined as the use of biological systems, living organisms or derivatives thereof, to make or modify products or processes for specific use. Apart from referring to a type of energy metabolism, fermentation in the industrial sense is regarded as any process for the production of various chemical or pharmaceutical compounds by means of the mass cultivation of micro-organisms.

The biochemical engineering approach to fermentation has been significant as the engineers have always been engaged in commercial fermentation operations. Perhaps the easiest way to assess and illustrate the role of biochemical engineering in fermentation technology is to first summarise its contributions in various aspects. Biochemical engineering contributions to fermentation technology can be looked at in many different ways. We can go through the characteristic fermentation process flowsheet and look at the main stages: (i) medium preparation and sterilisation, (ii) inoculum preparation, (iii) reaction (fermentation), and (iv) pretreatment for recovery. Alternatively we can adopt a unit operations approach and collectively examine all activities which have a common basis, heat sterilisation of media, aseptic transfer of fluids, mass transfer (aeration) and so forth.

This reference textbook on *Fermentation and Biochemical Engineering* is divided in two volumes. First volume contains seven sections and 1 to 33 chapters.

Section I discusses general considerations and biological aspects. Chapter 1 is devoted to historical perspective of fermentation and discusses various chronological development from the early 1700 when wooden vats of 1500 barrel capacity was introduced. Chapter 2 deals with fermentation feedstocks such as micro-organisms and nutrient sources. Chapter 3 concentrates on fermentation biotechnology: an overview and discusses basic concept of fermetation. Chapter 4 focuses on microbial growth kinetics. Microbial growth is described as an orderly increase in all chemical components in the presence of suitable medium and the culture environment. There are four types of microbial growth: bacteria grow by binary fission, yeast divide by budding, fungi divided by chain elongation and branching and viruses grow intracellularly in host cells.

Section II discusses industrial fermentation and solid state fermentation. Chapter 5 is devoted to microbiological aspects of solid substrate fermentation. Solid state fermentations (SSF) have attracted a renewed interest and attention from researchers due to recent developments in the field of microbial biotechnology. Chapter 6 deal with solid state fermentation for bioconversion of biomass and agricultural. This chapter discusses some important aspects of solid-state cultivation system, including the variety of substrates and micro-organisms used in SSF for the production of various end products, and the performance control of system by regulation of important factors. Chapter 7 focuses on engineering aspects of solid state fermentation. This chapter discusses the various micro- and macro- level engineering problems associated with SSF and some possible solutions for its full commercial realisation. Chapter 8 concentrates on microbial solid state fermentation for future biorefineries. Today's biorefinery technologies would be almost unthinkable without biotechnology. Novel biorefinery processes using solid state fermentation (SSF) technology have been developed as an alternative to conventional processing routes, leading to the production of added-value products from agriculture and food industry raw materials. Chapter 9 explains industrial fermentation processes.

Section III discusses biological basis of productivity in fermentation. Chapter 10 is devoted to isolation, preservation and improvement of important micro-organisms. The term isolation refers to the separation of a strain from a natural, mixed population of living microbes, as present in the environment, for example in water or soil flora, or from living beings with skin flora, oral flora or gut flora, in order to identify the microbe(s) of interest. Chapter 11 deal with industrial media and the nutrition of industrial organism. The use of a good, adequate, and industrially usable medium is as important as the deployment of a suitable micro-organism in industrial microbiology. Unless the medium is adequate, no matter how innately productive the organism is, it will not be possible to harness the organism's full industrial potentials. Chapter 12 focuses on sterilisation techniques in fermentation processes. Sterilisation is a technique to make anything free from organisms either by removing them or killing them. The removal or killing of all the organisms from fermentation medium is the main aim of the sterilisation process or else the contaminant will deteriorate the process. Chapter 13 concentrates on development of inocula for industrial fermentations. Inoculum is a small amount of material containing bacteria, viruses, or other micro-organisms that is used to start a culture.

Section IV discusses designing aspects of fermentator. Chapter 14 is devoted to fermentation monitoring and optimisation. Chapter 15 deals with designing parameters of fermentor. The function of the fermenter or bioreactor is to provide a suitable environment in which an organism can efficiently produce a target product—the target product might be cell biomass, metabolite and bioconversion product. It must be so designed that it is able to provide the optimum environments or conditions that will allow supporting the growth of the micro-organisms. Chapter 16 focuses on bioreactor design. In any fermentation process, the bioreactor plays a central role in determining the process efficiency. Even with recombinant products where stringent quality control implies that downstream processing is the major cost component, it is the bioreactor performance which determines product yields. Chapter 17 concentrates on aeration and agitation. The main function of aeration is to supply enough oxygen to the microbes in submerge culture technique for proper metabolism, while agitation provides proper mixing of the nutrient so that each and every organisms get proper nutrients. The main aim of the agitator is to provide homogenous environment all over the fermenter. It is also used for mixing of different phases, oxygen and heat transport.

Section V discusses biosensors and instrumentation and control systems. Chapter 18 provides information biosensors and nanobiosensors: design and applications. Biosensors are the device in which there is a coupling of biological sensing element with a detector system using a transducer. In comparison with any other currently available diagnostic device, biosensors are much higher in performance in terms of sensitivity and selectivity both. Advances in nanotechnology have led to the development of nanoscale biosensors that have exquisite sensitivity and versatility. The ultimate goal of nanobiosensors is to detect any biochemical and biophysical signal associated with a specific disease at the level of a single molecule or cell. They can be integrated into other technologies such as lab-on-a-chip to facilitate molecular diagnostics. Chapter 19 explains instrumentation and control systems and discusses the methods of measuring process variables along with on-line analysis of other chemical factors and control systems.

Section VI discusses enzymes and their importance in bioprocesses. Chapter 20 is devoted to characteristics of enzymes. Chapter 21 deals with production of industrial enzymes. Fermentation is a method of generating enzymes for industrial purposes. Fermentation involves the use of micro-organisms, like bacteria and yeast to produce the enzymes. Chapter 22 concentrates on fungal laccase enzyme for biotechnological application. Laccase belongs to the small group of enzymes called the blue multi copper oxidases. Chapter 23 focuses on enzymes in biosynthesis of nanoparticles. While a large number of microbial species are capable of producing metal nanoparticles (NPs), mechanism of nanoparticle biosynthesis is very important. Chapter 24 concentrates on nanoparticles in enzyme immobilisation. Chitosan nano-particles due to their highest specific surface area are much proper for immobilisation of higher amount of enzymes.

Section VII discusses recovery and purification of fermentation products. Chapter 25 is devoted to downstream processing: a review. Downstream processing refers to the recovery and purification of biosynthetic products, particularly pharmaceuticals, from natural sources such as animal or plant tissue or fermentation broth, including the recycling of salvageable components and the proper treatment and disposal of waste. Chapter 26 deals with solid-liquid separation which are used for clarification of liquids, solid recovery, dewatering of solids, thickening of slurries and washing of solids. Chapter 27 focuses on aqueous two-phase extraction systems which offers a suitable environment for protein separation because of the significant presence of water at all stages of the process, an important requirement for the maintenance of enzymic activity. Chapter 28 concentrates on chromatography which lies at the core of all biotechnology purification processes.

Chapter 29 explains membrane separation processes. Chapter 30 is devoted to affinity precipitation which is a relatively simple, convenient and reproducible technique that results in high target molecule recovery at high specificity. Chapter 31 deals with solvent extraction which is usually used to recover a component from either a solid or liquid. Chapter 32 concentrates on drying and crystallisation which involves removal of moisture from solids, solutions, slurries and pastes to give solid products, which often after drying are final products to be packed. Chapter 33 explains electrokinetic separation processes for biochemical products. The major recent developments in using electrokinetic processes for separation in biological system appear to have centred on the recovery and separation of high molecular weight, low volume, high value products such as therapeutic and diagnostic enzymes and proteins by some form of electrophoretic process.

Diagrams, figures, tables and index supplement the text. All topics have been covered in a cogent and lucid style to help the reader grasp the information quickly and easily.

It may not be wrong to hold that the present reference textbook of *Fermentation and Biochemical Engineering* is a complete treatise on this subject. It is essential reading for B Tech (environmental biotechnology/microbiology/food microbiology/biomedical and biochemical engineering) and students pursuing BSc/MSc course in Biotechnology and Microbiology. Besides students, this book will prove useful to industrialists and consultants in the respective fields.

This reference textbook also caters to the requirement of the syllabus prescribed by various universities for undergraduate and postgraduate courses in the above subjects. It has been prepared with meticulous care, aiming at making the book error-free. Constructive suggestions are always welcome from users of this book.

K M Richard

S R Durbin

Contents

Section III
BIOLOGICAL BASIS OF PRODUCTIVITY IN FERMENTATION

Section IV
DESIGNING ASPECTS OF FERMENTATOR

Section VII
RECOVERY AND PURIFICATION OF FERMENTATION PRODUCTS

SECTION I

General Consideration and Biological Aspects

Chapter 1

Historical Perspective of Fermentation

INTRODUCTION

Fermentation is a metabolic process that produces chemical changes in organic substrates through the action of enzymes. In biochemistry, it is narrowly defined as the extraction of energy from carbohydrates in the absence of oxygen. In the context of food production, it may more broadly refer to any process in which the activity of micro-organisms brings about a desirable change to a foodstuff or beverage.

The science of fermentation is known as zymology. In micro-organisms, fermentation is the primary means of producing ATP by the degradation of organic nutrients anaerobically. Humans have used fermentation to produce foodstuffs and beverages since the Neolithic age. For example, fermentation is used for preservation in a process that produces lactic acid found in such sour foods as pickled cucumbers, kimchi, and yogurt, as well as for producing alcoholic beverages such as wine and beer. Fermentation occurs within the gastrointestinal tracts of all animals, including humans.

BIOCHEMICAL ENGINEERING

Biochemical engineering, also known as bioprocess engineering, is a field of study with roots stemming from chemical engineering and biological engineering. It mainly deals with the design, construction, and advancement of unit processes that involve biological organisms or organic molecules and has various applications in areas of interest such as biofuels, food, pharmaceuticals, biotechnology, and water treatment processes. The role of a biochemical engineer is to take findings developed by biologists and chemists in a laboratory and translate that to a large-scale manufacturing process.

CHRONOLOGICAL DEVELOPMENT OF THE FERMENTATION INDUSTRY

The chronological development of the fermentation industry may be represented as five overlapping stages as illustrated in Table 1.1. The development of the industry prior to 1900 is represented by stage 1, where the products were confined to potable alcohol and vinegar. Although beer was first brewed by the ancient Egyptians, the first true large-scale breweries date from the early 1700s when wooden vats of 1500 barrels capacity were introduced. Even some process control was attempted in these early breweries, as indicated by the recorded use of thermometers in 1757 and the development of primitive heat

Table 1.1: The stages in the chronological development of the fermentation industry.

Stage	Main products	Vessels	Process control	Culture method	Quality control	Pilot plant	Strain selection
1. Pre-1900	Alcohol	Wooden, up to 1500 barrels capacity Copper used in later breweries	Use of thermometer, hydrometer and heat exchangers	Batch	Virtually nil	Nil	Pure yeast cultures used at the Carlsberg brewery (1886)
	Vinegar	Barrels, shallow trays, trickle filters		Batch	Virtually nil	Nil	Fermentations inoculated with 'good' vinegar
2. 1900–1940	Bakers' yeast glycerol, citric acid, lactic acid and acetone/ butanol	Steel vessels of up to 200 m³ for acetone/butanol Air spargers used for bakers' yeast. Mechanical stirring used in small vessels	pH electrodes with off-line control Temperature control	Batch and fed-batch system	Virtually nil	Virtually nil	Pure cultures used
3. 1940–date	Penicilling streptomycin, other antibiotics, gibberelin, amino acids, nucleotides, transformations, enzymes	Mechanically aerated vessels, operated aseptically—true fermenters	Sterilisable pH and oxygen electrodes. Use of control loops which were later computerised	Batch and fed-batch common. Continuous culture introduced for brewing and some primary metabolites	Very important	Becomes common	Mutation and selection programmes essential
4. 1964–date	Single-cell protein using hydrocarbon and other feedstocks	Pressure cycle and pressure jet vessels developed to overcome gas and heat exchange problems	Use of computer linked control loops	Continuous culture with medium recycle	Very important	Very important	Genetic engineering of producer strains attempted
5. 1979–date	Production of heterologous proteins by microbial and animal cells. Monoclonal antibodies produced by animal cells	Fermenters developed in stages 3 and 4. Animal cells reactors developed	Control and sensors developed in stages 3 and 4	Batch, fed-batch or continuous. Continuous perfusion developed for animal cell processes	Very important	Very important	Introduction of foreing genes into microbial and animal cell hosts. In vitro recombinant DNA techniques used in the improvement of stage 3 products

exchangers in 1806. By the mid-1800s the role of yeasts in alcoholic fermentation had been demonstrated independently by Schwann and Kutzing but it was Pasteur who eventually convinced the scientific world of the obligatory role of these micro-organisms in the process. During the late 1800s Hansen started his pioneering work at the Carlsberg brewery and developed methods for isolating and propagating single yeast cells to produce pure cultures and established sophisticated techniques for the production of starter cultures. However, use of pure cultures did not spread to the British ale breweries and it is true to say that many of the small, traditional, ale-producing breweries still use mixed yeast cultures at the present time but, nevertheless, succeed in producing high quality products.

Vinegar: Vinegar was originally produced by leaving wine in shallow bowls or partially filled barrels where it was slowly oxidised to vinegar by the development of a natural flora. The appreciation of the importance of air in the process eventually led to the development of the 'generator' which consisted of a vessel packed with an inert material (such as coke, charcoal and various types of wood shavings) over which the wine or beer was allowed to trickle. The vinegar generator may be considered as the first 'aerobic' fermenter to be developed. By the late 1800s to early 1900s the initial medium was being pasteurised and inoculated with 10 per cent good vinegar to make it acidic and therefore resistant to contamination, as well as providing a good inoculum. Thus, by the beginning of the twentieth century the concepts of process control were well established in both the brewing and vinegar industries.

Between the years 1900 and 1940 the main new products were yeast biomass, glycerol, citric acid, lactic acid, acetone and butanol. Probably the most important advances during this period were the developments in the bakers' yeast and solvent fermentations. The production of bakers' yeast is an aerobic process and it was soon recognised that the rapid growth of yeast cells in a rich wort led to oxygen depletion in the medium which, in turn, resulted in ethanol production at the expense of biomass formation. The problem was minimised by restricting the initial wort concentration such that the growth of the cells was limited by the availability of the carbon source rather than oxygen. Subsequent growth of the culture was then controlled by adding further wort in small amounts. This technique is now called fed-batch culture and is widely used in the fermentation industry to avoid conditions of oxygen limitation. The aeration of these early yeast cultures was also improved by the introduction of air through sparging tubes which could be steam cleaned.

Acetone-butanol fermentation: The development of the acetone-butanol fermentation during the First World War by the pioneering efforts of Weizmann led to the establishment of the first truly aseptic fermentation. All the processes discussed so far could be conducted with relatively little contamination provided that a good inoculum was used and reasonable standards of hygiene employed. However, the anaerobic butanol fermentation was susceptible to contamination in the early stages by aerobic bacteria and by acid-producing anaerobic ones, once anaerobic conditions had been established in the later stages of the process. The fermenters employed were vertical cylinders with hemispherical tops and bottoms constructed from mild steel. They could be steam sterilised under pressure and were constructed to minimise the possibility of contamination. Two-thousand-hecta-litre fermenters were commissioned which presented the problems fermentation technology and paved the way for the successful introduction of aseptic aerobic processes in the 1940s.

Penicillin: The third stage of the development of the fermentation industry arose as a result of the wartime need to produce penicillin in submerged culture under aseptic conditions. The production of penicillin is an aerobic process which is very vulnerable to contamination. Thus, although the knowledge gained from the solvent fermentations was exceptionally valuable, the problems of sparging a culture with large volumes of sterile air and mixing a highly viscous broth had to be overcome. Also, unlike the solvent fermentations, penicillin was synthesised in very small quantities by the initial isolates and this resulted in the establishment of strain-improvement programmes which became a dominant feature of the industry in subsequent years. Process development was also aided by the introduction of pilot-plant facilities which enabled the testing of new techniques on a semi-production scale. The development of a large-scale extraction process for the recovery of penicillin was another major advance at this time. The technology established for penicillin fermentation provided the basis for the development of a wide range of new processes. This was probably the stage when the most significant changes in fermentation technology took place resulting in the establishment of many new processes over the period, including other antibiotics, vitamins, gibberellin, amino acids, enzymes and steroid transformations. From 1960s onwards microbial products were screened for activities other than simply antimicrobial properties and screens became more and more sophisticated. These screens have evolved into those operating today utilising miniaturised culture systems, robotic automation and elegant assays.

Microbial biomass: In the early 1960s the decisions of several multi-national companies to investigate the production of microbial biomass as a source of feed protein led to a number of developments which may be regarded as the fourth stage in the progress of the industry. The largest mechanically stirred fermentation vessels developed during stage 3 were in the range 80,000 to 1,50,000 dm^3. However, the relatively low selling price of microbial biomass necessitated its production in much larger quantities than other fermentation products in order for the process to be profitable. Also, hydrocarbons were considered as potential carbon sources which would result in increased oxygen demands and high heat outputs by these fermentations. These requirements led to the development of the pressure jet and pressure cycle fermenters which eliminated the need for mechanical stirring. Another feature of these potential process was that they would have to be operated continuously if they were to be economic. At this time batch and fed-batch processes were common in the industry but the technique of growing an organism continuously by adding fresh medium to the vessel and removing culture fluid had been applied only to a very limited extent on a large scale. The brewers were also investigating the potential of continuous culture at this time, but its application in that industry was short-lived. Several companies persevered in the biomass field and a few processes came to fruition, of which the most long-lived was the ICI Pruteen animal feed process which utilised a continuous 30,00,000 dm^3 pressure cycle fermenter for the culture of *Methylophilus methylotrophus* with methanol as carbon source.

The operation of an extremely large continuous fermenter for time periods in excess of 100 days presented a considerable aseptic operation problem, far greater than that faced by the antibiotic industry in the 1940s. The aseptic operation of fermenters of this type was achieved as a result of the high standards of fermenter construction, the continuous sterilisation of feed streams and the utilisation of computer systems to control the sterilisation and operation cycles, thus minimising the possibility of human error. However, although the Pruteen process was a technological triumph it became an economic failure because the product was out-priced by soyabean and fishmeal. Eventually, in 1989, the plant was demolished, marking the end of a short, but very exciting, era in the fermentation industry. Whilst biomass is a very low-value, high-volume product, the fifth stage in the progress of the industry resulted in the establishment of very high-value, low-volume products. The development in *in vitro* genetic manipulation, commonly known as genetic engineering, enabled the expression of human and mammalian genes in micro-organisms, thereby enabling the large scale production of human proteins which could then be used therapeutically. According to Dykes, it was the small, venture-capital biotechnology companies that pioneered the development of heterologous proteins for therapeutic use. The established pharmaceutical companies used the new genetic engineering techniques to help in the discovery of natural products and in the rational design of drugs, for example, mammalian receptor proteins have been cloned and used in *in vitro* detection systems.

Insulin and human growth hormone have been the two most successful products but other products have far greater potential. Erythropoietin and the myeloid colony stimulating factors (CSFs) control the product of blood cells by stimulating the proliferation, differentiation and activation of specific cell types. Erythropoietin has been used to treat renal-failure anaemia and may have application in the treatment of the platelet deficiency associated with cancer chemotherapy, it is expected to become the top-selling therapeutic protein. Granulocyte-colony stimulating factor (G-CSF) is used during cancer chemotherapy. A number of different growth factors are involved in wound healing and recombinant forms of these proteins would be expected to yield significant returns in the coming years.

Recombinant proteins: The commercial exploitation of recombinant proteins has necessitated the design of contained production facilities. Thus, these processes are drawing on the experience of vaccine fermentations where pathogenic organisms have been grown on relatively large scales. Also, recombinant proteins have been classified as biologicals, not as drugs and thus come under the same regulatory authorities as do vaccines. The major difference between the approval of drugs and biologicals is that the process for the production of a biological must be precisely specified and carried out in a facility that has been inspected and licensed by the regulatory authority, which is not the case for the production of drugs (antibiotics, for example).

Thus, any changes which a manufacturer wishes to incorporate into a licensed process must receive regulatory approval. For drugs, only major changes require approval prior to implementation. The result of these containment and regulatory requirements is that the cost of developing a recombinant protein process is extremely high. Buckland illustrated this point in his claim that 'It now costs as much to build a 3000 dm^3 scale facility for Biologics as for a 200,000 dm^3 scale facility for an antibiotic. Also, even though titres are now reasonable for a recombinant protein (1 g dm^{-3}), the cost of manufacture kg^{-1} of bulk drug is about two orders of magnitude higher than that of an antibiotic at 10 g dm^{-3} titre'. Also, the development time for a recombinant protein is considerably longer than that for an antibiotic. For example, Bader claimed that it is feasible for an antibiotic plant to begin production four years after the initiation of the plant design whereas seven years would be required before a recombinant protein could be produced.

Genetic engineering: The exploitation of genetic engineering approximately coincided with another major development in biotechnology which influenced the progress of the fermentation industry—the production of monoclonal antibiodies. The availability of monoclonal antibodies opened the door to sophisticated analytical techniques and raised hopes for their use as therapeutic agents. Although the promise of therapeutic agents has yet to be realised (only one monoclonal antibody has been licensed for clinical use, OKT3, used in the treatment of acute renal allograft rejection, their use as tools in biological research has increased exponentially. Thus, animal cell culture processes were established to produce monoclonals on a commercial scale.

Animal cells: Subsequently, animal cells were also used as hosts for the production of some human proteins, especially where post-translational modification was essential for protein activity. Although these animal cell processes were based on microbial fermentation technology a number of novel problems had to be solved—animal cells are extremely fragile compared with microbial cells, the achievable cell density is very much less than in a microbial process and the media are very complex.

Recombinant fermentations: The outstanding developments in recombinant fermentations (stage 5) have tended to overshadow the progress which has been made in recent years in establishing new fermentations based on conventional microbial products (the continuing development of stage 4). However, the appreciation by the pharmaceutical industry that the activity of microbial metabolites extended well beyond antibacterials has resulted in a number of new microbial products reaching the marketplace in early 1990s. Buckland listed four secondary metabolites such as, cyclosporin, an immunoregulant used to control rejection of transplated organs, imipenem, a modified carbapenem which has the widest antimicrobial spectrum of any antibiotic, lovastatin, a drug used for reducing cholesterol levels and ivermectin, an anti-parasitic drug which has been used to prevent 'African River Blindness' as well as in veterinary practice. Buckland summarised these developments succinctly, 'One of the best kept secrets (unintentionally kept as a secret) in biochemical engineering was that working on secondary metabolites was a fascinating, important and rewarding experience. Furthermore the four products listed added together have higher sales than all of the recombinant products added together'. Thus, it is still relevant to heed Foster's warning 'never underestimate the power of the microbe'.

FERMENTATION

The fermentation is the chemical transformation of organic substances into simpler compounds by the action of enzymes, complex organic catalysts, which are produced by micro-organisms such as molds, yeasts, or bacteria. Enzymes act by hydrolysis, a process of breaking down or predigesting complex organic molecules to form smaller (and in the case of foods, more easily digestible) compounds and nutrients. For example, the enzyme protease (all enzyme names have the suffix - ase) breaks down huge protein molecules first into polypeptides and peptides, then into numerous amino acids, which are readily assimilated by the body. The enzyme amylase works on carbohydrates, reducing starches and complex sugars to simple sugars. And the enzyme lipase hydrolyses complex fat molecules into simpler free fatty acids. These are but three of the more important enzymes. There are thousands more, both inside and outside of our bodies. In some fermentations, important by-products such as alcohol or various gases are also produced. The word 'fermentation' is derived from the Latin meaning 'to boil,' since the bubbling and foaming of early fermenting beverages seemed closely akin to boiling.

Fermented foods: Fermented foods often have numerous advantages over the raw materials from which they are made. As applied to soyafoods, fermentation not only makes the end product more digestible, it can also create improved (in many cases meatlike) flavour and texture, appearance and aroma, synthesise vitamins (including B12, which is difficult to get in vegetarian diets), destroy or mask undesirable or beany flavours, reduce or eliminate carbohydrates believed to cause flatulence, decrease the required cooking time, increase storage life, transform what might otherwise be agricultural wastes (such as okara) into tasty and nutritious human foods (such as okara tempeh), and replenish intestinal microflora (as with miso or Acidophilus soyamilk).

Molds, yeasts, or bacteria: Most fermentations are activated by either molds, yeasts, or bacteria, working singularly or together. The great majority of these micro-organisms come from a relatively small number of genera, roughly eight genera of molds, five of yeasts, and six of bacteria. An even smaller number are used to make fermented soyafoods: the molds are *Aspergillus, Rhizopus, Mucor, Actinomucor,* and *Neurospora* species, the yeasts are *Saccharomyces* species, and the bacteria are *Bacillus* and *Pediococcus* species plus any or all of the species used to make fermented milk products. Molds and yeasts belong to the fungus kingdom, the study of which is called mycology. Fungi are as distinct from true plants as they are from animals. The study of all micro-organisms is called microbiology. While micro-organisms are the most intimate friends of the food industry, they are also its ceaseless adversaries. They have long been used to make foods and beverages, yet they can also cause them to spoil. When used wisely and creatively, however, micro-organisms are an unexploitable working class, whose very nature is to labour tirelessly day and night, never striking or complaining, ceaselessly providing human beings with new foods. Like human beings, but unlike plants, micro-organisms cannot make carbohydrates from carbon dioxide, water, and sunlight. They need a substrate to feed and grow on. The fermented foods they make are created incidentally as they live and grow.

BIOCHEMICAL ENGINEERING APPROACH TO FERMENTATION

The biochemical engineering approach to fermentation has been significant as the engineers have always been engaged in commercial fermentation operations. Perhaps the easiest way to assess and illustrate the role of biochemical engineering in fermentation technology is to first summarise its contributions in various aspects. Biochemical engineering contributions to fermentation technology can be looked at in many different ways. We can go through the characteristic fermentation process flowsheet and look at the main stages: (i) medium preparation and sterilisation, (ii) inoculum preparation, (iii) reaction (fermentation), and (iv) pretreatment for recovery. Alternatively we can adopt a unit operations approach and collectively examine all activities which have a common basis, heat sterilisation of media, aseptic transfer of fluids, mass transfer (aeration) and so forth.

Scale-up: Scale-up is an inherent part of process development, in fact the terms are virtually synonymous for many people. Scale-up has been successfully achieved when yields and productivities, have been produced in larger capacity units. This basic definition holds equally well for microbiological processes, though the problems encountered are markedly different.

Scale-up in fermentation: The real problem in fermentation process scale-up is the complete reproduction, in large capacity equipment, of those conditions for cell-medium interaction which have been established as optimum by small volume experiments. Many variables affect this interaction and oxygen supply is only one of them. Generally speaking, the most critical factors are considered to be:

Inoculum: Type, age, and amount.

Medium: Initial composition, pH, redox potential, changes during sterilisation, and additions during fermentation.

Conditions: Temperature, pressure, oxygen supply, and agitation/mixing.

The real key to successful scale-up of fermentation processes lies in a more complete understanding of the reactions taking place. Fermentation is, above all, a chemical process, although an extremely complex one. When the chemical mechanisms controlling the formation of desired products are reasonably understood, scale-up on a rational, but undoubtedly empirical, basis will be possible.

Thus, it seems quite clear that the bulk of biochemical engineering work so far lies in the areas of: (i) calibration and measurement and (ii) equipment development. The process design contributions, interestingly, are almost all from the biochemist-microbiologist, not the chemical engineer.

The key to future progress in fermentation is a more detailed knowledge of: (i) the biochemical mechanisms involved in growth and product formation and (ii) the factors which influence the organisms ability to catalyse these reactions. It follows that there will be a preponderance of biochemist-microbiologists, undoubtedly possessing increasing skills in physical chemistry and mathematics, in fermentation process development. The biochemical engineers role will be to translate this increased understanding by empirical means in most cases into increased productivities. Toward this end he can profitably spend more time on empirical analysis of process kinetics and improved methods for experimental design and analysis, particular computers for data analysis (digital) and process stimulation (analog) and scale-up techniques.

Chapter 2

Fermentation Feedstocks

INTRODUCTION

Fermentation is a chemical process by which molecules such as glucose are broken down anaerobically. More broadly, fermentation is the foaming that occurs during the manufacture of wine and beer, a process at least 10,000 years old. Generally the feedstocks of fermentation are micro-organisms and nutrient sources.

MICRO-ORGANISMS

Micro-organisms are those living things that are visible as individual organisms only with the aid of magnification. Micro-organisms are components of every ecosystem on Earth. Micro-organisms range in complexity from single to multicellular organisms. Most micro-organisms do not cause disease and many are beneficial. Micro-organisms require food, water, air, ways to dispose of waste, and an environment in which they can live. Investigation of micro-organisms is accomplished by observing organisms using direct observation with the aid of magnification, observation of colonies of these organisms and their waste, and observation of micro-organisms effects on an environment and other organisms. All micro-organisms are living things or organisms. Micro-organisms may be unicellular or single-celled, any living thing that has only one cell, the smallest unit of life. Some micro-organisms are multicellular, having more than one cell. Micro-organisms require food, air, water, ways to dispose of waste and an environment in which they can live. Some micro-organisms are producers, living things that make their own food from simple substances usually using sunlight, as plants do. Some micro-organisms eat other organisms to get their food. Most micro-organisms do not cause disease and many are helpful. There are many different kinds of micro-organisms. Scientists observe and classify micro-organisms just as they do plants and animals. These classifications are determined by the micro-organisms shape, structure, how they get food, where they live and how they move.

Bacteria are microscopic, single-celled organisms that exist all around you and inside you. Although they can cause sickness and disease, they are very important to life on Earth. We depend on bacteria to help in the digestion of food, for plant growth, and to help us make foods and medicines. Bacteria are an important part of the soil. They are able to capture some nutrients that plants cannot. When living things die, bacteria play a very important role as decomposers, bacteria and fungi feeding on and breaking down plant and animal matter. Without these decomposers, the bodies of all organisms that have ever lived would still remain. This would be messy. When bacteria break down the dead organisms, they release substances that can be used by other organisms in the ecosystem.

Bacteria can affect our bodies in several ways. Harmful bacteria can make us sick, but fortunately, our bodies will fight back. When streptococcus bacteria give us strep throat we can take medicine to help us get well faster. Some bacteria always live in our bodies. They are found in digestive systems and help digest food. Other bacteria are in our food. When you eat yogurt or cheese, you eat bacteria.

Bacteria are the smallest micro-organisms. You can see them when there are thousands of them growing together in a colony. To see bacteria as a single organism, requires a microscope with very high magnification. Bacteria live in almost every place on Earth. Scientists can culture, grow micro-organisms in a specially prepared nutrient medium. The drawing shows how colonies of bacteria look when cultured on a plate. The colonies vary in size and colour depending on the type of bacteria.

Fungi are organisms that are neither plant nor animal, yet have characteristics of both, and absorb food from whatever source they are growing on. A common fungus is a mushroom. It looks like a plant but is not green. Mushrooms cannot make their own food and must live on a food source. Some are poisonous, and only an expert can identify them. Another fungus, yeast, is used to make bread rise and give it flavour. Athletes foot is caused by a fungus. Some types of fungi rot wood in homes. Fungi also like warm moist places to grow. A good way to prevent fungus is to keep things, like your toes, dry.

Protozoans are microscopic organisms that usually live in water. They move through the water with tiny hair-like arms called cillia. The cillia are located all around the sack-like body of the protozoa and wave back and forth to move the protozoa through the water. Some protists are producers like plants.

Others must eat smaller things like bacteria or molds. Protozoa are an important food source for many pond creatures. Some protozoans are harmful to people. You may have heard that it is not a good idea to drink water from a stream. Streams sometimes contain a protozoan called Giardia that can make you sick. Algae are a type of protist that usually live in water and can produce their own food. Some algae are very large, while others are microscopic. Algae can be red, brown, yellow or green.

Some of the largest algae are kelp. They can grow to be 60 meters long. Algae are an important part of the oceans ecosystem. They provide food for fish, whales and many other sea animals. Phyloplaukton provide over half of oxygen on Earth. Algae are also eaten by people. In fact, algae are in ice cream.

NUTRIENTS AND THEIR SOURCES

Based on the amount of the nutrients that each person needs to consume on a daily basis, these nutrients are categorised into two groups. These are macronutrients, which should be consumed in fairly large amounts, and micronutrients, which are only required in small amounts.

Macronutrients

'Macro' means large; as their name suggests these are nutrients which people need to eat regularly and in a fairly large amount. They include carbohydrates, fats, proteins, fibre and water. These substances are needed for the supply of energy and growth, for metabolism and other body functions. Metabolism means the process involved in the generation of energy and all the 'building blocks' required to maintain the body and its functions. Macronutrients provide a lot of calories but the amount of calories provided varies, depending on the food source. For example, each gram of carbohydrate or protein provides four calories, while fat provides nine calories for each gram.

Micronutrients

As their name indicates ('micro' means small) micronutrients are substances which people need in their diet in only small amounts. These include minerals and vitamins. Although most foods are mixtures of nutrients, many of them contain a lot of one nutrient and a little of the other nutrients. Foods are often grouped according to the nutrient that they contain in abundance.

This chapter discusses in detail lignocellulose as feedstocks in fermentation processes.

LIGNOCELLULOSE AS FEEDSTOCKS IN FERMENTATION PROCESSES

Lignocellulose in the form of forestry, agricultural, and agro-industrial wastes is accumulated in large quantities every year. These materials are mainly composed of three groups of polymers, namely cellulose, hemicellulose, and lignin. Cellulose and hemicellulose are sugar rich fractions of interest for use in fermentation processes, since micro-organisms may use the sugars for growth and production of value added compounds such as ethanol, food additives, organic acids, enzymes, and others. Submerged and solid-state fermentation systems have been used to produce compounds of industrial interest from lignocellulose, as an alternative for valorisation of these wastes and also to solve environmental problems caused by their disposal. When submerged fermentation systems are used, a previous stage of hydrolysis for separation of the lignocellulose constituents is required. This section discusses the potential uses of lignocellulosic materials in fermentation processes. Aspects related to submerged and solid-state fermentation systems will be described focusing on the raw materials, hydrolysis processes, fermentation conditions, micro-organisms, and products that can be obtained.

Lignocellulose Structure

Lignocellulosic biomass comprising forestry, agricultural and agro-industrial wastes are abundant, renewable and inexpensive energy sources. Such wastes include a variety of materials such as sawdust, poplar trees, sugarcane bagasse, waste paper, brewers spent grains, switchgrass, and straws, stems. stalks, leaves, husks, shells and peels from cereals like rice, wheat, corn, sorghum and barley, among others. Lignocellulose wastes are accumulated every year in large quantities, causing environmental problems. However, due to their chemical composition based on sugars and other compounds of interest, they could be utilised for the production of a number of value added products, such as ethanol, food additives, organic acids, enzymes, and others. Therefore, besides the environmental problems caused by their accumulation in the nature, the non-use of these materials constitutes a loss of potentially valuable sources. The major constituents of lignocellulose are cellulose, hemicellulose, and lignin, polymers that are closely associated with each other constituting the cellular complex of the vegetal biomass. Basically, cellulose forms a skeleton which is surrounded by hemicellulose and lignin (Fig. 2.1).

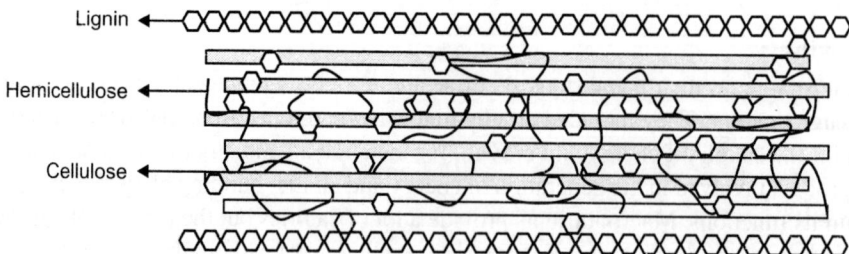

Fig. 2.1: Representation of lignocellulose structure showing cellulose, hemicellulose and lignin fractions.

Cellulose is a high molecular weight linear homopolymer of repeated units of cellobiose (two anhydrous glucose rings joined via a β-1,4 glycosidic linkage). The long-chain cellulose polymers are linked together by hydrogen and van der Walls bonds, which cause the cellulose to be packed into microfibrils. By forming these hydrogen bounds, the chains tend to arrange in parallel and form a crystalline structure. Therefore, cellulose microfibrils have both highly crystalline regions (around 2/3 of the total cellulose) and less-ordered amorphous regions. More ordered or crystalline cellulose is less soluble and less degradable. Hemicellulose is a linear and branched heterogeneous polymer typically made up of five different sugars - Larabinose, D-galactose, D-glucose, D-mannose, and D-xylose - as well as other components such as acetic, glucuronic, and ferulic acids. The backbone of the chains of hemicelluloses can be a homopolymer (generally consisting of single sugar repeat unit) or a heteropolymer (mixture of different sugars). According to the main sugar residue in the backbone, hemicellulose has different classifications, e.g., xylans, mannans, glucans, glucuronoxylans, arabinoxylans, glucomannans, galactomannans, galactoglucomannans, β-glucans, and xyloglucans. When compared to cellulose, hemicelluloses differ thus by composition of sugar units, by presence of shorter chains, by a branching of main chain molecules, and to be amorphous, which made its structure easier to hydrolyse than cellulose.

Lignin is a very complex molecule constructed of phenylpropane units linked in a large three-dimensional structure. Three phenyl propionic alcohols exist as monomers of lignin: p-coumaryl alcohol, coniferyl alcohol and sinapyl alcohol. Lignin is closely bound to cellulose and hemicellulose and its function is to provide rigidity and cohesion to the material cell wall, to confer water impermeability to xylem vessels, and to form a physico-chemical barrier against microbial attack. Due to its molecular configuration, lignins are extremely resistant to chemical and enzymatic degradation.

The amounts of carbohydrate polymers and lignin vary from one plant specie to another. In addition, the ratios between various constituents in a single plant may also vary with age, stage of growth, and other conditions. However, cellulose is usually the dominant structural polysaccharide of plant cell walls (3550%), followed by hemicellulose (2035%) and lignin (1025%). Average values of the main components in some lignocellulose wastes are shown in Table 2.1.

Table 2.1: Main components of lignocellulose wastes.

Lignocellulose waste	Cellulose (wt %)	Hemicellulose (wt %)	Lignin (wt %)
Barley straw	33.8	21.9	13.8
Corn cobs	33.7	31.9	6.1
Corn stalks	35.0	16.8	7.0
Cotton stalks	58.5	14.4	21.5
Oat straw	39.4	27.1	17.5
Rice straw	36.2	19.0	9.9
Rye straw	37.6	30.5	19.0
Soya stalks	34.5	24.8	19.8
Sugarcane bagasse	40.0	27.0	10.0
Sunflower stalks	42.1	29.7	13.4
Wheat straw	32.9	24.0	8.9

Pre-treatments for Selective Separation of Lignocellulose Constituents

Selective and effective separation of the main fractions present in lignocellulose biomass is of great interest, mainly when submerged fermentation process will be subsequently performed, because the by-products obtained during hydrolysis of these materials may affect the fermentation yield, and also because valuable sources are lost if other fractions are partially hydrolysed. Due to the structural differences among these fractions, separation of cellulose, hemicellulose and lignin from lignocellulose biomass requires the use of specific processes, which may be physical, physico-chemical, chemical or biological. There are variety of methods that have been proved to be efficient for the lignocellulose hydrolysis. The most commonly used are briefly described below.

Cellulose hydrolysis

Chemical and enzymatic methods are the most common techniques for hydrolysing cellulose. The chemical method, also known as concentrated acid hydrolysis, is conducted with mineral acids such as H_2SO_4 or HCl (in the range of 10–30%), at temperatures of about 160°C and pressures of about 10 atmosphere. These harsh conditions (high temperature and acid concentration) are needed to liberate glucose from the tightly associated chains, because most cellulose is crystalline. In this process, acid concentration, temperature and time are crucial factors, and must be controlled to avoid the sugars and lignin degradation to by-products. If present in the raw material to be hydrolysed, hemicellulose is also removed by using concentrated acids.

Enzymatic hydrolysis has attracted increasing attention as an alternative to concentrated acid hydrolysis because the process is highly specific, can be performed under milder reaction conditions (pH around 5 and temperature less than 50°C) with lower energy consumption and lower environmental impact. In addition, it does not present corrosion problems, and gives high yield of pure glucose with low formation of by-products that is favourable for the subsequent hydrolysate use in fermentation processes. Enzymatic hydrolysis of cellulose is a reaction carried out by cellulase enzymes, which correspond to a mixture of several enzymes, among which at least three major groups are involved in the hydrolysis of cellulose: (i) β-1-4-endoglucanase (EC 3.2.1.4.), which attacks regions of low crystallinity in the cellulose fibre creating free chain ends, (ii) β-1-4-exoglucanase or cellobiohydrolase (EC 3.2.1.91.), which degrades the molecule further by removing cellobiose units from the free chain ends, (iii) β-glucosidase (EC 3.2.1.21.), which hydrolyses cellobiose to produce glucose. A wide variety of cellulolytic fungi and bacteria have been reported. Among them, the enzyme system secreted by the filamentous fungus *Trichoderma reesei* has been intensively studied. Usually, lignocellulosic materials must be pretreated prior to the enzymatic hydrolysis in order to make the cellulose more accessible to enzymes.

Hemicellulose hydrolysis

Dilute acid hydrolysis is the most employed technique for the hemicellulose breakdown. In this process, the use of diluted acids (1–4%), under moderate temperatures (120 to 160°C), has proven to be adequate for hemicellulose hydrolysis, promoting little sugar decomposition. H_2SO_4 is the usual acid employed although HCl, HNO_3, and H_3PO_4 are also used. The mechanism involved in the acid hydrolysis of hemicellulose is similar to that for cellulose: acid catalyses the breakdown of long hemicellulose chains to form shorter chain oligomers and then to sugar monomers. However, because hemicellulose is amorphous, less severe conditions are required to release hemicellulose sugars. Other important advantages are the generation of lower degradation products as well much less corrosion problems in hydrolysis tanks.

Steam explosion is also a method commonly used for hemicellulose hydrolysis. In this method, the biomass is heated using high-pressure saturated steam (0.69–4.83 MPa, 160–260°C) for a short period (from seconds to few minutes). Steam condenses under high pressure, thereby wetting the material, and then the pressure is suddenly reduced, which makes the material undergo an explosive decompression.

Combination of acetic acid with sudden depressurisation, promote the hemicellulose hydrolysis and solubilisation. Although avoiding acid catalysts is stated as an advantage of this method, addition of an acid catalyst improves hemicellulose hydrolysis and decreases production of degradation compounds. Limitations of steam explosion include an incomplete disruption of the lignin carbohydrate matrix, and generation of compounds that may be inhibitory to micro-organisms. Autohydrolysis is a process similar to the steam explosion, but in this case, the explosion does not occur. This process uses compressed liquid hot water ($\approx 200°C$, pressure > saturation point) and the acids resulting from hydrolysis of acetyl and uronic groups, originally present in hemicelluloses, catalyse hydrolysis of links between hemicellulose and lignin as well as between the carbohydrates. This process is able to hydrolyse hemicellulose in minutes, with high yield, low by-products formation and no significant lignin solubilisation.

Specific micro-organisms or a suitable cocktail of enzymes known as hemicellulases can also promote the hemicellulose hydrolysis. Hemicellulases are produced by many species of bacteria and fungi, as well as by several plants. A number of hemicellulases, including xylanases and mannanases, have been identified in *Trichoderma reesei*. Today, most commercial hemicellulase preparations are produced by genetically modified *Trichoderma* or *Aspergillus* strains. Biological treatments have some advantages such as the high specificity, low energy consumption, no chemical requirement, and mild environmental conditions thus avoiding sugar degradation and resulting in high sugar yields.

Lignin hydrolysis

Alkaline treatments, ozonolysis, peroxide and organosolv treatments are some of the methods usually employed for lignin removal from lignocellulose biomass. Such methods are effective for lignin solubilisation but in most of them, part of the hemicellulose is also hydrolysed.

Alkali treatments refers to the application of alkaline solutions such as NaOH, $Ca(OH)_2$ or ammonia. Among these, treatment with NaOH is one of the most used for delignification of agricultural residues. The alkali treatment causes swelling, leading to an increase in internal surface area, a decrease in the degree of polymerisation, a decrease in crystallinity, separation of structural linkages between lignin and carbohydrates, and disruption of the lignin structure. As a consequence, the lignin is dissolved from the raw material, being separated in the form of a liquor rich in phenolic compounds that represents the process effluent. The inconvenient of this technique is that it also degrades part of the hemicellulose.

Hydrogen peroxide treatment utilises alkaline solutions at temperatures higher than 100°C, which promote a fast decomposition of H_2O_2. As a consequence, more reactive radicals such as hydroxyl radicals (HO) and superoxide anions (O_2^-) are produced, which are responsible for lignin degradation. This technique is commonly used in paper and pulp industries for bleaching and delignification purposes (to improve the brightness of pulp as it reacts with coloured carbonyl-containing structures in the lignin). However, delignification by this process on a large scale can be costly. Similarly to the alkaline treatment, during the lignin solubilisation by hydrogen peroxide part of the hemicellulose is also removed from the material structure.

Treatment with organo solvents involves the use of an organic liquid (for example, methanol, ethanol, acetone, ethylene glycol or triethylene glycol) and water, with or without addition of catalysts such as oxalic, salicylic, and acetylsalicylic acid. This mixture hydrolyses lignin bonds and lignin-carbohydrate bonds, but many of the carbohydrate bonds in the hemicellulose components are also broken. Lignin is dissolved as a result of the solvent action and the cellulose remains in the residual solid material.

Ozone treatment is another way of reducing the lignin content of lignocellulosic materials. Lignin attacks as a scavenger during this pre-treatment because it consumes most of ozone during the degradation of the carbohydrate content. As a consequence, low ozone amounts are available for cellulose degradation. However, this treatment may also attacks the cellulose and hemicellulose components besides the lignin molecule. Cellulose degradation has been attributed partly to a direct reaction of ozone with the glycosidic linkage and partly to a free radical mediated oxidation of hydroxyl groups in glucose. Some advantages of this treatment are that ozonolysis are carried out at room temperature and normal pressure. Furthermore, since ozone can be easily decomposed by using a catalytic bed or increasing the temperature, this process can be designed to minimise environmental pollution.

Biological treatments, based on the use of brown-, white- and soft-rot fungi have been commonly used to degrade the lignin, being considered a cheap and effective method of delignification. Degradation of lignin by fungi such as *Phanerochaete chrysosporium, Trametes versicolour, Trametes hirsuta* and *Bjerkandera adusta*, may be used to allow better access to the cellulose and hemicellulose components, besides to be also considered as an effective biological detoxification alternative. Main problems in using biological methods are that fungi may also attacks cellulose and hemicellulose, and hydrolysis rate in most biological materials is very low.

Three main enzymes are involved in the lignin biodegradation, namely lignin peroxidise, manganese peroxidise, and laccase. These enzymes have gained large attention by their industrial applications in pulp and paper industries, for biochemical pulping and decolourisation of bleach plant effluent.

Although each treatment is more efficient to remove a specific fraction of lignocellulose, in some cases, a combination of them is required to achieve an efficient conversion of the fraction of interest. Biological or enzymatic treatments, for example, cannot be applied directly on the raw materials because lignin hinders the attack of fungi or enzymes to the material cell wall. Therefore, for the enzymatic hydrolysis of cellulose is crucial a previous hydrolysis step to promote a partial removal of lignin and hemicellulose, so that the cellulose fibres become more accessible to the enzymes attack. In addition, the best method and conditions of treatment depend greatly on the type of lignocelluloses, varying thus, with the raw material used. It is important emphasising that the selection of a treatment method affects the cost and performance in the subsequent hydrolysis and fermentation stages. Ideal hydrolysis process should achieve high yields of fermentable reducing sugars, avoid degradation or loss of yielded sugars and the formation of inhibitors to the subsequent fermentation, and require minimal energy, chemicals and equipments.

Fermentation Processes for Value Added Compounds from Lignocellulose

Submerged fermentation systems

Submerged fermentation (SmF) systems can be defined as the cultivation of micro-organisms in a liquid medium containing soluble carbon source and nutrients, maintained or not under agitation. Several characteristics make these systems attractive for the micro-organisms cultivation and production of biological products, which include: (i) the mixture of the medium is easy and allows uniform conditions for the micro-organism growth, (ii) modification of the cultivation conditions like pH, dissolved oxygen, temperature, agitation, and nutrient concentration are easy and fast, (iii) the temperature control is favoured by the high specific heat and thermal conductivity, (iv) efficient technologies have already been developed, with high automation grade, diversity and availability of equipments. Countless micro-organism strains have been used in fermentation processes by submerged cultivation, including bacteria, yeasts, fungi and algae. Fermentation media used in these systems may be synthetically formulated or produced by hydrolysis of lignocellulose.

Sugars released from cellulose and hemicellulose structures can be microbially converted into various products of industrial interest by SmF systems, including organic acids, ethanol, glycerol, food additives, butanol, etc. However, bioconversion of these hydrolysates is not a simple process, but requires considerable attention in many points so that high yields and productivities may be attained. One of these points is related to the hydrolysates composition. Usually the hydrolysis processes not only extract sugars from the lignocellulose structure, but a broad range of compounds proceeding from the lignin or originated from the sugars degradation can also be found in the hydrolysates, being their concentration dependent of the raw material and hydrolysis process used.

Such compounds are toxic for the micro-organisms and therefore, previous the use as fermentation medium, lignocellulosic hydrolysates should be submitted to a detoxification process to become a more suitable medium for the micro-organisms cultivation. A number of detoxification methods, including biological, physical, and chemical ones, have been proposed to transform inhibitors into inactive compounds or to reduce their concentration. The effectiveness of a detoxification method depends both on the type of hemicellulose hydrolysate and on the species of micro-organism employed, because each type of hydrolysate has a different degree of toxicity, and each species of micro-organism has a different degree of tolerance to inhibitors. In some cases, the addition of nutrient sources to the hydrolysate may be necessary to improve the bioconversion yield. Another point requiring considerable attention is when hemicellulosic hydrolysates are used as fermentation medium. Such hydrolysates contain a mixture of pentose and hexoses sugars, and thus, for an efficient bioconversion process, the fermenting micro-organism should be able to consume and/or convert both kinds of sugars.

Cellulose bioconversion

Currently, ethanol production is one of the most studied and promising alternative for cellulosic biomass conversion, due to the large incentive that has been given to biofuels use in replacement of gasoline. In addition, ethanol production from lignocellulose wastes is very attractive because of their low cost and abundance, and non-competition with foodstuffs. Therefore, a large variety of lignocellulosic materials including wood, straws, agricultural wastes, and crop residues have been evaluated for use in this bioconversion process. The basic process steps in producing ethanol from cellulose biomass consist in an initial treatment (for example, diluted acid, alkaline or steam explosion) to render cellulose more accessible to the subsequent step of enzymatic hydrolysis, which break down polysaccharides to simple sugars. The glucose solution obtained is fermented to ethanol by micro-organisms; *Saccharomyces cerevisiae* being the most currently used since it gives high ethanol yields from glucose. Together with bioethanol, biohydrogen production from lignocellulose biomass has shown enormous potentialities for sustainable bioenergy production. The H_2 production from lignocellulosic materials can be performed by anaerobic fermentation process. However, direct conversion of lignocellulosic biomass to hydrogen needs a previous treatment to hydrolyse the cellulose crystalline structure. The cellulose present in the raw material may be hydrolysed by using acids such as HCl, for example. The hydrolysate obtained (rich in glucose and some hemicellulose sugars) is thus converted to biohydrogen by natural anaerobic microflora. Butyrate and acetate may be obtained as by-products of the hydrogen fermentation. Biogas rich in H_2 and CH_4 may be also produced by fermentation of glucose from lignocellulose. In this case, it is necessary to treat the raw material with chemicals such as NaOH solution to destroy and remove the solid lignin. Then, the released cellulose is hydrolysed with cellulase into reducing sugars, which can be fermented with anaerobic bacteria to produce biogas.

Besides biofuels, several organic acids, including lactic, citric, acetic, and succinic acids, may be produced by cellulose conversion. Lactic acid may be produced from lignocellulose materials by sequential steps of chemical processing (in order to make the cellulose more accessible to the enzymes), enzymatic saccharification (for obtaining solutions containing glucose as main sugar) and finally, the hydrolysate fermentation by micro-organisms, especially *Lactobacillus species*. The conventional process for cellulosic biomass conversion to acetic acid includes also an initial stage of acid or enzymatic hydrolysis of the substrate, followed by yeast fermentation and oxidation to acetic acid by *Acetobactor* sp. Similarly to these two cases, citric acid can also be produced by fermentation of glucose rich hydrolysates produced from cellulose. Several molds such as *Penicillium luteum*, *P. citrinum*, *Aspergillus niger*, *A. wentii*, *A. clavatus*, *Mucor piriformis*, *Citromyces pfefferianus*, *Paecilomyces divaricatum*, and *Trichoderma viride*, and yeasts such as *Yarrowia lipolytica* and *Candida guilliermondii* are able of producing this acid. Succinic acid is another acid that can be produced by fermentation of cellulosic hydrolysates. Many agricultural and industrial wastes, for example, whey, wheat, straw and wood, have been reported as raw carbon materials for the production of succinic acid. Wood hydrolysate prepared by steam explosion followed by enzymatic hydrolysis of the cellulose rich residue was successfully fermented to succinic acid by *Mannheimia succiniciproducens* bacteria. Corn stalk and cotton stalk hydrolysates produced by a sequence of steam explosion and enzymatic hydrolysis treatments were fermented to succinic acid by *Actinobacillus succinogenes*. Several micro-organisms have ability to produce this acid, among of which, *Anaerobiospirillum succiniciproduens* is considered the most efficient producer. This micro-organism only utilises glucose as carbon source, being not able of utilising xylose at all. *Mannheimia succiniciproducens* can use glucose as well as xylose, being thus more interesting for use in bioconversion processes from hemicellulosic hydrolysates. Production of carotenoids (natural pigments with a variety of applications in food technology) such as astaxanthin has already been performed by fermentation of wood hydrolysate. Hydrolysates containing glucose and cellobiose or glucose only, were obtained by alkali- or acid-processed wood treated with a cellulase complex. Such hydrolysates were suitable as fermentation media for carotenoids production by *Xanthophyllomyces dendrorhous*.

Hemicellulose bioconversion

Fermentation of hemicellulosic hydrolysates is usually more complicated than fermentation of cellulosic hydrolysates, due to the presence of pentoses sugars such as xylose, which is not metabolised by large number of micro-organisms. Even though, efforts have been directed to the use of this fraction from lignocellulose. Currently, one of the technologies that have been strongly investigated for conversion of hemicellulose sugars is for ethanol production. During the process for ethanol production from cellulose, a hemicellulosic hydrolysate is generated in the first step of raw material treatment (usually performed with dilute acids or steam explosion). Utilisation of the sugars released during this stage (hemicellulose sugars) is considered essential for efficient and cost-effective conversion of lignocellulose to ethanol. The main challenge of this process is that *Saccharomyces cerevisiae*, which is the most widely used micro-organism for ethanol production, does not utilise pentose sugars. *Pichia stipitis* has been described as a promising micro-organism for this bioprocess since this yeast is able of transforming both pentoses and hexoses sugars into ethanol, which is an important advantage since both kinds of sugars are currently found in hemicellulosic hydrolysates. Some bacteria, such as *Escherichia coli*, *Klebsiella*, *Erwinia*, *Lactobacillus*, *Bacillus*, and *Clostridia*, can utilise mixed sugars but produce no, or only a limited quantity of ethanol. Some attempts have also been made to genetically modify *S. cerevisiae* and other micro-organisms in order to produce ethanol with high yields from both, hexoses and pentoses.

Butanol is another biofuel that can be produced from hemicellulose bioconversion. As a product of ABE (acetone butanol ethanol) fermentation, butanol is an excellent feedstock chemical and a superior fuel compared to ethanol. This compound can be produced by fermentation of dilute sulphuric acid hemicellulosic hydrolysates produced from agricultural wastes such as barley straw, corn stover and switchgrass. A variety of micro-organisms are able of converting hemicellulose sugars to butanol, but the most commonly used strains are *Clostridium acetobutylicum* and *Clostridium beijerinckii*.

Xylitol, a five-carbon sugar alcohol that can be used as a natural food sweetener, a dental caries reducer, and as a sugar substitute for diabetics, may be produced by fermentation of xylose present in hemicellulosic hydrolysates. Production of this polyalcohol may be carried out by fermentation of hemicellulosic hydrolysates generated by acid hydrolysis or autohydrolysis from different raw materials, including rice and wheat straw, sugarcane bagasse, eucalyptus wood, corn cobs, brewers spent grains and several others. Many micro-organisms are able of producing xylitol from xylose, among of which, *Candida guilliermondii* yeast has been one of the most employed due to its high conversion efficiency. Besides xylitol, arabitol (another polyalcohol) may also be produced by fermentation of hemicellulosic hydrolysates. Yeasts like *Candida entomaea* and *Pichia guilliermondii* have demonstrated ability to convert both pentoses, xylose and arabinose, to xylitol and arabitol respectively, when cultivated in hemicellulosic hydrolysates. However, xylose is assimilated preferentially than arabinose. The arabinose utilisation by micro-organisms is of great interest for the economic conversion of lignocellulose biomass, since this sugar is present in significant amounts in many hemicelluloses.

2,3-Butanediol, also known as 2,3-butylene glycol, is a valuable chemical feedstock because of its application as a solvent, liquid fuel, and as a precursor of many synthetic polymers and resins. This compound may be produced by fermentation of hemicellulosic hydrolysates by several micro-organisms, including *Bacillus polymyxa, Klebsiella pneumoniae (Aerobacter aerogenes), Bacillus subtilis, Seratia marcescens* and *Aerobacter hydrophia*.

Among these strains, two bacterial species, namely *Bacillus polymyxa* and *Klebsiella pneumoniae,* have demonstrated potential for butanediol fermentation on a commercial scale. *Klebsiella pneumoniae* is also a micro-organism of interest due to its ability of utilising all the major sugars (hexoses, pentoses, certain disaccharides) and uronic acid derived from the hydrolysates of hemicellulosic and cellulosic materials. Sugars present in wood hemicelluloses could be efficiently utilised by this micro-organism after they had been released by either acidic or enzymatic hydrolysis.

Lactic acid production from lignocellulosic materials has been mostly reported using the cellulosic fraction, because there are few micro-organisms capable of fermenting hemicellulosic sugars. However, *Lactobacillus pentosus* has been reported as able to produce lactic acid by fermentation of hemicellulosic sugars (mainly xylose, arabinose, and glucose) from acid hydrolysates of most agricultural residues. Hemicellulosic hydrolysate of wet-oxidised wheat straw was successfully used as substrate for lactic acid production by *Lactobacillus brevis* and *Lactobacillus pentosus*. A lactic acid yield of 95% and complete substrate utilisation were achieved in this process using a mixed culture of these micro-organisms. Sugars from hemicellulosic hydrolysates may also be used for citric acid production by *Aspergillus niger.* Operational variables including agitation, aeration, and the time of spores cultivation influence this bioconversion process and must be taken into consideration to maximise the product yield. Besides lactic and citric acids, butyric acid can also be produced by fermentation of hemicellulosic hydrolysates. *Clostridial* species are known to ferment xylose, and are thus of interest for the production of butyric acid from plant biomass. Conversion of corn fibre hydrolysate produced by dilute acid hydrolysis, for example, gave high butyric acid yield when fermented with *Clostridium tyrobutyricum*.

Solid State Fermentation Systems

Solid State Fermentation (SSF) can be defined as any fermentation process allowing the growth of micro-organisms on moist solid materials in the absence of free-flowing water. The low moisture content means that fermentation can only be carried out by a limited number of micro-organisms, mainly yeasts and fungi, although some bacteria have also been used. Among these micro-organisms, fungi have been considered to be the most adapted to SSF because their hyphae can grow on particle surfaces and penetrate into the inter particle spaces and thereby colonising solid substrates.

In the last decades there has been an increasing trend towards the utilisation of these systems instead of SmF, because they have promoted higher yields and better product characteristics than cultivations in liquid medium. In addition, costs are much lower due to the efficient utilisation and value-addition of wastes. The main drawback of this type of cultivation concerns the scale up of the process, largely due to heat transfer and culture homogeneity problems. Several researches have been directed towards the development of bioreactors for SSF systems, however, the available information has not indicated an ideal bioreactor yet. Although this obstacle to be overcome, many studies have been performed using SSF systems for the production of different compounds of interest, including flavours, organic acids, enzymes and others.

The nature of the solid substrate is one of the most important factor affecting SSF processes and its selection depends on several factors mainly related with cost and availability. Therefore, many studies have been performed involving the screening of several agro-industrial wastes as solid substrate in SSF. Some examples of products obtained by SSF of lignocellulosic materials are described below.

Enzymes

SSF systems have been considered as a promising technology for the production enzymes and therefore, this is a topic that has been strongly studied. Many authors have reported the production of a variety of enzymes by SSF, using different micro-organisms and solid substrates. Some examples of enzymes that may be produced by SSF from lignocellulosic materials include α-amylase, cellulase, xylanase, protease, fructosyl transferase, chitinase, pectinase, among others.

Production of α-amylase is usually performed by bacteria strains such as *Bacillus subtilis, B. polymyxia, B. mesentericus, B. vulgarus, B. megaterium and B. licheniformis*, however, some fungi strains are also able of producing this enzyme. A variety of lignocellulosic wastes have been used as substrate materials for α-amylase production by SSF. For example, rice husks were used as substrates for the production of α-amylase by *B. subtilis*.

Rice straw, rice bran, red gram husk, jowar straw, jowar spathe and wheat bran were used as substrates for the production of α-amylase by the fungus *Gibberella fujikuroi*, and spent brewing grains were used for α-amylase production by *Aspergillus oryzae*.

Lignocellulolitic enzymes, more specifically cellulases and xylanases, have been produced by SSF on various agricultural residues. Cellulases production has been extensively studied using sugarcane bagasse as raw material. Production of this enzyme by a strain of *Streptomyces* sp. HM29 occurred with higher yields when the strain was grown on bagasse in comparison to rice straw, rye straw or corncobs. Production of xylanases by *Aspergillus terreus* and *Aspergillus niger* was performed in different lignocellulosic wastes, including sugarcane bagasse, rice straw and soyabean hulls. *Thermoascus aurantiacus* was able to produce a high level of thermostable xylanase from sugarcane bagasse, while *Bacillus* sp. was good producer of this enzyme from rice bran. Protease production has also been studied using different lignocellulosic wastes as solid substrate. In a study on the production of this enzyme by

Aspergillus oryzae from wheat bran, rice husk, rice bran and spent brewing grains, wheat bran provided the best production results. Green gram, chick pea, red gram and black gram husks, and wheat bran were used as solid substrate for protease production by *Bacillus* sp. Among the substrates, green gram husk gave the maximum protease production.

Fructosyl transferase, enzyme responsible for the fructooligosaccharides production from sucrose, was produced by *Aspergillus oryzae* using corn cobs, coffee husk, spent coffee, spent tea, sugarcane bagasse, cassava bagasse and cereal brans (from wheat, rice and oat) as solid substrates. Production of chitinase has been reported using *Penicillium aculeatum* grown on wheat brancrude chitin mixture, whereas the production of tannase has been reported using *Aspergillus* sp. grown on different material wastes, including ber leaves (*Zyzyphus mauritiana*), jamun leaves (*Syzygium cumini*), amla leaves (*Phyllanthus emblica*), jawar leaves (*Sorghum vulgaris*), wheat bran, rice bran, sawdust, rice straw dust and sugarcane pith. Most of agro-industrial wastes have also been reported as containing inducers of laccase synthesis, and therefore, they ensure an efficient production of laccase by SSF. Furthermore, agro-wastes have shown to produce higher laccase activities than inert supports for the same fungal strain and culture conditions. Use of SSF for pectinase production has been proposed using different agro-industrial wastes, such as coffee husk and sugarcane bagasse. Commercial pectinase preparation is produced from fungal micro-organisms, mainly by *Aspergillus niger* strains.

Flavours

Natural flavours are chemical substances with aroma properties that are produced from feedstock of plant or animal origin by means of physical, enzymatic or microbiological processing. Flavour synthesis by biotechnological processes plays an increasing role in the food, feed, cosmetic, chemical and pharmaceutical industries. SSF has been used for the production of flavour compounds by cultivating yeasts and fungi. The production of flavour compounds is related to the low oxygen availability in the medium, which results in production of odour compounds including alcohols, aldehydes and ketones. Coffee husk is one of the lignocellulosic materials that have already been used for the production of flavours. Cultivation of *Ceratocystis fimbriata* in coffee husk under SSF conditions proportioned the production of fruity aroma, with ethyl acetate, ethanol and acetaldehyde being the major compounds produced. When cultivated in cassava bagasse, wheat bran, and sugarcane bagasse, this micro-organism produced banana flavour and fruity complex flavours. *Kluyveromyces marxianus* produced aroma compounds, such as monoterpene alcohols and isoamyl acetate (responsible for fruity aromas), in SSF using cassava bagasse or giant palm bran as a substrate.

Organic acids

Several organic acids have already been produced by SSF from lignocellulose wastes. Citric acid is one of these examples. Almost the entire production of this acid has been obtained using crops and crop residues as substrates and *Aspergillus niger* as production strain. When comparing sugarcane bagasse, coffee husk and cassava bagasse as solid substrate for citric acid production by *Aspergillus niger*, cassava bagasse led to the highest production results. An important point to be considered in the production of citric acid by SSF is that the synthesis appears to be directly influenced by the nitrogen source and its consumption leads to a pH decrease. It has also been generally found that addition of methanol increases citric acid production in SSF. Lactic acid may be produced by SSF using fungal as well as bacterial cultures. Strains of *Rhizopus* sp. have been common among the fungal cultures and that of *Lactobacillus* sp. among the bacterial cultures. When cassava bagasse and sugarcane bagasse were used to produce

L(+) lactic acid by *Lactobacillus delbrueckii*, more than 99% of the total sugar available was converted into this acid. L(+)-lactic acid production by *Rhizopous oryzae* in solid-state conditions operating with sugarcane bagasse as a support gave slightly higher productivity than in SmF.

Single cell protein

Some processes have focused on the direct conversion of lignocellulosic wastes to single-cell protein. The micro-organisms strains *Chaetomium cellulolyticum* mutant, *Pleurotus sajor-caju*, strains of *Aspergillus* and *Penicillium* spp., may be used for this purpose. This co-cultivation of fungi has the ability to utilise cellulose and hemicellulose, after lignin degradation, for single cell protein production. There is no need for any treatment in the raw material before use in this system as together these fungi are capable of separating lignocellulose into its individual components. The cellulose obtained may be also used for paper production or as single cell protein for animal or human feed.

Bioactive compounds

Several bioactive compounds may be produced by SSF from different lignocellulose wastes. Some examples include: (i) the production of gibberellic acid by *Giberella fujikuroi* and *Fusarium moniliforme* from corn cobs, (ii) the production of tetracycline from cellulosic substrates, (iii) production of oxytetracycline by *Streptomyces rimosus* from corn cobs, (iv) the production of destrucxins (cyclodepsipeptides) by *Metarhizium anisopliae* from rice husk, and (v) production of ellagic acid by *Aspergillus niger* from pomegranate peel and creosote bush leaves.

Thus, lignocellulosic materials including forestry, agricultural and agro-industrial wastes contain several high value substances such as sugars, minerals and protein. Disposal of these wastes to the soil or landfill causes serious environmental problems, besides to constitute loss of these value added substances. Therefore, the development of processes for reuse of these wastes is of great interest. Since these wastes are rich in sugars, which are easily assimilated by micro-organisms, they are very appropriate for use as raw materials in the production of industrially relevant compounds by fermentation. Products of interest for food, pharmaceutical and biofuels industries may be produced by fermentation of lignocellulose, both by submerged or solid-state fermentation systems. Submerged fermentation systems require a previous step for fractionation of the lignocellulose to produce a sugar rich hydrolysate fermentable by micro-organisms. Different treatments may be used for the lignocellulose fractionation and the selection of the best one must be done considering the product that will be produced and economical aspects. Solid-state fermentation systems are other interesting option to reuse lignocellulose and have as advantage the no need of raw material fractionation previous the use in the fermentation stage. Besides to serve as low-cost raw materials for the production of important metabolites, the lignocellulose reuse in fermentation processes is an environment friendly method of waste management.

Chapter 3

Fermentation Biotechnology: An Overview

INTRODUCTION

'Biotechnology', the short form of biological technology, defies precise definition. The term biotechnology came into general use in the mid 1970s, gradually superseding the more ambiguous 'bioengineering', which was variously used, to describe chemical engineering processes using organisms and/or their products, particularly fermenter design, control, product recovery and purification. Most scientists agree that all processes that utilise biological organisms constitute biotechnology, but what is disputed is which processes do not. Ancient fermented food processes, such as making bread, wine, cheese, curds, idli, dosa, etc. some of which are some 6,000 yr old, and developed long before man had any knowledge of the existence of the micro-organisms involved, also genuinely constitute biotechnology. However, for the sake of convenience, many people exclude these traditional processes from the realm of biotechnology. Conventional agriculture is a well-developed industry in its own right, but in practice, this is not included in biotechnology. Aspects of 'modern biotechnology' may have significant effects on 'traditional biotechnology'. Genetic manipulation to improve brewing and baking yeasts or to introduce new characteristics in crops, biological control of plant pests, and new methods of diagnosing and preventing plant, human and animal disease, are all now realisable. Whatever the definition, experimental production of new varieties of organisms, is one of the important objectives of biotechnology.

Biotechnology deals with the introduction of biological methods within the framework of technical processes and industrial production. It involves the application of microbiology and biochemistry together with technical chemistry and process engineering.

FERMENTATION TECHNOLOGY

Fermentation technology is the oldest of all biotechnological processes. The term is derived from the Latin verb fevere, to boil- the appearance of fruit extracts or malted grain acted upon by yeast, during the production of alcohol. Fermentation is a process of chemical change caused by organisms or their products, usually producing effervescence and heat. Microbiologists consider fermentation as 'any process for the production of a product by means of mass culture of micro-organisms'. Biochemists consider fermentation as 'an energy-generating process in which organic compounds act both as electron donors and acceptors', hence fermentation is 'an anaerobic process where energy is produced without the participation of oxygen or other inorganic electron acceptors'. In biotechnology, the microbiological concept is widely used.

MICRO-ORGANISMS

Several species belonging to the following categories of micro-organisms are used in fermentation processes:

Prokaryotic	Unicellular: Bacteria, cyanobacteria
	Multicellular: Byanobacteria
Eukaryotic	Unicellular: Yeasts, algae
	Multicellular: Fungi, algae

Unicellular and micro-fauna are rarely a part of fermentation processes, while isolated cells of multicellular animals are frequently cultured.

MICROBIAL GROWTH

Requirements for Artificial Culture

The growth of organisms involves complex energy based processes. The rate of growth of micro-organisms is dependent upon several culture conditions, which should provide for the energy required for various chemical reactions.

The production of a specific compound requires very precise cultural conditions at a particular growth rate. Many systems now operate under computer control. The rate of growth of micro-organisms and hence the synthesis of various chemical compounds under artificial culture, requires organism specific chemical compounds as the growth (nutrient) medium. The kinds and relative concentrations of the ingredients of the medium, the pH, temperature, purity of the cultured organism, etc. influence microbial growth and hence the production of biomass (the total mass of cells or the organism being cultured), and the synthesis of various compounds.

The nutrient sources for industrial fermentation are given in Table 3.1.

Table 3.1: Nutrient sources for industrial fermentation.

Nutrient	Raw material
Carbon source	
Glucose	Corn sugar, Starch, Cellulose
Sucrose	Sugarcane, Sugar beet molasses
Lactose	Milk whey
Fats	Vegetable oils
Hydrocarbons	Petroleum fractions
Nitrogen source	
Protein	Soyabean meal, Cornsteep liquor, Distillers solubles
Ammonia	Pure ammonia or ammonium salts
Nitrate	Nitrate salts
Nitrogen	Air
Phosphorous source	Phosphate salts

Phases of Microbial Growth

When a particular organism is introduced into a selected growth medium, the medium is inoculated with the particular organism. Growth of the inoculum does not occur immediately, but takes a little while. This is the period of adaptation, called the lag phase.

Following the lag phase, the rate of growth of the organism steadily increases, for a certain period-this period is the log or exponential phase. After a certain time of exponential phase, the rate of growth slows down, due to the continuously falling concentrations of nutrients and/or a continuously increasing (accumulating) concentrations of toxic substances. This phase, where the increase of the rate of growth is checked, is the deceleration phase. After the deceleration phase, growth ceases and the culture enters a stationary phase or a steady state. The biomass remains constant, except when certain accumulated chemicals in the culture lyse the cells (chemolysis). Unless other micro-organisms contaminate the culture, the chemical constitution remains unchanged. Mutation of the organism in the culture can also be a source of contamination, called internal contamination.

FERMENTERS AND BIOREACTORS

A fermenter is the set up to carry out the process of fermentation. The fermenters vary from laboratory experimental models of one or two litres capacity, to industrial models of several hundred litres capacity, which refers to the volume of the main fermenting vessel. A bioreactor differs from a fermenter in that the former is used for the mass culture of plant or animal cells, instead of micro-organisms. The chemical compounds synthesised by these cultured cells, such as therapeutic agents, can be extracted easily from the cell biomass. The design engineering and operational parameters of both fermenters and bioreactors are identical. With the involvement of micro-organisms as elicitors in some situations, the distinction between the two concepts is being gradually obliterated.

DESIGN OF INDUSTRIAL FERMENTATION PROCESS

The fermentation process requires the following:

1. A pure culture of the chosen organism, in sufficient quantity and in the correct physiological state.
2. Sterilised, carefully composed medium for growth of the organism.
3. A seed fermenter, a mini-model of production fermenter to develop an inoculum to initiate the process in the main fermenter.
4. A production fermenter, the functional large model.
5. Equipment for (i) drawing the culture medium in steady state, (ii) cell separation, (iii) collection of cell free supernatant, (iv) product purification, and (v) effluent treatment.

Items (1) to (3) above constitute the upstream and (5) constitutes the downstream, of the fermentation process. Fermenters/bioreactors are equipped with an aerator to supply oxygen in aerobic processes, a stirrer to keep the concentration of the medium uniform, and a thermostat to regulate temperature, a pH detector and similar control devices.

TYPES OF CULTURE SYSTEMS

Batch Processing or Culture

At about the onset of the stationary phase, the culture is disbanded for the recovery of its biomass (cells, organism) or the compounds that accumulated in the medium (alcohol, amino acids), and a new batch is set up. This is batch processing or batch culture. The best advantage of batch processing is the optimum levels of product recovery. The disadvantages are the wastage of unused nutrients, the peaked input of labour and the time lost between batches.

Continuous Processing or Culture

The culture medium may be designed such that growth is limited by the availability of one or two components of the medium. When the initial quantity of this component is exhausted, growth ceases and a steady state is reached, but growth is renewed by the addition of the limiting component. A certain amount of the whole culture medium (aliquot) can also be added periodically, at the time when steady state sets in. The addition of nutrients will increase the volume of the medium in the fermentation vessel. It is so arranged that the increased volume will drain off as an overflow, which is collected and used for recovery of products. At each step of addition of the medium, the medium becomes dilute both in terms of the concentration of the biomass and the products. New growth, stimulated by the added medium, will increase the biomass and the products, till another steady state sets in; and another aliquot of medium will reverse the process. This is continuous culture or processing. Since the growth of the organism is controlled by the availability of growth limiting chemical component of the medium, this system is called a chemostat. The rate at which aliquots are added is the dilution rate that is in effect the factor that dictates the rate of growth. The events in a continuous culture are:

1. The growth rate of cells will be less than the dilution rate and they will be washed out of the vessel at a rate greater than they are being produced, resulting in a decrease of biomass concentration both within the vessel and in the overflow.

2. The substrate concentration in the vessel will rise because fewer cells are left in the vessel to consume it.

3. The increased substrate concentration in the vessel will result in the cells growing at a rate greater than the dilution rate and biomass concentration will increase.

4. The steady state will be re-established.

Hence, a chemostat is a nutrient limited self-balancing culture system, which may be maintained in a steady state over a wide range of sub-maximum specific growth rates. The continuous processing offers the most control over the growth of cells.

Commercial adaptation of continuous processing is confined to biomass production, and to a limited extent to the production of potable and industrial alcohol. The steady state of continuous processing is advantageous as the system is far easier to control. During batch processing, heat output, acid or alkali production, and oxygen consumption will range from very low rates at the start to very high rates during the late exponential phase. The control of the environmental factors of the system becomes difficult. In the continuous processing, the rates of consumption of nutrients and those of the output chemicals are maintainable at optimal levels. Besides, the labour demand is also more uniform.

Continuous processing may suffer from contamination, both from within and outside. The fermenter design, along with strict operational control, should actually take care of this problem.

The production of growth associated products like ethanol is more efficient in continuous processing, particularly for industrial use. Continuous culturing is highly selective and favours the propagation of the best-adapted organism in culture. A commercial organism is highly mutated such that it will produce very high amounts of the desired product. But physiologically such strains are inefficient and give way in culture to inferior producers-a kind of contamination from within.

Fed-Batch Culture or Processing

In the fed-batch system, a fresh aliquot of the medium is continuously or periodically added, without the removal of the culture fluid. The fermenter is designed to accommodate the increasing volumes. The system is always at a quasi-steady state.

Fed-batch achieved some appreciable degree of process and product control. A low but constantly replenished medium has the following advantages:

1. Maintaining conditions in the culture within the aeration capacity of the fermenter.
2. Removing the repressive effects of medium components such as rapidly used carbon and nitrogen sources and phosphate.
3. Avoiding the toxic effects of a medium component.
4. Providing limiting level of a required nutrient for an auxotrophic strain.

Production of baker's yeast is mostly by fed-batch culture, where biomass is the desired product. Diluting the culture with a batch of fresh medium prevents the production of ethanol, at the expense of biomass; the moment traces of ethanol were detected in the exhaust gas.

The production of penicillin, a secondary metabolite, is also by fed-batch method. Penicillin process has two stages: an initial growth phase followed by the production phase called the 'idiophase'. The culture is maintained at low levels of biomass and phenyl acetic acid, the precursor of penicillin, is fed into the fermenter continuously, but at a low rate, as the precursor is toxic to the organism at higher concentrations.

PRODUCTS OF FERMENTATION PROCESSES

The growth of micro-organisms or other cells results in a wide range of products. Each culture operation has one or few set objectives. The process has to be monitored carefully and continuously, to maintain the precise conditions needed and recover optimum levels of products. Accordingly, fermentation processes aim at one or more of the following:

1. Production of cells (biomass) such as yeasts.
2. Extraction of metabolic products such amino acids, proteins (including enzymes), vitamins, alcohol, etc. for human and/or animal consumption or industrial use such as fertiliser production.
3. Modification of compounds (through the mediation of elicitors or through biotransformation).
4. Production of recombinant products.

Microbial Biomass

Microbial biomass is produced commercially as single cell protein (SCP) using such unicellular algae as species of Chlorella or Spirulina for human or animal consumption, or viable yeast cells needed for the baking industry, which was also used as human feed at one time. Bacterial biomass is used as animal feed. The biomass of *Fusarium graminearum* is also produced, for a similar use.

Microbial Metabolites

Primary metabolites

During the log or exponential phase organisms produce a variety of substances that are essential for their growth, such as nucleotides, nucleic acids, amino acids, proteins, carbohydrates, lipids, etc. or by-products of energy yielding metabolism such as ethanol, acetone, butanol, etc. This phase is described as the tropophase, and the products are usually called primary metabolites. Commercial examples of such products are given in Table 3.2.

Table 3.2: Examples of commercially produced primary metabolites.

Primary metabolite	Organism	Significance
Ethanol	Saccharomyces cerevisiae, Kluyveromyces fragilis	Alcoholic beverages
Citric acid	Aspergillus niger	Food industry
Acetone and butanol	Clostridium acetobutyricum	Solvents
Lysine Glutamic acid	Corynebacterium glutamacium	Nutritional additive flavour enhancer
Riboflavin	Ashbya gossipii, Eremothecium ashbyi	Nutritional
Vitamin B12	Pseudomonas denitrificans, Propionibacterium shermanii	Nutritional
Dextran	Leuconostoc mesenteroides	Industrial
Xanthan gum	Xanthomonas campestris	Industrial

Secondary metabolites

Organisms produce a number of products, other than the primary metabolites. The phase, during which products that have no obvious role in metabolism of the culture organisms are produced, is called the idiophase, and the products are called secondary metabolites. In reality, the distinction between the primary and secondary metabolites is not a straightjacket situation. Many secondary metabolites are produced from intermediates and end products of secondary metabolism. Some like those of the *Enterobacteriaceae* do not undergo secondary metabolism. Examples of secondary metabolites are given in Table 3.3.

Table 3.3: Examples of commercially produced secondary metabolites.

Metabolite	Species	Significance
Penicillin	Penicillium chrysogenum	Antibiotic
Erythromycin	Streptomyces erythreus	Antibiotic
Streptomycin	Streptomyces griseus	Antibiotic
Cephalosporin	Cephalosporium acrimonium	Antibiotic
Griseofulvin	Penicillium griseofulvin	Antifungal antibiotic
Cyclosporin A	Tolypocladium inflatum	Immunosuppressant
Gibberellin	Gibberella fujikuroi	Plant growth regulator

Secondary metabolism may be repressed in certain cases. Glucose represses the production of actinomycin, penicillin, neomycin and streptomycin; phosphate represses streptomycin and tetracyclin production. Hence, the culture medium for secondary metabolite production should be carefully chosen.

Production of Enzymes

Industrial production of enzymes is needed for the commercial production of food and beverages. Enzymes are also used in clinical or industrial analysis and now they are even added to washing powders (cellulase, protease, lipase). Enzymes may be produced by microbial, plant or animal cultures. Even plant and animal enzymes can be produced by microbial fermentation. While most enzymes are produced in the tropophase, some like the amylases (by *Bacillus stearothermophilus*) are produced in the idiophase, and hence are secondary metabolites. Examples of enzymes produced through fermentation processes are given in Table 3.4.

Table 3.4: Examples of commercially produced enzymes.

Organism	Enzyme
Aspergillus oryzae	Amylases
Aspergillus niger	Glucamylase
Trichoderma reesii	Cellulase
Saccharomyces cerevisiea	Invertase
Kluyveromyces fragilis	Lactase
Saccharomycopsis lipolytica	Lipase
Aspergillus species	Pectinases and proteases
Bacillus species	Proteases
Mucor pusillus	Microbial rennet
Mucor meihei	Microbial rennet

Food Industray Products

A very wide range of innumerable products of the food industry, such as sour cream, yoghurt, cheeses, fermented meats, bread and other bakery products, alcoholic beverages, vinegar, fermented vegetables and pickles, etc. are produced through microbial fermentation processes. The efficiency of the strains of the organisms used, and the processes are being continuously improved to market quality products at more reasonable costs.

Recombinant Products

Recombinant DNA technology has made it possible to introduce genes from any organism into micro-organisms and vice versa, resulting in transgenic organisms and the latter are made to produce the gene product. Genetically manipulated *Escherichia coli*, *Saccharomyces cerevisiae*, other yeasts and even filamentous fungi are now being used to produce interferon, insulin, human serum albumin, and several other products.

Biotransformation

Production of a structurally similar compound from a particular one, during the fermentation process is transformation, or biotransformation, or bioconversion. The oldest instance of this process is the production of acetic acid from ethanol.

Immobilised plant cells may be used for biotransformation. Using alginate as the immobilising polymer, digitoxin from Digitalis lanata was converted into digoxin, which is a therapeutic agent in great demand. Similarly, codeinone was converted into codeine and tyrosine from Mucuna pruriens was converted into Dihydroxyphenylalanine (DOPA).

Elicitors

It is possible to induce production or enhance production of a compound in cultures by using elicitors, which may be micro-organisms. For example, *Saccharomyces cerevisiae* was an efficient elicitor in the production of glyceollin (Glycine max) and berberine (*Thalictrum rugosum*). *Rhizopus arrhizus* trebled diosgenin production by *Dioscorea deltoidea*. The production of morphine and codeine by *Papaver somniferum* was increased 18 times by Verticillium dahliae.

GENETIC IMPROVEMENT OF FERMENTATION PROCESSES

The genome of the organism ultimately controls its metabolism. Although improved fermenter engineering design and optimal cultural conditions can quantitatively enhance the microbial products, this will only be up to a limit. Genetic improvement of the organism is fundamental to the success of fermentation technology. Mutation and recombination are the two ways to meet this end.

Mutation

A certain amount of mutational change in the genome occurs as a natural process, though the probability is small. Exposing a culture of a micro-organism to UV light, ionising radiation or certain chemicals, enhances the rate of occurrence of mutations. But it is a tremendous task for the industrial geneticist to screen the very large number of randomly produced mutants and to select the ones with the desired qualities. The synthesis of a number of products of cell metabolism is controlled by a 'feed-back inhibition'. When a compound reaches a particular level of accumulation, its synthesis is stopped. Synthesis starts again when the level of the compound falls below the specific level. If a mutant is produced, in which the feedback signalling is suppressed, the product is synthesised continuously. By such a manipulation, a high producing strain of *Corynebacterium glutamacium* was developed to recover very high quantities of lysine. Such strains that do not produce controlling end products are called auxotrophs.

Recombination

Recombination is defined as any process that brings together genes from different sources. A strain of *Brevibacterium flavum* is a high producer of lysine, but is limited by its poor capacity to absorb glucose. Another strain of the bacterium, which is an efficient absorber of glucose but which does not produce lysine, was used to develop a recombinant strain, through protoplast fusion. The new strain utilises high levels of glucose and yields higher levels of lysine.

A gene for the synthesis of phenylalanine was transferred to a chosen strain of *Escherichia coli*, which was a non-producer, but a good experimental and production tool. Transformation of a high cephalosporin producing strain of *Cephalosporium acremonium* with a plasmid containing the gene REXH has significantly increased the titre. A number of human proteins, such as insulin, human growth hormone, bone growth factor, alpha, beta and gamma interferons, interleukin-2, tumour necrosis factor, tissue plasminogen activator, blood clotting factor VIII, epidermal growth factor, granulocyte colony stimulating factor, erythropoietin, etc. are being produced through recombinant micro-organisms.

DNA Manipulation

In vitro DNA technology was used to increase the number of copies of a critical pathway gene (operon), as for example the production of threonine in *Escherichia coli*, at rates 40 to 50 times higher than usual.

Thus, fermentation technology is a very vibrant and fast growing area of biotechnology, absorbing an ever increasing processes and products. With a longer history than any area of biological sciences, fermentation technology has a longer and brighter future, in the service of mankind, covering such important areas as food and medicine.

Chapter 4

Microbial Growth Kinetics

INTRODUCTION

Kinetics, is a branch of natural science that deals with the rates and mechanisms of any processes—physical, chemical, or biological. Kinetic studies in microbiology cover all dynamic manifestations of microbial life: growth itself, survival and death, product formation, adaptations, mutations, cell cycles, environmental effects, and biological interactions. Kinetics provides a theoretical framework for optimal design in biotechnologies based on fermentation and enzyme catalysis, as well as on employment of outdoor activity of natural microbial populations (waste-water treatment, soil bioremediation, etc.).

Contrary to simple rates measurements, kinetic studies require the perception of the underlying basic mechanisms of studied processes. Kinetic mechanistic studies are those that interpret some complex process as an interplay of several simpler reactions, for example, cell growth can be explained through activity of enzymes and microbial community dynamics can be interpreted through behaviour of individual cells and populations. Ideally, mechanistic studies infer the coupling of experimental measurements with analysis of simulating mathematical models. The models formalise postulated mechanisms, so that the comparison of observations and the model's predictions allows one to discard an incorrect hypotheses.

The quantitative studies in microbiology often involve the assessment of growth stoichiometry. Stoichiometry is the quantitative relationship between reactants and products in a chemical reaction. In microbiology, stoichiometry stands for a quantitative relationship between substrates and products of microbial processes, including biomass formation (the consequence of complying with mass and energy conservation laws). In practical terms, kinetic and stoichiometry are tightly linked to each other, but stoichiometry mainly addresses problems of a static nature (how much? in what proportion?), whereas kinetics considers the dynamics questions (at what rate? by which mechanism?).

It is well known that micro-organisms fuel their lives by performing oxidation/reduction reactions that generate the energy and reducing power needed to construct and maintain themselves. Because redox reactions are nearly always very slow unless catalysed, micro-organisms produce enzyme catalysts that increase the kinetics of their essential reactions to rates fast enough for them to exploit the chemical resources available in their environment. Engineers want to take advantage of these microbially catalysed reactions, because the chemical resources of the micro-organisms usually are the pollutants that the engineers must control. For example, the biochemical oxygen demand (BOD) is an organic electron donor for heterotrophic bacteria, NH_4^+–N is an inorganic electron donor for nitrifying bacteria, NO_3^-–N is an electron acceptor for denitrifying bacteria, and PO_4^{3-} is a nutrient for all micro-organisms. In trying to employ micro-organisms for pollution control, engineers must recognise two interrelated principles: First, metabolically active micro-organisms catalyse the pollutant-removing reactions. The rate of pollutant removal depends on the concentration of the catalyst, or the active biomass. Second, the active biomass is grown and sustained through the utilisation of its energy- and electron-generating primary substrates, which are its electron donor and electron acceptor. The rate of production of active biomass is proportional to the utilisation rate of the primary substrates.

The connection between the active biomass (the catalyst) and the primary substrates is the most fundamental factor needed for understanding and exploiting microbial systems for pollution control. Because those connections must be made systematically and quantitatively for engineering design and operation, mass-balance modelling is an essential tool.

MICROBIAL GROWTH KINETICS

Microbial growth is described as an orderly increase in all chemical components in the presence of suitable medium and the culture environment. There are four types of microbial growth: bacteria grow by binary fission, yeast divide by budding, fungi divided by chain elongation and branching and viruses grow intracellularly in host cells. Growth of the cell mass or cell number can be described quantitatively as a doubling of the cell number per unit time for bacteria and yeasts or a doubling of biomass per unit time for filamentous organisms such as fungi. Measuring techniques involve direct counts, visually or using instruments and indirect cell counts.

The first method is to measure the dry weight of the cell material in a fixed volume of the culture by measuring the dry weight of the cell material in a given volume of the culture. The cells need to be removed from the medium and dried. Another method is to use the spectrophotometer to estimate absorbance of cell suspensions. The absorbance at a particular wavelength is proportional to the cell concentration. By plotting a standard curve of absorbance versus cell concentration, the cell concentration of an unknown sample can be calculated. Direct microscopic counts using a counting chamber can be used but this technique has limitations as the dead cells cannot be distinguished from living cells. Electronic counting chambers count numbers and measure size distribution of cells. These are more often used to count eucaryotic cells like blood cells. Indirect viable cell counts also called plate counts may be used. This involves plating out dilutions of a culture on nutrient agar. Each viable unit will form a colony and each colony that can be counted is called a colony forming unit (CFU) and the number of CUFs is related to the viable count in the sample. Turbidity measurement is a fast and nondestructive method especially for counting large numbers of bacteria in clear liquid media and broths – but cannot detect cell densities less than 10^7 cells per ml. The biochemical activity may also be measured, e.g. O_2 uptake, CO_2 production, ATP production. This method requires a fixed standard to relate chemical activity to cell mass and/or volume. Bacterial growth rates during the phase of exponential growth, under standard nutritional conditions (culture medium, temperature, pH, etc.), define the bacterium's generation time. Generation times for bacteria vary from about 12 minutes to 24 hr. The generation time for *E. coli* in the laboratory is 15–20 min. Symbionts such as *Rhizobium* tend to have a longer generation time. Some pathogenic bacteria, e.g., *Mycobacterium tuberculosis* have especially long generation times and this is thought to be an advantage to their virulence. When growing exponentially by binary fission, the increase in a bacterial population is by geometric progression.

The generation time is the time interval required for cells (or population) to divide:

$$G = t/n$$

where, G is generation time, n is number of generations and t is time in min/hr.

Growth Medium

A growth medium or culture medium is a liquid or gel designed to support the growth of micro-organisms or cells. There are different types of media for growing different types of cells. There are two major types of growth media: those used for cell culture, which use specific cell types derived from plants or animals, and microbiological culture, which are used for growing micro-organisms, such as bacteria or yeast. The most common growth media for micro-organisms are nutrient broths and agar plates; specialised media are sometimes required for micro-organism and cell culture growth. Some organisms, termed fastidious organisms, require specialised environments due to complex nutritional requirements. Viruses, for example, are obligate intracellular parasites and require a growth medium containing living cells.

Types of growth media

The most common growth media for micro-organisms are nutrient broths (liquid nutrient medium) or lysogeny broth (LB) medium. Liquid media are often mixed with agar and poured into *Petri* dishes to solidify. These agar plates provide a solid medium on which microbes may be cultured. They remain solid, as very few bacteria are able to decompose agar. Bacteria grown in liquid cultures often form colloidal suspensions. The differences between growth media used for cell culture and those used for microbiological culture are because cells derived from whole organisms and grown in culture often cannot grow without the addition of, for instance, hormones or growth factors which usually occur *in vivo*. In the case of animal cells, this difficulty is often addressed by the addition of blood serum or a synthetic serum replacement to the medium. In the case of micro-organisms, there are no such limitations, as they are often unicellular organisms. One other major difference is that animal cells in culture are often grown on a flat surface to which they attach, and the medium is provided in a liquid form, which covers the cells. In contrast, bacteria such as *Escherichia coli* may be grown on solid media or in liquid media. An important distinction between growth media types is that of defined versus undefined media. A defined medium will have known quantities of all ingredients. For micro-organisms, they consist of providing trace elements and vitamins required by the microbe and especially a defined carbon source and nitrogen source. Glucose or glycerol are often used as carbon sources, and ammonium salts or nitrates as inorganic nitrogen sources.

A process, which employs micro-organisms, animal cells and/or plant cells for the production of materials, is a bioprocess. Most biotechnical products are produced by fermentation. In fermentation, the products are formed by catalysts that catalyse their own synthesis. Enzymes are biological catalysts and are produced as secondary metabolites of enzyme fermentation.

There are many aspects that complicate the modelling of the bioprocesses. A fermentation process has both nonlinear and dynamic properties. The metabolic processes of the micro-organisms are very complicated and cannot be modelled precisely. Because of these reasons, traditional modelling methods fail to model bioprocesses accurately. The modelling is further complicated because the fermentation runs are usually quite short and large differences exist between different runs.

Fermentations can be operated in batch, fed-batch or continuous reactors. In batch reactor all components, except gaseous substrates such as oxygen, pH-controlling substances and antifoaming agents, are placed in the reactor in the beginning of the fermentation. During process there is no input nor output flows. In fed-batch process, nothing is removed from the reactor during the process, but one substrate component is added in order to control the reaction rate by its concentration. There are both input and output flows in a continuous process, but the reaction volume is kept constant.

CELLULAR PROCESS

Quantitative description of cellular processes is an indispensable tool in the design of fermentation processes. The two most important quantitative design parameters, yield and productivity, are quantitative measures that specify how the cells convert the substrates to the product. The yield specifies the amount of product obtained from the substrate, and it is of particular importance when the raw material costs make up a large fraction of the total costs, as exemplified in the production of solvents, antibiotics, alcohol, and other primary metabolites. The productivity specifies the rate of product formation, and is particularly important when the capital investments play an important role, such as in a growing market where there is an increasing demand for producing the product by a given capacity (or factory). These two design parameters can easily be derived from experimental data but, what is more difficult to predict, is how they change with the operating conditions, e.g., if the medium composition changes or the temperature changes. To do this it is necessary to set up a mathematical model.

A model is a set of relationships between the variables in the system being studied. These relationships are normally expressed in the form of mathematical equations, but they may also be specified as logic expressions (or cause/effect relationships) which are used in the operation of a process. The variables include any property that are of importance for the process, such as the agitation rate, the feed rate, pH, temperature, concentrations of substrates, metabolic products and biomass and the state of the biomass — often represented by the concentration of a set of key intracellular compounds. To set up a mathematical model it is necessary to specify a control volume wherein all the variables of interest are taken to be uniform. For fermentation processes the control volume is typically the whole bioreactor, but for large bioreactors the medium may be nonhomogeneous due to mixing problems and here it is necessary to divide the bioreactor into several control volumes. When the control volume is the whole bioreactor it may either be of constant volume or it may change with time depending on the operation of the bioprocess.

When the control volume has been defined, a set of balance equations can be specified for the variables of interest. These balance equations specify how material is flowing in and out of the control volume and how material is converted within the control volume. Rate equations (or kinetic expressions) specify the conversion of material within the control volume. They may be anything from a simple empirical correlation that specifies the product formation rate as a function of the medium composition to a complex model that accounts for all the major cellular reactions involved in the conversion of the substrates to the product. Independent of the model structure, the process of defining a quantitative description of a fermentation process involves a number of steps, as shown in Fig. 4.1.

A key aspect in setting up a model is to specify the model complexity. This depends on what the model is going to be used for. Specification of the model complexity involves defining the number of reactions to be considered in the model, and specification of the stoichiometry for these reactions. When the model complexity has been specified, rates of the cellular reactions considered in the model are described with mathematical expressions, i.e., the rates are specified as functions of the variables; namely the concentration of the substrates (and in some cases the metabolic products). These functions are normally referred to as kinetic expressions, since they specify the kinetics of the reactions considered in the model. This is an important step in the overall modelling cycle and in many cases, different kinetic expressions have to be examined before a satisfactory model is obtained.

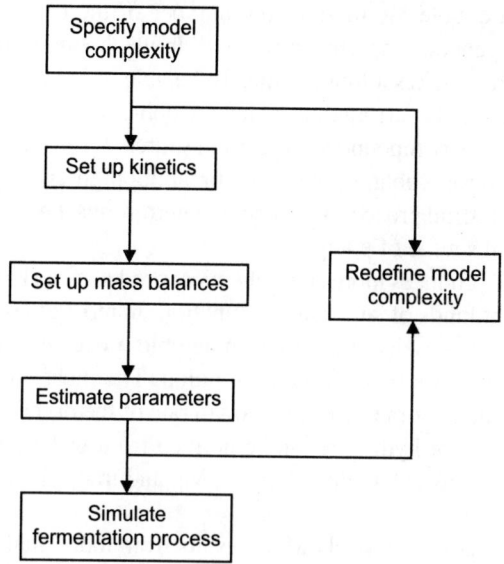

Fig. 4.1: Different steps in quantitative description of fermentation processes.

The next step in the modelling process is to combine the kinetics of the cellular reactions with a model for the reactor in which the cellular process occurs. Such a model specifies how the concentrations of substrates, biomass, and metabolic products change with time, and what flows in and out of the bioreactor. These bioreactor models are normally represented in terms of simple mass balances over the whole reactor, but more detailed reactor models may also be applied, if inhomogeneity of the medium is likely to play a role. The combination of the kinetic and the reactor model makes up a complete mathematical description of the fermentation process and this model can be used to simulate the profile of the different variables of the process, e.g., the substrate and product concentrations. However, before this can be done it is necessary to assign values to the parameters of the model. Some of these parameters are operating parameters, which are dependent on how the process is operated, e.g., the volumetric flow in and out of the bioreactor, whereas others are kinetic parameters which are associated with the cellular system. To assign values to these parameters, it is necessary to compare model simulations with experimental data and hereby estimate a parameter set that gives the best fit of the model to the experimental data. This is referred to as parameter estimation. The evaluation of the fit of the model to the experimental data can be done by simple visual inspection of the fit, but generally it is preferential to use a more rational procedure, such as minimising the sum of squared errors between the model and the experimental data.

In the following section we will consider the two different elements needed for setting up a bioprocess model, namely kinetic modelling and mass balances. This will lead to a description of different types of bioreactor operation, and hereby simple design problems can be illustrated.

KINETIC MODELLING OF CELL GROWTH

Every cell in nature has a finite lifetime and in order to maintain the species the continuous growth of the organisms is needed. A bacterial cell is able to duplicate itself. The duplication process is quite complicated and includes as many as 2000 different chemical reactions. The generation time, that is the time needed for the cells to double the mass or the number of the cells, depends on the number of factors, both nutritional and genetic. For *Escherichia coli* in ideal conditions the doubling time can be as short as 20 min. but usually it takes a longer time. To be able to live, reproduce and make products, a cell must obtain nutrients from its surroundings. Heterotrophic micro-organisms, which include most of the bacteria, require an organic compound as the carbon source. A cell can use either light or chemicals as its energy source. A chemotroph obtains energy by breaking high-energy bonds of chemicals. Most organisms that are used in industrial processes are chemoheterotrophs, i.e., organisms that use an organic carbon source and a chemical source of energy.

Cell as an open system and produces more cells, chemical products and heat from chemical substrates. A cell requires many different kinds of substrates to function. In most cases carbon is supplied as sugar or some other carbohydrate. Glucose is often used. In aerobic processes oxygen is a vital component. Oxygen can be fed into the process by continuous aeration. The most common source of nitrogen is ammonia or an ammonium salt. In some cases the growth rate of the organisms increases if amino acids are supplied. Required amounts of hydrogen can be derived from water and organic substrates. Other compounds that are needed for growth include P, S, K, Mg and trace elements, which are added in the growth media as inorganic salts.

All researchers in life sciences use models when results from individual experiments are interpreted and when results from several different experiments are compared with the aim of setting up a model that may explain the different observations. During the last 20 years there has been a revolution in experimental techniques applied in life sciences, and this has made possible far more detailed modelling of cellular processes. Furthermore, the availability of powerful computers has made it possible to solve complex numerical problems with a reasonable computational time; even complex mathematical models for biological processes can be handled and experimentally verified. However, often such detailed (or mechanistic) models are of little use in the design of a bioprocess, whereas they mainly serve a purpose in fundamental research of biological phenomena. In this section, we will focus on models which are useful for design of bioprocesses, but in order to give an overview of the different mathematical models applied to describe biological processes we start the presentation of kinetic models with a discussion of model complexity.

Model Structure and Model Complexity

Biological processes are *per se* extremely complex. Cell growth and metabolite formation are the result of a very large number of cellular reactions and events like gene expression, translation of mRNA into functional proteins, further processing of proteins into functional enzymes or structural proteins, and sequences of biochemical reactions leading to building blocks needed for synthesis of cellular components.

It is clear that a complete description of all these reactions and events cannot possibly be included in a mathematical model. In fermentation processes, where there is a large population of cells, non-homogeneity of the cells with respect to activity and function may add further to the complexity. In setting up fermentation models lumping of cellular reactions and events is, therefore, always done but the detail level considered in the model, i.e., the degree of lumping, depends on the aim of the modelling.

Fermentation models can roughly be divided into four groups depending on the detail level included in the model, (Fig. 4.2). The simplest description is the so-called unstructured models where the biomass is described by a single variable (often the total biomass concentration) and where no segregation in the cell population is considered.

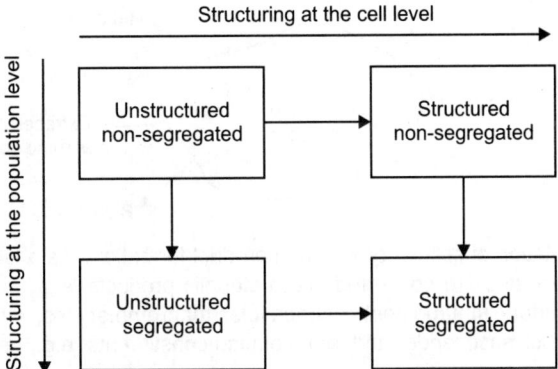

Fig. 4.2: Different types of model complexity, with increasing complexity going from the upper left corner to the lower right corner. When there is structuring at the cell level, specific intracellular events or reactions are considered in the model, and the biomass is structured into two or more variables. When there is structuring at the population level, segregation of the population is considered, i.e., it is accounted for that not all the cells in the population are identical.

These models can be combined with a segregated population model, where the individual cells in the population are described by a single variable, e.g., the cell mass or cell age, but often it is relevant to add further structure to the model when segregation in the cell population is considered. In the so-called structured models the biomass is described with more than one variable, i.e., structure in the biomass is considered. This structure may be anything from a few compartments to a detailed structuring into individual enzymes and macromolecular pools. It is clear that a very important element in mathematical modelling of fermentation processes is defining the model structure (or specifying the complexity of the model), and for this, a general rule can be stated: As simple as possible but not simpler. This rule implies that the basic mechanisms always should be included and that the model structure depends on the aim of the modelling exercise.

Definitions of Rates and Yield Coefficients

Before we turn to describing different unstructured models, a few definitions are needed. Figure 4.3 is a representation of the overall conversion of substrates into metabolic products and biomass components (or total biomass). The rates of substrate consumption can be determined during a fermentation process by measuring the concentration of these substrates in the medium. Similarly, the rates of formation of metabolic products and biomass can be determined from measurements of the corresponding concentrations. It is, therefore, possible to determine what flows into the total pool of cells and what flows out of this pool.

The inflow of a substrate is normally referred to as the substrate uptake rate and the outflow of a metabolic product is normally referred to as the product formation rate. From the direct measurements of the concentrations, one obtains so-called volumetric rates. Often it is convenient to normalise the rates with respect to the amount of biomass present, since the rates hereby easily can be compared between fermentation experiments, even when the amount of biomass changes. Such normalised rates are referred to as specific rates, and these are often represented as r_i, where the subscript indicates whether it is a substrate (s) or a metabolic product (p). The specific growth rate of the total biomass is also a very important variable, and it is generally designated μ.

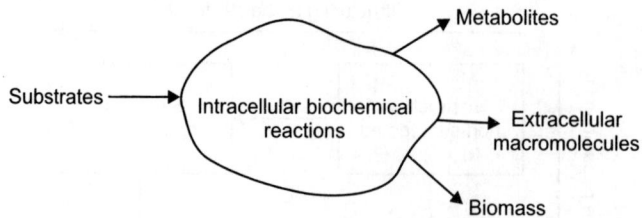

Fig. 4.3: General representation of cellular growth and product formation. Via a large number of intracellular biochemical reactions, substrates are converted into metabolic products, e.g., ethanol, acetate, lactate, or penicillin (and other secondary metabolites), extracellular macromolecules, e.g., a secreted enzyme, a heterologous protein, or a polysaccharide, and into biomass constituents, e.g., cellular protein, lipids, RNA, DNA, and carbohydrates.

The specific growth rate is related to the doubling time t_d (hr) of the biomass through:

$$t_d = \frac{\ln 2}{\mu} \qquad \qquad ...(4.1)$$

The doubling time t_d is equal to the generation time for a cell, i.e., the length of a cell cycle for unicellular organisms, which is frequently used by life scientists to quantify the rate of cell growth.

The specific rates, or the flow in and out of the cell, are very important design parameters since they are related to the productivity of the cell. Thus, the specific productivity of a given metabolite directly indicates the capacity of the cells to synthesise this metabolite. Furthermore, if the specific rate is multiplied by the biomass concentration in the bioreactor one obtains the volumetric productivity, or the capacity of the biomass population per reactor volume. In simple kinetic models the specific rates are specified as functions of the variables in the system, e.g., the substrate concentrations. In more complex models where the rates of the intracellular reactions are specified as functions of the variables in the system, the substrate uptake rates and product formation rates are given as functions of the intracellular reaction rates. Another class of very important design parameters are the yield coefficients, which quantify the amount of substrate recovered in biomass and the metabolic products. The yield coefficients are given as ratios of the specific rates, e.g., for the yield of biomass on a substrate:

$$Y_{sx} = \frac{\mu}{r_s} \qquad \qquad ...(4.2)$$

and similarly for the yield of a metabolic product on a substrate:

$$Y_{sp} = \frac{r_p}{r_s} \qquad \qquad ...(4.3)$$

The yield coefficients are clearly determined by how the carbon in the substrate is distributed among the different cellular pathways towards the end products of the catabolic and anabolic routes. These parameters can be considered as an overall determination of metabolic fluxes, a key aspect in modern physiological studies where methods to quantify intracellular, metabolic fluxes have become an important tool in defining the activity of the different pathways within the complete metabolic network. In the production of low-value added products, e.g., ethanol, antibiotics, amino acids and baker's yeast, it is generally of utmost importance to optimise the yield of product on the substrate and the target is, therefore, to direct as much carbon as possible towards the product and minimise the carbon flow to by-products (including biomass in metabolite production processes). In this process the yield coefficient is the most important design parameter, both for characterising different mutants and for characterising different fermentation schemes. For aerobic processes the yield of CO_2 from O_2 is often used to characterise the metabolism of the cells. This yield coefficient is referred to as the respiratory quotient (RQ). With complete respiration the RQ is close to 1 whereas if a metabolite is formed it deviates from 1.

The yield coefficients are always given with a double index that indicates the direction of the conversion, i.e., the yield for the conversion of substrate to biomass ($s \rightarrow x$) has the index sx. Thus, the yield coefficient Y_{xs} specifies the amount of substrate converted per unit biomass formed and, similarly, the yield coefficient, Y_{xp}, specifies the amount of product formed per unit biomass formed.

Effect of Temperature and pH

The reaction temperature and the pH of the growth medium are other process conditions with a bearing on the growth kinetics. It is normally desired to keep both of these variables constant (and at their optimal values) throughout the cultivation process, hence they are often called culture parameters to distinguish them from other variables such as reactant concentrations, stirring rate, oxygen supply rate, etc., which can change dramatically from the start to the end of a cultivation. The influence of temperature and pH on individual cell processes can be very different, and since the growth process is the result of many enzymatic processes the influence of both variables (or culture parameters) on the overall bioreaction is quite complex.

The influence of temperature on the maximum specific growth rate of a micro-organism is similar to that observed for the activity of an enzyme: an increase with increasing temperature up to a certain point where protein denaturation starts, and a rapid decrease beyond this temperature. For temperatures below the onset of protein denaturation the maximum specific growth rate increases in much the same way as for a normal chemical rate constant.

The influence of pH on the cellular activity is determined by the sensitivity of the individual enzymes to changes in the pH. Enzymes are normally only active within a certain pH range, and the total enzyme activity of the cell is, therefore, a complex function of the environmental pH. As an example, we shall consider the influence of pH on a single enzyme which is taken to represent the cell activity.

MASS BALANCES FOR IDEAL BIOREACTORS

The last step in modelling of fermentation processes is to combine the kinetic model with a model for the bioreactor. A bioreactor model is normally represented by a set of dynamic mass balances for the substrates, the metabolic products and the biomass, which describes the change in time of the concentration of these state variables. The bioreactor may be any type of device ranging from a test tube or a shake flask to a well-instrumented bioreactor. The feed is normally assumed to be sterile, i.e., the biomass concentration in the feed is zero. The bioreactor may be operated in three different modes: (i) batch, (ii) continuous, and (iii) fed-batch.

SECTION II

Industrial Fermentation and Solid State Fermentation

Chapter 5

Microbiological Aspects of Solid Substrate Fermentation

INTRODUCTION

There are many biotechnological processes that involve the growth of organisms on solid substrates in the absence or near absence of free water. Solid state fermentation (SSF) deals with substrates that are solid and contain low moisture levels. The most regularly used solid substrates are cereal grains (rice, wheat, barley and corn), legume seeds, wheat bran, lignocellulose materials such as straws, sawdust or wood shavings, and a wide range of plant and animal materials.

Most of these compounds are polymeric molecules – insoluble or sparingly soluble in water – but most are cheap and easily obtainable and represent a concentrated source of nutrients for microbial growth.

PROPERTIES OF SOLID SUBSTRATE FERMENTATION

Aerobic microbial transformation of solid materials or 'Solid Substrate Fermentation' (SSF) can be defined in terms of the following properties:

- A solid porous matrix which can be biodegradable or not, but with a large surface area per unit volume, in the range of 10^3 to 10^6 m^2/cm^3, for a ready microbial growth on the solid/gas interface.
- The matrix should absorb water amounting one or several times its dry weight with a relatively high water activity on the solid/gas interface in order to allow high rates of biochemical processes.
- Air mixture of oxygen with other gases and aerosols should flow under a relatively low pressure and mix the fermenting mash.
- The solid/gas interface should be a good habitat for the fast development of specific cultures of molds, yeasts or bacteria, either in pure or mixed cultures.
- The mechanical properties of the solid matrix should stand compression or gentle stirring, as required for a given fermentation process. This requires small granular or fibrous particles, which do not tend to break or stick to each other.

43

- The solid matrix should not be contaminated by inhibitors of microbial activities and should be able to absorb or contain available microbial foodstuffs such as carbohydrates (cellulose, starch, sugars) nitrogen sources (ammonia, urea, peptides) and mineral salts.

Typical examples of SSF are traditional fermentations such as:

- Japanese 'koji' which uses steamed rice as solid substrate inoculated with solid strains of the mold *Aspergillus oryzae*.
- Indonesian 'tempeh' or Indian 'ragi' which use steamed and cracked legume seeds as solid substrate and a variety of non toxic molds as microbial seed.
- French 'blue cheese' which uses perforated fresh cheese as substrate and selected molds, such as *Penicillium roquefortii* as inoculum.
- Composting of lignocellulosic fibres, naturally contaminated by a large variety of organisms including cellulolytic bacteria, molds and *Streptomyces sp.*
- In addition to traditional fermentations, new versions of SSF have been invented. For example, it is estimated that nearly a third of industrial SSF and koji processes in Japan has been modernised for large scale production of citric and itaconic acids.

Furthermore, new applications of SSF have been suggested for the production of antibiotics, secondaries metabolites or enriched foodstuffs.

LARGE-SCALE INDUSTRIAL PROCESSES

Presently SSF has been applied to large-scale industrial processes mainly in Japan. Traditional *koji*, manufactured in small wooden and bamboo trays, has changed gradually to more sophisticated processes: fixed bed room fermentations, rotating drum processes and automated stainless steel chambers or trays with microprocessors, electronics sensors and servomechanical stirring, loading and discharging. The usual scale in *sake* or *miso* factories is around 1 or 2 MT batch, but reactors can be made and delivered by engineering firms to a capacity as large as 20 T.

Outside Japan, Smith has reported medium scale production of enzymes, such as pectinases, in India. *Koji* type processes are widely used in small factories of the Far East and *koji* fermentation has been adapted to local conditions in United States (USA) and other Western countries, including Cuba. In France, a new firm was recently created to commercialise a process for pectinase production from sugarbeet pulp. Blue cheese production in France is being modernised with improvements on the mechanical conditioning of cheeses, production of mold spores and control of environmental conditions.

Composting, which was developed for small-scale production of mushrooms, has been modernised and scaled up in Europe and USA. Also, various firms in Europe and USA produce mushroom spawn by cultivating Agaricus, Pleurotus or Shii-Take aseptically on sterile grains in static conditions.

New versions for SSF reactors have been developed in France, Cuba, Chile and fundamental studies on process engineering are being conducted in Mexico.

SSF is a batch process: SSF is a batch process using natural heterogeneous materials, containing complex polymers like lignin, pectin, lignocellulose. SSF has been focused mainly to the production of feed, hydrolytic enzymes, organic acids, gibberelins, flavours and biopesticides. Most of the recent research activity on SSF is being done in developing nations as a possible alternative for conventional submerged fermentations, which are the main process in pharmaceutical and food industries in industrialised nations. SSF seems to have theoretical advantages over liquid substrate fermentation (LSF). Nevertheless, SSF has several important limitations.

Table 5.1 shows advantages and disadvantages of SSF compared to LSF. Table 5.2 presents a list of SSF processes in economical sectors of agro-industry, agriculture and fermentation industry. Most of the processes are commercialised in South-East Asian, African, and Latin American countries. Nevertheless, a resurgence of interest has occurred in Western and European countries over the last 10 years. The future potentials and applications of SSF for specific processes are discussed later.

Table 5.1: Comparison between liquid and solid substrate fermentations.

Factor	Liquid substrate fermentation	Solid substrate fermentation
Substrates	Soluble substrates (sugars)	Polymer insoluble substrates Starch cellulose pectines lignin
Aseptic conditions	Heat sterilisation and aseptic control	Vapour treatment, non sterile conditions
Water	High volumes of water consumed and effluents discarded	Limited consumption of water, low a_w. No effluent
Metabolic heating	Easy control of temperature	Low heat transfer capacity easy aeration and high surface exchange air/substrate
Aeration	Limitation of soluble oxygen high level of air required	
pH control	Easy pH control	Buffered solid substrates
Mecanical agitation	Good homogenisation	Static conditions prefered
Scale up	Industrial equipments available	Need for engineering and new design equipment
Inoculation	Easy inoculation , continuous process	Spore inoculation, batch
Contamination	Risks of contamination for single strain bacteria	Risk of contamination for low rate growth fungi
Energetic consideration	High energy consuming	Low energy consuming
Volume of equipment	High volumes and high cost technology	Low volumes and low costs of equipments
Effluent and pollution	High volumes of polluting effluents	No effluents, less pollution
Concentration S/Products	30–80 g/l	100/300g/l

Table 5.2: Main applications of SSF processes in various economical sectors.

Economical sector	Applications	Examples
Agro-food industry	Traditional food fermentations	Koji, tcznpch, rae, attickc, fermented cheeses
	Mushroom production and spawn	Agaricus, pleurotus, shn-take
	Bioconversion by-products	Sugar pulp bagass Coffee pulp
	Silage composting, detoxication	
	Food additives	Flavours, dyestuffs, essential fat and organic acids
Agriculture	Biocontrol, bioinsecticide	Beauveria metarhizium, tricho derma
	Plant growth, hormones	Gioberellins, rhizobium, inchoderma
	Mycorhisation, wild mushroom	Plant inoctiation,
Industrial fermentation	Enzymes production	Amylases, cellulases proteases, pectinases, xylanases
	Artibiotic prduction	Penecillin, feed and probiotics
	Organic acid production	Ciric acid, fumaric acid, gallic acid, lactic acid
	Ethanol prodixtion	Schwanniomyces sp. sbrch, malting and brewing
	Fungal metabolites	Hormones alcaloides

The following considerations summarise the present status of SSF:

- Potentially many high value products, as enzymes, primary and secondary metabolites, could be produced in SSF. But improvements in engineering and socio-economic aspects are required because processes must use cheap substrates locally available, low technology applicable in rural areas, and processes therefore must be simplified.

- Potential exists in developed countries, but close co-operation and exchange between developing and industrialised countries are required for further application of SSF.

- The greatest socio-economical potential of SSF is the raising of living standards through the production of protein rich foods for human consumption. Protein deficiency is a major cause of malnutrition and the problem will become worse with further increases in world population. Two alternatives can be explored to tackle this problem.

- Production of protein-enriched fermented foods for direct human consumption. This alternative involves starchy substrates for its initial nutritional caloric value. Successful production of such foods will require demonstration of economical feasibility, safety, significant nutritional improvement, and cultural acceptability.

- Production of fermented materials for animal feeding. Starchy substrates protein-enriched by SSF could be fed to monogastric animals or poultry. Fermented lignocellulosic substrates, by increasing its fibre digestibility, could be fed to ruminants. In this case, the economical feasibility should be favourable in comparison to the common model using protein of soyabean cake, a by-product of soyabean oil, product of soyabean oil.

The Orstom group has been working on solid fermentation process for improving protein content of cassava and other tropical starchy substrates using fungi, specially from *Aspergillus* group, in order to transform starch and mineral salts into fungal proteins. More recently, C. Soccol is working in Orstom laboratory in Montpellier, obtained good results with fungi of the *Rhizopus* group, of special interest in human traditional fermented foods. Increasing knowledge about specificity of strains of *Rhizopus* able to degrade crude granules of starch has been recently gathered at Orstom, which will drastically simplify the process of SSF. Recently the scientist are working on the following aspects:

- Protein enrichment of cassava and starchy substrates.
- Production of organic acids or ethanol by SSF from starchy substrate and cassava.
- Digestibility of fibres and lignocellulosic materials for animal feeding.
- Degradation of caffein in coffee pulp and ensiling for conservation and detoxification.
- Production of enzymes and fungal metabolites by SSF using sugarcane bagasse.

MICRO-ORGANISMS IN SSF PROCESSES

Bacteria, yeasts and fungi can grow on solid substrates, and find application in SSF processes. Filamentous fungi are the best adapted for SSF and dominate in research works. Some examples of SSF processes for each category of micro-organisms are reported in Table 5.3. Bacteria are mainly involved in composting, ensiling and some food processes. Yeasts can be used for ethanol and food or feed production. But filamentous fungi are the most important group of micro-organisms used in SSF process owing to their physiological, enzymological and biochemical properties. The hyphal mode of fungal growth and their good tolerance to low water activity (a_w) and high osmotic pressure conditions make fungi efficient and competitive in natural microflora for bioconversion of solid substrates.

Table 5.3: Main groups of micro-organisms involved in SSF processes.

Microflora	SSF process
Bacteria	
Bacillus sp.	Composting, natto, amylase
Pseudomonas sp.	Composting
Serratia sp.	Composting
Streptoccus sp.	Composting
Lactobacillus sp.	Ensiling, food
Clostidrium sp.	Ensiling, food
Yeast	
Endomicopsis burtonii	Tape, cassava, rice
Saccharomyces cerevisiae	Food, ethanol
Schwanniomyces castelli	Ethanol, amylase
Fungi	
Altemaria sp.	Composting
Aspergillus sp.	Composting, industrial, food
Fusarium sp.	Composting, gibberellins
Monilia sp.	Composting
Mucor sp.	Composting, food, enzyme
Rhizopus sp.	Composting, food, enzymes, organic acids
Phanerochaete chrysosporium	Composting, lignin degradation
Trichoderma sp.	Composting biological control, bioinsecticide
Beauveria sp., Metharizium sp.	Biological control, bioinsecticide
Amylomyces rouxii	Tape cassava, rice
Aspergillus oryzae	Koji, food, citric acid
Rhizopus oligosporus	Tempeh, soyabean, amylase, lipase
Aspergillus niger	Feed, proteins, amylase, ctric acid
Pleurotus oestreatus, sajor-caju	Mushroom
Lentinus edodes	Shii-take mushroom
Penicilium notatum, roquefortii	Penicillin, Cheese

Koji and *Tempeh* are the two most important applications of SSF with filamentous fungi. *Aspergillus oryzae* is grown on wheat bran and soyabean for *Koji* production, which is the first step of soyasauce or citric acid fermentation. Koji is a concentrated hydrolytic enzyme medium required in further steps of the fermentation process. Tempeh is an Indonesian fermented food produced by the growth of *Rhizopus oligosporus* on soyabeans. People consume the fermented product after cooking or toasting. The fungal fermentation allows better nutritive quality and degrades some antinutritional compounds contained in the crude soyabean.

The hyphal mode of growth gives a major advantage to filamentous fungi over unicellular micro-organisms in the colonisation of solid substrates and for the utilisation of available nutrients. The basic mode of fungal growth is a combination of apical extension of hyphal tips and the generation of new hyphal tips through branching. An important feature is that, although extension occurs only at the tip at a linear and constant rate, the frequency of branching makes the kinetic growth pattern of biomass exponential, mainly in the first steps of the vegetative stage. That point is important for growth modelling and will be discussed further. The hyphal mode of growth gives the filamentous fungi the power to penetrate into the solid substrates. The cell wall structure attached to the tip and the branching of the mycelium ensure a firm and solid structure. The hydrolytic enzymes are excreted at the hyphal tip, without large dilution like in the case of LSF, what makes the action of hydrolytic enzymes very efficient and allows penetration into most solid substrates. Penetration increases the accessibility of all available nutrients within particles.

Fungi cannot transport macromolecular substrates, but the hyphal growth allows a close contact between hyphae and substrate surface. The fungal mycelium synthetises and excretes high quantities of hydrolytic exoenzymes. The resulting contact catalysis is very efficient and the simple products are in close contact to enter the mycelium across the cell membrane to promote biosynthesis and fungal metabolic activities. This contact catalysis by enzymes can explain the logistic model of fungal growth commonly observed.

Substrates of SSF

All solid substrates have a common feature and have macromolecular structure. In general, substrates for SSF are composite and heterogeneous products from agriculture or by-products of agro-industry. This basic macromolecular structure (e.g. cellulose, starch, pectin, lignocellulose, fibres, etc.), confers the properties of a solid to the substrate. The structural macromolecule may simply provide an inert matrix (sugarcane bagasse, inert fibres, resins) within which the carbon and energy source (sugars, lipids, organic acids) are adsorbed. But generally, the macromolecular matrix represents the substrate and provides also the carbon and energy source.

Preparation and pre-treatment represent the necessary steps to convert the raw substrate into a form suitable for use, that include:

- Size reduction by grinding, rasping or chopping.
- Physical, chemical or enzymatic hydrolysis of polymers to increase substrate availability by the fungus.
- Supplementation with nutrients (phosphorus, nitrogen, salts) and setting the pH and moisture content, through a mineral solution.
- Cooking or vapour treatment for macromolecular structure pre-degradation and elimination of major contaminants. Pre-treatments will be discussed under individual applications.

The most significant problem of SSF is the high heterogeneity, which makes difficult to focus one category of hydrolytic processes, and leads to poor trials of modelling. This heterogeneity is of different nature:

- Non-uniform substrate structure (mixture of starch, lignocellulose, pectin).
- Variability between batches of substrates, limiting the reproducibility.
- Difficulty of mixing solid mass in fermentation, in order to avoid compaction, which causes non uniform growth, gradients of temperature, pH and moisture, that makes representative samples almost impossible to obtain.

Each macromolecular type of substrate presents different kind of heterogeneity. Lignocellulose occurs within plant cell walls, which consist of cellulose microfibrils embedded in lignin, hemicellulose and pectin. Each category of plant material contains variable proportion of each chemical compound.

Two major problems can limit lignocellulose breakdown:

- Cellulose exists in four recognised crystal structures known as celluloses I, II, III and IV. Various chemical or thermal treatments can change the structure from crystalline to amorphous.
- Different enzymes are necessary in order degrade cellulose, e.g. endo and exo-cellulases plus cellobiase.

Pectins are polymers of galacturonic acid with different ratio of methylation and branching. Exo-and endo pectinases and demethylases that hydrolyse pectin into galacturonic acid and methanol. Hemicelluloses are divided in major three groups: xylans, mannans and galactans. Most of hemicelluloses are heteropolymers containing two to four different types of sugar residues.

Lignin represents between 26 to 29% of lignocellulose, and is strongly bounded to cellulose and hemicellulose, hiding them and protecting them from the hydrolase attack. Lignin peroxidase is the major enzyme involved in lignin degradation. *Phanerochaete chrysosporium* is the most recognised fungi for lignin degradation.

So, lignocellulose hydrolysis is a very complex process. Effective cellulose hydrolysis requires the synergetic action of several cellulases, hemicellulases and lignin peroxidases. Despite this, lignocellulose is a very abundant and cheap natural renewable material, so a lot of work has been conducted on its microbial breakdown, specially with fungal species.

Starch is another very important and abundant natural solid substrate. Many micro-organisms are capable to hydrolyse starch, but generally its efficient hydrolysis requires previous gelatinisation. Some recent works concern the hydrolysis of the raw (crude or native) starch as it occurs naturally.

The chemical structure of starch is relatively simple compared to lignocellulose substrates. Essentially starch is composed of two related polymers in different proportions according to its source: amylose (16–30%) and amylopectin (65–85%). Amylose is a polymer of glucose linked by α-1,4 bonds, mainly in linear chains. Amylopectin is a large highly branched polymer of glucose including also α-1,6 bonds at the branch points.

Within the plant, cell starch is stored in the form of granules located in amyloplasts, intracellular organelles surrounded by a lipoprotein membrane. Starch granules are highly variable in size and shape depending on the plant material. Granules contain both amorphous and crystalline internal regions in respective proportions of about 30/70. During the process of gelatinisation, starch granules swell when heated in the presence of water, which involves the breaking of hydrogen bonds, especially in the crystalline regions.

Many micro-organisms can hydrolyse starch, specially fungi which are then suitable for SSF application involving starchy substrates. Glucoamylase, α-amylase, β-amylase, pullulanase and isoamylase are involved in the processes of starch degradation. Mainly α-amylase and glucoamylase are of importance for SSF.

α-amylase is an endo-amylase attacking α-1,4 bonds in random fashion which rapidly reduce molecular size of starch and consequently its viscosity producing liquefaction. Glucoamylase occurs almost exclusively in fungi including *Aspergillus* and *Rhizopus* groups. This exo amylase produces glucose units from amylose and amylopectin chains.

Micro-organisms generally prefer gelatinised starch. But large quantity of energy is required for gelatinisation so it would be attractive to use organisms growing well on raw (ungelatinised) starch. Different works are dedicated to isolate fungi producing enzymes able to degrade raw starch.

In various laboratories, many studies concerning SSF of cassava, a very common tropical starchy crop, have been conducted with the purpose of upgrading protein content, both for animal feeding using *Aspergillus sp.* and for direct human consumption, using *Rhizopus*.

Recently good results were obtained by Soccol in the protein enrichment of cassava and cassava bagasse using selected strains of *Rhizopus*. Biotransformed starchy flours were produced containing 10–12% of good quality protein, comparable to cereal. Such biotransformed cassava flour can be used as cereal substitute for bread making up to a level of 20%, without any perceivable change by the consumer.

Biomass Measurement

Biomass is a fundamental parameter in the characterisation of microbial growth. Its measurement is essential for kinetic studies on SSF. Direct determination of biomass in SSF is very difficult due to the problem of separating the microbial biomass from the substrate. This is especially true for SSF processes involving fungi, because the fungal hyphae penetrate into and bind tightly to the substrate. On the other hand, for the calculation of growth rates and yields, it is the absolute amount of biomass which is important. Methods that have been used for biomass estimation in SSF belong to one of the following categories.

Direct evaluation of biomass

Complete recovery of fungal biomass is possible only under artificial circumstances in membrane filter culture, because the membrane filter prevents the penetration of the fungal hyphae into the substrate. The whole of the fungal mycelium can be recovered simply by peeling it off the membrane and weighing it directly or after drying. Obviously, this method cannot be used in actual SSF. However, it could find application in the calibration of indirect methods of biomass determination. Indirect biomass estimation methods should be calibrated under conditions as similar as possible to the actual situation in SSF. The global mycelium composition could be estimated through analysis of the mycelium cultivated in LSF in conditions as close as possible to SSF cultivation.

Microscopic observations can also represent a good way to estimate fungal growth in SSF. Naturally, optic examination is not possible at high magnitude but only at stereo microscope. Scanning Electron Microscope (SEM) is a useful tool to observe the pattern of growth in SSF. New approaches and researches are developed for image analysis by computer software in order to evaluate the total length or volume of mycelium on SEM photography. Another new very promising approach is the Confocal Microscopy, based on specific reaction of fungal biomass with specific fluorochrome probes. Resulting 3D images of biomass can open new ways to appreciate and measure biomass *in situ* in a near future.

Since direct measurement of exact biomass in SSF is a very difficult task, then other approaches have been preferred.

That for, the global stoichiometric equation of microbial growth can be considered:preferred. That for, the global stoichiometric equation of microbial growth can be considered:

$$\text{Carbon source} + \text{Water} + \text{Oxygen} + \text{Phosphorus} + \text{Nitrogen}$$

$$\text{Biomass} + CO_2 + \text{Metabolites} + \text{Heat}$$

Variation in each component is strictly related to the variation of others when all coefficients are maintained constants. Then, measuring one of them allows to determine the evolution of the others.

Metabolic measurement of biomass

Respiratory metabolism: Oxygen consumption and carbon dioxide release result from respiration, the metabolic process by which aerobic micro-organisms derive most of their energy for growth. These metabolic activities are therefore growth associated and can be used for the estimation of biomass synthesis. As carbon compounds within the substrate are metabolised, they are converted into biomass and carbon dioxide. Production of carbon dioxide causes the weight of fermenting substrate to decrease during growth, and the amount of weight lost can be correlated to the amount of growth that has occurred. Growth estimation based on carbon dioxide release or oxygen consumption assumes that the metabolism of these compounds is completely growth associated, which means that the amount of biomass produced per unit of gas metabolised must be constant. Sugama and Okazaki reported that the ratio of mg CO_2 evolved to mg dry mycelia formed by *Aspergillus oryzae* on rice ranged from 0.91 to 1.26 mg CO_2 per mg dry mycelium. A gradual increase in this ratio was observed late in growth due to endogenous respiration. Drastic changes can be observed for the respiratory quotient which commonly changes with the growth phase, i.e. germination, rapid and vegetative growth, secondary metabolism, conidiation and degeneration of the mycelium. Evolution of CO_2 and O_2 during SSF of *Rhizopus* on cassava.

The measurement of either carbon dioxide evolution or oxygen consumption is most powerful when coupled with the use of a correlation model. The term correlation model is used here to denote a model that correlates biomass with a measurable parameter. Correlation models are not growth models as such, since they make no predictions as to how the measurable parameter changes with time. The usefulness of correlation models is that a biomass profile can be constructed by following the profile of the parameter during growth.

Application of these correlation models involving prediction of growth from oxygen uptake rates or carbon dioxide evolution rates requires the use of numerical techniques to solve the differential equations. A computer and appropriate software is therefore essential. If both the monitoring and computational equipment is available, then these correlation models provide a powerful mean of biomass estimation since continuous on line measurements can be made. Other advantages of monitoring effluent gas concentrations with paramagnetic and infrared analysers include the ability to monitor the respiratory quotient to ensure optimal substrate oxidation, the ability to incorporate automated feedback control over the aeration rate, and the non-destructive nature of the measurement procedure. The metabolic activity in SSF is very important for studying all theoretical and practical aspects of respirometric measurement of fungal biomass cultivated in SSF. Various authors reported data concerning laboratory and scale up experiments on respirometric measurement in several applications. An international training course was recently dedicated to advances and techniques to study fungal growth on SSF.

Production of extracellular enzymes or primary metabolites: Another metabolic activity that may be growth associated is extracellular enzyme production. Okazaki and coworkers claimed that the α-amylase activity was directly proportional to mycelial weight for *Aspergillus oryzae* grown in SSF on steamed rice. For growth of *Agaricus bisporus* on mushroom compost, mycelial mass was directly proportional to the extracellular laccase activity for 70 days.

Smith and others observed a good correlation between growth and hydrolytic enzymes such as amylases, cellulases or pectinases. Also, it has been observed a good correlation between mycelial growth and organic acid production, which can be estimated from pH measurement or correlated by HPLC analysis on extracts. In the case of *Rhizopus*, Soccol demonstrated a close correlation between fungal protein (biomass) and organic acids (citric, fumaric, lactic or acetic).

Biomass components

The biomass can also be estimated from measurements of a specific component, as long as the composition of the biomass is constant and stable and the fraction of the component representative.

Protein content: The most readily measured biomass component is protein. Generally the protein content is used (as determined by the Lowry method) to measure the growth rate of *Aspergillus niger* on cassava meal. Growth of *Chaetomium cellulolyticum* on wheat straw was determined from TCA insoluble nitrogen using the Kjeldahl method, biomass protein was then calculated as 6,25 times this value. In all cases the protein content of the biomass was assumed to be constant. Values of biomass protein content measured by the Biuret method were consistent with those measured by the Kjeldahl method. But, unfortunately, the Biuret method was not suitable for application to SSF because of nonspecific interference by substrate starch. The Folin method is more sensitive and allowed a greater dilution of the sample, which avoided interference from the starch in the substrate. Therefore it is advisable to chose the Folin technique to measure protein enrichment in starchy substrates.

Nucleic acids: DNA content has been used to estimate the biomass of *Aspergillus oryzae* on rice. The method was calibrated using the DNA contents of fungal mycelia obtained in submerged culture. DNA contents were higher during early growth and then decreased, levelling off as the stationary phase was approached. The method was corrected for the DNA content of rice, which remained unchanged since *Aspergillus oryzae* did not produce extracellular DNase. Methods based on DNA or RNA determination are reliable only if there is little nucleic acid in the substrate and no interfering chemicals are present.

Glucosamine: Glucosamine is a useful compound for the estimation of fungal biomass, taking advantage of the presence of chitin, poly-Nacetylglucosamine, in the cell walls of many fungi. Interference with this method may occur when using complex agricultural substrates containing glucosamine in glucoproteins. The accuracy of the glucosamine method for determination of fungal biomass depends on establishing a reliable conversion factor relating glucosamine to mycelial dry weight. However, the proportion of chitin in the mycelium will vary with age and the environmental conditions. Mycelial glucosamine contents ranged from 67 to 126 mg per g dry mycelium. Another disadvantage of this method is the tedious extraction procedures and processing times over 24 hr.

Ergosterol: Ergosterol is the predominant sterol in fungi. Glucosamine estimation was therefore compared with the estimation of ergosterol for the determination of growth of *Agaricus bisporus*. In solid cultures, directly proportional relationships for glucosamine and ergosterol against linear extension of the mycelium were obtained. Determination of ergosterol was claimed to be more convenient than glucosamine. It could be recovered and separated by HPLC and quantified simply by spectrophotometry, providing a sensitive index of biomass at low levels of growth. HPLC was necessary to separate the ergosterol from sterols endogenous to the solid substrate. However, Nout and others showed that the ergosterol content of *Rhizopus oligosporus* varied from 2 to 24 micrograms per mg dry biomass, depending on the culture conditions, aeration and substrate composition, concluding that it was an unreliable method for following growth.

Physical measurement of biomass: Peñaloza evaluated mycelial growth, based on the difference in the electric conductivity between biomass and the substrate. Good correlation with biomass was obtained and a model was proposed. Auria and others monitored the pressure drop in a packed bed during SSF of *Aspergillus niger* on a model solid substrate consisting of ion exchange resin beads. Pressure drop was closely correlated with protein production. Pressure drop is a parameter that is simple to measure and can be measured on-line. Further studies are required to determine whether the use of pressure drop is generally applicable for monitoring growth in SSF bioreactors under forced aeration.

In conclusion, the measurement of biomass in SSF is important to follow the kinetics of growth in relation to the metabolic activity. Measurement of metabolic activity by carbon dioxide evolution or oxygen consumption can be generally applied, whereas extracellular enzyme production will only be useful when enzyme production is reasonably growth-associated.

Vital staining with fluorescein diacetate has potential in providing basic information as to the mode of growth of fungi on complex solid surfaces, as this method can show the distribution of metabolic activity within the mycelium. But it cannot be measured on line.

On the other hand, in the production of protein enriched feeds, the protein content itself is of greater importance than the actual biomass concentration, and the variation in biomass protein content during growth becomes less relevant.

Overall, oxygen uptake and carbon dioxide evolution methods are probably the most promising techniques for biomass estimation in aerobic SSF as they provide on-line information. The monitoring and computing equipment is relatively expensive and will not be suitable for low technology or rural applications. No method is ideally suited to all situations, so the method most appropriate to a particular SSF application must be chosen in each case on the basis of simplicity, cost and accuracy. A good strategy can be a combination of several techniques based on the determination of different parameters that can correlate actual biomass with the material balance.

Environmental Factors

Environmental factors such as temperature, pH, water activity, oxygen levels and concentrations of nutrients and products significantly affect microbial growth and product formation. In submerged stirred cultures, environmental control is relatively simple because of the homogeneity of the suspension of microbial cells and of the solution of nutrients and products in the liquid phase.

The low moisture content of SSF enables a smaller reactor volume per substrate mass than LSF and also simplifies product recovery. However, serious problems arise with respect to mixing, heat exchange, oxygen transfer, moisture control and gradients of pH, nutrient and product as a consequence of the heterogeneity of the culture. The latter characteristic of SSF render the measurement and control of the above mentioned parameters difficult, laborious and often inaccurate, thereby limiting the industrial potential of this technology. Due to these problems, the micro-organisms that have been selected for SSF are the more tolerant to a wide range of cultivation conditions.

Moisture content and water activity (a_w)

SSF process can be defined as microbial growth on solid particles without the presence of free water. The water present in SSF systems exists in a complexed form within the solid matrix or as a thin layer either absorbed to the surface of the particles or less tightly bound within the capillary regions of the solid. Free water will only occur once the saturation capacity of the solid matrix is exceeded. However, the moisture level at which free moisture becomes apparent varies considerably between substrates and is dependant upon their water binding characteristics. For example, free water is observed when the moisture content exceeds 40% in maple bark and 50–55% in rice and cassava. With most lignocellulosic substrates free water becomes apparent before the 80% moisture level is reached.

The moisture levels in SSF processes, which vary between 30 and 85%, has a marked effect on growth kinetics. The optimum moisture level for the cultivation of *Aspergillus niger* on rice was 40%, whereas on coffee pulp the level was 80%, which illustrates the unreliability of moisture level as a parameter for predicting microbial growth. It is now generally accepted that the water requirements of micro-organisms should be defined in terms of the water activity (a_w) rather than the water content of the solid substrate. a_w is a thermodynamic parameter defined in relation to the chemical potential of water. a_w is related to the condensed phase of absorbed water, but it is well correlated (less than 0.2% error) to the relative humidity (RH).

The reduction of a_w has a marked effect on microbial growth. Typically, a reduction in a_w extends the lag phase, decreases the specific growth rate, and results in low amount of biomass produced. In general, bacteria require higher values of a_w for growth than fungi, thereby enabling fungi to compete more successfully at the a_w values encountered in SSF processes. With the exception of halophilic bacteria, few others grow at a_w values below 0.9 and most bacteria require considerably higher minimum a_w values for growth. Some fungi, on the other hand, stop growing only at a_w values as low as 0.62 and a number of fungi used in SSF processes have minimum growth a_w values between 0.8 and 0.9.

The optimum moisture content for growth and substrate utilisation is between 40 and 70% but depends upon the organism and the substrate used for cultivation. For example, cultivation of *Aspergillus niger* on starchy substrates, such as cassava and wheat bran, was optimal at moisture levels considerably lower than on coffee pulp or sugarcane bagasse. This is probably because of the greater water holding capacity of the latter substrate. The optimum a_w for growth of a limited number of fungi used in SSF processes was at least 0.96, whereas the minimum a_w required for growth was generally greater than 0.9. This suggests that fungi used in SSF processes are not especially xerophilic. The optimum a_w values for sporulation in *Trichoderma viride* and *Penicillium roqueforti* were lower than those for growth. Maintenance of the a_w at the growth optimum would allow fungal biomass to be produced without sporulation.

Temperature and heat transfer

Stoichiometric global equation of respiration is highly exothermic and heat generation by high levels of fungal activity within the solids lead to thermal gradients because of the limited heat transfer capacity of solid substrates. In aerobic processes, heat generation may be approximated from the rate or CO_2 evolution or O_2 consumption. Each mole of CO_2 produced during the oxidation of carbohydrates releases 673 Kcal. Therefore, it is important to measure CO_2 evolution during SSF because it is directly related to the risk of temperature increase. Detailed calculations of the relation between respiration, metabolic heat and temperature were discussed in early works on SSF with *Aspergillus niger* growing on cassava or potato starch. The overall rate or heat transfer may be limited by the rates of intra- and inter-particle heat transfer and by the rate at which heat is transferred from the particle surface to the gas phase. The thermal characteristics of organic material and the low moisture content in SSF are especially difficult conditions for heat transfer. Saucedo-Castañeda and others developed a mathematical model for evaluating the fundamental heat transfer mechanism in static SSF and more specifically to assess the importance of convection and conduction in heat dissipation. This model can be used as a basis for automatic control of static bioreactors.

Heat removal is probably the most crucial factor in large scale SSF processes. Conventional convection or conductive cooling devices are inadequate for dissipating metabolic heat due to the poor thermal conductivity of most solid substrates and results in unacceptable temperature gradients. Only evaporative cooling devices provide sufficient heat elimination capacity. Although the primary function of aeration during aerobic solid state cultivations was to supply oxygen for cell growth and to flush out the produced carbon dioxide, it also serves a critical function in heat and moisture transfer between the solids and the gas phase. The most efficient process for temperature control is water evaporation.

Maintaining constant temperature and moisture content simultaneously in large scale SSF is generally difficult, but using the proper ancillary equipment can do this. The reactor type can have a large influence on the quality of temperature control achieved. It depends highly of the type of SSF: static on clay or vertical exchangers, drums or mechanically agitated.

Control of pH and risks of contamination

The pH of a culture may change in response to metabolic activities. The most obvious reason is the secretion of organic acids such as citric, acetic or lactic, which will cause the pH to decrease, in the same way than ammonium salts consumption. On the other hand, the assimilation of organic acids which may be present in certain media will lead to an increase in pH, and urea hydrolysis will result in alkalinisation. The kinetics of pH variation depends highly on the micro-organism. With *Aspergillus sp.*, *Penicillium sp.* , and *Rhizopus sp.* the pH can drop very quickly below 3.0, for other types of fungi, like *Trichoderma, Sporotrichum, Pleurotus sp.* the pH is more stable between 4 and 5. Besides, the nature of the substrate has a strong influence on pH kinetics, due to the buffering effect of lignocellulosic materials.

The mixture of ammonium salt and urea is used to control the pH decrease during growth of *A. niger on* starchy substrates. A degree of pH control may be obtained by using different ratios of ammonium salts and urea in the substrate. Hydrolysis of urea liberates ammonia, which counteracts the rapid acidification resulting from uptake of the ammonium ion. In this manner the optimal growth of *Aspergillus niger* on granulated cassava meal when using a 3:2 ratio (on a nitrogen basis) of ammonium to urea is obtained. It was observed that during the first stage of cultivation the pH increased as the urea was hydrolysed. During the subsequent rapid growth stage, ammonium assimilation exceeded the rate of urea hydrolysis and the pH decreased, but increased again in the stationary phase. Thus, during the first cultivation, the pH remained within the limits of pH 5 to pH 6.2, whereas a lower urea concentration resulted in a rapid decrease in pH.

In a similar way, pH adjustment during pilot plant cultivation of *Trichoderma viride* on sugar-beet pulp was effective by spraying with urea solutions due to the urease activity of the micro-organism that caused an increase in pH by producing ammonia. Finally, in fungal or yeast SSF, bacterial contamination may be minimised or prevented by employing a suitably low pH.

Oxygen uptake

Aeration fulfils four main functions in SSF, namely (i) to maintain aerobic conditions, (ii) to desorb carbon dioxide, (iii) to regulate the substrate temperature and (iv) to regulate the moisture level. The gas environment may significantly affect the relative levels of biomass and enzyme production.

In aerobic LSF oxygen supply is often the growth limiting factor due to the low solubility of oxygen in water. In contrast, a solid state process allows free access of atmospheric oxygen to the substrate. Therefore, aeration may be easier than in submerged cultivations because of the rapid rate of oxygen diffusion into the water film surrounding the insoluble substrate particles, and also because of the very high surface of contact between gas phase, substrate and aerial mycelium. The control of the gas phase and air flow is a simple and practical mean to regulate gas transfer and generally no oxygen limitation is observed in SSF when the solid substrate is particular. It is important to maintain a good balance between the three phases in SSF. By this very simple aeration process, it is also possible to induce metabolic reactions, either by water stress, heat stress or temperature changes, all processes that can drastically change biochemical or metabolic behaviour.

SSF is a well-adapted process for cultivation of fungi on vegetal materials which are breakdown by excreted hydrolytic enzymes. In contrast with LSF where water is in large excess, water activity is a limiting factor in SSF. On the other hand, oxygen is a limiting factor in LSF but not in SSF, where aeration is promoted by the porous and particular structure and by the high surface are of contact which facilitate mass transfer between gas and liquid phases. SSF are aerobic processes where respiration is fundamental for energy supply but, because respiratory metabolism is highly exothermic, severe limitation of growth can occur when heat transfer is not efficient enough to avoid temperature increase.

Chapter 6

Solid State Fermentation for Bioconversion of Biomass and Agricultural

INTRODUCTION

Solid State fermentations (SSF) have attracted a renewed interest and attention from researchers due to recent developments in the field of microbial biotechnology. Hence, for the practical, economical and environmentally-friendly bioconversion of agro-industrial wastes, solid state or substrate fermentation has been researched globally and proved to be the ideal technology for this purpose. In this chapter some important aspects of solid-state cultivation system have been discussed, including the variety of substrates and micro-organisms used in SSF for the production of various end products; and the performance control of system by regulation of important factors.

AGRO-RESIDUE BIOCONVERSION IN SSF

Commonly used substrates in SSF are natural agricultural products, as well as agro-industrial waste residues and by-products serve as a source of carbon in SSF (Table 6.1). Lignocellulosic materials of agriculture origin compose more than 60% of plant biomass produced annually through the process of photosynthesis. This vast resource is the potential and renewable source of biofuels, biofertilisers, animal feed and chemical feedstocks. Lignocellulose may be a substrate for the production of value-added products, such as biofuels, biochemicals, biopesticides, biopromoters, or may even be a product itself after biotransformation.

In all applications the primary requirement is the hydrolysis of lignocellulose into fermentable sugars by lignocellulolytic enzymes, or appropriate modification of the structure of lignocellulose. Economical and effective lignocellulolytic enzyme complexes, containing cellulases, hemicellulases, pectinases and ligninases may be prepared by SSF (Table 6.2). Lignocellulose is also the raw material of the paper industry. To fully utilise the potential of lignocellulose, it has to be converted by chemical and/or biological processes. Solid substrate fermentation (SSF) plays an important role, and has a great perspective for the bioconversion of plant biomass. Lignocellulose may be a good feedstock for the production of biofuels, enzymes and other biochemical products by SSF. Crop residues (straw, corn by-products,

Table 6.1: Diverse range of agro-residues utilisation in SSF technology.

Substrates for SSF	Micro-organisms used in SSF
Starchy raw materials	*Aspergillus* spp.
Bannana waste	*A. niger*
Barley Husk	*Bjkendra adusta*
Corn cob	*A. niger*
Citrus peel	*A. niger*
Sugarcane by-products	*A. terreus*
Cassava	*Rhizopus oryzae*
Sugarbeet pulp	*Trichoderma viride*
Cassava	*T. resei and yeast*
Wheat straw	*T. reesei and Endomycopsis fibuleger*
Wheat straw	*T. reesei, Chaeotominum*
Sugarbeet pulp	*T. reesei and Fusarium oxysporum*
Sugarcane bagasse	*Polyporus* spp.
Saccharum munja	*Pleurotus* spp.
Residues wheat straw	*Coprinus* spp.
Cassava	*Sporotrichum pulverulentum*
Straw	*Candida utilis*
Sweet potato	*Pichia bartonii*
Fodder beets	*Saccharomyces cerevisiae*

Table 6.2: Agro-residues used in SSF for enzyme production.

Substrates	Micro-organisms	Enzymes
Bagasse, sawdust, corn cobs	*A. niger*	Cellulose, beta glucosidase
Corn cobs	*A. niger*	Cellulose
Wheat bran	*A. niger*	Glucoamylase
Sugarbeet pulp	*A. phoenicis*	Beta glucosidase
Wheat bran	*A. flavus*	Protease
Wheat bran	*A. carbonarius*	Pectinase
Wheat bran	*A. niveus*	Catalase
Sugarbeet pulp	*T. viride and A. niger*	Cellulase and amylase
Wheat bran and rice straw	*Trichoderma* spp., *A. ustus*, *Botritis* spp., *S. pulverulentum*	Cellulose, betaglucosidase, Xylanse
Wheat bran	*Pencillium* spp., *Geotrichwn*, *Candidum*, *Mucor meihei* and 2, *Rhizopus* spp.	Lipase
Sugarbeet pulp	*P. capsulatum*	Enzymes
Citrus pulp-pellets	*P. charlesii*, *Talaromyces flavus*, *Tubercularia vulgaris*	Pectic enzymes
Citrus pulp	*T. vulgaris*	Pectic enzymes
Bagasse	*Polyporous* spp.	Cellulase and ligninase

(Cont'd...)

Substrates	Micro-organisms	Enzymes
Lignocellulosis	Lentinula edodus	Enzymes
Wheat bran	Bacillus licheniformis	Alpha amylase
Wheat bran	Bacillus subtilis	Protease
Straw	Neurospora crasse	Carboymethyl cellulase, beta glucosidase

bagasse, etc.) are particularly suitable for this purpose, since they are available in large quantities in processing facilities. Lignocellulose in wood may be transformed into good quality paper products with the help of SSF biopulping and biobleaching. Agricultural residues may be converted into animal feed enriched with microbial biomass, enzymes, biopromoters, and made more digestible by SSF. Lignocellulosic waste may be composted to targeted biofertiliser, biopesticide and biopromoter products. Post-harvest residue may be decomposed on site by filamentous fungi and recycled to the soil with improved biofertiliser and bioprotective properties.

Nature of Substrates

The major organic material available in nature are polymeric in nature, e.g. polysaccharides (cellulose, hemicellulose, pectins, and starch, etc.) lignin and protein, which can be metabolised by different micro-organisms as a source of energy. These substrates that are insoluble in water, absorb water onto their matrix, which provides required moisture in SSF system for the growth and metabolic activities of micro-organisms. Bacterial and yeast cultures grow on the surface of substrate fibrils and particles while fungal mycelia penetrate into the particles of substrate for nutrition.

The solid phase in SSF provides a rich and complex source of nutrients that may be sufficient or sometimes insufficient and incomplete with respect to the overall nutritional requirements of that particular micro-organism that is cultivated on that substrate. The constituents in the agricultural solids are approximately analysed in terms of total carbohydrates, proteins, lipids, various elements and ash content. The solid substrates generally contain some small carbon compounds whereas the bulk of total dry weight is a complex polymer. The polymeric forms require enzymatic hydrolysis for their mineralisation as carbon-energy sources in microbial metabolism. In comparison with liquid-state fermentation, which generally use less complex carbon energy sources, solid insoluble substrates provide mixed ingredients of high molecular weight carbon compounds. Such complex carbon compounds may contribute inhibition, induction, or repression mechanism in microbial metabolism during solid state cultivation.

BIOTECHNOLOGY OF SOLID STATE FERMENTATION

Solid substrate systems have been defined in several ways:

1. Solid substrate fermentation is the microbial transformation of biological materials in their natural state, in contrast with liquid or submerged fermentation that is carried out in dilute solutions.

2. Solid substrate fermentation is generally defined as the growth of micro-organisms on solid substrates or sometimes referred to as solid-state fermentation since the process taking place is in the absence or near-absence of free water in the system. The substrate however, must contain enough moisture, which exists in absorbed form within the solid substrate matrix and simulates the fermentation reaction occurring in nature. These moist solid substrates are insoluble in water and polymeric in nature, are a source of carbon and energy, vitamins, minerals, nutrients and also provide their absorbed water for microbial growth as well as anchorage.

3. Solid-state or solid-substrate fermentation means that the substrate is moistened, often with a thin layer of water on the surface of the particles, but there is not enough water present to make fluid mixture. Weight ratios of water to substrate in SSF are usually between 1:1 and 1:10.

4. SSF can be defined as a system with solid matrix particles, a liquid phase bound to them and a gaseous phase entrapped within the particles. The physical properties of this system such as the water potential and water holding capacity, (can be used as an index of aeration) and bulk density (which predicates the volume of pore space) help to define the conditions of solid-state fermentation.

ADVANTAGES OF SSF OVER CONVENTIONAL LIQUID FERMENTATION

Traditional SSF came about for two primary reasons:

1. The desire for more tasty food, as with Oriental fermented foods and moldripened cheese.

2. The need to dispose of agricultural and farm waste materials (as in composting).

A closer examination of SSF processes in recent years in several research centres throughout the world has led to the realisation of its numerous economical and practical advantages. The attraction of SSF comes from its simplicity and its closeness to the natural way of life for many micro-organisms. Since large amount of water are not added to the biological systems, fermenter volumes remain small, necessary manipulations become less expensive and the cost of water removal at the end of fermentation in minimised. This type of fermentation is especially suitable for growing mixed cultures of micro-organisms where symbiosis stimulates better growth and productivity. Solid-state fermentations are clearly distinguished from submerged cultures by the fact that microbial colonisaton occur at or near the surfaces of solid substrate, or in few cases the soluble substrate supported on the solid insoluble-matrix in the environment of low-moisture contents. In contrast to liquid fermentation, the substrates traditionally fermented in the solid-state are renewable agricultural products, such as wheat, rice, millet, barley, corn and soyabeans. The non-traditional substrates, which can be used in industrial process development, include an abundant availability of agricultural, forest and food-processing wastes. From an engineering point of view, SSF offers many attractive features in comparison to conventional stirred tank reactors or aerated liquid medium fermentations because no free water is present, this leads to many benefits.

Solid-state fermentations can be used to provide low-shear environments for the cultivation of shear-sensitive mycelial organisms. Solid state cultivations can be and have been used for mass production of spores, which can than be used for the transformation of organic compounds such as steroids, antibiotics, fatty acids, and carbohydrates. Fungal spores have applications in the production of food-flavours and insecticides. The advantage of solid state fermentation includes simplicity, yields and the homogeneity of spore preparations. The expected advantages of SSF over submerged fermentations are:

- Smaller fermenter volume, relative to the yield of the product, as there is no excess water taking space in the fermenter.
- Lower sterilisation energy costs, as less volume of water needs to be heated.
- Seed tanks are not necessary in all cases, as the spore inocula can be successfully used to inoculate the solid medium.
- Easier aeration, as air can circulate easily and freely between the substrate particles, and also because the liquid film covering the substrate has a large surface area compared to its volume. Aeration is facilitated by spaces between substrate particles and particle mixing.
- Reduced or eliminated capital and operating costs for stirring, since occasional stirring is sufficient.

- Lower costs of product recovery and drying; in many cases the product is concentrated in the substrate and can be used directly, e.g. Oriental foods and cheeses, or the products can be directly incorporated into animal feeds.
- If the product is to be extracted from the substrate, e.g. enzymes and other metabolites, then much less solvent is needed. The fermented solids may be extracted immediately by direct addition of solvents or maintained in frozen storage before extraction.
- Reduced or eliminated capital and operating costs for effluent treatment due to lower water content in the system.

The other benefits are:

1. The media are relatively simple; a natural, as opposed to a synthetic, medium is used.
2. A more natural environment for micro-organisms, e.g. agricultural wastes degrading organisms: many of these fungi grow and perform better under SSF than submerged conditions.
3. A less favourable environment for many bacteria, which require a high moisture level to survive, lowering the risk of contamination, therefore many SSF processes need no sterilisation.
4. SSF is adaptable to either continuous or batch process and the complexity of equipment is no greater than that required for submerged reactors.

Above described advantages are so attractive for the biological processing of agricultural by-products that most of the work has used SSF process. These advantages can outweigh the disadvantages of SSF, which are the slowness of fermentation and the difficulty of controlling the process precisely.

PERFORMANCE CONTROL OF SSF PROCESS

The difference in process control between SSF and SmF is mainly due to the use of solid substrates with a very low moisture content in system. The disadvantages of large-scale solid cultures are due to the problems of process-control, process scaleup and the major problem of heat build-up. Despite these drawbacks, large-scale SSF processes have been developed successfully in Japan for the manufacture of a variety of products, including fermented foods and food-products, enzymes, and organic acids. The drawbacks have been overcome by carrying these fermentations in stationary and rotary tray processes, where the temperature and humidity-controlled air is circulated through the stacked beds of fermenting solid substrate particles. These tray methods of cultivation have been used for centuries in the manufacture of traditional food products and the cultures experience the shear-sensitivity in some of these processes. These are main reasons of less frequent use of rotary drum-type fermenters.

Little information is available on the details of modern control systems in large-scale solid-state cultivations. The control of temperature and humidity within practical limits is exercised through water temperatures, which is used to humidify the circulating air. The humidified air is circulated at flow-rates to meet the requirements of heat and mass transfer. The gas environment has been found to significantly affect the rate and extent of culture colonisation and product formation in SSF. In the commercial production of amylase using rice substrate in SSF, oxygen pressures above atmospheric have been found to significantly stimulate the enzyme productivity, suggesting oxygen limitation at normal atmospheric pressure. The DNA measurements revealed that this only caused a little effect on biomass formation, but the carbon dioxide pressures above 0.01 atm severely affected the process through the inhibition in amylase productivity. In a protein production process by *Aspergillus* species using alfalfa residues, cellulase and pectinase activities have been found stimulated by oxygen and carbon dioxide pressures above atmospheric levels, and with no effect on biomass formation. These studies have been conducted in controlled gas environments at constant partial pressures, which is maintained by admitting pure oxygen on demand at pressures below a set point and purging carbon dioxide in 30% KOH at pressures above a set point in a closed aeration system. In another type of SSF performed for the degradation of natural birch lignin employing *Phanerochaete chrysosporium*, high oxygen pressures have been found to be stimulating, whereas the high carbon dioxide pressures have been found inhibiting the process. The stimulatory effect of oxygen on breakdown of lignins has been confirmed in laboratory studies by using labelled synthetic lignins and natural wood lignins.

Given the present state of the art, the most promising approach in solid state fermentation processes development happens to be the measurements and control of various parameters and process variables, similarly as in any liquid fermentation. In SSF processes, various methods are selected to analyse the temperature, pH, humidity, oxygen and carbon dioxide concentrations in gas phases, biochemical analysis of fermented and unfermented solids and their extracts. The manufacturing productivities of some industrial scale submerged liquid fermentations have increased significantly over years, e.g. antibiotic production. This development has been possible due to applied and basic research in microbial-biochemistry, microbial-physiology, and genetics. To some extent the contribution also goes to engineering research based on concepts of stoichiometry, kinetics, thermodynamics, and heat and mass transfer in control of the microbial fermentation process and its environment. Direct economic comparisons of solid-state and liquid-state fermentations are not possible, it is apparent that the large-scale solid-state fermentations (known as Koji in Orient) have been developed in Japan on an economic basis.

Potential economic advantages of such processes to employ suitable microbe-substrate system include:

1. Reduced thermal processing requirements, since many processes are not aseptic.
2. Reduced energy requirements for agitation, since surface-to-volume ratios for gas transfer are high and many processes do not require agitation due to their shear sensitivity.
3. High extracellular product concentrations, that can be efficiently recovered by superficial-extraction or leaching methods.

Performance Control by Particle Size of Agro Residues

SSF processes performance can be varied and controlled by changing physical and chemical factors. It has been reported that substrates with finer particles showed improved degradation due to an increase in surface area for enzymatic action. The greater growth of fungal cultures has been found stimulated by smaller particle size substrates. Higher enzyme productivity in SSF has been achieved with substrates, which contained particles of mixed sizes from 180 µm to 1.4 mm. Particles and kernels of grain must be of suitable size, but not be too small in order to avoid particle agglomeration.

The particle size must be in a limited size range to be maintained at relatively low moisture content to prevent contamination. The smaller particle size provides a larger surface area which facilitates heat transfer and gas exchange. Smaller particle sizes also distribute equivalent moisture concentrations in thinner films on external surfaces exposed to the gas environments, given the same void volume fraction (porosity) and pore size distribution. Internal pores maintain the same surface-to-volume ratios with respect to solid surfaces, based on geometric considerations of spherical particles. This results in higher surface nutrient concentrations and the diffusion of nutrients takes place via shorter pathways at the surfaces as well as in the pores of those substrates which have same tortuosity. Too small a particle size may result in closer packing densities of the substrates and the void space between particles becomes considerable reduced. The reduced space between particles tends to reduce the available area for heat transfer and gas-exchange with the surrounding environment. If such condition arises, densely packed particles in a cultivation system have to be sufficiently agitated to provide a better separation of particles for the exchanges of gases and heat transfer. There may be a lower limit in particle size at which the heat transfer or gas exchange becomes rate limiting and there may be an upper limit at which the nutrient transfer becomes limiting. Conclusively under any condition, the particle size of the substrate to be used is one of the major variables in the SSF-process development. Various methods are available to obtain particle sizes such as milling, grinding, chopping and sieving to obtain substrates of particular particle-sizes. In the case of lignocellulosic substrates, smaller particle size substrate is usually obtained through ball-milling.

Performance Control by Medium Preparation of Agro Residues

Some SSF systems do not require any nutritional supplements as do most of the traditional food fermentations. Medium supplementation is necessary in nontraditional SSF fermentations, as it induces enzyme-synthesis, provides balanced growth conditions for mycelial-colonisation and biomass formation, as well as prolonging the production of secondary metabolites. SSF employing brown-rot fungi, require an additional carbon source for the induction of enzymes for the cellulose utilisation. Certain fungi including *Lentinus lapidus*, *Poria monticola*, and *Lezites trabea* can be cultivated on lignin-containing natural wood substrates from aspen, pine and spruce, when the SSF medium is supplemented with glucose or cellobiose in smaller quantities of 0.5%, w/v, and an even smaller amount of peptone, asparagine and yeast extract. In unsupplemented media, growth of these fungi was very slow as negligible. A co-metabolite, such as glucose or cellulose, stimulates the lignin-degrading system in white-rot fungi such as *Phaenerochaete chrysosporium* and *Coriolus versicolor* when these organisms are cultivated on spruce lignin. Other supplementations of cellobiose, mannose, xylose, glycerol or succinate have been found less effective.

Studies for the nutritional requirements for a developmental microbe-substrate system to be used on a large-scale SSF, can be done in preliminary experiments in small-scale liquid or SSF on laboratory scale. There is a procedure for evaluating the effects of nutritional supplements on culture-growth and product formation, in which microbial-cultures and the solid substrate are contained in separate compartments divided by a membrane with a molecular-weight-cut-off. The membrane permits the passage of enzymes and small molecular weight compounds but restricts microbial and substrate solids. One of the major difficulties in the development of solid state fermentations has been the problem in separating microbial biomass from the solid substrate particles after the mycelial growth has covered the substrate surfaces. In solid culture cultivation the micro-organism and substrate are intimately associated making the analytical methods of limited value in stoichiometric analysis of SSF. The analysis of biomass yield and growth rate by the measurement of glucosamine, protein, RNA, DNA, oxygen consumption, and carbon dioxide or heat evolution, can not be accurately used in samples of SSF.

Solid cultures for the production of secondary metabolites may have another problem in that the nutrient, whose deficiency triggers the pathway leading to formation of secondary metabolite, may be available in excess when the microbial growth becomes limited by other nutrient. Therefore, the selection of a solid substrate and required-supplements is more critical for a SSF process for antibiotic production that for a SSF designed for enzyme and organic acid biosynthesis.

Performance Control by Moisture Content of Agro Residues

Solid state or solid substrate fermentation means that the substrate is moistened, often with a thin layer of water on the surface of the particles, although there is not enough water present to make a fluid mixture. Weight ratios of water to substrate in SSF are usually between 1:1 and 1:10. Since biological activity ceases below a moisture content of about 12%, this establishes the lower limit at which SSF can take place. The upper limit is a function of absorbency and hence, moisture content varies with the substrate material type.

Solid substrates may be viewed as gas-liquid-solid mixtures. The aqueous phase in such mixtures is intimately associated with solid surfaces in various states of sorption. The aqueous phase in a cultivation system is in contact with the gas phase continuous with the external gas environment. Different types of solid substrates can absorb different amounts of water. Depending on the moisture content of the solid; some of the water is tightly bound to solid surfaces, some amount of water is less tightly bound and remaining water may exist in a free state inside the capillary regions of the solid substrates. The gas-liquid interface provides a boundary for gaseous exchange between carbon dioxide and oxygen as well as for heat exchanges.

Water in biological materials exists in three states. The moisture isotherm measurements determines that the solids sorb or desorb water vapour in equilibrium with relative humidities in a gas phase (water activities), which can be maintained by saturated salt solutions at a constant temperature. Water is tightly bound to solid surfaces at the surface in a monolayer region. In case of agricultural residues, monolayer binding is generally 5 to 10 g per 100 g of dry solids. Beyond the surface monolayer in a multilayer region, water is less tightly bound in additional layers at progressively decreasing energy levels. Then beyond the multilayer region, free water exists in a region of capillary condensation. In terms of relationships between water activity and moisture content, the distinction between the multilayer and capillary regions is ambiguous.

The electric measurements of an agricultural residue containing high starch content has been used to determine the dividing line between multilayer and capillary regions. The dividing line was defined by amoisture content of about 25 to 30% by weight at a water activity of 80 to 85%, which is the lower limit for microbial growth except for some halophilic or osmophilic microbes.

The sorption isotherm may vary from one type of product to another, the hysteresis is seen in sorption and desorption isotherms. Water may exist in free state at moisture levels of interest in solid state fermentation, which is in contrast with general perception about SSF that the free water does not exist in such systems. Moisture is a critical factor in SSF of aflatoxin production on rice, the yields of aflatoxins have been found decreasing rapidly at moistures above 40%. The rice particles become sticky at moistures above 30 to 35%. Moisture content plays an important role on the growth of lactic acid bacteria on feed lot wastes liquids mixed with cracked corn; growth and acid production was limited at moisture level less than 35%, whereas the higher level above 42% in SSF-mixtures caused the contents to become gummy and aggregate. One of the secrets of a successful SSF-process is to keep the fermenting substrate moist enough for fungal-growth and colonisation and to avoid higher moisture level not to promote the unwanted bacterial growth. Therefore, the optimum moisture content for a particular type of SSF for its microbe-substrate system should be determined for a particular end-product and cultivation conditions of that SSF.

The level of moisture content affects the process productivity significantly in any SSF system, when available in lower or higher quantities than the optimum value. Hence, it should be in limited and required amounts in system. The presence of an optimum moisture content in SSF medium has been emphasised also for the cultivation of bacterial cultures. The process productivities are affected by water content because the physico-chemical properties of the solids depend and vary with moisture available to them. Therefore, the major key factors determining the outcome of the SSF-process are the moisture content and the relative humidity levels. Heat removal during fermentation is mostly achieved by evaporative cooling. This leads to an uneven distribution of water in system due to large quantities of water evaporation. Workers have practised various ways to maintain the moisture content of the solids.

Control of water activity factor in SSF

Water activity of the substrate has been proposed as the condition of growth and viability of the microbes and hence, the importance of a_w in SSF has widely been studied. Water activity is defined as the relative humidity of the gaseous atmosphere in equilibrium with the substrate and the water activity factor, a_w of the substrate quantitatively expresses the water requirement for microbial activity. The types of the micro-organisms that can grow in SSF systems are determined by the water activity factor, a_w. Bacteria mainly grow at higher a_w values while filamentous fungi and some yeasts can grow at lower a_w values (0.6–0.7). The micro-organisms capable of carrying out their metabolic activities at lower a_w values are suitable for SSF processes. High a_w favours sporulation in the course of growth in SSF, but low a_w favours spore germination and mycelial growth.

Numerous experiments have demonstrated the influence of a_w on microbial metabolism, such as, on growth rate and sporogenesis of filamentous fungi on enzyme biosynthesis by fungi, and on cheese aroma production. The a_w of the medium is a fundamental parameter for mass transfer of the water and solutes across the cell membrane. The control of this parameter could be used to modify the metabolic production or excretion of a micro-organism.

MICRO-ORGANISMS USED FOR AGRO RESIDUES BIOCONVERSION

Selection of a suitable micro-organisms is one of the most important criteria in SSF. The vast majority of wild type micro-organisms are incapable of producing commercially acceptable yields of the desired products. The unique characteristics of solid-state cultivations are their ability to provide a selective environment at lower concentrations of moisture ideal formycelial organisms. Themycelial organisms are capable of producing a range of extracellular enzymes required for the hydrolysis of complex, polymeric solid substrates. Such micro-organisms are able to colonise at high nutrient concentrations near solid surfaces. The mycelial organisms include a large number of filamentous fungi and a few bacteria of actinomycetes. The importance of micro-organisms can be seen from the fact that a culture of *Aspergillus niger* can produce as many as 19 types of enzymes, while enzyme alpha amylase can be produced by some 28 different types of cultures. SSF processes can be placed in two main classes based on the type of micro-organism involved:

1. *Natural (indigenous) SSF*: Ensiling and composting are SSF processes, that utilise natural microflora. In nature, SSF is often carried out by mixed cultures in which several micro-organisms show symbiotic cooperation.

2. *Pure culture SSF*: Known purified micro-organisms are used in such processes either singly or in mixed culture. SSF using a pure culture is known since antiquity, e.g. the Koji process with *Aspergillus oryzae*. A pure culture is necessary in industrial SSF process for improved rate of substrate utilisation and controlled product formation. A typical example of pure mixed culture SSF is the bioconversion of agricultural residues to fungal biomass (protein) using two pure cultures of *Chaetomium cellulolyticum* and *Candida utilis*.

Several micro-organisms have been employed in a wide range of SSF processes for various objectives. The cultivation of filamentous fungi on solid substrates has been widely used for different purposes at laboratory scale, e.g. for Koji fermentation, for lignocellulose fermentation, for fungal spores, and for mycotoxin production. For various purposes, among the filamentous fungi three classes, viz.

Phycomycetes (*Mucor* and *Rhizopus*), Ascomycetes (*Aspergillus* and *Penicillium*) and Basidiomycetes, have been most widely used. SSF has been most commonly used mploying *Aspergillus niger* for protein enrichment as well as for enzymes production, such as, cellulase, amylase, glucoamylase, beta glucosidase, and protease. Production of alcohols, ketones and aldehyde in rice fermentation was achieved by the use of *A. oryzae*. For protein enrichment and kinetic studies related to SSF process *Rhizopus oligosporus* has been employed.

Fungal rennet has been produced by *R. oligosporus* and *Mucor meihei*. For enzyme production and protein enrichment cultures of *Trichoderma* spp. have been employed in pure, single and mixed SSF. Lipase enzyme production has been reported using six species of *Penicillium*, two species of *Rhizopus, Geotrichum candidum* and *Mucor meihei*, whereas the maximum lipase activity was obtained with *P. candidum, P. camembertii* and *M. meihei*. For the production of several other enzymes, e.g. hydrolases and pectic enzymes several other species of *Penicillium* have been employed in SSF.

Production of the antibiotic penicillin was achieved in a non-sterile SSF process on sugar cane bagasse impregnated with culture medium using *Penicillium chrysogenum*. Protein enrichment of lignocellulosic substrates for animal feed production, lignin degradation, and cellulase and ligninase enzyme production have been obtained by white-rot cultures in SSF. Production of gibberellic acid has been reported using *Fusarium monoliforme* and *Gibberella fugikuroi*. Bacterial alpha amylase production is reported using *Bacillus licheniformis* in SSF. Several yeasts have been used for protein enrichment and ethanol fermentation in SSF. For protein enrichment of straw *Candida utilis* was used whereas *Saccharomyces cerevisiae* has most commonly been employed for ethanol production.

DESIGNING AND TYPES OF SSF

Fermenter Design for SSF

Several miscellaneous types of fermenters have been used in batch or continuous mode in SSF processes. Process parameters are very important factors and they have to be considered in a bioreactor design for any SSF. Design considerations in types of SS-fermenters used by various researchers are described by Aidoo and others. The engineering aspects, with major types of fermenters describing their advantages and drawbacks has been reviewed by Fernandez and others. Solid state cultivations are not as well characterised on a fundamental scientific or engineering basis, as are the liquid fermentation systems that are used in the West for the industrial production of microbial-metabolites. Solid-state fermentations are, however, widely used in the Orient and therefore, the old traditional methods of cultivation systems which have been used in food-processing for more than 2,000 years, have now been modernised and well characterised for their extended application to non-traditional products.Mitchell and others have described in detail the modelling aspects of SSF. The physical state of the substrate and the products to be produced in the system characterise the design-type of solid state cultivation process:

1. Low-moisture solids are fermented:
   - Without any agitation for the production of Tempeh and Natto.
   - By occasional stirring for the production of Miso and Soya sauce.
   - With continuous stirring for the production of Aflatoxin.
2. Suspended solids are fermented in packed bed columns:
   - Through which the liquid is circulated, as for the production of rice-wine.
   - Which contain stationary or agitated liquid media, for the production of Kaffir beer.

Types of SSF Systems

There are two types based on process design:

1. Fermentation in static reactor, e.g. Tray fermentations.
2. Fermentation with occasional or continuous agitation, e.g. Production of aflatoxin, ochtratoxin and enzymes.

Type two has 4 variations according to the need of process:

1. Occasional agitation, without forced aeration
2. Slow continuous agitation, without forced aeration
3. Occasional agitation with forced aeration
4. Continuous agitation with forced aeration.

SSF Bioreactors

Three basic groups of reactor exist for SSF, and these may be distinguished by the type of mixing and aeration used. In laboratory scale, SSF occurs mainly in flasks whereas following reactors are used for large-scale product-formation.

Tray bioreactors

Tray bioreactors tend to be very simple in design, with no forced aeration or mixing of the solid substrate. Such reactors are restrictive in the amount of substrate that can be fermented, as only thin layers can be used, so as to avoid overheating and maintain aerobic conditions. Tray undersides are perforated to allow aeration of the solid substrate, each arranged above each other. In such reactors, temperature and relative humidity are the only controllable external parameters.

Wooden trays were initially used for sauce production in Koji fermentations by *Aspergillus oryzae*. The use of tray fermenters in large-scale production is limited as they require a large operational area and tend to be labour intensive. The lack of adaptability of this type of fermenter makes it an unattractive design for any large-scale production.

Drum bioreactors

Drum bioreactors are designed to allow adequate aeration and mixing of the solid, whilst limiting the damage to the inoculum or product. As previously mentioned, mixing and aeration of the medium has been explored in two ways: by rotating the entire vessel or through the use of various agitation devices. Rotation or the use of agitation can be carried out on a continuous or periodic basis. In contrast to tray reactors, growth of the inoculum in drum bioreactors is considered to be better and more uniform. Increased sheer forces through mixing, can however, have a detrimental affect on the ultimate product yield. Although the mass heat transfer, aeration and mixing of the substrate is increased, damage to inoculum and heat build up through sheer forces may affect the final product yield. Application of drum reactors for large-scale fermentations also poses handling difficulties.

Packed bed bioreactors

Columns are usually constructed from glass or plastic with the solid substrate supported on a perforated base through which forced aeration is applied. They have been successfully used for the production of enzymes, organic acids and secondary metabolites. Forced aeration is generally applied at the bottom of the column, with the humidity of the air kept high to avoid desiccation of the substrate.

Disadvantages associated with packed bed column bioreactors for SSF include difficulties in retrieving the product, non-uniform growth, poor heat removal and scale-up problems.

SCALE-UP STAGES OF SSF

Scale-up of SSF has been defined in many ways. There are mainly four stages:

Flask Level

This is smallest scale using 50–1000 g substrate working capacity, and used for the selection of the organism, optimisation of the process and experimental variables in a short time and at low cost. The vessels used are conical flasks and beakers, jars, and glass tubes.

Laboratory Fermenter Level

This is next to flask scale using a 5–20 kg substrate working capacity. It is used for a selection of procedures such as, inoculum development, medium sterilisation, aeration, agitation and downstream processing. Standardisation of various parameters, selection of control strategies and instruments, evaluation of economics of the process and its commercial feasibility are also examined at this level. The fermenters used are glass incubators, column fermenters, polypropylene bags, and miscellaneous types of fermenters.

Pilot Fermenter Level

This scale is a stage before the commercial scale using 50–5000 kg of substrate. This level is necessary for the confirmation of laboratory data and selection of optimised procedures. It facilitates market trials of the product, physico-chemical characterisation and determination of viability of the process. Most large scale SSFs employ tray type fermenters as in the oldest soya sauce Koji process, horizontal paddle fermenters and mixed layer pilot plant fermenters. Durand and Chereau reported the use of a pilot reactor having a 1 T working capacity.

Production Fermenter Level

The commercial scale fermenter utilises 25–1000 T of substrate and is performed for streamlining of the developed process. Yokotsuka described deep trough methods and mechanical continuous equipment for Koji production generating 50–100 T of Koji per day.

FACTORS AFFECTING SSF

Each microbe-substrate system is unique and the process variables must be considered in terms of the physical properties and chemical composition of its substrate, growth characteristics and physiological properties of the micro-organisms to be cultivated in SSF. The nature of the product, if the process involves the synthesis of primary or secondary metabolite may be based on the synthesis of extracellular enzymes in growth-associated metabolism. The process variables affecting a solid state cultivation include, pretreatment of substrates, particle-size of substrates, medium-ingredients, supplementation of growth medium, sterilisation of SSF-medium, moisture-content, inoculum-density, temperature, pH, agitation and aeration. These variables should be considered in process-development of a SSF to be carried out for different purposes. Some of these variables have been discussed in some sections as above, the rest are discussed.

Significance of Aeration and Mixing in SSF

In any SSF-process an adequate supply of oxygen is required to maintain the aerobic conditions and for the transfer of excess carbon dioxide produced during metabolism. This requirement can be achieved through the process of aeration and mixing of the fermenting solids. In certain cases, the mixture can not be agitated vigorously or in some cases, at all, if the micro-organism used in SSF is shear sensitive. The shear sensitivity is attributed to disruption of mycelial-substrate contact, this is particularly concerned to those organisms which possess mycelial-bound enzymes required for the hydrolysis of solid substrate-polymers. Most Koji processes in Japan performed for the commercial production of enzymes do not involve great agitation. The fermenting substrate is gently turned periodically just to bring the bottom of Koji to the top. These processes have been developed in highly controlled environments, using automated systems for inoculum mixing, and turning of the fermenting substrate.

Most of the traditional food-fermentation in Japan use the rotary-tray method for SSF with the circulation of humidified air to create the conditions suitable for gas-exchange and heat-transfer. In the SSF for the production of certain secondary metabolites such as aflatoxin and ochratoxin, and in some processes for the enzyme production, mixing and particle separation are achieved by agitation on shakers or in rotating vessels with circulating conditioned air. Maximum rotation rates generally decrease with the size of the fermentation-vessel. Therefore, solid-state fermentations are ideal for the cultivation of those micro-organisms that are extremely sensitive to the shear rates of the impeller speeds required for stringent oxygen demand rates in liquid fermentaton. Such micro-organisms colonise the solid substrates by microbe-substrate attachment and there is no pellet formation in solid-state cultivation, which is added advantage to SSF.

Aeration plays an important role in solid state fermentations as compared to liquid fermentation where it only helps in gas transfer. Aeration facilitates in heat, gas and moisture transfer between the fermenting solid particles and the gas environment of the system. The temperature of the gas phase serves by supplying or removing heat, in maintaining the relative humidity in equilibrium with the liquid phase. In liquid fermentations the substrates are dissolved in at low substrate concentrations in large volumes of fluid, but in solid cultures with respect to moisture transfer, the loss or gain of moisture during SSF is extremely sensitive to the water activity of the gas-phase. Therefore, small changes in the relative humidity of the gas phase in equilibrium with the solids may cause the large changes of moisture content in the solid state, depending on the sorption-desorption characteristics of the solid substrate.

There are two main functions of the gas phase in SSF, the primary function is to supply oxygen and remove the carbon dioxide from the system. The secondary function of aeration is in heat and moisture transfer that is more important, when the rates of oxygen and carbon dioxide are not limiting. The gas phase can facilitate in the control of solid cultures, due to the fact that direct measurements can not be performed to estimate dissolved oxygen or carbon dioxide concentrations in low-moisture solids during the course of the fermentation on either a continuous or sampling basis. The methods of aerationmay cause the conditions of gas transfer being relatively stagnant. This condition may be responsible for the oxygen limitation at small penetration depths or may lead to inhibitory carbon dioxide concentrations in normal atmospheric environments. The gas phase in the SSF during the course of microbial metabolism, can be analysed for oxygen, and carbon dioxide pressures using analysers which function on thermal-conductivity, paramagnetism, or infrared absorption. The technique of gas chromatography can also be used for gas-analysis of the gas phase of a SSF.

Significance of Control of Temperature and pH in SSF

Two significant variables affecting any SSF are the incubation temperature and the pH of SSF-medium. Both variables are specific for each SSF process depending on the micro-oganisms to be cultivated and the product to be formed. Unlike submerged fermentation, these factors are difficult to control in SSF. These variables can not be directly measured in the liquid phase, as these are associated with the solids at lower moisture content without any free liquid in the fermenting medium. The other difficult situation arises when the growth temperature of cultivated micro-organism is different than the optimal temperature for the product formation. Such systems require a possible need for temperature profiling or shift in the later stages of fermentation. The thermal gradients may be induced within SSF-mixture due to the rate of heat generation in SSF-system at high levels of biological activity. This gradient may limit the heat transfer and may lead to sub-optimal conditions for microbial-biomass and product formation.

The local pH levels at solid surfaces near which the biological activity occurs, may be considerable different than the bulk pH of the liquid phase. This difference in pH levels happens due to surface charge effects and ionic equilibria modified by solute transport effects. There is no suitable method to measure the precise pH of fermenting solid residues in SSF. A general method used for measuring pH of solid agricultural residues involves mixing one part of fermented solids (dry weight) and three parts of freshly boiled and cooled water, and measuring the pH of the resultant liquid after five minutes using a glass electrode. This procedure can be used to monitor pH changes during fermentation on intervals using minimum one gram of the SSF-mixture.

It is easier to measure temperature of the fermenting SSF-mixture, in comparison to pH measurement. Temperature can be measured using thermistor or thermocouple probes at various depths of the SSF-mixture below the medium-surface. In various SSF-processes for the production of enzymes, mycelial-biomass or organic acids, total heat generation of up to 600 kcal per kilogram of fermenting solids has been observed. A study of composting of animal wastes and agricultural residue has revealed that such heat generations may lead to rapid temperature rise of the fermenting mass in the system limited by heat transfer. The study also revealed that the biological activity was considerably higher near the surface of the compost pile than in the depth of pile that was at lower oxygen pressure. This phenomenon happens due to a decrease in interior oxygen concentrations inside the SSF-mixture pile of compost. Thus the heat generation in such fermentations is coupled to conditions for heat as well as mass transfer.

Chapter 7

Engineering Aspects of Solid State Fermentation

INTRODUCTION

Solid substrate cultivation (SSC) or solid state fermentation (SSF) is envisioned as a prominent bioconversion technique to transform natural raw materials into a wide variety of chemical as well as biochemical products. This process involves the fermentation of solid substrate medium with micro-organism in the absence of free flowing water. Recent developments and concerted focus on SSF enabled it to evolve as a potential biotechnology as an alternative to the traditional chemical synthesis. SSF is being successfully exploited for food production, fuels, enzymes, antibiotics, animal feeds and also for dye degradation. This chapter discusses the various micro and macro level engineering problems associated with SSF and some possible solutions for its full commercial realisation.

Modern chemical synthesis aims at three E's. Energy, Economy and Environment. Any new chemical product today must be produced with minimum energy requirement at optimal cost with zero environmental pollution. As the biochemical processes are environment friendly, cost effective and carried out at ambient conditions, satisfy the above constraints of three E's for the production of a wide variety of chemical and biochemical products to serve the modern society. Hence, biosyntheses have emerged as the potential alternative to traditional chemical syntheses. There are two types of bioconversion methods in operation, one is the submerged fermentation (SmF) which is well established and the other is solid state fermentation (SSF), which is still in evolutionary state and under intensive research. There are several recent publications describing the solid state fermentation of agro-industrial residues such as rice bran, rice husk, potato wastes, cassava husk, wheat bran, sugar cane bagasse, sugar beet pulp, palm kernel cake, rice straw, cocoa pod, fruit wastes, etc. into bulk chemicals and value added fine products such as ethanol, enzymes, antibiotics, biofuel, mushrooms, organic acids, amino acids, biologically active secondary metabolites. More recently the focus has been the conversion of natural wastes into animal, poultry and fish meals through SSF. In principle SSF refers to the microbial growth on moist solid substrates or within the pores with out free flow of water. The required moisture for SSF exists in the solid as absorbed or complex form is more helpful for oxygen availability to the microbial population.

In SSF the microbe is in contact with atmospheric oxygen unlike in submerged fermentation (SmF). SSF is simpler and requires less processing energy. The basic differences between SSF and SmF are presented in Table 7.1 for better understanding. Some of the SSF processes developed are presented in Table 7.2. The low moisture content means that fermentation can only be carried out by a limited number of micro-organisms, mainly yeasts and fungi. Therefore, because of low moisture levels spoilage or contamination by unwanted bacteria is reduced, and more concentrated product is produced. The chapter also highlights some of the commercial applications of SSF.

Table 7.1: Basic differences in solid state fermentation and submerged.

Solid state fermentation	*Submerged fermentation*
Medium is not free-flowing	Medium free flowing
Shallow depth	Greater
Single solid substrate provides C, N_2, minerals and energy	Employed
Medium absorbs water, up-takes nutrients	Disolved in water
Gradients of T, pH, C_s, C_n	Uniform
Minimum water,(less volume)	More water, more volume
3 phase system	2 phases system
T, O_2, H_2O control (H_2O critical)	T, O_2 control
Inoculum ratio large	Low
Intra particle resistances	No such resistances
Bacterial and yeast cells adhere to solid and grow	Uniformly distributed
Highly concentrated product	Low concentration product

Table 7.2: Solid state fermentation processes developed.

Process/product	*Substrate*	*Microbe*
Protein enrichment	Cassava bagasse, cassava crude	*Rhizopus* sp.
Citric acid	Cassava bagasse	*Aspergillus niger*
Lactic acid	Sugarcane bagasse	*Rhizopus oryzae*
Mushrooms	Cassava bagasse, coffee residues	*Pleurotus ostrreatus, lentinus edodes, flamulina velutipes*
Aroma production	Cassava bagasse, coffee husks	*Ceratocystis, rhizopus* sp.
Detoxification	Coffee husks	*Aspergillus* sp.
Biopesticide	Potato waste	*Bauveria bassiana*
Hormones	Coffee husks	*Gibberella fugikuroi*
Xanthan gum	Sugarcane bagasse	*Xanthomonas campestri*
Plant cell culture	Sugarcane bagasse	*Molus prunifolia borkh*
Amylase	Casava bagasse	*Rhizopus arrhizus*
Protease	Soyabean defatted cake	*Penicillum* sp.

SSF PROCESS METHODOLOGY

In any SSF process the basic steps carried out are:

1. The preparation of a solid substrate (d_p, pH, C_n, C_s).
2. Sterilisation of the substrate.

3. Rising of suitable inoculum (Traditional or pure culture technique).
4. The inoculation of the moist substrate.
5. The incubation in appropriate culture of vessels or reactors.
6. Maintenance of optimal conditions. (pH, T, H_m, mixing, aeration, flow pattern, Q, N_A).
7. Harvesting of solids.
8. Drying/Extraction of product.
9. Further downstream processing if necessary.

The major problems encountered in the above sequential steps are mainly design of appropriate reactor which can maintain required moisture, temperature and microbe concentration on the solid substrate with no inter and intraparticle oxygen gradients.

MASS TRANSFER PROBLEMS

The efficiency, productivity and economy of SSF are affected by various factors like mass and heat transport phenomena at micro and macroscopic levels in the reactor. At micro level the mass transfer depends on the nature and growth pattern of the micro-organism and their response to local environment change. The growth of microbe depend on inter and intra particle diffusion of gases like O_2, CO_2 and enzymes, nutrients and products of metabolism in the substrate.

At the macro level, the bulk flow of air into and out of the reactor which affects the sensible heat and compositions of O_2, CO_2 and moisture. Problems in SSF occur in industrial reactors where, the problem of the lack of free water and generation of metabolic heat is greatly exaggerated as the system struggles to provide adequate agitation, aeration and cooling. As a result high temperature gradients may result. During forced aeration, natural or forced convection, diffusion and conduction of heat and mass take place in a direction normal to the flow of air. To ensure good heat and mass transfer in the reactor, proper flow rates, contact patters be adopted between the phases in the reactor. Excessive shear forces within the bioreactor due to mixing may affect the functions of the microbe. But reasonable shear will keep it hale and active during the process.

The particle size and shape of substrate may affect the flow pattern and porosity in the bed. Mixing and aeration provide good transfer of nutrients and product gases in the bed. To maintain adequate moisture in the bed, continuous monitoring of moisture in the inlet and outlet air is essential. In order to transfer O_2 to the Micro-organism growing on and in the particle it has to cross hydrodynamic boundary layers, interparticle space and diffusion within the particle. Overall seven resistance steps are involved in O_2 transfer to reach the microbe. These resistances will decide specially the highest resistance step, the overall rate of conversion in SSF reactor. The hydrodynamic conditions of air and a moisture across the bed will either increase or decrease the overall rate of transfer of oxygen as well as moisture in the bed. In dealing with intraparticle transfer of O_2, nutrients and enzymes secreted by growing micro-organisms, the effectiveness factor (η) is a very useful concept. It is the ratio of the observed reaction rate to that in the absence of any substrate concentration gradients. Though it is used for quantifying the diffusional limitations in catalysis also applicable to SSF as it is visualised as an heterogeneous system. An important parameter required for the evaluation of this is the Thiele modulus (Φ) which is the ratio of biochemical reaction rate to that of diffusional mass transfer rate within the solid. By making use of this concept, a criterion can be developed to evaluate intraparticle mass and heat transfer limitations to provide greater insight and understanding into the heat and mass transfer mechanisms in SSF reactors. Mitchel and others studied the diffusional limitations of glucoamylase in a gel substrate to convert to glucose. Its production was as low as 20% because they did not consider the O_2 diffusion and consumption at the intraparticle level. Mitchel and Moo-Young studied the diffusional limitations on SSF of different substrates, which hindered the growth rate of micro-organisms and product yields.

In systems with forced aeration, O_2 transfer is less likely to be rate limiting, but some nutrients transfer might affect the growth of the micro-organism. In the case of filamental fungi, the layer on the substrate depends on the intra particle oxygen transfer and the moisture content of the particle.

DIFFUSION OF ENZYMES

In majority of SSF processes the carbon and energy substrate is macromolecular hence micro-organisms cannot transport macromolecular substrate across the cell membrane, so the action of extracellular enzymes in degrading the solid state substrate into soluble fragments is a very important step. Sometimes enzyme diffusion in the SSF reactor could be the rate controlling step. Thus exoenzymes may experience diffusion and steric limitations depending on the porosity of the macromolecular structure. The diffusion of enzymes is caused by the open pore geometry of the substrate, but when the porosity of substrate is less the degradation occurs at the outer surface.

So, the particle size, shape, porosity, consistency and strength are some of the important parameters to be considered in industrial solid state fermentation as they affect profoundly the heat and mass transfer rates and maintenance of adequate moisture in the bed. However for optimum particle size and structure modifications before the fermentation are yet to be found experimentally though pretreatment techniques like steaming, puffing, extrusion, etc. can be adapted to enhance the interfacial area and accessibility of O_2, nutrients and enzymes to the microbes in the solid substrate. These mass transfer studies indicate forced circulation of air through bed can enhance O_2 transfer and CO_2 dissipation in the bed if fluidised bed is used as SSF.

HEAT TRANSFER PROBLEMS

The heat generation is directly proportional to the metabolic activity in the SSF reactor. As fermentation progresses O_2 diffuses and triggers the bioreactions, liberating heat which accumulates in the reactor due to poor transport property of substrate. Hot spots may develop with a temperature rise of as high as 70°C. This may affect porosity of the bed. The heat removal and regulation depends on the aeration of the fermentation system. High temperatures affect spore germination, growth, product formation and sporulation, where low temperatures affect adversely.

Low moisture, poor thermal conductivity of the substrate result in poor heat transfer in SSF. Hence it is very difficult to maintain favourable temperature in the reactor. Thus water addition with continuous mixing is favourable in SSF, which can be achieved in a properly designed fluidised bed bioreactor. The importance of evaporative cooling and moisture content of the substrate on the performance of SSF bioreactor has been highlighted in the literature to control the rising temperature. But this type of cooling may affect the reactor performance. Water activity could also drop due to build-up of solutes such as glucose, amino acids, etc. This could be prevented by spraying water on to the solid substrate coupled with mixing. Sufficient water supply must be made available to the growing spores or fungi, and for water activity of the substrate.

In industrial SSF reactors forced moist air or dry air circulation, cooling the external surface or with internal surface of the reactor with chilled water or by covering it with water soaked burlap are worth trying to control the fermentation temperature.

BIOREACTOR DESIGN

SSF processes could be operated in batches, fed batches or continuous modes. Shear sensitivity of the substrate and the micro-organism must be taken into account during reactor design. Over the period a good understanding of SSF led to design, operate and scale up SSF bioreactors. The process is aerobic in nature and contains a solid substrate bed with moisture and porosity. The SSF bioreactor system should fulfill the following requirements.

1. A suitable vessel for holding the solid substrate. The material of construction should be mechanically strong, non-toxic, corrosion resistant and less cost.
2. Environmental friendly. Should be able to control the bio-emissions during its operation.
3. Should be equipped with controls and regulators for effective aeration, mixing, heat removal and moisture control.
4. Sterilisation mechanisms (*in situ* or off-line).
5. Safe loading and unloading and product recovery systems.

In contrast to submerged fermentation systems, SSF bioreactor systems are yet to reach sophistication and perfection. To overcome the problems of heat and mass transfer phenomena and easy diffusion of O_2, CO_2 and other metabolites, a suitable bioreactor design is yet to be achieved. Temperature and conducive moisture maintenance in the reactor have become the main topics for research. Many bioreactor designs like tray type, packed bed, rotating, rocking drum, stirred type have been proposed recently.

Mass diffusion mechanisms and limitations, in the interstices and inter particle transfer of by-product gases and O_2 are to be understood clearly for efficient design.

MEASUREMENT AND CONTROL OF SSF PARAMETERS

Measurement and control of state or operating variables is crucial for better performance of the reactors. Moisture, temperature, oxygen and gaseous products like CO_2 concentrations are to be accurately measured on-line and controlled for good fermentation and product yield at optimum levels. The operating variables which can be manipulated to achieve optimum conditions depend on the type of reactors and the operating variables like flow rates, humidity of inlet air, frequency and intensity of agitation. Thermocouples can be used for online measurement of temperatures. Sensors will be effective to minimise the error in measurements so that control becomes easy and accurate. On-line sensors to measure relative humidity, pH, and pressure and concentration gradients across the SSF reactors are the best bet for the accuracy and consistency.

Smoothing algorithms may be used to account for noise in the measurement of reactor variables. Bio mass estimation can be carried out based on oxygen consumption or CO_2 evolution during the fermentation process. Applications of off-line measurement techniques for water activity, pH and biomass provide a check on online measurements as they suffer from less noise.

SCALE-UP PROBLEMS OF SSF BIOREACTORS

Scale-up of bioreactors can be done based on chemical engineering fundamental principles of mass and energy balances on moisture, temperature, aeration and oxygen transfer and dynamic conditions of the substrate in the reactor. The important factor here is to find operating conditions for the bioreactor that will allow the water and the energy balances to remain at a constant value as scale increases. Theory without experiment is dry and experiment without theory is sterile. So, both theory and formulating the model and experiment in verifying their validity is essential. The models proposed should be simple and easily able to incorporate the complexities of SSF processes into the model equation to get better insight into the understanding of the growth kinetics and transport phenomena of heat and mass transfer. Most of the models proposed in the literature to date need improvement and experimental validation. An ideal model of SSF bioreactors should represent the following features.

1. Distribution of substrate particles in the reactor and their dynamics under agitation.
2. Heat generation, transfer and its effect on growth of microbes.
3. O_2 and CO_2 diffusion and their effect on growth.
4. Exoenzyme production and its diffusion.
5. Substrate degradation and uptake.
6. Biomass production and its dynamics.
7. pH and water activity changes.
8. Change in physiology of the biomass.

Such an ideal model may never be achieved but some of the critical parameters can be embedded in the model to describe the true characteristics of SSF processes thereby maximizing their economic performance. The kinetic and transport models of SSF system will help in the design, development and operation of commercial reactors. Geometric and dynamic similarity approach will be a rationale and realistic one for scale-up of SSF reactors. Dimensional analysis of the parameters affecting the overall performance of the rector must be done critically for a reliable scale-up. The dimension less numbers like Reynolds, Nusselts, Dam Koehler, Weber, Prandtle, etc. must be adapted to describe the simulation of the conditions in industrial reactors through laboratory reactor data to ensure adaptability in the designing of commercial reactors. Pressure drop criterion is one of the possible scale-up factors to be explored. There is no information on characterisation of heat and mass transfer coefficient in SSF reactors up to date. There is need for experimental work to formulate the models on pressure drop, mass and heat transfer coefficients in SSF labscale bioreactors to adopt in commercial units.

Applications of SSF

Following the global trends on SSF research the potential applications of SSF are classified as follows.

1. Agro-industrial residues conversion into value added and protein enriched end products for poultry and cattle feed. Residues like coffee pulp and husk, soyabeans, cassava husk and bagasse, sugarcane bagasse, sugarbeet pulp, fruit wastes, palm tree wastes, etc. bioconverted into ethanol, single cell protein, organic acids like citric and lactic acids, aminoacids, pigments, antibiotics, mushrooms, biopesticides, gibberellic acid, flavour and aroma compounds, etc. The fungus culture on coffee husk produced a strong alcoholic aroma with fruity flavour compounds such as acetal dehyde, ethanol, etyl acetate, were the major compounds produced. The head space of the cultures is composed of the compounds. Citric acid is probably the only product produced on large scale by fermentation which is used in food and pharmaceutical industry. Many bacteria, yeast and fungi are capable of growth on solid substrates, but filamentous fungi are the best adapted for SSF process and dominates the research presently due to their physiological capabilities and hyphal mode of growth under low moisture. Filamentous fungi is extensively used for protein enrichment of starch substrates such as cassava. sago and banana wastes as well as of cellulosic substrates such as wheat straw, corn straw and sugar beet pulp. All these SSF products are aimed for animal feed and animal feed supplementation.

2. SSF is increasingly applied in environmental control and monitoring. Bioremediation and biodegradation of hazardous compounds. Biological detoxification of industrial wastes are the latest. Bioinsecticides for pest control in crops are looked at with promise by SSF.

3. Enzyme production. Enzymes like amylase, protease, xylaxase, lipase, etc. were produced by SSF using cheap carbon sources like cassava bagasse and soyabean defatted cake, etc. Development of a lab-scale reactor for tannase production from coffee industrial waste is reported by some of the researchers in recent literature. Laccase and Manganese-peroxidase are produced from malted barley waste using *Lentinus edodes*. Amyloglucosidase and lignin peroxidase were also produced from SSF process.

4. Biopulping of wheat straw using *Phanerochaete chrysosporium* was reported by Chen and others. Work is also being carried out-on enzyme inhibitors and bio molecules production through SSF.

To sum up, SSF could be perfected for value-addition and utilisation of these products and their residues to boost the economy of the nation. Hence the research on SSF to develop commercial processes with techno-economic feasibility is worth continuing. The intricacies in solid state fermentation technology is to be understood clearly through modelling, kinetics of growth of microbes, control of parameters, etc. and finally scale up and commercialisation of SSF processes are essential for establishing the SSF technologies to apply in divergent areas.

Chapter 8

Microbial Solid State Fermentation for Future Biorefineries

INTRODUCTION

Today's biorefinery technologies would be almost unthinkable without biotechnology. This is a growing trend and biorefineries have also increased in importance in agriculture and the food industry. Novel biorefinery processes using solid state fermentation (SSF) technology have been developed as alternative to conventional processing routes, leading to the production of added-value products from agriculture and food industry raw materials. SSF involves the growth of micro-organisms on moist solid substrate in the absence of free-flowing water. Future biorefineries based on SSF aim to exploit the vast complexity of the technology to modify biomass produced by agriculture and the food industry for valuable by-products through microbial bioconversion. This chapter discusses microbial SSF technology for future biorefineries for the production of many added value products ranging from feedstock for the fermentation process and biodegradable plastics to fuels and chemicals.

There is a strong need to produce new products (food and non-food, chemicals) from food and agro-industrial residues in a more sustainable way than are currently realised or possible. Sustainable production is defined here as production that is efficient and effective in the use of raw materials and energy, generates a minimum of waste or low-value streams, and leads to new and existing products and/or materials with the required functionality, safety and integrity. Solid state fermentation (SSF) and related technologies offer alternative production routes for such biotechnology-based products.

SSF has been used in the world for a long time. This technology is commonly known in the East, for traditional manufacture of fermented foods, and in the West, for mold-ripened cheese. It can be defined as a system in which the growth of selected micro-organism(s) occurs on solid materials with a low moisture content and has been identified as a potentially important methodology and technique in biotechnology. Now-a-days, SSF is an economically viable, practically acceptable technology for large-scale microbial bioconversion and biodegradation processes. Development of sustainable SSF and bioprocess technology is an emerging, multidisciplinary field with possible application to the production of food, enzymes, animal feed, chemicals, cosmeceutical products, bioethanol and pharmaceuticals/

nutraceuticals. During recent years, SSF has received fresh attention from researchers and industries all over the world. This is due to several major advantages that it offers over submerged fermentation (SmF), particularly in the area of solid waste treatment. Apart from the production of food and feed, SSF shows tremendous potential in applications to produce high-value low-volume products such as enzymes, biologically active secondary metabolites, and chemicals. SSF offers many advantages over conventional SmF such as simple and inexpensive substrates, elimination of the need for solubilisation of nutrients from within solid substrates, elimination of the need for rigorous control of many parameters during fermentation, mostly higher product yields, lower energy requirements, less waste water produced, no foam generation, and relatively easy recovery of end products. SSF provides flexibility in terms of the raw materials to be used and their capability to produce various value-added products by microbial bioconversion. Thus, modern SSF holds the highest potential for biorefinery targets for the production of many value-added products.

DEFINITION OF SOLID STATE FERMENTATION

Over the last four decades, various terms have been used as synonyms of SSF. The most popular term is 'solid state fermentation' itself, but terms such as 'solid substrate fermentation', 'solid state bioprocessing', 'solid substrate cultivation', 'solid substrate process', 'solid state digestion', 'solid state cultivation', 'solid state culture', 'surface cultivation', and 'surface culture' have also been used to describe the same process. The term 'solid state fermentation' is the most commonly used term followed by 'solid substrate fermentation'. Later, Botella and others introduced a new term 'particulate bioprocessing', a novel process strategy developed for biorefineries based on SSF. In this case, they described particulate bioprocessing as confined to those systems that involve the growth of micro-organisms on moist solid materials in a particulate state. Amore and Faraco introduced the the term consolidated bioprocessing (CBP) as the main route for lignocellulosic degradation that makes fungi alternative and better candidates. Here, they introduced cellulolytic fungi as candidates with great potential to provide saccharolytic enzymes to digest lignocellulose efficiently and produce sugars, then later convert those sugars to ethanol. The proposed activities provide an important contribution to reducing ethanol production cost and showing that the fungi naturally possess all pathways required for the conversion of lignocellulose to bioethanol.

SSF has been defined in many ways. Many researchers in the field have introduced their own ways to define SSF. For example, Viniegra-Gonzalez defined SSF as a microbial process occurring mostly on the surface of solid materials that have the property to absorb or contain water, with or without soluble nutrients. Mitchell and others described SSF as any processes in which substrates in a solid particulate state are utilised, while Pandey and others defined SSF as the cultivation of micro-organisms on moist solid supports, either on inert carriers or on insoluble substrates that can also be used as carbon and energy source. Rahardjo and others came out with a definition that SSF is the growth of micro-organisms on moistened solid substrate, in which enough moisture is present to maintain microbial growth and metabolism, but where there is no free-moving water and air is the continuous phase. Rosales and others gave a simple definition of SSF where the growth of micro-organisms is on solid or semisolid substrates or support. Later, Mitchell and others defined SSF as a process that involves the growth of micro-organisms on moist particles of solid materials in beds in which the spaces between the particles are filled with a continuous gas phase. In the latest definition, Thomas and others defined SSF as a three-phase, heterogeneous process, comprising solid, liquid, and gaseous phases, which offers potential benefits for the microbial cultivation for bioprocess and products development. Whatever the definition, we can understand that SSF is referring to the microbial fermentation, which takes place in the absence or near absence of free water, thus being close to the natural environment to which the selected micro-organisms, especially fungi, are naturally adapted.

SOLID STATE FERMENTATION – CURRENT STATE AND PERSPECTIVES

Industrial practice in SSF for secondary metabolite production has been led by companies such as Biocon Ltd. Biocon India developed technology based on SSF as a low-cost, low-energy option for the production of specialty enzymes. The company started as early as 1990 on an 8-year research and development programme to create a novel bioreactor capable of conducting SSF with comparable levels of automation and specialisation as those associated with SmF. Cristobal and others mentioned that the great success of SSF is not only related to the notable increase in research in this area, but also to significant industrial developments especially in enzyme production. The last two decades witnessed an unprecedented increase in interest in SSF. There has been a concerted effort to understand the issues involved in SSF and to apply them to a wide range of new products. The majority of publications are about the use of microbial fermentation and the possibility of utilising different solid waste products as raw material. Through SSF, solid waste either from food or agro-industry can be used as commercially desirable substrates.

Many research works have so far focused on the general applicability of SSF for the production of enzymes, metabolites, and spores. Food and agro-industry provide many different solid waste products suitable as valuable solid substrates, which have been combined with many different micro-organisms and resulted in a wide range of fermentation processes. For example, enzyme production by SSF is a growing field due to the simplicity of the processes, high productivity, and generation of concentrated products. Another important factor that influences the development of SSF is that both food and agro-industrial waste is rich in carbohydrates and other nutrients so that it can serve as a substrate for the production of bulk chemicals and enzymes. In addition, SSF is a good alternative to help in solving pollution problems, rather than disposing of waste into the land and causing environmental harm. Research studies on microbial growth in SSF are limited compared to those in SmF. The majority of published research focuses on the optimisation of environmental conditions to achieve maximum production and substrate utilisation rates. With the advances of biotechnology and bioprocessing now-a-days, for example in the area of enzyme and fermentation technology, many new avenues have opened for their utilisation in SSF. Growth and product formation kinetics, bioreactor design, and process control in SSF are becoming popular research subjects. With the increased interest in SSF, progress is being made now-a-days with the goal of developing industrially applicable SSF systems. Research on kinetic studies has become one of the most popular and critical subjects to be explored, and with developments in computer and software technology, such studies have become easier. The prediction of microbial growth and product formation could be achieved with simulations and parameter estimation if we had a better understanding of how micro-organisms grow and produce the desired products in SSF. Accurate modeling and determination of process variables in SSF such as moisture, temperature, pH, colour changes, biomass determination by oxygen uptake rate, carbon dioxide production rate, glucosamine, fungal growth and enzyme production related to kinetic studies, and modeling for optimisation of SSF have been investigated extensively. Bioreactor design for SSF has been studied extensively and is another important area for development. Developments during the past 15 years suggest that there is now considerable interest in studying various aspects of SSF bioreactor design. Bioreactors such as packed beds, multi-layer packed beds, rotating drums bioreactor, column bioreactors, column-tray bioreactor, magnetic rotating biological contactors, fixed beds, immersion bioreactors, tray systems, horizontal stirred tanks and other bioreactors have been reported. The biotransformation and biological upgrading of food and agro-industry waste for improved nutritional qualities can be achieved through SSF technology. This has been the most important area where the potential of SSF has been recognised to offer economically feasible technology and provide the possibility of a continuous operation for new value added products. Various high-value biotechnological products such as enzymes, primary and secondary metabolites, antibiotics, and chemicals can be produced through SSF because it uses cheap solid substrates which are available locally and are rich in carbohydrates and other nutrients. In the case of wheat bran, research has been developed for its utilisation as an SSF substrate for added value products. As examples, the production of the antifungal antibiotic iturin, nigerloxin, meroparamycin, γ-aminobutyric acid (GABA), antioxidants, monacolin K and biopigments, clavulanic acid, lovastatin, gallic acid, L (+) lactic acid and bioactive phenolic compounds have been reported.

Many published research articles refer to the advantages of SSF in enzyme production. These advantages include higher enzyme titers, higher productivity levels, stability of excreted enzymes, a low level of catabolic repression, and short fermentation time. Production of enzymes can be stimulated by a high sugar concentration in SSF but in SmF, such high concentrations have an inhibitory effect because of catabolic repression. SSF appears to be more robust than SmF with regard to catabolic repression and can therefore, be more productive with a wider variety of substrate mixtures. According to Castilho and others, SSF has many advantages over SmF for lipase production by *Penicillium candidum* when it is grown on wheat bran substrate. The SSF process is very attractive from an economic point of view as well. Studies on economic analysis of lipase production showed that total capital investment needed for SmF was 78% higher than that needed for the SSF process. Thus, there has been much development of SSF in various biotechnology applications and in product development. The last two decades have changed the perception of SSF as 'low technology' and it is becoming a promising technology for the production of added value, 'low volume and high cost' products. The majority of publications indicate that filamentous fungi are the most suitable organisms for growing under SSF conditions.

CONCEPTS OF BIOREFINERY

A biorefinery can use all kinds of biomass that can be upgraded to one or more valuable products such as agricultural by-products (wheat bran, rapeseed meal, straw, corn stover, bagasse), waste from the food industry (including kitchen and household waste), grains/cereals (wheat, maize, corn, soyabean), starch and sugars, aquatic biomass (algae and seaweeds), as well as wood and lignocellulosic materials. A biorefinery is not a completely new concept. According to Berntsson and others, the term 'biorefinery' became visible in the 1990s in response to at least four industry trends. First, there was a growing awareness in the industry to use biomass resources sustainably for economic and environmental sustainability. Second, there was an increase in interest in upgrading low-quality lignocellulosic biomass into highly valuable products. Third, there was a concern in starch production for energy applications. Finally, there was a need to develop high value-added products to face global competition by adopting the biomass produced from industry. Biorefineries can provide a significant contribution to sustainable development, generating added value to sustainable biomass use, and producing a range of bio-based products (food, feed, materials, chemicals) and bioenergy (fuels, power, and/or heat) at the same time.

CLASSIFICATION OF BIOREFINERIES

In the past, biorefineries were classified based on a variety of different bases, such as:

1. Type of main intermediates produced: Syngas platform biorefineries and sugar platform biorefineries.

2. Technological implementation status: Conventional and advanced biorefineries; first, second, and third generation biorefineries.

3. Type of raw materials used: Whole crop biorefineries, oleochemical biorefineries, lignocellulosic feedstock biorefineries, green biorefineries, and marine biorefineries.

5. Main type of conversion process applied: Thermochemical biorefineries, biochemical biorefineries, and two-platform concept biorefineries.

However, according to Lange, biorefineries can be classified into six categories based on the availability of biomass:

1. The yellow biorefinery: Straw, corn stover, wood.

2. The green biorefinery: Fresh green biomass, grass for protein-rich feed.

3. The blue biorefinery: Fish by-catch/cut-offs, fish discards and innards, mussels as biomass, brown seaweed, red and green algae, invertebrates such as sea cucumber.

4. The red biorefinery: Slaughterhouse waste identified as a new resource for upgrade to higher value products, for example food ingredients.

5. The white biorefinery: Agro-industry side streams.

6. The brown biorefinery: Sludge and household waste.

MICROBIAL FERMENTATION

A wide range of substrates is used for SSF. These include waste products from the agricultural and food industries. There is a need to decrease food and agro-industry processing waste because of environmental and economic problems. Instead, better use of raw materials and more efficient processing of food and agro-industry solid materials into added-value products is needed. Now-a-days, scientists are using SSF to develop new process strategies for biorefineries. SSF bioprocessing in a biorefinery concept offers the versatile possibility to convert the sugar-containing polymers starch, cellulose and hemicellulose into a range of products. Currently, commercially viable SSF processes involve solid substrates for producing products ranging from biofuels, bioethanol, biomethanol, biogas, and biodegradable plastics to commodity, platform, and specialty chemicals like succinic acid and pharmaceutical products. This is achieved by microbial bioconversions or enzymatic biotransformation. Using food and agro-industry solid materials provides almost complete nutrient sources. This is an advantage because during fermentation, solid materials can be used with/without supplementation.

Processes using micro-organisms are very efficient and can be controlled. Another advantage is that the micro-organisms act as specific catalysts that can produce a range of targeted products. An example includes using the fungus *Aspergillus niger* in the bioconversion of apple pomace into a multi-enzyme bio-feed. Another example is making use of agro-industry waste (coconut husks, apple pomace, orange and lemon peel) as these materials can be used as inert carriers in SSF and allows good micro-organism growth. As found by many researchers globally, solid materials from food and agro-industry waste can be used as feedstocks for fermentation processes. They can be applied to biological activities such as producing various enzymes, animal feed, food ingredients, primary and secondary metabolites or nutraceutical, and pharmaceutical products. Emerging technologies have high potentials because of their application in producing non-food materials. Examples include bioethanol and biofuels produced by food and agro-industry waste, biopulping, biological control (bioinsecticides), biodegradable plastic (poly (3-hydroxybutyrate): PHB), the production of functional chemicals, and also production of feedstock for sequel fermentation processes (Fig. 8.1). Koutinas and others and Webb and others proposed a

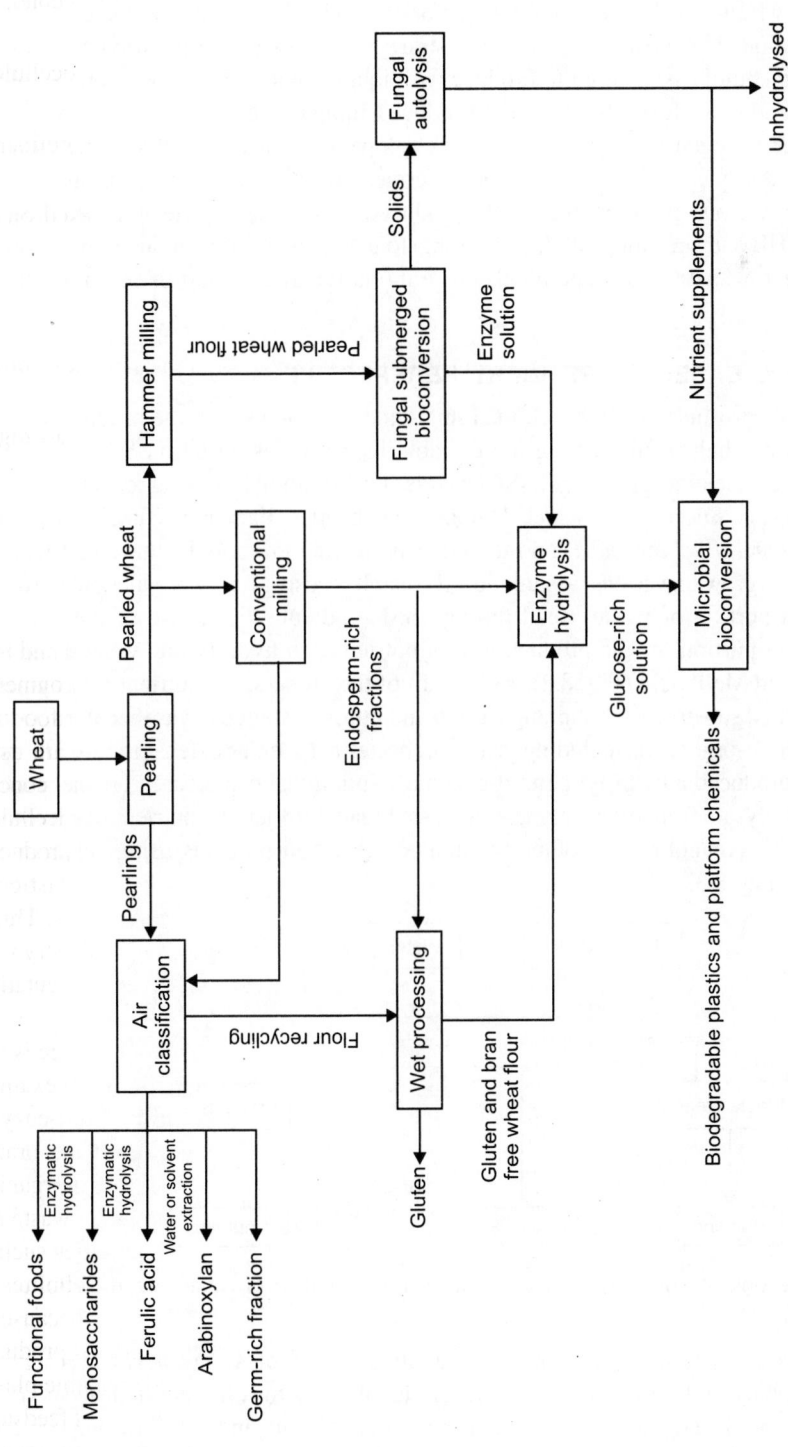

Fig. 8.1: Schematic diagram of microbial fermentations proposed in a possible biorefinery utilising wheat for the production of PHB and succinic acid.

novel cost-competitive wheat-based biorefining strategy for the production of nutrient-complete feedstock for microbial fermentation. This upstream processing strategy exploits (i) the hydrolytic capability of fungi and (ii) the natural autolysis of fungi to produce two liquid streams. Strategy (i) produced wheat hydrolysate rich in the glucose form. Strategy (ii) produced fungal autolysate rich in supplementing nutrients, especially the nitrogen form. These strategies compensate for the most suitable carbon and nitrogen ratio for the subsequent submerged microbial fermentation process. Through this study, they proved the utilisation of various microbial feedstocks produced from wheat via the proposed biorefinery for the production of PHB and succinic acid. It is desirable to use food and agro-industry waste solids as a renewable resource for sustainable chemical and non-chemical production through microbial bioconversions.

SSF BIOPROCESSING-BASED BIOREFINERY PERSPECTIVE

Biomass from food and agro-industry (through microbial bioconversion or for the production of more added-value products) has helped in creating more viable biorefineries based on SSF. Generating a generic fermentation feedstock through fungal SSF provides the possibility of using residual solid waste from food and agro-industry. Successful research demonstrated that two filamentous fungi, one producing amylolytic enzymes (*Aspergillus awamori*) and the other producing proteolytic enzymes (*Aspergillus oryzae*), were both able to grow on various waste solids. Research focusing on the production of hydrolytic enzymes from a small portion of waste bread through SSF, and subsequent use of the enzymes to hydrolyse-the remaining portion for the production of a nutrient-rich hydrolysate, was carried out by Melikoglu and others and Melikoglu (Fig. 8.2). As a result of the process, the nutrient-rich hydrolysate taken from both fermented substrates (containing carbon and nitrogen sources) was then used to support growth of a range of micro-organisms including yeast and bacteria. Later, as extension from the project, Melikoglu and others produced a hydrolytic multi-enzyme solution that can be used for the production of a monomer rich hydrolysate from other segment of waste bread. Further, the nutrient rich hydrolysate can be converted into bioethanol or any other potential value-added products using selected micro-organisms as shown in Fig. 8.3.

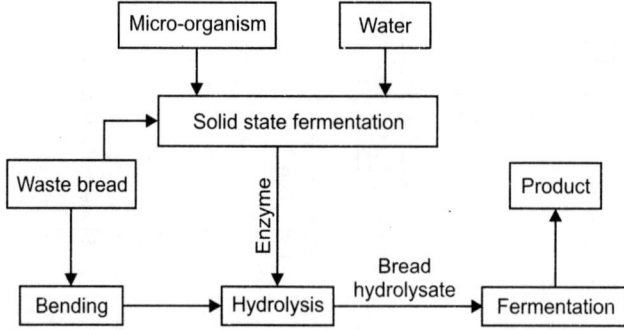

Fig. 8.2: Schematic presentation of a novel bioprocess for the utilisation of waste bread based on multi-enzyme producing SSF.

Within their continuous studies, Du and others also applied an SSF of *A. awamori* and *A. oryzae* on wheat bran to produce amylolytic-rich and proteolytic-rich solutions for efficient hydrolysis of wheat starch and protein. This produced glucose-rich and nitrogen-rich streams. In the following year, a wheat bran fraction was used as the only solid medium in two SSF processes of *A. awamori* and *A. oryzae* that

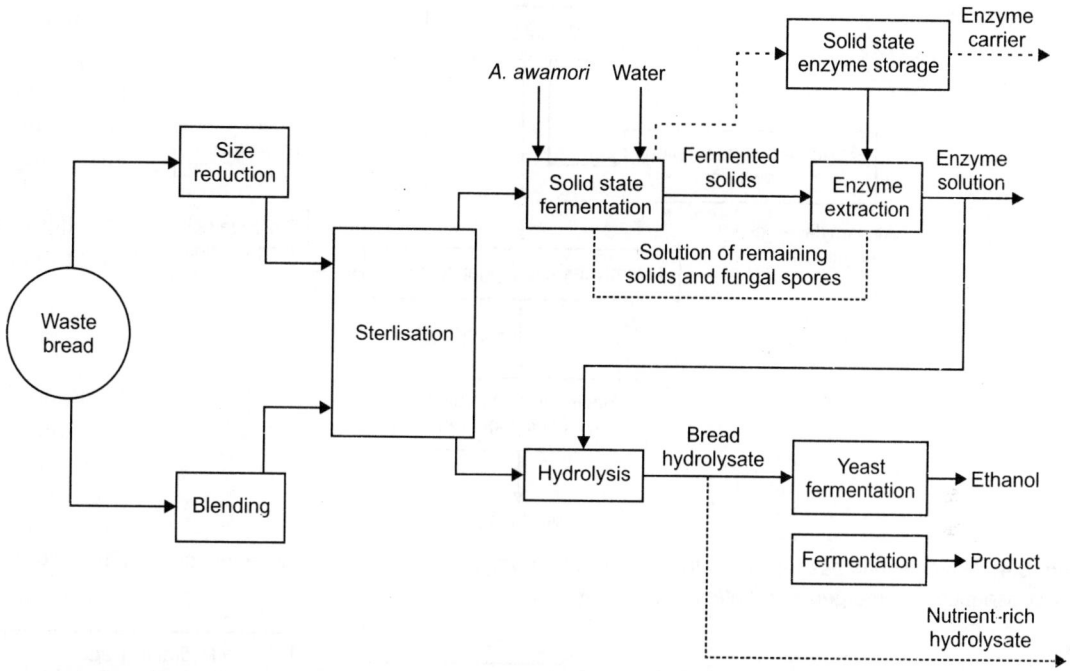

Fig.8.3: Proposed bioprocess for the production of bioethanol and other potential value added products from waste bread by *A. awamori* in SSF. Dotted lines indicate optional steps in the process.

produced enzyme complexes rich in amylolytic and proteolytic enzymes, respectively. In a research work, Leung and others reported how a glucoamylase- and protease-rich multi-enzyme solution was formed from waste bread using the fungi *A. awamori* and *A. oryzae*. Their SSF-based biorefining could be divided into three important steps as shown in Fig. 8.4: (i) SSF of a selected fungus was carried out on solid waste material to obtain enzyme-rich fungal solids, (ii) the fermented solids were subsequently added to a media suspension to produce a nutrient-rich hydrolysate with optimal C/N source, and (iii) another subsequent bacterial fermentation was carried out using the waste solid hydrolysate for the production of chemicals or targeted products.

This strategy in the subsequent *Actinobacillus succinogenes* and recombinant *Escherichia coli* fermentation to produce succinic acid. Prior to that, Koutinas and others designed an integrated biorefinery that produced high-quality wheat flour and upgraded the by-product stream into a nutrient-enrichment animal feed and added-value chemicals through microbial bioconversion. Figure 8.4 shows this process based on the bioproduction of succinic acid. One of the processes presented in Fig. 8.5 used wheat bran in fungal SSF for enzyme production. This was then used in hydrolysis reactions to produce nutrient-rich hydrolysates from wheat macromolecules (e.g. starch, protein). The nutrient-rich hydrolysate was then converted into the desired products by proper fermentation as shown in Figs 8.3 and 8.4. Another work by Botella explained how nutrient-rich hydrolysate was produced *via* SSF of wheat grains using *A. awamori* and followed next by SmF by *Wautersia eutropha* to produce the biodegradable plastic PHB and by *Saccharomyces cerevisiae* for ethanol production. Salakkam worked with nutrient-rich hydrolysate derived from rapeseed meal *via* SSF by *A. oryzae* followed by the hydrolysis of fermented solids to produce hydrolysate, which later was used for PHB production through SmF by *Cupriavidus necator*.

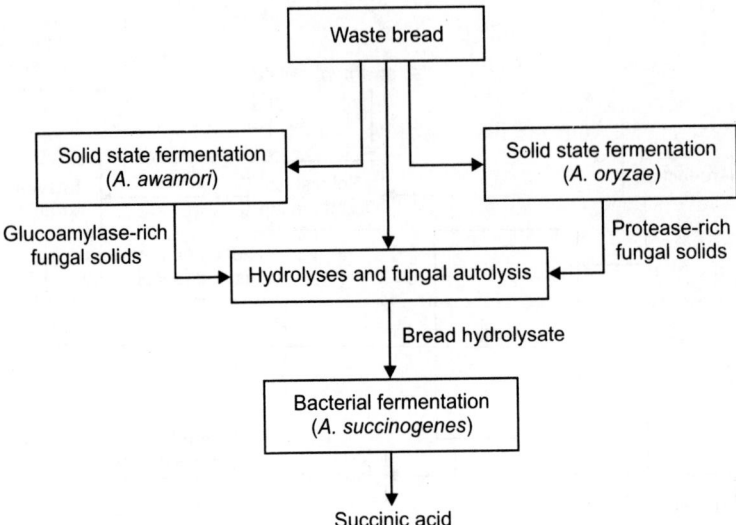

Fig. 8.4: Novel bioprocess based on the production of amylolytic and proteolytic enzymes *via* SSF followed by subsequent submerged fermentation.

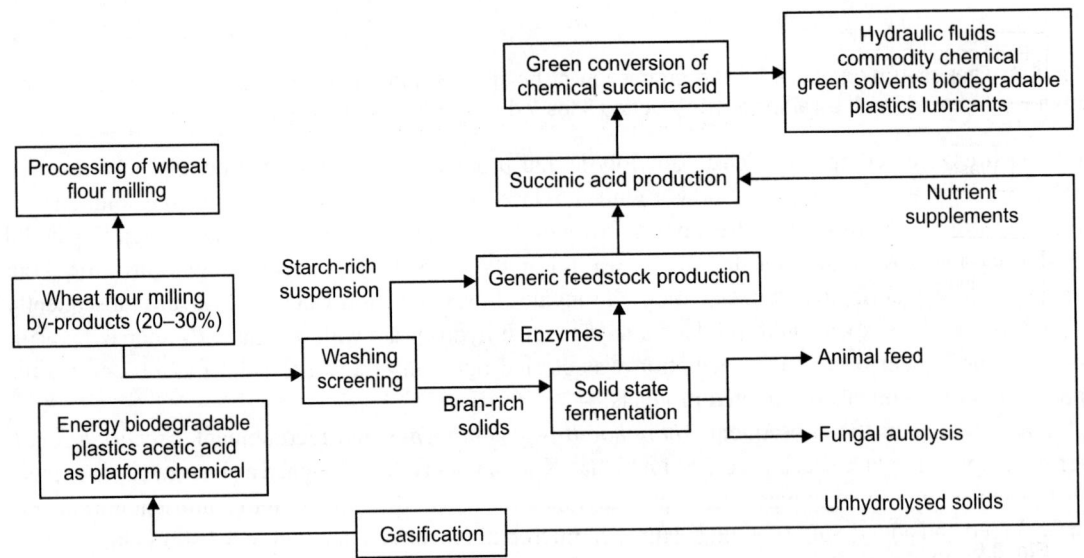

Fig. 8.5: Schematic diagram of SSF rules in a possible biorefinery utilising wheat flour milling by-products to produce added-value products.

Added-value technology through microbial bioconversion technology will have a powerful impact on food and agro-industry waste processing and create exciting opportunities for addressing environmental issues and economic concerns. SSF with the use of micro-organisms could lower operating costs by eliminating the need to purchase unnecessarily purified commercial enzymatic products. In addition, nutrient-rich hydrolysates with high glucose and free amino nitrogen (FAN) levels were shown to have great potential to replace more expensive synthetic media. Kwan and others demonstrated the bioconversion of mixed food waste and bakery waste through fungal SSF by *A. awamori* 14331. After fungal SSF, fungal hydrolyses were performed in order to recover nutrients such as glucose, fructose, and FAN which are essential requirements for the subsequent lactic acid fermentation of *Lactobacillus casei* Shirota to produce lactic acid. Later, Kwan and others developed further process to investigate the behavior and lactic acid production by *Streptococcus thermophilus* YI-B1 and *L. casei* Shirota during fermentation process using semi-defined medium and food waste derived media and found that the varied composition of food waste, enzyme loading, and solid-to-liquid ratio led to different nutrient composition and levels of the hydrolysate.

A potential integrated biorefinery based on food and agricultural industry waste solids can be summarised in the schematic diagram in Fig. 8.6. At the end of SSF, hydrolysates are mixed with water and could be directly used for subsequent fermentation. However, nutrient-complete generic fermentation can be achieved with another step, continuing with simultaneous hydrolyses and fungal autolysis. In this step, enzyme activity, such as glucoamylase and protease activity, continues to degrade the solids from the fermented substrates to produce glucose and FAN.

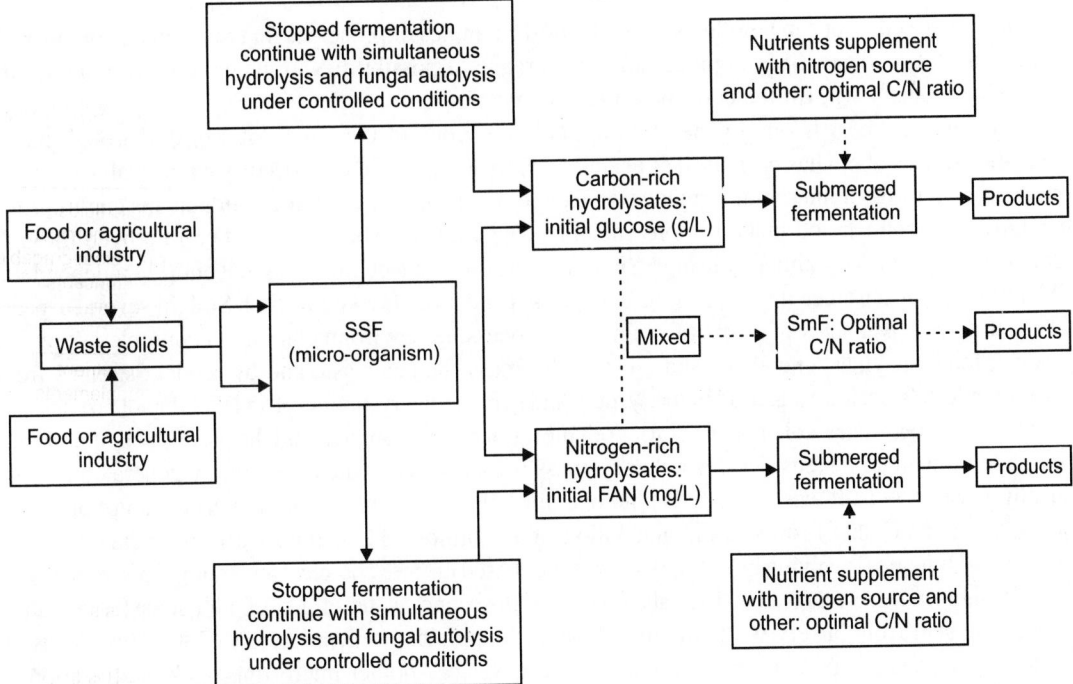

Fig. 8.6: Summary of a novel process strategy for biorefineries based on SSF to produce generic fermentation feedstocks.

Research has shown that hydrolysates can either be used separately or in a mixture. Nutrient supplements might need to be added if separate feedstocks are used. With the respect to this, if carbon-rich hydrolysates from one micro-organism are used, nutrients with a nitrogen source should be added. Likewise, if hydrolysates rich in nitrogen from a second micro-organism are used, nutrients with a carbon source might be added. As a result, it has been proved that the performance of micro-organisms is extremely efficient when both generic fermentation feedstocks are used together in a mixture. This is due to achieving optimal nutrient media with carbon and nitrogen sources (C/N ratio) to support growth of micro-organisms for selected fermentations. Such biorefining strategy provides complete feedstocks for subsequent microbial fermentation and leads to the production of chemicals and of targeted added-value products. According to Koutinas and others, three steps are required to produce a nutrient-complete medium for microbial fermentation: (i) on-site enzyme production through SSF or SmF, (ii) enzymatic hydrolysis of substrate suspensions, and (iii) fungal autolysis for production of hydrolysate with a high nutrient composition.

It is predicted that SSF technology will appeal to countries that have a high level of agro-industry waste, as this can be used as inexpensive raw material. On top of that, advanced biorefining strategies have been restructured in order to reduce environmental impact, improve overall economics, and meet market and societal needs. The promising results taken from these research activities have led to the conclusion that biomass from renewable resources, such as food and agro-industry waste, has practical benefits and should be further taken advantage of within future biorefining strategies.

INVESTING IN A GREEN SUSTAINABLE FUTURE

The biorefinery concept can be filled with real world examples of processes that make use of biomass to produce useful products. SSF bioprocessing and a range of possibilities to integrate biorefining in the processing industry will fill the concept with some meaning.

For example, wheat is one of the main agricultural crops in the whole of world. Through many generations, wheat bran has been used as an animal feed and can be categorised as waste. Value addition of wheat bran through microbial fermentation technology using selected micro-organisms can improve the nutritional content and make it more valuable. The value of wheat bran can be improved through fermentation technology either by using SSF or SmF. These two techniques are not entirely independent. SSF has been chosen since it appears to be exceptional and more favourable than SmF in several aspects where it gives advantages in terms of biological, processing, environmental and economic aspects to produce food, enzymes, chemicals and bio-oils. Biomass such as waste and by-product streams from existing industrial sectors (e.g. food industry, pulp and paper industry, biodiesel and bioethanol production) can also be used as renewable resources for both biorefinery development and the production of nutrient-complete fermentation feedstocks. Another biomass is food waste, which is currently generated in huge quantities worldwide. It was estimated about one third of food produced for human consumption is lost or wasted globally that results in the generation of 1.3 billion MT of food waste per year. Proposed processing strategies involving biorefinery approaches based on both chemical and biological technologies will definitely help. As reported by Lam and others, for the economic feasibility of pilot-scale fermentation of succinic acid from bakery waste in Hong Kong, it is possible to generate US$ 380,000/year (as on 2015) overall revenue by converting 1 T/d of bakery waste. Another interesting work is utilisation of different feedstock formulation strategies based on the utilisation of microbial feedstocks produced from bakery waste through the proposed Starbucks biorefinery development in Hong Kong for the production of biodegradable plastics, succinic acid, and multi-enzyme solutions. Other than that is spent ground coffee obtained after brewing the coffee beans which is an abundant by-product of the coffee industry making it a promising and interesting feedstock for a biorefinery. This is important for attracting investment and industrialisation interest in the biorefinery process using domestic waste as raw material.

As mentioned earlier, SSF offers a more favourable environment for fungal growth, yielding higher productivity in a relatively low-cost process by using nutrient-rich agro-industrial residues as substrates.

Note-worthy to say, with the increasing interest in SSF now-a-days, researchers are keen to discover as many new ways to explore the usage of this technology as possible to develop new added-value materials from by-products. The technology know-how, protocol, and manufacturing process will be obtained and optimised taking into account raw materials, technologies, processing routes, products, technical, economical, and environmental aspects. The use of renewable resources will make an essential contribution towards sustainable development while, at the same time; the generation of pollutants or harmful waste during product manufacture can be minimised.

- Future biorefinery studies can be carried out to use the possible solid state bioreactor for the production of nutrient-rich hydrolysate for sequel fermentation for the production of valuable chemicals. The productivity of the systems can be enhanced by modification and adding several control parameters for better control of SSF processes.

- SSF has proved to be excellent technology for solid waste treatment. It can be suggested that food and agro-industry solid waste can be utilised for the production of various added-value products. Industrial bodies should consider this scenario as part of their strategy for tackling the food and agro-industry waste problem and for the environmentally friendly production of biomaterial, enzymes, secondary metabolites, chemicals, and even biofuels. In many countries food waste for example, is currently landfilled or incinerated together with other combustible wastes for possible energy recovery, while through circular economy, it could serve as a promising source of energy and value-added products, especially in developing countries. Therefore, the issue of solid waste being thrown onto the land and the subsequent environmental problems could be overcome.

Overall, the value of food and agro-industrial waste can be improved through microbial SSF technology and thus, SSF can be an ideal platform for biomass biochemical conversion for bio-based products. Improvements in the quality of food and agro-industrial waste can be associated with the improved value of food and agro-industrial waste. In another word, these approaches may lead to associate improvement in food and agro-industrial waste quality and to manage the issue of solid wastes in a green approach. The micro-organism chosen must also be capable of utilising all the nutrients contained in food and agro-industrial waste and provide unique microenvironments conducive to microbial growth and metabolic activity. The quality of food and agro-industrial waste may be improved either in a nutritional way or by improving its processing properties. Therefore, the microbial bioconversion technology involved may allow novel alteration of nutritional quality. Potential value-added by-products from food and agro-industrial waste fractions could be then used for the production of various co-products, such as methane, hydrogen, ethanol, butanol, cosmeceutical, pharmaceutical and nutraceutical products, enzymes, organic acids, natural biopigments, feed and food ingredients, prebiotics for animals, natural polymers (e.g. arabinoxylans, β-glucans), monomers (e.g. glucose, xylose, arabinose, succinic acid, lactic acid), or oil components (e.g. triglycerides, sterols). Definitely, the main benefit anticipated by using microbial bioconversion technology is to add value to food and agro-industrial waste, which can be identified, from the product quality and safety of the final fermented products. It is desirable to use this as a renewable resource for sustainable production of value-added products through microbial fermentation technology and SSF is a promising technology, however, rising concerns related to scale up and reproducibility in a productive process, which should be addressed in the future.

To sum up, advances in the understanding of microbiology and of the composition of targeted products and their raw materials (biomass), as well as the development of advanced SSF bioreactors, allow more consistent research and development in SSF. Thus, we arrive at the modern day bioprocessing and microbial fermentation processes. Despite many limitations, many industrial facilities worldwide successfully operate SSF processes, although some of them produce relatively small quantities of high added-value products that do not require large-scale bioreactors. In other cases, the bioreactors do not operate optimally, or bioreactors that are not easily adaptable for different processes are used. In the near future flexible and optimum performance large-scale SSF bioreactors will be designed, built, and operated successfully, although more engineering research is needed. These bioreactors will use low-cost and reliable instruments, especially designed for SSF, and sophisticated control strategies that will include advanced control techniques such as expert- and model-based controlled systems. Indeed, the biorefinery idea and prospect can be pervaded with real world examples to produce functional and valuable products. On top of that, it will add value to the sustainable use of biomass and make a significant contribution to sustainable development. Last but not least, the ability of micro-organisms, especially filamentous fungi, to convert biomass through SSF bioconversion will have a great impact on food and agro-industry in every aspect of life from food and medicine to fuel. This is a future that undoubtedly will present challenges, but one that we believe we should embrace.

Chapter 9

Industrial Fermentation Processes

INTRODUCTION

The most common meaning of fermentation is the conversion of a sugar into an organic acid or an alcohol. Fermentation occurs naturally in many foods and humans have intentionally used it since ancient times to improve both the preservation and organoleptic properties of food. However, the term 'fermentation' is also used in a broader sense for the intentional use of micro-organisms such as bacteria, yeast, and fungi to make products useful to humans (biomass, enzymes, primary and secondary metabolites, recombinant products, and products of biotransformation) on an industrial scale.

Modern industrial fermentation processes used in the food and beverage industry can be described according to different perspectives. In the center of these processes are usually bioreactors, which can be classified with respect to the feeding of the bioreactor (batch, fed-batch, and continuous mode of operation), immobilisation of the biocatalyst (free or immobilised cells/enzymes), the characteristic state of matter in the system (submerged or solid substrate fermentations), single strain/mixed culture processes, mixing of the bioreactor (mechanical, pneumatic, and hydraulic agitation), or the availability of oxygen (aerobic, micro aerobic, and anaerobic processes). The decision as to which bioreactor or fermentation process should be implemented in any particular application involves considering the advantages and disadvantages of each setup. This includes examining the properties and availability of the primary raw materials, any necessary investment and operating costs, sustainability, availability of a competent workforce, as well as the desired productivity and return on investment. Since in large-scale applications, each fermentation system needs to operate efficiently and reliably, the major criterion for the selection of a bioreactor/fermentation process remains the minimum for capital costs per unit of product recovered. Simultaneously, with efficient design and operation, in largescale processes, the issues concerning by-product and wastewater management are inevitable.

TYPES OF FERMENTATION PROCESSES

Submerged Cultivation

Submerged cultivation of microbial cells in bioreactors guarantees a controlled environment for the efficient production of high-quality end products and to achieve optimum productivity and yield. Industrial bioreactors operated in batch, fed-batch, or continuous mode are utilised to culture different types of micro-organisms producing a wide range of products. In the following sections, different approaches to submerged cultivation of micro-organisms in bioreactors are discussed briefly and the typical features, benefits, and drawbacks of each cultivation mode are highlighted. Finally, the relevant applications for batch, fed-batch, and continuous cultivation of micro-organisms in liquid media used in the production of different types of food industry products are also discussed.

Batch cultivation

Batch culture represents a closed system in which the medium, nutrients, and inoculum are added to the bioreactor, mostly under aseptic conditions, at the beginning of cultivation, that is, the volume of the culture broth in the bioreactor is theoretically constant during cultivation (practically, small deviations in culture volume are caused by a low feed rate of acid/base solutions to keep the pH at a desired level and by sampling or introducing air/gas into the culture; on balance, such changes are usually ignored due to their small value relative to the total working volume of the bioreactor).

Typically, at the beginning of batch cultivation, a known number of viable cells are inoculated into the bioreactor that is already filled with sterilised medium containing all nutrients. After inoculation, the cell culture follows the classical growth curve described by Monod, which is divided into four main phases. As the lag phase is an 'inefficient' stage of culture (even though the cells are metabolically active—they are adapting their enzymatic apparatus to a new environment, no significant increases in biomass concentration, substrate consumption, or product synthesis are observed), it is desirable to shorten it as much as possible. The length of the lag phase is influenced mainly by the concentration of cells in the inoculum and their physiological state, the composition of the inoculation and cultivation medium (mainly the source of carbon and energy, pH, and temperature), and the size of the inoculum.

The exponential (logarithmic growth) phase is characterised by rapid cell proliferation (biomass concentration is an exponential function of time), constant specific growth rate, which is equal to the maximum specific growth rate of the culture under conditions of absence of growth limitation (growth rate is not limited because all nutrients are present in excess, while also not attaining growth-inhibiting concentrations), fast consumption of the source of carbon and energy, and a high rate of primary metabolite production. The depletion of nutrients by the end of the exponential phase (in the case of aerobically grown cultures, these are signalled by a rapid increase in dissolved oxygen concentration) causes a progressive reduction in the specific growth rate and a transition to the stationary phase, characterised by the stagnation of growth and utilisation of endogenous reserves of carbon and energy, this phase is important for the synthesis of secondary metabolites.

Most industrial bioreactors are operated in batch mode due to the relative simplicity of this process. The whole batch operation consists of several steps, including medium formulation, filling the bioreactor, sterilisation in place (SIP systems), inoculation, cultivation, product harvesting, and bioreactor cleaning in place (CIP systems). For efficient performance of batch operation, it is important to minimise all nonproductive steps (all steps listed above except cultivation), achieve a high rate of product synthesis, optimise productivity, and maximise the yield of the end product. The performance of any particular batch operation is thus influenced by the type of end product—an extension of exponential growth is advantageous for the efficient production of biomass (baker's yeasts, feed biomass) or primary metabolites (ethanol, acetic, citric, or lactic acids), whereas in the case of secondary metabolite production, the exponential phase is shortened (by the limitation of one nutrient, usually the source of nitrogen) and the stationary phase is prolonged to achieve the maximum yield of the product.

Submerged batch cultivation can be used for the production of alcoholic beverages (beer, wine, and distilled spirits such as whisky, brandy, rum, and others), organic acids used in the food industry either as acidifiers or as preservatives (citric, acetic (vinegar), and lactic acids), and amino acids used as flavour enhancers (e.g. monosodium glutamate) or sweeteners (e.g. aspartate).

For distilled spirits, the fermentation of wort during Scotch whisky production is taken as an example. Washbacks, simple cylindrical fermentation vessels (volume 250–500 m^3) for the production of distilled spirits are made either from wood or from stainless steel. Although wood washbacks are difficult to clean and sanitise, they are still used, especially in malt whisky distilleries. Wort to be fermented is pumped to the washback, cooled to 20°C, and inoculated with either fresh or dried yeast cells.

The 90% of citric acid is produced by microbial (*Aspergillus niger*) synthesis from sugar- or starch-containing materials (sugar beet, sugarcane molasses, and corn) and about 60% of this amount was consumed in the food industry. Although citric acid can be produced at an industrial scale using surface liquid cultivation, solidstate cultivation, or submerged liquid cultivation, now-a-days, the latter predominates. Submerged cultivation is carried out in stirred bioreactors (capacity 150–200 m^3) or bubble columns (capacity up to 1000 m^3), usually operating aerobically for 4–10 days until the citric acid concentration reaches 10–15% w/v.

Fed-batch cultivation

Fed-batch culture represents a semi-open system in which one or more nutrients are aseptically and gradually added to the bioreactor while the product is retained inside, that is, the volume of the culture broth in the bioreactor increases within this time. The main advantages of fed-batch over batch cultures are: (i) the possibility to prolong product synthesis, (ii) the ability to achieve higher cell densities and thus increase the amount of the product, which is usually proportional to the concentration of the biomass, (iii) the capacity to enhance yield or productivity by controlled sequential addition of nutrients, and (iv) the feature of prolonged productive cultivation over the 'unprofitable periods' when the bioreactor would normally be prepared for a new batch.

Fed-batch is advantageously used in processes: (i) where substrate inhibition or catabolic repression is expected, this problem can be overcome by using a 'safe' concentration of the substrate in batch mode followed by feeding the remaining substrate within fed-batch operation, (ii) where a Crabtree effect (repression of yeast respiratory enzymes by high concentrations of glucose) is expected, by gradual feeding of the substrate, the production of ethanol by yeasts can be eliminated under aerobic conditions, (iii) where a high cell density is required, a high and constant specific growth rate can be maintained by exponential feeding of the substrate, (iv) where a high production rate should be achieved; cell metabolism can be regulated by precise sequential feeding of nutrients, and (v) where a high viscosity of culture broth is expected (e.g. production of dextran or xanthan); a gradual dilution of the medium can overcome the problems of mixing and oxygen transfer.

There are many methods of adding a substrate to the bioreactor (either as a concentrated solution of a sole carbon and energy source or as a medium containing carbon plus other nutrients); the proper choice of the nutrient feeding rate can enhance the culture performance considerably since it influences cellular growth rate, cell physiology, and the rate of product formation. The common feeding strategies are: (i) discontinuous feeding, achieved by regular or irregular pulses of substrates and (ii) regular continuous feeding of nutrients designed according to a precalculated profile (Fig. 9.1) or based on the feedback control of online measured variables associated with cell growth and metabolism, for example, dissolved oxygen concentration, pH, CO_2, evolution rate, and biomass concentration.

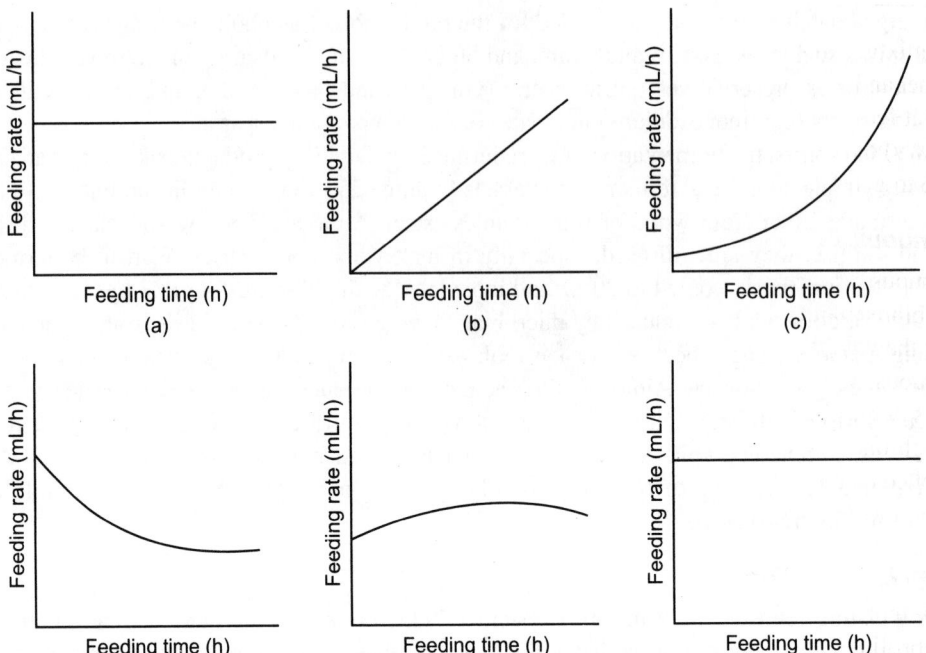

Fig. 9.1: Relationship between precalculated profiles of substrate feeding and specific growth rates of cell culture. (a) Constant feeding rate, (b) linearly increasing feeding rate, and (c) exponential feeding rate.

The typical food fed-batch fermentations are large-scale production of baker's yeast, pure ethanol, which is further utilised for alcoholic beverages produced by mixing ingredients such as liquors or cordials, and submerged acetification for vinegar production.

Baker's yeast (*Saccharomyces cerevisiae*), which is distributed as compressed, dried, or instant biomass, is valued for its dough-leavening ability. Yeast production is the only technology in which the respiratory metabolism of *S. cerevisiae*, leading to high biomass yield, is stressed. Nevertheless, due to the Crabtree effect, that is, the formation of ethanol under aerobic conditions in the presence of excess substrate, the only alternative for producing baker's yeast is fed-batch cultivation. The main bottleneck of large-scale baker's yeast production is the control of nutrient medium inflow, which was traditionally based on empirical data. Currently, many feeding methods utilising different approaches have been developed, for example, a logistic feeding profile, evolutionary optimisation of genetic feeding algorithms, fuzzy control, pulsed feeding optimisation, and others.

Human knowledge of vinegar production dates back to ancient history and several distinct production methods have been used, for example, surface oxidation (Orleans process, slow vinegar production), the quick vinegar process using a trickle-bed bioreactor, or submerged acetification. The most modern process is submerged acetification, in which acetic acid (vinegar) is produced by *Acetobacter*-mediated oxidation of ethanol, and takes place in special 'acetator' bioreactors. The most common acetator, a Frings Acetator®, differs from the usual bioreactors by using a special rotor/stator turbine aerator. The aerator with a double function (aeration and mixing), consists of a rotor placed under the bioreactor and connected to an air-suction pipe surrounded by a stator. The aerator is self respiring, that is, during rotation (speed about 1500 rpm), it sucks air and pumps liquid, which causes the formation of an air–

liquid mixture that is radially injected into the culture medium. The foam is broken by a mechanical defoamer and because oxidation is an exothermal process, cooling is necessary. The acetators are operated in repeat fed-batch mode and one cultivation cycle in a single acetator can produce vinegar containing 15% (w/v) of acetic acid. In addition, a dual-stage high-strength fermentation process, which allows the culture to generate up to 20.5% (w/v) acetic acid, has been developed.

Continuous cultivation

Continuous culture represents an open system in which nutrients are aseptically and continuously added to the bioreactor, and the culture broth (containing cells and metabolites) is removed at the same time, that is, the volume of the culture broth is constant due to a constant feed-in and feed-out rate. Frequently, continuous culture is used as a synonym for a chemostat, represented by a constant specific growth rate of cells, which is equal to the dilution rate and is controlled by the availability of the limiting nutrient, although other types of continuous operation such as turbidistat (a constant concentration of biomass controlled by the dilution rate) or nutristat (a constant parameter related to cell growth controlled by the dilution rate) can be employed. The main advantages of continuous culture (chemostat) over the batch mode are: (i) the possibility to set up optimum conditions for maximum and long-term product synthesis, (ii) the ability to achieve stable product quality (the steady state is characterised by a homogeneous cell culture represented by a constant concentration of biomass and metabolites), and (iii) a distinct reduction in 'unprofitable' periods of the bioreactor operation.

In spite of these advantages, there are also several problems that hamper the extensive utilisation of continuous operation on a large scale. These include: (i) increased risk of contamination due to the pumping of the medium in and out of the bioreactor, (ii) the danger of genetic mutations in the production strain in a long-term operation, and (iii) additional investments may be required for technical facilities.

Despite the producer's proclamations, only a few food or feed production systems employing micro-organisms are operated in a genuine continuous mode, where continuous fermentation is defined as a process running in one or more bioreactors at a stable dilution rate. One of the rare examples of this is fodder yeast (*Kluyveromyces fragilis*) production using spent sulphite liquor as the substrate, which is operated in the Czech Republic. The main reason why this process can be performed continuously is the low sensitivity of the production medium to possible contamination. At the same time, this process can be taken as a rare example of industrial-scale production using a highly valued substrate, lignocellulosic hydrolysate, which is obtained as a waste product from pulp production.

In other food applications such as in modern distilleries, a semicontinuous fermentation operated in a series of fermentors is usually used instead of genuine continuous fermentation. The current estimates are that only about 16% of ethanol in North America is produced in a continuous mode due to problems with contamination. Nevertheless, both batch and semicontinuous mode fermentations permit continuous distillation, which is the reason why the whole distillery production is often considered as continuous.

Continuous fermentation systems based on immobilised cell technology have also been studied in beer production. Although continuous beer fermentation has been tested as a promising technology for several decades, the number of industrial applications is still limited. The reasons include engineering problems (excess biomass and problems with CO_2 removal, optimisation of operating conditions, clogging, and channelling of the reactor) and unrealised cost benefits (carrier price, complex, and unstable operation). The major obstacle hindering the extensive industrial exploitation of this technology is the difficulty in achieving the correct balance of sensory compounds in a short time typical for continuous systems. However, recent developments in reactor design and in our understanding of immobilised cell physiology, together with applications of novel carrier materials, could provide a new stimulus to research and potential applications of this promising technology.

INTENSIFICATION OF FERMENTATION PROCESSES

There are several strategies available for the intensification of bioprocesses. Some of them focus on engineering aspects, whereas others exploit the tools of physiological modulation (selection or adaptation of micro-organisms), mutagenesis, or genetic manipulation to improve the production strains. Examples of areas where a significant improvement of the fermentation processes can be achieved by engineering approaches are improved mass and heat transfer, reduction of power consumption, high-density cell cultures, and low-shear mixing. The performance of the bioprocess is both individually and synergistically influenced by all components of the production unit and related know-how (strain, bioreactor, media composition, feeding strategy, etc.). In addition, the biological elements of the process (micro-organisms, animal and plant cells, and enzymes) are subject to many processing constraints (fragility, temperature, pH range, and hygienic design of the equipment). These facts place important practical limitations on the bioreactor and bioreaction engineering.

In the last few decades, there has been a significant progress in the area of process control and instrumentation for bioreactors. This has an economic importance because the optimum operation of a fermentation process is associated with improved productivity (high product concentration, high production rate) and savings in product separation. The ability to operate a process at high productivity requires a sound understanding of the biological requirements, process kinetics (limitation, inhibition), and transport phenomena. The following sections provide the principles and examples of some bioprocess intensification methods.

Immobilised Cell Technology

The increased productivity of bioprocesses can be achieved through controlled contact of substrates with a high concentration of the active biocatalyst, enzyme, or microbial cells. These high-cell-density cultures can be created by feeding strategies, cell retention/recycle, or cell immobilisation. Among the strategies to create high cell-density cultures, cell immobilisation is the most widely studied and applied in the food and beverage industries.

The maximum immobilised biomass concentration achieved in continuous beer fermentation was up to 10 times greater than the free cell concentration at the end of the conventional batch fermentation. However, the immobilisation of micro-organisms provokes different physiological responses when compared to low-cell-density cultures of free cells and therefore, their application has to be carefully considered in terms of product quality. For example, the application of immobilised micro-organisms in fermentation processes induces modifications in cell physiology due to mass transfer limitations, concentration gradients created by an immobilisation matrix, and by ageing of the immobilised biomass.

Immobilised Cell Physiology

An important factor influencing the growth and metabolic activity of immobilised cells is the microenvironment of the solid immobilisation matrix, represented by parameters such as water activity, pH, oxygen, substrate and product concentration gradients, and mechanical stress. The interplay between the appropriate production strains and immobilisation methods is very important in immobilised cell reactors and their suitable combination can improve both system performance and product quality. The importance of careful matching of the chosen yeast strain with the immobilisation method and the suitable reactor arrangement was demonstrated in beer production. Although there are a variety of methods for investigating the metabolic state of immobilised cells (monitoring of cellular activity, microscopic, noninvasive, and destructive methods), acquiring the reliable data is still the limiting factor for process optimisation. In addition, the data concerning the physiological conditions of immobilised microbial cells are rather complex, due to different matrices and variable system configurations, and therefore their interpretation is difficult. Immobilisation has been reported to activate some metabolic functions (substrate uptake, product formation, enzyme expression, and activity) of microbial cells. According to some authors, the enhanced metabolic activity can also be attributed to surface-sensing responses in immobilised microbial cells but the reasons are still a matter of controversy. Overall, conclusions should be very carefully drawn from the results, since sampling and sample treatment may also influence the measurements of immobilised cell physiology.

It has been shown that immobilised cells exhibit increased levels of deoxyribonucleic acid (DNA), structural carbohydrates, glycogen, and fatty acids, as well as modifications of cell proteome, cell wall, and cell membrane composition. Not surprisingly, alterations in plasma membrane composition have a profound impact on several enzymes, sensor proteins, transporters, and membrane fluidity. Many reports also underline increased stress resistance of immobilised cells. The increased resistance to inhibit substances (ethanol, pollutants, antimicrobial agents, etc.) can be ascribed to changes in the composition and organisation at the level of the cell wall and plasma membrane and/or to the protective effect of the immobilisation support.

Mass transfer in immobilised cell systems

The diffusional resistance to substrate transport from the bulk solution to the biocatalyst and the hindered diffusion of products in the opposite direction may represent the most significant mass transfer limitations arising from the use of immobilised cell technology. These mass transfer limitations constitute the most evident hypothesis to explain the often-observed decrease in immobilised cell growth rate and specific productivities as compared to free cell cultures. The typical immobilisation materials exhibiting internal mass transfer limitations are polymeric matrices. In these materials used for cell entrapment, the internal mass transfer limitations of cells by nutrients can be further influenced by the position of the cells, bead size, and structure of the polymer. Mass transfer limitations are crucial in immobilised cell systems when oxygen supply to cells and the removal of carbon dioxide are required. Oxygen transfer from the gas phase to the immobilised biocatalyst has long been recognised as the major rate-limiting step in aerobic immobilised cell processes. The most common option to improve mass transfer in these systems is to reduce the bead diameter.

Unlike polymeric matrices, the preformed porous (sintered glass) and nonporous (DEAE-cellulose, wood chips, and spent grains) carriers do not have the additional gel-diffusion barrier. However, depending on the porosity of the carrier and on the amount of biomass adsorbed in the pores, internal mass transfer limitations may also occur. In the case of nonporous carriers, internal mass transfer problems vary with the thickness of the cell layer (biofilm). The yeast adhered in a single layer of DEAE-cellulose showed similar metabolic activities whereas multilayers of yeast attached to spent grains had a significantly lower specific sugar consumption rate as compared to free cells.

Engineering Aspects of Process Intensification

Fermentation processes can be divided into two main categories based on the characteristic state of matter of the medium: solid substrate fermentation and submerged fermentation. Among the latter, the most common bioreactor configurations used in food and beverage applications are stationary particle bioreactors, such as packed-bed/fibrous-bed, trickle-bed reactors, and mixed (particle) bioreactors, such as fluidised bed, gas lift, bubble column, and stirred tank. The stationary particle bioreactors are either operated with immobilised cells (enzymes) or a mixture of free and immobilised cells (enzymes), and their typical internal mass transfer issues. Mixed bioreactors may contain solely free or immobilised cells as well as their mixture. There are also bioreactor configurations that do not fit into the two previous categories. For example, rotating biological contactors (RBCs), also classified among moving surface reactors, where a biofilm grows on rotating disks partially or completely immersed in a liquid medium. The use of RBCs in food applications is rare, and is limited to the production of citric acid.

The selection of a suitable reactor design from numerous available types and configurations is a complex task and depends on various factors (Table 9.1). The importance of individual factors may change depending on the process requirements and the product characteristics. However, there is a need to have a fundamental understanding of the kinetics and transport limitations when a bioreactor is selected or when a new bioreactor is designed and constructed. Two phase bioreactors are generally limited to anaerobic processes or to processes where gas–liquid mass transfer plays a marginal role. Conversely, in three-phase bioreactors, efficient mass transfer usually requires an intimate mixing of all three phases. Three-phase bioreactor design is an area in which significant process intensification can be achieved through the enhancement of gas–liquid mass transfer.

Table 9.1: Factors influencing the selection of bioreactor type.

Free cell bioreactors	
Nature of substrate and product	Biological requirements
Kinetics of product formation	Process conditions (T, pH, and Δp)
Mass transfer considerations	Hygienic considerations
Heat transfer considerations	Scale-up considerations
Hydrodynamic considerations	Ease of fabrication and reactor costs
Process control	Running costs
Immobilized cell bioreactors (additional factors)	
Method of immobilisation	Internal mass transfer in the biocatalyst
Carrier costs	Biocatalyst replacement/regeneration

Gas–liquid mass transfer considerations

In aerobic bioprocesses, oxygen is the key substrate due to its low solubility in aqueous media. Consequently, a continuous supply of oxygen into aerobic bioreactors is often needed. Therefore, the oxygen transfer rate (OTR) should be predicted prior to the choice/design and scale-up of a bioreactor. Many studies have been conducted to estimate the efficiency of oxygen transfer in different bioreactors and these have been reviewed in various works. Another area where gas–liquid mass transfer rate is crucially important (CO_2 supply and O_2 removal) is the construction and operation of photobioreactors used for the cultivation of photoautotrophic micro-organisms (micro algae) with nutritional potential.

The dissolved oxygen concentration in aerobic cultures depends on the rate of oxygen transfer from the gas phase (usually air bubbles) to the liquid, on the rate at which oxygen is transported into the cells, and on rate of oxygen uptake by the micro-organism. The transport of oxygen from air bubbles to the site of oxygen consumption can be described in a number of steps, among which oxygen diffusion through the liquid film surrounding the bubble shows the greatest resistance. The gas–liquid mass transfer rate is usually modelled according to the two-film theory and is characterised by the volumetric (gas–liquid) mass (oxygen) transfer coefficient (k_La), while the driving force of the process is the difference between the concentration of oxygen at the interface (C^*) and that in the bulk liquid (C_L). In the case of large microbial pellets, immobilised cells, or fungal hyphae, the resistance in the liquid film surrounding the solid can also be significant. Oxygen transfer in aerobic bioprocesses is strongly influenced by the hydrodynamic conditions in the bioreactors. These conditions are known to be affected by the operational conditions (stirrer speed, superficial gas velocity, liquid circulation velocity, etc.) physico-chemical properties of the culture (viscosity, density, and surface tension), bioreactor geometry, and also by the presence of oxygen-consuming cells.

Stirred tank bioreactors (STBRs) are widely used in a large variety of bioprocesses taking advantage of free cell (enzyme) suspensions. An industrial-scale STBR usually consists of a stainless-steel vessel, motor-driven impeller, and gas sparger positioned below the impeller. Aerated STBRs generally have high mass and heat transfer coefficients, good homogenisation, and the capability of handling a wide range of superficial gas velocities. Mass transfer and mixing in STBRs are most significantly affected by stirrer speed, type and number of stirrers, and the gas flow rate.

In bioreactors with a height-to-diameter ratio (H/T) above two, standard single-impeller systems were often found to have unsuitable operating parameters. Oxygen transfer in these geometries can be improved by using multiple-impeller configurations (approximately one impeller per each H/T = 1) that exhibit efficient gas distribution, increased gas holdup, superior liquid-flow characteristics, and lower power consumption per impeller as compared to single-impeller systems.

Pneumatic bioreactors consist of a cylindrical vessel, into the bottom (usually) of which air (gas) is introduced to ensure aeration, mixing, and liquid circulation, without any moving mechanical parts. In pneumatically agitated reactors such as bubble columns (random liquid circulation) and airlift reactors (streamlined liquid circulation), the homogeneous shear environment compared to the local shear extremes in STBRs has enabled the successful cultivation of shear-sensitive cells such as mammalian and plant cells or mycelial fungi. In contrast, the lack of mechanical agitation can cause poor mixing in a highly viscous medium and serious foaming under high aeration. Airlift and stirred tank reactors exhibit comparable mass transfer capacities; however, airlift reactors can be superior to STBRs in terms of operating costs because of lower power consumption. A further increase in the overall volumetric gas–liquid oxygen transfer coefficient (k_La) in bubble column and airlift reactors was achieved by the installation of static mixers into the draft tubes and riser sections, respectively. The improvement of OTR achieved by static mixers is a result of air bubble breakup increasing the specific gas–liquid interfacial area (a). Industrial applications include the cultivation of a filamentous mold and ethanol production in an airlift reactor. However, in the second example, increased ethanol productivity was also achieved as a consequence of size reduction of yeast flocs, and thus improved liquid–solid mass transfer, provoked by the new riser design. The predictions of OTR determined by a dynamic method in sterile culture medium in the absence of biomass, often underestimates the k_La value for the real bioprocess. In fermentation processes, an enhancement of OTR was found to be due to oxygen consumed by the micro-organisms, leading to a lower dissolved oxygen concentration in the bulk liquid (C_L). Simultaneously, mass transfer enhancement was also attributed to the presence of a dispersed phase (micro-organisms) adsorbed onto the gas–liquid interface, influencing the oxygen adsorption rate and gas–liquid interfacial area. The extent of this enhancement was expressed as the biological enhancement factor (E). According to some studies, the E value of k_La can be up to 1.3 times that of the mass transfer coefficient determined for the system without microbial cells.

FUTURE PROSPECTS OF FERMENTATION PROCESSES

The development of fermentation processes for the food and beverage industries aims at improving the productivity and product quality by means of process design, strain selection/construction, and process monitoring. In all these areas, there have emerged some very innovative ideas that could lead to economically attractive solutions.

With regard to (online) process monitoring, significant progress is required, particularly in the area of advanced instrumentation and sensor development, for solid substrate fermentations. The innovative techniques described so far include different sensor technologies, respirometry, X-rays, image analysis, infrared spectrometry, magnetic resonance imaging, and so on. However, for some of them, the main drawback is high cost, which makes these techniques unsuitable for large-scale applications.

One of the significant challenges in the bioreactor design is the improvement of large-scale photobioreactors and phycocultures (seaweed farms) for the production of micro- and macro algae and algae-derived food products. Another prospective strategy to increase the metabolic productivity in bioprocesses is the use of suitably controlled ultrasonication. The beneficial effects of ultrasound can be exploited at the level of biocatalysts (cells and enzymes) and their function (e.g. cross-membrane ion fluxes, stimulated sterol synthesis, altered cell morphology, and increased enzyme activity) and sonobioreactor performance (mass transfer enhancement).

The potential for genetic engineering in the field of food fermentation is indisputable and has been reviewed. However, the nutritional status of fermented foods can also be improved by the rational choice of food-fermenting microbes based on the understanding of their interaction with diet and human gastrointestinal microbiota. In this respect, fermented foods can be regarded as an extension of the food digestion and fermentation processes and can be steered toward beneficial health attributes.

SECTION III

Biological Basis of Productivity in Fermentation

Chapter 10

Isolation, Preservation and Improvement of Important Micro-organisms

INTRODUCTION

The term isolation refers to the separation of a strain from a natural, mixed population of living microbes, as present in the environment, for example in water or soil flora, or from living beings with skin flora, oral flora or gut flora, in order to identify the microbe(s) of interest. Historically, the laboratory techniques of isolation first developed in the field of bacteriology and parasitology (during the 19th century), before those in virology during the 20th century.

Methods of microbial isolation have drastically changed over the past 50 years, from a labour perspective with increasing mechanisation, and in regard to the technology involved, and hence speed and accuracy. In order to isolate a microbe from a natural, mixed population of living microbes, as present in the environment, for example in water or soil flora, or from living beings with skin flora, oral flora or gut flora, one has to separate it from the mix.

Traditionally microbes have been cultured in order to identify the microbe(s) of interest based on its growth characteristics. Depending on the expected density and viability of microbes present in a liquid sample, physical methods to increase the gradient as for example serial dilution or centrifugation may be chosen. In order to isolate organisms in materials with high microbial content, such as sewage, soil or stool, serial dilutions will increase the chance of separating a mixture. In a liquid medium with few or no expected organisms, from an area that is normally sterile (such as CSF, blood inside the circulatory system) centrifugation, decanting the supernatant and using only the sediment will increase the chance to grow and isolate bacteria or the usually cell-associated viruses.

If one expects or looks for a particularly fastidious organism, the microbiological culture and isolation techniques will have to be geared towards that microbe. For example, a bacterium that dies when exposed to air, can only be isolated if the sample is carried and processed under airless or anaerobic conditions. A bacterium that dies when exposed to room temperature (thermophilic) requires a pre-warmed transport container, and a microbe that dries and dies when carried on a cotton swab will need a viral transport medium before it can be cultured successfully.

BACTERIAL AND FUNGAL CULTURE

Inoculation

Laboratory technicians inoculate the sample onto certain solid agar plates with the streak plate method or into liquid culture medium, depending what the objective of the isolation is:

- If one wants to isolate only a particular group of bacteria, such as Group A *Streptococcus* from a throat swab, one can use a selective medium that will suppress the growth of concomitant bacteria expected in the mix (by antibiotics present in the agar), so that only *Streptococci* are 'selected', i.e. visibly stand out. To isolate fungi, Sabouraud agar can be used. Alternatively, lethal conditions for *streptococci* and gram negative bacteria like high salt concentrations in Mannitol salt agar favour survival of any *staphylococci* present in a sample of gut bacteria, and phenol red in the agar acts as a ph indicator showing if the bacteria are able to ferment mannitol by excreting acid into the medium. In other agar substances are added to exploit an organisms ability to produce a visible pigment (e.g. granada medium for Group B *Streptococcus*) which changes the bacterial colony's colour, or to dissolve blood agar by hemolysis so that they can be more easily spotted. Some bacteria like *Legionella* species require particular nutrients or toxin binding as in charcoal to grow and therefore media such as Buffered charcoal yeast extract agar must be used.

- If one wants to isolate as many or all strains possible, different nutrient media as well as enriched media, such as blood agar and chocolate agar and anaerobic culture media such as thioglycolate broth need to be inoculated. To enumerate the growth, bacteria can be suspended in molten agar before it becomes solid, and then poured into petri dishes, the so-called 'pour plate method' which is used in environmental microbiology and food microbiology (e.g. dairy testing) to establish the so-called 'aerobic plate count'.

Incubation

After the sample is inoculated into or onto the choice media, they are incubated under the appropriate atmospheric settings, such as aerobic, anaerobic or microaerophilic conditions or with added carbon dioxide (5%), at different temperature settings, for example 37°C in an incubator or in a refrigerator for cold enrichment, under appropriate light, for example strictly without light wrapped in paper or in a dark bottle for scotochromogen mycobacteria, and for different lengths of time, because different bacteria grow at a different speed, varying from hours (*Escherichia coli*) to weeks (e.g. mycobacteria).

At regular, serial intervals laboratory technicians and microbiologists inspect the media for signs of visible growth and record it. The inspection again has to occur under conditions favouring the isolate's survival, i.e. in an 'anaerobic chamber' for anaerobe bacteria for example, and under conditions that do not threaten the person looking at the plates from being infected by a particularly infectious microbe, i.e. under a biological safety cabinet for Yersinia pestis (plague) or *Bacillus anthracis* (anthrax) for example.

Identification

When bacteria have visibly grown, they are often still mixed. The identification of a microbe depends upon the isolation of an individual colony, as biochemical testing of a microbe to determine its different physiological features depends on a pure culture. To make a subculture, one again works in aseptic technique in microbiology, lifting a single colony off the agar surface with a loop and streaks the material into the 4 quadrants of an agar plate or all over if the colony was singular and did not look mixed.

Gram staining the raw sample before incubation or staining freshly grown colony material helps to determine if a colony consists of uniformly appearing bacteria or is mixed, and the colour, and shape of bacteria allow a first classification based on morphology. In clinical microbiology numerous other staining techniques for particular organisms are used (acid fast bacterial stain for mycobacteria). Immunological staining techniques, such as direct immunofluorescence have been developed for medically important pathogens that are slow growing (Auramine-rhodamine stain for mycobacteria) or difficult to grow (such as *Legionella pneumophila* species) and where the test result would alter standard management and empirical therapy. Biochemical testing of bacteria involves a set of agars in vials to separate motile from non-motile bacteria.

ISOLATION AND SCREENING OF MICRO-ORGANISMS

The success of an industrial fermentation process chiefly depends on the micro-organism strain used. An ideal producer or economically important strain should have the following characteristics.

1. It should be pure, and free from phage.
2. It should be genetically stable, but amenable to genetic modification.
3. It should produce both vegetative cells and spores; species producing only mycelium are rarely used.
4. It should grow vigorously after inoculation in seed stage vessels.
5. It should produce a single valuable product, and no toxic by-products.
6. Product should be produced in a short time, e.g. 3 days.
7. It should be amenable to long term conservation.
8. The risk of contamination should be minimal under the optimum performance conditions.

Screening of Micro-organisms for New Products

The next step after isolation of micro-organisms is their screening. A set of highly selective procedures, which allows the detection and isolation of micro-organisms producing the desired metabolite, constitutes primary screening. Ideally, primary screening should be rapid, inexpensive, predictive, specific but effective for a broad range of compounds and applicable on a large scale.

Primary screening is time consuming and labour intensive since a large number of isolates have to be screened to identify a few potential ones. However, this is possibly the most critical step since it eliminates the large bulk of unwanted useless isolates, which are either non producers or producers of known compounds. Computer based databases play an important role by instantaneously providing detailed information about the already known microbial antibiotic compounds.

Rapid and effective screening techniques have been devised for a variety of microbial products, which utilise either a property of the product or that of its biosynthetic pathway for detection of desirable isolates. Some of the screening techniques are relatively simple. e.g. for extracellular enzymes and enzyme inhibitors.

However, for most microbial products of high value, the screening is usually complex and tedious, and often may involve two or more steps, e.g. for antimicrobials. In some cases, it may be desirable to concentrate on a group of organisms expected to yield new products. For example, the search for new antibiotics now focusses on rare *Actinomycetes*, i.e. *Actinomycetes* other than those belonging to the genus *Streptomyces*. Suitably designed specialised screening techniques may be used to detect compounds having various pharmacological activities other than antibiotics.

Strategies for Isolation of Industrially Important Microbes

- The diversity of micro-organisms may be exploited still by searching for strains from the neutral environment able to produce products of commercial value.
- The first stage in the screening of micro-organisms of potential industrial is their 'isolation'.
- Isolation involves obtaining either pure or mixed cultures followed by their assessment to determine which carry out the desired reaction or produce the desired product.
- In some cases it is possible to design the isolation procedure in such a way that the growth of producers is encouraged or that they may be recognised at the isolation stage, whereas in other cases organisms must be isolated and producers recognised at a subsequent stage.
- It should be remembered that the isolate must carry out the process economically and therefore the selection of the culture to be used is a compromise between the productivity of the organism and the economic constraints of the process.

Criteria used for Choice of Organisms

- The nutritional characteristics of the organism: Organism should be capable to utilise the ingredients present in the medium to produce interested product.
- The optimum temperature of the organisms: For instance, the use of an organism having an optimistic temperature above 40°C considerably reduces the cooling costs of a large-scale fermentation, and therefore, the use of such a temperature in the isolation procedure may be beneficial.
- The reaction of the organism with the equipment to be employed.
- The stability of the organism and its amenability to genetic manipulation.
- The productivity of the organism, measured in its ability to convert substrate into product and to give a high yield of product per unit time.
- The easy product recovery from the cultures.
- It should have stable biochemical characteristics.
- It should not produce undesirable substances.
- It should be easily cultivated on a large scale.
- The ideal isolation procedure commences with an environmental source (frequently soil), which is highly profitable to be rich in the desired types.
- Selective pressure may be used in the isolation of organism that will grow on particular substrates in the presence of certain compounds or under agricultural conditions adverse in their types.
- If it is not possible to apply selective pressure for the desired character it may be possible to design a procedure to select for a microbial taxon which is known to show the characteristics at a relatively high frequency, e.g. the production of antibiotic by Streptomycin.
- Alternately, the isolation procedure may be designed to exclude certain microbial 'weeds' and to encourage the growth of more novel types.
- The advantages in the taxonomic description of taxa have allowed the rational design of procedures for the isolation of strains that may have a high probability of being productive or are representatives of unusual groups.
- The advances in pharmacology and molecular biology have also enabled the design of more effective screening tests to identify productive strains amongst the isolated organisms.

PRIMARY AND SECONDARY SCREENING OF INDUSTRIALLY IMPORTANT MICROBES

Screening

The use of highly selective procedures to allow the detection and isolation of only those micro-organisms which are of interest from among a large microbial population.

Screening allows the discarding of many valueless micro-organisms, at the same time it allows the easy detection of the useful micro-organisms that are present in the population in very less number.

Primary Screening

Primary screening allows the detection and isolation of micro-organisms that possess potentially interesting industrial application. Primary screening separate out only a few micro-organisms having real commercial value.

Primary screening determines which micro-organisms are able to produce a compound without providing much idea of the production or yield potential of the organisms.

1. Primary screening of organic acid producing micro-organisms:
   - Incorporation of a pH indicating dye such as neutral red or bromothymol blue into a poorly buffered agar medium.
   - Greater buffer capacity of medium screen microbes having capability to produce considerable quantities of the acid.
   - Incorporation of calcium carbonate in the medium is also used to screen organic acid producing microbes on the basis cleared zone of dissolved calcium carbonate around the colony.
   - These screening approaches do not give idea that which organic acid has been produced.
   - Thus the colonies of micro-organisms showing the potential to produce any fermentation product should immediately be purified and sub-cultured into appropriate medium to be maintained as stock cultures for further testing.

2. Primary screening of antibiotic producing micro-organisms:
   - The simplest screening technique for antibiotic producers is: Crowded Plate technique.
   - The technique is used to find out the micro-organisms that produce an antibiotic without giving much information of sensitivity towards other micro-organisms.
   - Procedure include dilution and spreading or pouring of soil samples that give 300 or 400 or more colonies per plate.
   - Colonies producing antibiotic activity are indicated by an area of agar around the colony.
   - Such a colony is sub-cultured to a similar medium and purified by streaking, before making stock cultures. The purified culture is then tested to find what types of micro-organisms are sensitive in the presence of these the antibiotics, i.e. Microbial Inhibition Spectrum (MIS).
   - The crowded plate procedure also does not necessarily select an antibiotic producing micro-organism, because the inhibition area around the colony sometimes can be due to other reason like: (i) marked change in the pH of the medium resulted due the metabolism of the colony and (ii) rapid utilisation of critical nutrients in the vicinity of the colony, etc.
   - Thus further testing is required to confirm the inhibitory activity associated with a micro-organisms is whether attributed to the presence of an antibiotic or not.
   - Screening of antibiotic producing micro-organisms can be improved by using a 'test organism.

3. Primary screening of extracellular metabolites (vitamins, amino acids and growth factors) producing micro-organisms.

4. Primary screening of micro-organisms utilising specific Carbon and Nitrogen sources secondary screening.

Secondary screening allows further sorting out of micro-organisms obtained from PS having real value for industrial processes and discarding of those lacking this potential

1. SS is conducted on agar plates, in flasks or small fermenter containing liquid media.

2. SS can be qualitative or quantitative in its approach.

3. Secondary screening should give information about the evaluation of the true potential of the micro-organisms for industrial usage.

4. SS should determine whether micro-organisms are actually producing new chemical compounds not previously described.

5. SS should reveal whether there is pH, aeration or other critical requirements associated with particular micro-organisms, both for the growth of the organism and for the formation of chemical products.

6. SS should also detect gross genetic instability in microbial cultures.

7. SS should show whether certain medium constituents are missing or possibly, are toxic to the growth of the organisms or its ability to accumulate fermentation products.

8. SS should determine whether the product has a simple, complex, or even a macromolecular structure, if this information is not already available.

9. SS should show something of the chemical stability of the product and of the product's solubility picture on various organic solvents.

10. SS should show whether the product possesses physical properties such as UV light absorption or fluorescence or chemical properties that can be employed to detect the compound during the use of paper chromatography or other analytical methods and which also might be of value in predicting the structure of the compound.

11. In some case, for certain kinds of fermentation product determinations should be made as to whether gross animal, plant or human toxicity can be attributed to the fermentation product, particularly if it is utilised (as are antibiotics) in disease treatment.

12. SS should reveal whether a product resulting from a microbial fermentation occurs in the culture broth in more than one chemical form and whether it is an optically or biologically active material.

13. SS should reveal whether the micro-organisms are able to chemically alter or even destroy their own fermentation products.

14. Secondary screening helps in predicting the approaches to be utilised in conducting further research on the micro-organisms and its fermentation processes.

TYPES OF METABOLITES

Primary Metabolite

A primary metabolite, is directly involved in normal growth, development, and reproduction. Microbial production of primary metabolites contributes significantly to the quality of life. Through fermentation, micro-organisms growing on inexpensive carbon sources can produce valuable products such as amino acids, nucleotides, organic acids, and vitamins which can be added to food to enhance its flavour or increase its nutritive value. The contribution of micro-organisms will go well beyond the food industry with the renewed interest in solvent fermentations. Micro-organisms have the potential to provide many petroleum derived products as well as the ethanol necessary for liquid fuel. The role of primary metabolites and the microbes which produce them will certainly increase in importance.

Conversely, a secondary metabolite is not directly involved in those processes, but usually has an important ecological function.

Secondary Metabolite

Secondary metabolites are organic compounds that are not directly involved in the normal growth, development, or reproduction of organisms. Unlike primary metabolites, absence of secondary metabolities does not result in immediate death, but rather in longterm impairment of the organism's survivability, fecundity, or aesthetics, or perhaps in no significant change at all. Secondary metabolites are often restricted to a narrow set of species within a phylogenetic group.

Categories

Most of the secondary metabolites of interest to humankind fit into categories which classify secondary metabolites based on their biosynthetic origin. Since secondary metabolites are often created by modified primary metabolite synthases, or borrow substrates of primary metabolite origin, these categories should not be interpreted as saying that all molecules in the category are secondary metabolites (for example the steroid category), but rather that there are secondary metabolites in these categories.

Small molecules can have a variety of biological functions, serving as cell signalling molecules, as tools in molecular biology, as drugs in medicine, and in countless other roles. These compounds can be natural (such as secondary metabolites) or artificial (such as antiviral drugs), they may have a beneficial effect against a disease (such as drugs) or may be detrimental (such as teratogens and carcinogens). Biopolymers such as nucleic acids, proteins, and polysaccharides (such as starch or cellulose) are not small molecules, although their constituent monomers—ribo or deoxyribonucleotides, amino acids, and monosaccharides, respectively—are often considered to be. Very small oligomers are also usually considered small molecules, such as dinucleotides, peptides such as the antioxidant glutathione, and disaccharides such as sucrose.

SCREENING FOR ACTIVITIES

Microbes are exceptionally rich, diverse, and easily accessible sources of novel metabolites that can inhibit enzyme pathways related to disease targets. These metabolites vary enormously in structural complexity and biological activity. To discover therapeutically useful metabolites, it is critical not only to design suitable and sensitive assays for screening microbial extracts but also to test extracts that contain most or all of the metabolites from culture broths with a minimum of interference.

In addition to sensitive assays, novel and diverse producing micro-organisms are critical to the success of any natural products programme, implicit in this statement is that biological diversity may lead to chemical diversity.

STRAIN IMPROVEMENT

After an organism producing a valuable product is identified, it becomes necessary to increase the product yield from fermentation to minimise production costs. Product yields can be increased by: (i) developing a suitable medium for fermentation, (ii) refining the fermentation process and (iii) improving the productivity of the strain. Generally, major improvements arise from the last approach; all fermentation enterprises place a considerable emphasis on this activity. The techniques and approaches used to genetically modify strains to increase the production of the desired product is called strain improvement or strain development. Strain improvement is based on the following three approaches: (i) mutant selection, (ii) recombination, and (iii) recombinant DNA technology.

Mutant Selection

Large scale mutant selection programmes begin when favourable reports of clinical trials are obtained. In the early stages, selection of spontaneous mutants may be helpful, but induced mutations are the most common sources of improvements. Mutations occurring without any specific treatment are called spontaneous mutation, while those resulting due to a treatment with certain agents are known as induced mutations, such agents are referred to as mutagens. Either physical and chemical mutagens can be employed. Usually, the frequency of mutants with desirable phenotype is quite low; hence the major bottleneck is the identification and isolation of such cells from among the large number of non-mutant/undesirable mutant cells. Many mutations (a sudden and heritable change in the traits of an organism) bring about marked changes in a biochemical character of practical interest; these are called major mutations. Some major mutations can be useful in strain improvement. For example, the original strain of *Streptomyces griseus* produced small amounts of streptomycin and large amounts of mannosidostreptomycin which has low antibiotic activity. A major mutant isolated from this strain produced negligible amounts of mannosidostreptomycin and much larger quantities of streptomycin. Similarly, a mutant strain (S-604) of *Streptomyces aureofaciens* produces 6-demethyl tetracycline in place of tetracycline; this demethylated form of tetracycline is the major commercial form of tetracycline.

In contrast, most improvements in biochemical production have been due to the stepwise accumulation of so called minor genes. These genes lead to small increases (or decreases) in the antibiotic or other biochemical production, and selection may be expected to result in a 10–15 per cent increase in yield. The selected strains are usually subjected to successive cycles of mutagenesis and selection, and after several cycles large increases is yields are likely to be obtained. Application of mutagens to induce mutations is called mutagenesis. In some cases, improvements have been obtained even without the use of mutagens. Mutants of *Penicillium chrysogenum* were selected for increased penicillin production; each cycle of selection was preceded by mutagen (chemical) treatment and resulted in only small changes in penicillin yield. But after several (about dozen) cycles of selection, a strain (E 15-1) was obtained that yielded 55 per cent more penicillin than the original strain (Fleming strain).

Selective isolation of mutants

A majority of desirable mutants, especially the 'minor gene' mutants showing increased production, are isolated by screening a large number of clones surviving the mutagen treatment; this is called secondary screening. But this approach requires a large amount of work. Therefore, efforts have increasingly focused on developing techniques for the isolation of particular classes of mutants which are likely to be over producers (Table 10.1).

Some of the strategies are briefly summarised below, the selection for these classes of mutants is simple, easy and effective.

1. Isolation of auxotrophic mutants is the basis for commercial amino acid production in Japan from the bacterium *Corynebacterium glutamicus*. An auxotrophic mutant has a defect in one of its biosynthetic pathways so that it requires a specific biochemical for normal growth and development.

Table 10.1: A summary of different approaches in utilisation of mutation and genetic recombination for strain improvement.

Approach	Chief feature	Example/Remark
Mutant selection: Types		The main approach to strain improvement, produces new alleles of existing genes
Spontaneous mutations	Occurs without any treatment with a mutagen	Used in the initial stages of strain improvement, also for maintenance of improved strains
Induced mutations	Induced by chemical (mainly) or physical mutagens	Mutagenesis followed by selection, several cycles employed
Major mutations	Affect the pattern of metabolite production	Production of 6-demethyl tetracycline in place of tetracycline by *S. griseus*
Minor mutations	Affect the rate of metabolite production	Small gains in each cycle of selection, substantial improvement after several cycles
Mutant selection: Strategies		
Auxotrophic mutants	Defective biosynthesis of a biochemical	Enhanced production of an amino acid, e.g. *phe⁻* mutants accumulate ⸌yrosine
Analogue-resistant mutants	Feed-back insensitive enzymes	Overproduction of metabolites, e.g. amino acids by *C. glutamicus*
Revertants of nonproducing mutants		Some mutants are high producers, e.g. chlortetracycline by *S. viridifaciens*
Revertants of auxotrophic mutants		Some are high producers, e.g. chlortetracycline by *S. viridifaciens*
Resistance to the antibiotic produced by the organism itself		Increased production, e.g. chlortetracyline by *S. aureofaciens*
Recombination		Produces new combinations of existing alleles
Sexual reproduction	Conjugation; fusion of gametes	Some bacteria and Actinomycetes, fungi and yeast
Heterokaryosis	Nuclear fusion followed by mitotic recombination and mitotic reduction	Fungi
Protoplast fusion	Protoplasts produced by lytic enzymes; fusion by PEG, recombinant recovery	Bacteria, Actinomycetes, fungi, quite successful

For example, *phe⁻* mutants require phenylalanine for growth; such mutants of *C. glutamicus* accumulate tyrosine. Similarly, *tyr⁻* mutants accumulate phenylalanine, while phe-+ tyr- mutants accumulate tryptophan.

2. Many analogue-resistant mutants have feedback insensitive enzymes of the biosynthetic pathway the analogue of whose product was used for selection of such cells. In feedback inhibition, activity of an enzyme is inhibited by the end-product of the biosynthetic pathway in which the enzyme participate. For example, when *tyr⁻* mutants of *C. glutamicus* were selected for resistance to 50 mg/l *p*-fluorophenylalanine (analogue of phenylalanine), there was a nearly seven-fold increase in phenylalanine accumulation over that of the *tyr⁻* mutant.

3. Sometimes revertants from non producing mutants of a strain are high producers, e.g. one such reversion mutant of *Streptomyces viridifaciens* showed over 6-fold increase in chlortetracycline production over the original strain from which the nonproducing mutant was obtained. When a mutant mutates back to its original phenotype it is called reversion, and the mutant is known as revertant, e.g. nonproducer mutant mutating to back producer.

4. Reversion mutants of appropriate auxotrophs may often be high producers, e.g. in case of *S. viridifaciens* reversion mutants of an auxotrophic mutant requiring homocysteine showed 28 per cent more chlortetracycline yield than the original strain.

5. In some cases, selection for resistance to the antibiotic produced by the organism itself may lead to increased yields. For example, *Streptomyces aureofaciens* mutants selected for resistance to 200–400 mg/l chlortetracycline showed a four-fold increase in the production of this antibiotic.

6. Sometimes, mutants with altered cell membrane permeability show high production of some metabolites. A mutant *E. coli* strain has defective lysine transport; it actively excretes L-lysine into the medium to 5-times as high concentration as that within its cells.

7. Mutants have been selected to produce altered metabolites, especially in case of aminogycoside antibiotics. For example, *Pseudomonas aureofaciens* produces the antibiotic pyrrolnitrin; a mutant of this fungus yields 4′-fluoropyrrolnitrin.

The above and many other approaches for selection of mutants can be most profitably used when the biosynthetic pathway for the concerned product is known, as are the precursors and the regulatory mechanisms. Mutant selection has been the most successful approach for strain improvement, but major advances are being made in the exploitation of other strategies, i.e. recombination and recombinant DNA technology.

Strain Improvement Methods

Strain improvement can generally be described as the use of any scientific techniques that allow the isolation of cultures exhibiting a desired phenotype. The technology has bean utilised for more than 50 years in conjunction with modern submerged culture fermentations and perhaps in a less systematic way for as long as fermented products have been made by humans. Most commonly, the ability of a strain to exhibit increased product accretion is the desired phenotype. However, the spectrum of improvements can include other traits, such as the elimination of toxic cometabolites or those problematic in downstream processing, the ability to degrade complex waste materials, or greater genetic stability of recombinant hosts.

The utility of strain improvement arises because of the existence of rate limiting steps within all metabolic pathways. Most of these events are not readily measurable owing to the nature of metabolic transients at reaction fluxes during the course of an industrial fermentation. The metabolic flux involved in the biosynthesis of a secondary metabolite generally includes numerous specific reactions from the biosynthetic enzymes, primary metabolism for the supply of precursors to growth and secondary metabolites, and regulatory circuits involved in cell growth and differentiation. Likewise, heterologous protein expression in bacterial or fungal systems offers a significantly complex pathway. Because the rate limiting enzymatic reactions or flux nodes are often unknown, an empirical process such as classical strain improvement is well suited to manipulation of the pathway. A screening programme can be initiated with limited knowledge of the physiology or genetics associated with production of the molecule of interest. Classical mutagenesis and screening, also referred to as nonrecombinant strain improvement, can thus offer a significant advantage over genetic engineering approaches alone by yielding gains with minimal startup time and sustaining such gains over years despite a lack of detailed knowledge concerning the physiology of the producing micro-organism. This empirical approach has a long history of success, as best exemplified by the improvements achieved for penicillin production in which reported penicillin titers are 50 g/l, an improvement of at least 4000 fold over the original parent.

Examples of fungal or actinomycaete cultures capable of overproducing metabolites in quantities as high as 80 g/l can be found in the literature. Thus, application of strain improvement to new fermentation processes continues to be documented in the literature despite the age of the technology. One part of this continued interest in strain improvement is the marriage of classical techniques and molecular genetics to create a synergistic effect for process improvement. Fermentation processes for products as diverse as antibiotics and human proteins have benefited from this combination of approaches. A second area of interest has been the application of new approaches or technology to strain improvement. The greater availability of userfriendly equipment and enhanced detection limits for mass spectroscopy and high pressure liquid chromatography (HPLC) have made their use more common. In addition to using design to improve media during fermentation development, statistical analysis can also lead to enhancements in screening programmes, as will be discussed. The availability of userfriendly software such as JMP (JMP Statistical Discovery Software; SAS Institute Inc., Cary, NC) has allowed the wider use of such analyses by the scientist. Finally, the growing field of metabolic flux analysis or quantitative physiology will likely become a tool in directing screening work or explaining the success of such work.

Regardless of the methods of strategy, strain improvement relies on the iteration of three operations: genetic alteration, fermentation, and assay.

Genetic Alterations

Mutagenesis

The first key step is the generation of mutants. This can be accomplished by using either chemical or physical treatments to modify the genome of the target organism. It should be stressed that safety of the handler should be a consideration before starting such work. All mutagens should be considered potential carcinogens, and care should be taken to avoid exposure. Biosafety cabinets and protective equipment should be used, surfaces should be decontaminated, and used equipment should be decontaminated or disposed of by incineration.

There are many excellent reviews that list protocols for mutagenesis with a variety of different mutagens. Different mutagens are presumed to have different mechanisms of action, such as genetic alteration by base transitions or by frameshifts. During a longterm strain improvement programme, it is advisable to change mutagens periodically to take advantage of these different mechanisms of action. The detailed procedure for the isolation of mutations and the sensitivity of an organism to a particular mutagen will vary considerably from organism to organism. For example, a highly pigmented organism will show increased resistance to the killing effects of UV light exposure. Alkalophilic organisms need to be harvested and resuspended in a neutral pH buffer before treatment with chemical mutagens because of the inactivation of the mutagen at high pH.

The degree of killing versus the frequency of observed mutants will vary with different organisms. This can easily be verified by using antibiotic resistance as an indicator of mutation frequency. It is recommended that the multiplicity of mutation be determined in addition to monitoring survival rates as a more meaningful measure of efficiency.

In addition to vegetative cells and spore preparations, protoplasts can be used as starting material for mutagenesis. This is especially useful for basidiomycetes and other mycelial organisms. The nature of the protoplasting condition itself may prove to be mutagenic, and this can be enhanced by exposure of protoplasts to N-methyl-N'-nitro-N-nitrosoguanidine (MNNG) or UV light.

Protoplast fusion

Protoplast fusion is another tool to be used to achieve genetic alterations in the lineage of industrial micro-organisms. The technique can offer a means of combining favourable traits from two lineages or parental cultures. Fusion unfortunately does not allow the scientist to direct specific genes or DNA segments and thus, like mutagenesis, relies on empirical measurements to determine the success at combining two or more traits. The need for genetic markers thus becomes important in measuring efficiency of the approach. Phenotypic determinants such as auxotrophy, extracellular enzyme production, morphological differences, levels of antibiotic production, or antibiotic resistance can offer selectable traits. However, the use of auxotrophic markers may not be desirable for industrial micro-organisms because of the cost of supplementing medium at the production scale. Protoplast fusion also offers the advantage of not requiring significant knowledge of the genetics of a particular culture. In addition, fusion is considered natural or homologous recombination and thus can avoid the regulatory constraints of fermenting the resulting strains at largescale.

Operational Considerations

The ultimate success of a strain improvement programme charged with developing and improving a fermentation process will be based largely on resource allocation. The key labour intensive steps in strain improvement include the segregation and isolation of individual clones, preparation and dispensing of sterile media, transfer of the isolates in order to initiate the vegetative and fermentation stages, and assay of the fermentation broth from individual flasks or other containers.

In general, the number of isolates screened will determine the success of detecting improved strains. If a manual operation is employed, the number of strains examined will be roughly proportional to the number of workers available. Alternative approaches toward minimising the number of manipulations include bioassays or selective agents (see rational screening below). However, such methods have limited applications in a lineage and may not be effective over the longer lifetimes of some processes. Alternatively, automating the key steps in the process can drive throughput higher without adding labour.

Automated screening

Toward the goal of increasing throughput without adding significant labour, automation of the key steps is a useful approach. It is generally desirable to miniaturise where possible to reduce the cost of equipment required as well as to reduce volumes of solvents and fermentation waste streams in the laboratory, which have come under more stringent control by environmental regulations. Efforts can range from automating single steps in a process to construction of an integrated system. In this industrial system, both agar and liquid media are charged into a vessel or fermenter, where they are sterilised under conditions that can simulate those at pilotscale. Media are robotically dispensed in sterile laminar flow hoods into plates or fermentation bottles.

Manual screening

A manual screen had advantages in that improved yield can be obtained readily with more limited capital investment. If a manual operation is employed, the number of strains examined will roughly be proportional to the number of workers available. Thus, labour tends to be the key driver of cost. The desired target of a programme will dictate approaches. For a single protein product, it is often advisable to employ recombinant techniques as a first step. There may be multiple regulatory steps, precursor availability, export functions, and end product sensitivity that must be addressed, however, for complex antibiotic pathways.

Selection and rational screening

Use of a selection strategy can greatly increase the efficiency of a strain improvement project. In random screening, a high percentage of putative mutants examined will be carried over as survivors from the mutagenesis and will exhibit the same yields or lower yields than the parent strain. However, direct selection on plates, for example, allows for a higher throughput, and only mutants are examined in shake flask experiments. Rational screening requires some knowledge or inferred knowledge of the biosynthetic pathway to a product.

FUTURE POTENTIAL AND NEEDS OF SCREENING PROCESSES

The future of the screening processes for the selection of the best inoculum is promising. This is due to the development of knowledge and expertise in the fields of protein research and genetic engineering which will have a significant impact. The field of Industrial microbiology have been significantly impacted by protein and genetic engineering specifically at the point of maintaining the optimisation of the product. Optimisation of the product is maintained by bringing about significant changes in the protein, i.e. increasing the stability of the protein. Different studies proved the increased stability of the subtilisin protein (alkaline protease) after some alterations were brought about as a result of genetically engineered. In another study, a site directed mutation was made in the protein subtilisin. As a result of this mutation the activity of the subtilisin protein was enhanced. In order to surge the product formation, the target is to enhance the gene expression. In order to achieve this many firms are aiming to develop enhanced genetic promoters which would increase the expression of the gene and thus the product formation. Sayler used colony hybridisation technique to identify the exact DNA sequence which was specific for the catabolism of hydrocarbons. The gene expression of the identified DNA sequence was enhanced to increase the catabolic activity.

Later techniques aiming at the unique DNA sequences of organisms can be employed for the purpose of screening. A common example is dot blot. It involves the hybridisation of the sequence of the microbes with the probes. So, one can look for microbes with unique sequence using this technique involving hybridisation of DNA from environmental samples with the designed probes. The screening potential of the microbes is greatly dependent upon the diversity of the microbes present in the environment. There are a number of microbes yet to be discovered and they are living around us. We just need to develop appropriate techniques to isolate them and then select the best ones for industrial purposes. However, major issue is the deficiency of research programmes and lack of collaboration because this is multidisciplinary task and thus it needs cooperation from different disciplines, i.e. chemists, engineers and microbiologists.

IMPROVEMENT OF YIELDING POTENTIAL OF INDUSTRIALLY IMPORTANT MICROBES

Industrial Products produced by the wild strains isolated from nature. They are selected based on particular product produced by strain, e.g. *Bacillus* is isolated from soil sample for the production of amylase. (Amylase can act on starch to give glucose). These naturally isolated organisms usually produce commercially important products in very low concentrations and therefore it is essential to increase the productivity of the selected organism. Yield of the desired product, can be increased by optimising medium and growth conditions. Medium and growth optimisation is having limited effect on increase in the product due to organism's maximum ability to synthesise the product, which is control by its genome. Thus if one want to increase the product then one should modify the genome. After the creation of desired genome for the increased desired product, then cultural requirements of the modified organisms is tested. Modified genome further modifie d forimprovement. Thus the process of strain improvement require continuous changing of genes of given organism with changes in the medium components. Genetic modification achieved by

1. Mutation
2. Recombination

Only yield increase is not the criteria for the improvement of microbes but their stability, resistance to infection, medium components, non-foaming capabilities, tolerance to low oxygen tension, no undesirable product formation is also taken into consideration.

Mutation

Mutation can be spontaneous or induced. Spontaneous mutation can lead to the change in the genetic make-up of the organisms but chances are there that culture may have problem of yield degeneration because variants are usually inferior producers. Sometime natural variants can also give better yield but it is not reliable one so other techniques are employinglike induced mutation and recombination.

Induced mutation

Variants or mutants obtain giving better yield by inducing mutation to the wild strain. Mutation Induced by various physical and chemical agents, known as mutagens. Mutation can creates so many mutants of which some are superior producers and some are inferior producers of the product. It is not easy to select them because of one or two criteria but superior mutant is selected using more than one criterion for the formation of product.

Selection of mutants giving better primary metabolites

Any metabolite is produce by particular micro-organisms via certain pathway. When concentration of particular metabolite increases then organisms have regulatory mechanism to control over the production of those metabolites. If such control is block by mutation, than improvement in product is possible. Example *Corynebacterium glutamicum* produces glutamic acid; this product increased by mutation. Kinoshita isolated mutant of *C glutamicum*, which is deficient in the production of biotin as well as in the synthesis of enzyme α-ketogluterate dehydrogenase. Biotin deficient strain do not produce proper membrane and thus deficient in selective permeability. While defect in the production of enzyme α-ketogluterate dehydrogenase will not allow the formation of succinic acid from the α-ketoglutaric acid, and α-ketogluteratediverted to glutamic acid synthesis. Organisms used for the commercial production of primary metabolites rarely modified at only one genetic site, sometime it is necessary to alter several control sites to produce desire product in high quantity.

Selection of mutants producing improved levels of secondary metabolites

The design of producers for the isolation of mutants overproducing secondary metabolites is more difficult due to the fact that far less information is available on the control of production and, also, that the end products of secondary metabolism are not required for the growth. Screening achieves considerable success in selecting mutant for the production of secondary metabolites. Several workers have obtained improved secondary metabolite producing strains by isolating auxotrophic mutants. In many cases, there is no correlation between the compound and the secondary metabolites produced. Possible explanation for this may be that they are double mutants and their auxotrophy not directly related to the improved productivity. Organisms exploited by using Nitrosoguanidine (NTG) as a mutagen. NTG causes clusters of mutations around the replication fork of the bacterial chromosome.

Thus, if one of the mutations were selectable it may be possible to isolate a strain containing the selectable mutation which is close by, for this one should require the accurate knowledge about the positions of the genes important in secondary metabolites. The technique of selecting mutants resistant to inhibitory analogues has found some application in the selection of secondary metabolites overproducers. For example, Elander and other isolated tryptophan analogue resistant mutant of *Pseudomonas aureofaciens*, which overproduced antibiotic pyrrolnitrin. Tryptophan is precursor for this pyrrolnitrin and resistant mutant can produce more of this limiting precursor.

Recombination

Recombination is process of creating new combination of genes in given organisms. Recombination process is not that much successful as the use of induced mutant and selection, this is due to success of mutation programme. Now recombination technique are used for the strain improvement after the different techniques available, which help us to use this technique more conveniently. Recombination process carried out naturally or artificially. Recombination that is occurring naturally is applicable to few organisms, due to limitation of genetic exchange between these organisms. With artificial recombination,insertion of any gene is possible in any cell.

Natural recombination by parasexual cycle in some fungi

In Fungi, imperfect nuclear fusion and gene segregation could take place outside the sexual organ. This process is parasexual cycle. For this process genetically unlike nuclei must be present in one of the fungi. After the fusion of genetic material of two different organisms, heterokaryon is produce. These heterokaryons contain genetic information of two different organisms or recombinant gene. However, this technique is not useful in strain improvement is the establishment of heterokaryons.

Natural recombination by conjugation

Conjugation is the process in bacteria whereby genetic information is transfer from one cell to another by cell-to-cell contact. The chromosome of the 'donor' cell mobilises by the integration of a normally extrachromosomal DNA particle into the recipient. This technique used in the preparation of strain producing particular compound in excess, or producing compound, which is not previously present in recipient organism. Conjugation demonstrated in Streptomyces, which have enormous industrial significance. However, the disadvantage of the conjugation is that considerable genetic knowledge of the organism is required to perform the cross effectively.

Natural recombination by protoplast fusion

Protoplasts are the cells devoid of cell wall. It can be prepared by subjecting organisms to wall degrading enzymes. Cell fusion, followed by nuclear fusion will occur between protoplasts that would not otherwise fuse and resulting in fused protoplast may generate cell wall and grow into mutant cell. In this technique whole genome of one fused with genome of other organism thus chances of mutations are more in this technique compared to conjugation. Fusion of two protoplast result in the formation of heterokaryons, where the limitation of one cell with different genome of traditional parasexual cycle is over in this technique.

In vitro recombinant DNA technology

In this technique, Host's chromosomal or extrachromosomal DNA cut and the desired small DNA inserted into it, this lead to the recombination of the Host DNA.

The majority of the recombinant DNA prepared by this technique is use for the improvement of organisms producing primary metabolites. The efficiency of the organisms used in the single cell protein process, *Methylophilusmethylotrophus*, improved by the incorporation of a plasmid containing the glutamate dehydrogenase from *E. coli*.

Techniques of genetic manipulation improve production of commercially important enzyme. Enzyme yield increases by incorporating the chromosomal gene coding for the enzyme into a plasmid, which then introduced into the original strain and maintained at high copy number. This technique is also available for bacteria, yeast, and fungi. However, it is limited to improvement of organism for primary metabolites, due to lack of information regarding basic genetics of secondary metabolites production.

Chapter 11

Industrial Media and the Nutrition of Industrial Organism

INTRODUCTION

The use of a good, adequate, and industrially usable medium is as important as the deployment of a suitable micro-organism in industrial microbiology. Unless the medium is adequate, no matter how innately productive the organism is, it will not be possible to harness the organism's full industrial potentials. Indeed not only may the production of the desired product be reduced but toxic materials may be produced. Liquid media are generally employed in industry because they require less space, are more amenable to engineering processes, and eliminate the cost of providing agar and other solid agents.

BASIC NUTRIENT REQUIREMENTS OF INDUSTRIAL MEDIA

All microbiological media, whether for industrial or for laboratory purposes must satisfy the needs of the organism in terms of carbon, nitrogen, minerals, growth factors, and water. In addition they must not contain materials which are inhibitory to growth. Ideally it would be essential to perform a complete analysis of the organism to be grown in order to decide how much of the various elements should be added to the medium. However, approximate figures for the three major groups of heterotrophic organisms usually grown on an industrial scale are available and may be used in such calculations.

Carbon or energy requirements are usually met from carbohydrates, notably (in laboratory experiments) from glucose. It must be borne in mind that more complex carbohydrates such as starch or cellulose may be utilised by some organisms. Furthermore, energy sources need not be limited to carbohydrates, but may include hydrocarbons, alcohols, or even organic acids.

In composing an industrial medium the carbon content must be adequate for the production of cells. For most organisms the weight of organism produced from a given weight of carbohydrates (known as the yield constant) under aerobic conditions is about 0.5 gm of dry cells per gram of glucose. This means that carbohydrates are at least twice the expected weight of the cells and must be put as glucose or its equivalent compound. Nitrogen is found in proteins including enzymes as well as in nucleic acids hence it is a key element in the cell. Most cells would use ammonia or other nitrogen salts. The quantity of nitrogen to be added in a fermentation can be calculated from the expected cell mass and the average composition of the micro-organisms used. For bacteria the average N content is 12.5%. Therefore to produce 5 gm of bacterial cells per liter would require about 625 mg N (Table 11.1).

Table 11.1: Average composition of micro-organisms (% dry weight).

Component	Bacteria	Yeast	Molds
Carbon	48 (46–52)	48 (46–52)	48 (45–55)
Nitrogen	12.5 (10–14)	7.5 (6–8.5)	6 (4–7)
Protein	55 (50 –60)	40 (35–45)	32 (25–40)
Carbohydrates	9 (6–15)	38 (30–45)	49 (40–55)
Lipids	7 (5–10)	8 (5–10)	8 (5–10)
Nucleic acids	23 (15–25)	8 (5–10)	5 (2–8)
Ash	6 (4–10)	6 (4–10)	4 (4–10)
Minerals (same for all three organisms)			
Phosphorus	1.0–2.5		
Sulphur, magnesium	0.3–1.0		
Potassium, sodium	0.1–0.5		
Iron	0.01–0.1		
Zinc, copper, manganese	0.001–0.01		

Any nitrogen compound which the organism cannot synthesise must be added. *Minerals* form component portions of some enzymes in the cell and must be present in the medium. The major mineral elements needed include P, S, Mg and Fe. Trace elements required include manganese, boron, zinc, copper and molybdenum. Growth factors include vitamins, amino acids and nucleotides and must be added to the medium if the organism cannot manufacture them.

Under laboratory conditions, it is possible to meet the organism's requirement by the use of purified chemicals since microbial growth is generally usually limited to a few liters. However, on an industrial scale, the volume of the fermentation could be in the order of thousands of liters. Therefore, pure chemicals are not usually used because of their high expense, unless the cost of the finished material justifies their use. Pure chemicals are however used when industrial media are being developed at the laboratory level. The results of such studies are used in composing the final industrial medium, which is usually made with unpurified raw materials. The extraneous materials present in these unpurified raw materials are not always a disadvantage and may indeed be responsible for the final and distinctive property of the product. Thus, although alcohol appears to be the desired material for most beer drinkers, the other materials extraneous to the maltose (from which yeasts ferment alcohol) help confer on beer its distinctive flavour.

CRITERIA FOR THE CHOICE OF RAW MATERIALS USED IN INDUSTRIAL MEDIA

In deciding the raw materials to be used in the production of given products using designated micro-organism(s) the following factors should be taken into account.

Cost of the material: The cheaper the raw materials the more competitive the selling price of the final product will be. No matter, therefore, how suitable a nutrient raw materials is, it will not usually be employed in an industrial process if its cost is so high that the selling price of the final product is not economic. Thus, although lactose is more suitable than glucose in some processes (e.g. penicillin production) because of the slow rate of its utilisation, it is usually replaced by the cheaper glucose. When used, glucose is added only in small quantities intermittently in order to decelerate acid production. Due to these economic considerations the raw materials used in many industrial media are usually waste products from other processes. Corn steep liquor and molasses are, for example, waste products from the starch and sugar industries, respectively. They will be discussed more fully below.

Ready availability of the raw material: The raw material must be readily available in order not to halt production. If it is seasonal or imported, then it must be possible to store it for a reasonable period. Many industrial establishments keep large stocks of their raw materials for this purpose. Large stocks help beat the ever rising cost of raw materials; nevertheless large stocks mean that money which could have found use elsewhere is spent in constructing large warehouses or storage depots and in ensuring that the raw materials are not attacked during storage by micro-organisms, rodents, insects, etc. There is also the important implication, which is not always easy to realise, that the material being used must be capable of long-term storage without concomitant deterioration in quality.

Transportation costs: Proximity of the user-industry to the site of production of the raw materials is a factor of great importance, because the cost of the raw materials and of the finished material and hence its competitiveness on the market can all be affected by the transportation costs. The closer the source of the raw material to the point of use the more suitable it is for use, if all other conditions are satisfactory.

Ease of disposal of wastes resulting from the raw materials: The disposal of industrial waste is rigidly controlled in many countries. Waste materials often find use as raw materials for other industries. Thus, spent grains from breweries can be used as animal feed. But in some cases no further use may be found for the waste from an industry. Its disposal especially where government regulatory intervention is rigid could be expensive. When choosing a raw material therefore the cost, if any, of treating its waste must be considered.

Uniformity in the quality of the raw material and ease of standardisation: The quality of the raw material in terms of its composition must be reasonably constant in order to ensure uniformity of quality in the final product and the satisfaction of the customer and his/her expectations. In cases where producers are plentiful, they usually compete to ensure the maintenance of the constant quality requirement demanded by the user. Thus, in the beer industry information is available on the quality of the barley malt before it is purchased. This is because a large number of barley malt producers exist, and the producers attempt to meet the special needs of the brewery industry, their main customer. On the other hand molasses, which is a major source of nutrient for industrial micro-organisms, is a by-product of the sugar industry, where it is regarded as a waste product. The sugar industry is not as concerned with the constancy of the quality of molasses, as it is with that of sugar. Each batch of molasses must therefore be chemically analysed before being used in a fermentation industry in order to ascertain how much of the various nutrients must be added. A raw material with extremes of variability in quality is clearly undesirable as extra costs are needed, not only for the analysis of the raw material, but for the nutrients which may need to be added to attain the usual and expected quality in the medium.

Adequate chemical composition of medium: As has been discussed already, the medium must have adequate amounts of carbon, nitrogen, minerals and vitamins in the appropriate quantities and proportions necessary for the optimum production of the commodity in question. The demands of the micro-organisms must also be met in terms of the compounds they can utilise. Thus most yeasts utilise hexose sugars, whereas only a few will utilise lactose; cellulose is not easily attacked and is utilised only by a limited number of organisms. Some organisms grow better in one or the other substrate. Fungi will for instance readily grow in corn steep liquor while actinomycetes will grow more readily on soya bean cake.

Presence of relevant precursors: The raw material must contain the precursors necessary for the synthesis of the finished product. Precursors often stimulate production of secondary metabolites either by increasing the amount of a limiting metabolite, by inducing a biosynthetic enzyme or both. These are usually amino acids but other small molecules also function as inducers. The nature of the finished product in many cases depends to some extent on the components of the medium. Thus dark beers such as stout are produced by caramelised (or over-roasted) barley malt which introduce the dark colour into these beers. Similarly for penicillin G to be produced the medium must contain a phenyl compound. Corn steep liquor which is the standard component of the penicillin medium contains phenyl precursors needed for penicillin G. Other precursors are cobalt in media for Vitamin B_{12} production and chlorine for the chlorine containing antibiotics, chlortetracycline, and griseofulvin.

Satisfaction of growth and production requirements of the micro-organisms: Many industrial organisms have two phases of growth in batch cultivation: the phase of growth, or the trophophase, and the phase of production, or the idiophase. In the first phase cell multiplication takes place rapidly, with little or no production of the desired material. It is in the second phase that production of the material takes place, usually with no cell multiplication and following the elaboration of new enzymes. Often these two phases require different nutrients or different proportions of the same nutrients. The medium must be complete and be able to cater for these requirements. For example high levels of glucose and phosphate inhibit the onset of the idiophase in the production of a number of secondary metabolites of industrial importance. The levels of the components added must be such that they do not adversely affect production.

SOME RAW MATERIALS USED IN COMPOUNDING INDUSTRIAL MEDIA

The raw materials to be discussed are used because of the properties mentioned above: Cheapness, ready availability, constancy of chemical quality, etc. A raw material which is cheap in one country or even in a different part of the same country may however not be cheap in another, especially if it has already found use in some other production process. In such cases suitable substitutes must be found if the goods must be produced in the new location. The use of local substitutes where possible is advantageous in reducing the transportation costs and even creating some employment in the local population. Prior experimentation may however be necessary if such new local materials differ substantially in composition from those already being used. Some well-known raw materials will now be discussed. In addition, some of potential useability will also be examined.

Corn steep liquor: This is a by-product of starch manufacture from maize. Sulphur dioxide is added to the water in which maize is steeped. The lowered pH inhibits most other organisms, but encourages the development of naturally occurring lactic acid bacteria especially homofermentative thermophilic *Lactobacillus* spp. which raise the temperature to 38–55°C. Under these conditions, much of the protein present in maize is converted to peptides which along with sugars leach out of the maize and provide nourishment for the lactic acid bacteria. Lactic fermentation stops when the SO_2 concentration reaches about 0.04% and the concentration of lactic acid between 1.0 and 1.5%. At this time the pH is about 4. Acid conditions soften the kernels and the resulting maize grains mill better while the gel-forming property of the starch is not hindered. The supernatant drained from the maize steep is corn steep liquor.

Before use, the liquor is usually filtered and concentrated by heat to about 50% solid concentration. The heating process kills the bacteria. As a nutrient for most industrial organisms corn steep liquor is considered adequate, being rich in carbohydrates, nitrogen, vitamins, and minerals. Its composition is highly variable and would depend on the maize variety, conditions of steeping, extent of boiling, etc. The composition of a typical sample of corn steep liquor is given in Table 11.2. As corn steep liquor is highly acidic, it must be neutralised (usually with $CaCO_3$) before use.

Table 11.2: Approximate composition of corn steep liquor (%).

Lactose	3.0–4.0
Glucose	0–0.5
Non-reducing carbohydrates (mainly starch)	1.5
Acetic acid	0.05
Glucose lactic acid	0.5
Phenylethylamine	0.05
Amino aids (peptides, mines)	0.5
Total solids	80–90
Total nitrogen	0.15–0.2%

Pharmamedia: Also known as proflo, this is a yellow fine powder made from cotton-seed embryo. It is used in the manufacture of tetracycline and some semi-synthetic penicillins. It is rich in protein, (56% w/v) and contains 24% carbohydrate, 5% oil, and 4% ash, the last of which is rich in calcium, iron, chloride, phosphorous, and sulphate.

Distillers solubles: This is a by-product of the distillation of alcohol from fermented grain. It is prepared by filtering away the solids from the material left after distilling fermented cereals (maize or barley) for whiskey or grain alcohol. The filtrate is then concentrated to about one-third solid content to give a syrup which is then drum-dried to give distillers soluble. It is rich in nitrogen, minerals, and growth factors (Table 11.3).

Table 11.3: Composition of maize distillers soluble.

	%
Moisture	5
Protein	27
Lipid	9
Fibre	5
Carbohydrate	43
Ash (mainly K, Na, Mg, CO_3, and P)	11

Soyabean meal: Soyabeans (soja) (*Glycine max*), is an annual legume which is widely cultivated throughout the world in tropical, sub-tropical and temperate regions between 50°N and 40°S. The seeds are heated before being extracted for oil that is used for food, as an antifoam in industrial fermentations, or used for the manufacture of margarine. The resulting dried material, soyabean meal, has about 11% nitrogen, and 30% carbohydrate and may be used as animal feed. Its nitrogen is more complex than that found in corn steep liquor and is not readily available to most micro-organisms, except actinomycetes. It is used particularly in tetracycline and streptomycin fermentations.

Molasses: Molasses is a source of sugar, and is used in many fermentation industries including the production of potable and industrial alcohol, acetone, citric acid, glycerol, and yeasts. It is a by-product of the sugar industry. There are two types of molasses depending on whether the sugar is produced from the tropical crop, sugar cane (*Saccharum officinarum*) or the temperate crop, beet, (*Beta alba*).

Four stages are involved in the manufacture of cane sugar. After *crushing*, a clear greenish dilute sugar solution known as 'mixed juice' is expressed from the canes. During the second stage known as *clarification* the mixed juice is heated with lime. Addition of lime changes the pH of the juice to alkaline and thus stops further hydrolysis (or inversion) or the cane sugar (sucrose), while heating coagulates proteins and other undesirable soluble portions of the mixed juice to form 'mud'. The supernatant juice is then *concentrated* (in the third stage) by heating under high vacuum and increasing low pressures in a series of evaporators. In the fourth and final stage of *crystallisation,* sugar crystals begin to form with increasing heat and under vacuum, yielding a thick brown syrup which contains the crystals, and which is known as 'massecuite'. (In the beet industry it is known as 'fillmass'.) The massecuite is centrifuged to remove the sugar crystals and the remaining liquid is known as molasses. The first sugar so collected is 'A' and the liquid is 'A' molasses. 'A' molasses is further boiled to extract sugar crystals to yield 'B' sugar and 'B' molasses. Two or more boilings may be required before it is no longer profitable to attempt further extractions. This final molasses is known as 'blackstrap molasses'. The sugar yielded with the production of black strap molasses is low-grade and brown in colour, and known as raw sugar, cargo sugar, or refining sugar. This raw sugar is further refined, in a separate factory, to remove miscellaneous impurities including the brown colour (due to caramel) to yield the white sugar used at the table. The heavy liquid discarded from the refining of sugar is known in the sugar refining industry as 'syrup' and corresponds to molasses in the raw sugar industry. The above description has been of cane sugar molasses. In the beet sugar industry the processes used in raw refined sugar manufacture are similar, but the names of the different fractions recovered during purification differ. Cane and beet molasses differ slightly in composition (Table 11.4). Beet molasses is alkaline while cane molasses is acid.

Table 11.4: Average composition of beet and cane molasses.

	Beet molasses % (W/W)	Cane molasses % (W/W)
Water	16.5	20.0
Sugars	53.0	64.0
Sucrose	51.0	32.0
Fructose	1.0	15.0
Glucose	–	14.0
Raffinose	1.0	–
Non-sugar (nitrogeneous materials, acids, gums, etc.)	19.0	10.0
Ash	11.5	8.0

Even within same type of molasses – beet or cane – composition varies from year to year and from one locality to another. The user industry selects the batch with a suitable composition and usually buys up a year's supply. For the production of cells the variability in molasses quality is not critical, but for metabolites such as citric acid, it is very important as minor components of the molasses may affect the production of these metabolites. 'High test' molasses (also known as inverted molasses) is a brown thick syrup liquid used in the distilling industry and containing about 75% total sugars (sucrose and reducing sugars) and about 18% moisture. Strictly speaking, it is not molasses at all but invert sugar, (i.e. reducing sugars resulting from sucrose hydrolysis). It is produced by the hydrolysis of the concentrated juice with acid. In the so called Cuban method, invertase is used for the hydrolysis. Sometimes 'A' sugar may be inverted and mixed with 'A' molasses.

Sulphite liquor: Sulphite liquor (also called waste sulphite liquor, sulphite waste liquor or spent sulphite liquor) is the aqueous effluent resulting from the sulphite process for manufacturing cellulose or pulp from wood. Depending on the type, most woods contain about 50% cellulose, about 25% lignins and about 25% of hemicelluloses. During the sulphite process, hemicelluloses hydrolyse and dissolve to yield the hexose sugars, glucose, mannose, galactose, fructose and the pentose sugars, xylose, and arabinsoe. The acid reagent breaks the chemical bonds between lignin and cellulose; subsequently they dissolve the lignin. Depending on the severity of the treatment some of the cellulose will continue to exist as fibres and can be recovered as pulp. The presence of calcium ions provides a buffer and helps neutralise the strong lignin sulphonic acid. The degradation of cellulose yields glucose. Portions of the various sugars are converted to sugar sulphonic acids, which are not fermentable. Variable but sometimes large amounts of acetic, formic and glactronic acids are also produced.

Sulphite liquor of various compositions are produced, depending on the severity of the treatment and the type of wood. The more intense the treatment the more likely it is that the sugars produced by the more easily hydrolysed hemicellulose will be converted to sulphonic acids; at the same time the more intense the treatment the more will glucose be released from the more stable cellulose. Hardwoods not only yield a higher amount of sugar (up to 3% dry weight of liquor) but the sugars are largely pentose, in the form of xylose. Hardwood hydrolysates also contains a higher amount of acetic acid. Soft woods yield a product with about 75% hexose, mainly mannose. Sulphite liquor is used as a medium for the growth of micro-organisms after being suitably neutralised with $CaCO_3$ and enriched with ammonium salts or urea, and other nutrients. It has been used for the manufacture of yeasts and alcohol. Some samples do not contain enough assaimilable carbonaceous materials for some modern fermentations. They are therefore often enriched with malt extract, yeast autolysate, etc.

Other substrates: Other substrates used as raw materials in fermentations are alcohol, acetic acid, methanol, methane, and fractions of crude petroleum.

GROWTH FACTORS

Growth factors are materials which are not synthesised by the organism and therefore must be added to the medium. They usually function as cofactors of enzymes and may be vitamins, nucleotides, etc. The pure forms are usually too expensive for use in industrial media and materials containing the required growth factors are used to compound the medium. Growth factors are required only in small amounts. Table 11.5 gives some sources of growth factors.

Table 11.5: Some sources of growth factors.

Growth factor	Source
Vitamin B	Rice polishing, wheat germ, yeasts
Vitamin B_2	Cereals, corn steep liquor
Vitamin B_6	Corn steep liquor, yeasts
Nicotinamide	Liver, penicillin spent liquor
Panthothenic	Acid corn steep liquor
VitaminB_{12}	Liver, silage, meat

WATER

Water is a raw material of vital importance in industrial microbiology, though this importance is often overlooked. It is required as a major component of the fermentation medium, as well as for cooling, and for washing and cleaning. It is therefore used in rather large quantities, and measured in thousands of liters a day depending on the industry. In some industries such as the beer industry the quality of the product depends to some extent on the water. In order to ensure constancy of product quality the water must be regularly analysed for minerals, colour, pH, etc. and adjusted as may be necessary. Due to the importance of water, in situations where municipal water supplies are likely to be unreliable, industries set up their own supplies.

SOME POTENTIAL SOURCES OF COMPONENTS OF INDUSTRIAL MEDIA

The materials to be discussed are mostly found in the tropical countries, including those in Africa, the Caribbean, and elsewhere in the world. Any microbiological industries to be sited in these countries must, if they are not to run into difficulties discussed above, use the locally available substrates. It is in this context that the following are discussed.

Carbohydrate Sources

These are all polysaccharides and have to be hydrolysed to sugar before being used.

Cassava (manioc)

The roots of the cassava-plant *Manihot esculenta* Crantz serve mainly as a source of carbohydrate for human (and sometimes animal) food in many parts of the tropical world. Its great advantage is that it is high yielding, requires little attention when cultivated, and the roots can keep in the ground for many months without deterioration before harvest. The inner fleshy portion is a rich source of starch and has served, after hydrolysis, as a carbon source for single cell protein, ethanol, and even beer. In Brazil it is one of the sources of fermentation alcohol which is blended with petrol to form gasohol for driving motor vehicles.

Sweet potato

Sweet potatoes *Ipomca batatas* is a warm-climate crop although it can be grown also in sub tropical regions. There are a large number of cultivars, which vary in the colours of the tuber flesh and of the skin; they also differ in the tuber size, time of maturity, yield, and sweetness. They are widely grown in the world and are found in South America, the USA, Africa and Asia. They are regarded as minor sources of carbohydrates in comparison with maize, wheat, or cassava, but they have the advantage that they do not require much agronomic attention. They have been used as sources of sugar on a semi-commercial basis because the fleshy roots contain saccharolytic enzymes. The syrup made from boiling the tubers has been used as a carbohydrate (sugar) source in compounding industrial media. Butyl alcohol, acetone and ethanol have been produced from such a syrup, and in quantities higher than the amounts produced from maize syrup of the same concentration.

Since sweet potatoes are not widely consumed as food, it is possible that it may be profitable to grow them for use, after hydrolysis, in industrial microbiology media as well as for the starch industry. It is reported that a variety has been developed which yields up to 40 T/hectare, a much higher yield than cassava or maize.

Yams

Yams (*Dioscorea spp*) are widely consumed in the tropics. Compared to other tropical roots however, their cultivation is tedious; in any case enough of this tuber is not produced even for human food. It is therefore almost inconceivable to suggest that the crop should be grown solely for use in compounding industrial media. Nevertheless yams have been employed in producing various products such as yam flour and yam flakes. If the production of these materials is carried out on a sufficiently large scale it is to be expected that the waste materials resulting from peeling the yams could yield substantial amounts of materials which on hydrolysis will be available as components of industrial microbiological media.

Cocoyam

Cocoyam is a blanket name for several edible members of the monocotyledonous (single seed-leaf) plant of the family *Araceae* (the aroids), the best known two genera of which are *Colocasia* (tano) and *Xanthosoma* (tannia). They are grown and eaten all over the tropical world. As they are laborious to cultivate, require large quantities of moisture and do not store well they are not the main source of carbohydrates in regions where they are grown.

However, this relative unimportance may well be of significance in regions where for reasons of climate they can be suitably cultivated. Cocoyam starch has been found to be of acceptable quality for pharmaceutical purposes. Should it find use in that area, starchy by-products could be hydrolysed to provide components of industrial microbiological media.

Millets

This is a collective name for several cereals whose seeds are small in comparison with those of maize, sorghum, rice, etc. The plants are also generally smaller. They are classified as the minor cereals not because of their smaller sizes but because they generally do not form major components of human food. They are however hardy and will tolerate great drought and heat, grow on poor soil and mature quickly. Attention is being turned to them for this reason in some parts of the world. It is for this reason also that millets could become potential sources of cereal for use in industrial microbiology media. Millets are grown all over the world in the tropical and sub-tropical regions and belong to various genera: *Pennisetum americanum* (pearl or bulrush millet), *Setaria italica* (foxtail millet), *Panicum miliaceum* (yard millet), *Echinochloa frumentacea* (Japanese yard millet) and *Eleusine corcana* (finger millet). Millet starch has been hydrolysed by malting for alcohol production on an experimental basis as far back as 50 years ago and the available information should be helpful in exploiting these grains for use as industrial media components.

Rice

Rice, *Oryza sativa* is one of the leading food corps of the world being produced in all five continents, but especially in the tropical areas. Although it is high-cost commodity, it has the advantage of ease of mechanisation, storability, and the availability of improved seeds through the efforts of the International Rice Research Institute, Philippines and other such bodies. The result is that this food crop is likely in the near future to displace, as a carbohydrate source, such other starch sources as yams, and to a lesser extent cassava in tropical countries. The increase in rice production is expected to become so efficient in many countries that the crop would yield substrates cheap enough for industrial microbiological use. Rice is used as brewing adjuncts and has been malted experimentally for beer brewing.

Sorghum

Sorghum, *Sorghum bicolor,* is the fourth in term of quantity of production of the world's cereals, after wheat, rice, and corn. It is used for the production of special beers in various parts of the world. It has been mechanised and has one of the greatest potential among cereals for use as a source of carbohydrate in industrial media in regions of the world where it thrives. It has been successfully malted and used in an all-sorghum lager beer which compared favourably with barley lager beer.

Jerusalem artichoke

Jerusalem artichoke, *Helianthus tuberosus,* is a member of the plant family compositae, where the storage carbohydrate is not starch, but inulin a polymer of fructose into which it can be hydrolysed. It is a root-crop and grows in temperate, semi-tropical and tropical regions.

Protein Sources

Peanut (groundnut) meal

Various leguminous seeds may be used as a source for the supply of nitrogen in industrial media. Only peanuts (groundnuts) *Arachis hypogea* will be discussed. The nuts are rich in liquids and proteins. The groundnut cake left after the nuts have been freed of oil is often used as animal feed. But just as is the case with soya bean, oil from peanuts may be used as anti-foam while the press-cake could be used for a source of protein. The nuts and the cake are rich in protein.

Blood meal

Blood consists of about 82% water, 0.1% carbohydrate, 0.6% fat, 16.4% nitrogen, and 0.7% ash. It is a waste product in abattoirs although it is sometimes used as animal feed. Drying is achieved by passing live steam through the blood until the temperature reaches about 100°C. This treatment sterilises it and also causes it to clot. It is then drained, pressed to remove serum, further dried and ground. The resulting blood-meal is chocolate-coloured and contains about 80% protein and small amounts of ash and lipids. Where sufficient blood is available blood meal could form an important source of proteins for industrial media.

Fish meal

Fish meal is used for feeding farm animals. It is rich in protein (about 65%) and, minerals (about 21% calcium 8%, and phosphorous 3.5%) and may therefore be used for industrial microbiological media production. Fish meal is made by drying fish with steam either aided by vacuum or by simple drying. Alternatively hot air may be passed over the fish placed in revolving drums. It is then ground into a fine powder.

USES OF PLANT WASTE MATERIALS IN INDUSTRIAL MICROBIOLOGY MEDIA: SACCHARIFICATION OF POLYSACCHARIDES

The great recommendation of plant agricultural wastes as sources of industrial microbiological media is that they are not only plentiful but that in contrast with petroleum, a major source of chemicals, they are also renewable. Serious consideration has therefore been given, in some studies, to the possibility of deriving industrial microbiological raw materials not just from wastes, but from crops grown deliberately for the purpose. However, plant materials in general contain large amounts of polysaccharides which are not immediately utilizable by industrial micro-organisms and which will therefore need to be hydrolysed or saccharified to provide the more available sugars. Thereafter the sugars may be fermented to ethyl alcohol for use as a chemical feed stock. The plant polysaccharides whose hydrolysis will be discussed in this section are starch, cellulose and hemicelluloses.

Starch

Starch is a mixture of two polymers of glucose: amylose and amylopectin. Amylose is a linear $(1\rightarrow 4)$ $\propto$– D glucan usually having a degree of polymerisation (DP, i.e. number of glucose molecules) of about 400 and having a few branched residues linked with $(1\rightarrow 6)$ bondings. Amylopectin is a branched D glucan with predominantly $\propto$–D $(1\rightarrow 4)$ linkages and with about 4% of the $\propto$–D $(1\rightarrow 6)$ type.

Starches from various sources differ in their proportion of amylopectin and amylose. The more commonly grown type of maize, for example, has about 26% of amylose and 74% of amylopectin. Others may have 100% amylopectin and still others may have 80–85% of amylose.

Saccharification of Starch

Starch occurs in discrete crystalline granules in plants, and in this form is highly resistant to enzyme action. However when heated to about 55–82°C depending on the type, starch gelatinises and dissolves in water and becomes subject to attack by various enzymes.

Before saccharification, the starch or ground cereal is mixed with water and heated to gelatinise the starch and expose it to attack by the saccharifying agents. The saccharifying agents used are dilute acids and enzymes from malt or micro-organisms.

Saccharification of starch with acid

The starch-containing material to be hydrolysed is ground and mixed with dilute hydrochloric acid, sulphuric acid or even sulphurous acid. When sulphurous acid is used it can be introduced merely by pumping sulphur dioxide into the mash. The concentrations of the mash and the acid, length of time and temperature of the heating have to be worked out for each starch source. During the hydrolysis the starch is broken down from starch (about 2,000 glucose molecules) through compounds of decreasing numbers of glucose moieties to glucose. The actual composition of the hydrolysate will depend on the factors mentioned above. Starch concentration is particularly important: if it is too high, side reactions may occur leading to a reduction in the yield of sugar.

At the end of the reaction the acid is neutralised. If it is desired to ferment the hydrolysate for ethanol, yeast or single cell production, ammonium salts may be used as they can be used by many micro-organisms.

Uses of enzymes

Enzymes hydrolysing starch used to be called collectively diastase. With increased knowledge about them, they are now called amylases. Enzymatic hydrolysis has several advantages over the use of acid: (i) since the pH for enzyme hydrolysis is about neutral, there is no need for special vessels which must stand the high temperature, pressure, and corrosion of acid hydrolysis, (ii) enzymes are more specific and hence there are fewer side reactions leading therefore to higher yields, (iii) acid hydrolysis often yields salts which may have to be removed constantly or periodically thereby increasing cost, (iv) it is possible to use higher concentrations of the substrates with enzymes than with acids because of enzyme specificity, and reduced possibility of side reactions.

Enzymes involved in the hydrolysis of starch

Several enzymes are important in the hydrolysis of starch. They are divisible into six groups.

1. Enzymes that hydrolyse α-1,4 bonds and by-pass α-1,6 bonding: The typical example is α-amylase. This enzyme hydrolyses randomly the inner (1→4)-α-D-glucosidic bonds of amylose and amylopectin. The cleavage can occur anywhere as long as there are at least six glucose residues on one side and at least three on the other side of the bond to be broken. The result is a mixture of branched α-limit dextrins (i.e. fragments resistant to hydrolysis and contain the α-D (1,6) linkage derived from amylopectin) and linear glucose residues especially maltohexoses, maltoheptoses and maltotrioses. α- amylases are found in virtually every living cell and the property and substrate pattern of α-amylases vary according to their source. Thus, animal α-amylases in saliva and pancreatic juice completely hydrolyse starch to maltose and D-glucose. Among microbial α-amylases some can withstand temperatures near 100°C.

2. Enzymes that hydrolyse the α-1,4 bonding, but cannot by-pass the α-1,6 bonds: Beta *amylase:* This was originally found only in plants but has now been isolated from micro-organisms. Beta amylase hydrolyses alternate α-1,4 bonds sequentially from the non-reducing end (i.e. the end without a hydroxyl group at the C–1 position) to yield maltose. Beta amylase has different actions on amylose and amylopectin, because it cannot by-pass the α-1:6 – branch points in amylopectin. Therefore, while amylose is completely hydrolysed to maltose, amylopectin is only hydrolysed to within two or three glucose units of the α-1.6 – branch point to yield maltose and a 'beta-limit' dextrin which is the parent amylopectin with the ends trimmed off. Debranching enzymes (see below) are able to open up the α–1:6 bonds and thus convert beta-limit dextrins to yield a mixture of linear chains of varying lengths; beta amylase then hydrolyses these linear chains. Those chains with an odd number of glucose molecules are hydrolysed to maltose, and one glucose unit per chain. The even numbered residues are completely hydrolysed to maltose. In practice there is a very large population of chains and hence one glucose residue is produced for every two chains present in the original starch.

3. Enzymes that hydrolyse (α-1,4 and α-1:6 bonds: The typical example of these enzymes is amyloglucosidase or glucoamylase. This enzyme hydrolyses α-D-(1→ 4)-D–glucosidic bonds from the non-reducing ends to yield D–glucose molecules. When the sequential removal of glucose reaches the point of branching in amylopectin, the hydrolysis continues on the (1→6) bonding but more slowly than on the (1→4) bonding. Maltose is attacked only very slowly. The end product is glucose.

4. De-branching enzymes: At least two de-branching enzymes are known: pullulanase and iso-amylase. Pullulanase: This is a de-branching enzyme which causes the hydrolysis of α-D–(1→ 6) linkages in amylopectin or in amylopectin previsouly attacked by alphaamylase. It does not attack α-D (1→4) bonds. However, there must be at least two glucose units in the group attached to the rest of the molecules through an α-D-(1→ 6) bonding. Iso-amylase: This is also a de-branching enzyme but differs from pullulanase in that three glucose units in the group must be attached to the rest of the molecules through an α-D–(1→6) bonding for it to function.

5. Enzymes that preferentially attack α-1,4 linkages: Examples of this group are glucosidases. The maltodextrins and maltose produced by other enzymes are cleaved to glucose by α-glucosidases. They may however sometime attack unaltered polysaccharides but only very slowly.

6. Enzymes which hydrolyse starch to non-reducing cyclic D-glucose polymers known as cyclodextrins or Schardinger dextrins: Cyclic sugar residues are produced by Bacillus macerans. They are not acted upon by most amylases although enzymes in Takadiastase produced by Aspergillus oryzae can degrade the residues.

Industrial saccharification of starch by enzymes

In industry the extent of the conversion of starch to sugar is measured in terms of dextrose equivalent (D.E.). This is a measure of the reducing sugar content, expressed in terms of dextrose, determined under defined conditions involving Fehling's solution. The D.E is calculated as percentage of the total solids. For the saccharification of starch in industry acid is being replaced more and more by enzymes. Sometimes acid is used only initially and enzymes employed at a later stage. Acid saccharification has a practical upper limit of 55 D.E. Beyond this, breakdown products begin to accumulate. Furthermore, with acid hydrolysis reversion reactions occur among the sugar produced. These two deficiencies are avoided when enzymes are utilised. Besides, by selecting enzymes specific sugars can be produced. Starch-splitting enzymes used in industry are produced in germinated seeds and by micro-organisms. Barley malt is widely used for the saccharification of starch. It contains large amounts of various enzymes notably β-amylase and α-glucosidase which further split saccharides to glucose. All the enzymes discussed above are produced by different micro-organisms and many of these enzymes are available commercially. The most commonly encountered organisms producing these enzymes are *Bacillus* spp, *Streptomyeces* spp, *Aspergillus* spp, *Penicillium* spp, *Mucor* spp and *Rhizopus* spp.

Cellulose, Hemi-celluloses and Lignin in Plant Materials

Cellulose

Cellulose is the most abundant organic matter on earth. Unfortunately it does not exist pure in nature and even the purest natural form (that found in cotton fibres) contains about 6% of other materials. Three major components, cellulose, hemi-cellulose and lignin occur roughly in the ratio of 4:3:3 in wood. Before looking more closely at cellulose, the other two major components of plant materials will be briefly discussed.

Hemicelluloses

These are an ill-defined group of carbohydrates whose main and common characteristic is that they are soluble in, and hence can be extracted with, dilute alkali. They can then be precipitated with acid and ethanol. They are very easily hydrolysed by chemical or biological means. The nature of the hemicellulose varies from one plant to another. In cotton the hemicelluloses are pectic substances, which are polymers of galactose. In wood, they consist of short (DP less than 200) branched heteropolymers of glucose, xylose, galactose, mannose and arabinose as well as uronic acids of glucose and galactose linked by 1-3, 1-6 and 1-4 glycosidic bonding.

Lignin

Lignin is a complex three-dimensional polymer formed from cyclic alcohols. It is important because it protects cellulose from hydrolysis. Cellulose is found in plant cell-walls which are held together by a porous material known as middle lamella. In wood the middle lamella is heavily impregnated with lignin which is highly resistant and thus protects the cell from attack by enzymes or acid.

Pretreatment of cellulose-containing materials before saccharification

In order to expose lignocellulosics to attack, a number of physical and chemical methods are in use, or are being studied, for altering the fine structure of cellulose and/or breaking the lignin-carbohydrate complex. Chemical methods include the use of swelling agents such a NaOH, some amines, concentrated H_2SO_4 or HCI or proprietary cellulose solvents such as 'cadoxen' (tris thylene-diamine cadmium hydroxide). These agents introduce water between or within the cellulose crystals making subsequent hydrolysis, easier. Steam has also been used as a swelling agent. The lignin may be removed by treatment with dilute H_2SO_4 at high temperature. Physical methods of pretreatment include grinding, irradiation and simply heating the wood.

Hydrolysis of cellulose

Following pretreatment, wood may be hydrolysed with dilute HCI, H_2SO_4 or sulphites of calcium, magnesium or sodium under high temperature and pressure as described for sulphite liquor production in paper manufacture. When, however, the aim is to hydrolyse wood to sugars, the treatment is continued for longer than is done for paper manufacture.

A lot of experimental work has been done recently on the possible use of cellulolytic enzymes for digesting cellulose. The advantage of the use of enzymes rather than harsh chemicals methods have been discussed already. Fungi have been the main source of cellulolytic enzymes. *Trichoderma viride* and *T. koningii* have been the most efficient cellulase producers. *Penicillicum funiculosum* and *Fusarium solani* have also been shown to possess equally potent cellulases. Cellulase has been resolved into at least three components: C_1, C_x, and β-glucosidases. The C_1 component attacks crystalline cellulose and loosens the cellulose chain, after which the other enzymes can attack cellulose. The C_x enzymes are β-(1→4) glucanases and hydrolyse soluble derivatives of cellulose or swoollen or partially degraded cellulose. Their attack on the cellulose molecule is random and cellobiose (2-sugar) and cellotroise (3-sugar) are the major products of their actions. There is evidence that the enzymes may also act by removing successive glucose units from the end of a cellulose molecule. β-glucosidases hydrolyse cellobiose and shortchain oligo-saccharides derived from cellulose to glucose, but do not attack cellulose.

They are able to attack cellobiose and cellotriose rapidly. Many organisms described in the literature as 'cellulolytic' produce only C_x and β-glucosidases because they were isolated initially using partially degraded cellulose. The four organisms mentioned above produce all three members of the complex.

Molecular structure of cellulose

Cellulose is a linear polymer of D-glucose linked in the Beta-1,4 glucosidic bondage. The bonding is theoretically as vulnerable to hydrolysis as the one in starch. However, cellulose – containing materials such as wood are difficult to hydrolyse because of: (i) the secondary and tertiary arrangement of cellulose molecules which confers a high crystallinity on them and (ii) the presence of lignin.

The degree of polymerisation (DP) of cellulose molecule is variable, but ranges from about 500 in wood pulp to about 10,000 in native cellulose. When cellulose is hydrolysed with acid, a portion known as the amorphous portion which makes up 15% is easily and quickly hydrolysed leaving a highly crystalline residue (85%) whose DP is constant at 100–200. The crystalline portion occurs as small rod-like particles which can be hydrolysed only with strong acid.

Chapter 12

Sterilisation Techniques in Fermentation Processes

INTRODUCTION

Sterilisation is technique to make anything free from organisms either by removing them or killing them. The removal or killing all organisms from fermentation medium is the main aimof the sterilisation process or else the contaminant will deteriorate the process.

Contaminant can enter into the fermenter from various points or sight like:

- Improper sterilisation of media.
- Partial sterilisation of air.
- Cooling water system.
- Through out-lets, in-lets or other openings.
- Faulty inoculum procedure.
- Due to faulty process used for pretreatment of crude ingredients.
- Due to leakage in design of fermenter.

NEED FOR STERILISATION

If any foreign micro-organism invades, the fermentation process then product will not be produce. Thus, it is very necessary to sterilise the medium and other materials so that only required organisms, which we inoculate, will grow and give high yield of fermentation product. Thus, process of sterilisation plays vital role in the fermentation processes.

Following are the consequences of contamination of fermenter.

1. The medium would have to support the growth of both the production organism and the contaminant resulting in a loss of productivity.

2. If the fermentation is a continuous one then the contaminant may outgrow the production organism and displace it from the fermentation.

3. Contamination of a bacterial fermentation with phage could result in the lysis of the culture.

4. The foreign organism may contaminate the final product for example single-cell protein where the cells separated from the broth constitute the product.

5. The contaminant may produce compounds, which make subsequent extraction of the final product difficult.

6. The contaminant may degrade the desired product this is common in bacterial contamination of antibiotic fermentations where the contaminant would have to be resistant to the normal inhibitory effects of the antibiotic and degradation of the antibiotic is a common resistance mechanism for example, the degradation of beta lactam antibiotics by beta lactamase producing bacteria.

Contamination may be avoid in Fermenter by using following Precautions

- Using a pure inoculum to start the fermentation.
- Sterilising the medium to be employed.
- Sterilising the fermenter vessel.
- Sterilising everything that is used during the process.
- Maintaining aseptic conditions during the fermentation.

Principles of Sterilisation

Micro-organisms will be removed by sterilisation process which otherwise will create problem in fermentation process. There are two main methods for the sterilisation.

1. Destruction of micro-organism.
2. Removal of micro-organism.

Methods of sterilisation/disinfection are shown in Fig. 12.1.

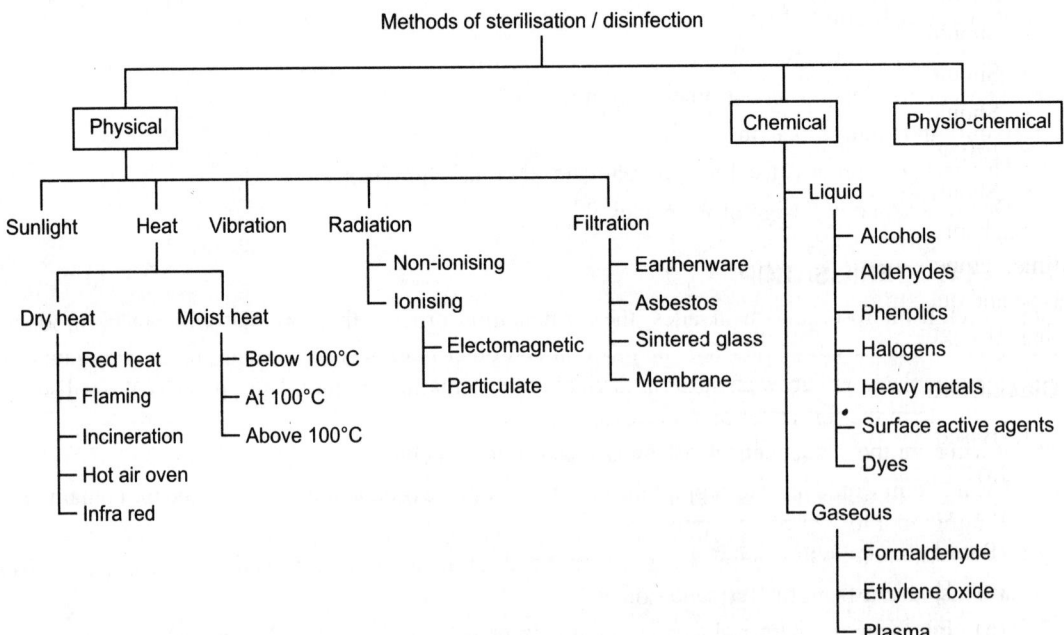

Fig. 12.1: Methods of sterilisation/disinfection.

SONIC AND ULTRASONIC VIBRATIONS

Sound waves of frequency >20,000 cycle/second kills bacteria and some viruses on exposing for one hour. Microwaves are not particularly antimicrobial in themselves, rather the killing effect of microwaves are largely due to the heat that they generate. High frequency sound waves disrupt cells. They are used to clean and disinfect instruments as well as to reduce microbial load. This method is not reliable since many viruses and phages are not affected by these waves.

CHEMICAL METHODS OF DISINFECTION

Disinfectants are those chemicals that destroy pathogenic bacteria from inanimate surfaces. Some chemical have very narrow spectrum of activity and some have very wide. Those chemicals that can sterilise are called chemisterilants. Those chemicals that can be safely applied over skin and mucus membranes are called antiseptics.

An ideal antiseptic or disinfectant should have following properties:

- Should have wide spectrum of activity.
- Should be able to destroy microbes within practical period of time.
- Should be active in the presence of organic matter.
- Should make effective contact and be wettable.
- Should be active in any pH.
- Should be stable.
- Should have long shelf life.
- Should be speedy.
- Should have high penetrating power.
- Should be non-toxic, non-allergenic, non-irritative or non-corrosive.
- Should not have bad odour.
- Should not leave non-volatile residue or stain.
- Efficacy should not be lost on reasonable dilution.
- Should not be expensive and must be available easily.

Such an ideal disinfectant is not yet available. The level of disinfection achieved depends on contact time, temperature, type and concentration of the active ingredient, the presence of organic matter, the type and quantum of microbial load. The chemical disinfectants at working concentrations rapidly lose their strength on standing.

Classification of Disinfectants

1. Based on consistency:
    (a) Liquid (e.g. alcohols, phenols).
    (b) Gaseous (formaldehyde vapour, ethylene oxide).
2. Based on spectrum of activity:
    (a) High level.
    (b) Intermediate level.
    (c) Low level.

3. Based on mechanism of action:
   (a) Action on membrane (e.g. alcohol, detergent).
   (b) Denaturation of cellular proteins (e.g. alcohol, phenol).

Alcohols

Mode of action

Alcohols dehydrate cells, disrupt membranes and cause coagulation of protein.

Examples: Ethyl alcohol, isopropyl alcohol and methyl alcohol.

Application: A 70% aqueous solution is more effective at killing microbes than absolute alcohols. 70% ethyl alcohol (spirit) is used as antiseptic on skin. Isopropyl alcohol is preferred to ethanol. It can also be used to disinfect surfaces. It is used to disinfect clinical thermometers. Methyl alcohol kills fungal spores, hence is useful in disinfecting inoculation hoods.

Disadvantages: Skin irritant, volatile (evaporates rapidly), inflammable.

Aldehydes

Mode of action

Acts through alkylation of amino-, carboxyl- or hydroxyl group, and probably damages nucleic acids. It kills all micro-organisms, including spores.

Examples: Formaldehyde, Gluteraldehyde.

Application: 40% Formaldehyde (formalin) is used for surface disinfection and fumigation of rooms, chambers, operation theatres, biological safety cabinets, wards, sick rooms, etc. Fumigation is achieved by boiling formalin, heating paraformaldehyde or treating formalin with potassium permanganate. It also sterilises bedding, furniture and books. 10% formalin with 0.5% tetraborate sterilises clean metal instruments. 2% gluteraldehyde is used to sterilise thermometers, cystoscopes, bronchoscopes, centrifuges, anasethetic equipments, etc. An exposure of at least 3 hr at alkaline pH is required for action by gluteraldehyde. 2% formaldehyde at 40°C for 20 minutes is used to disinfect wool and 0.25% at 60°C for 6 hr to disinfect animal hair and bristles.

Disadvantages: Vapours are irritating (must be neutralised by ammonia), has poor penetration, leaves non-volatile residue, activity is reduced in the presence of protein. Gluteraldehyde requires alkaline pH and only those articles that are wettable can be sterilised.

Phenol

Mode of action

Act by disruption of membranes, precipitation of proteins and inactivation of enzymes.

Examples: 5% phenol, 1–5% cresol, 5% lysol (a saponified cresol), hexachlorophene, chlorhexidine, chloroxylenol (Dettol).

Applications: Joseph Lister used it to prevent infection of surgical wounds. Phenols are coal-tar derivatives. They act as disinfectants at high concentration and as antiseptics at low concentrations. They are bactericidal, fungicidal, mycobactericidal but are inactive against spores and most viruses. They are not readily inactivated by organic matter. The corrosive phenolics are used for disinfection of ward floors, in discarding jars in laboratories and disinfection of bedpans. Chlorhexidine can be used in an isopropanol solution for skin disinfection, or as an aqueous solution for wound irrigation. It is often used as an antiseptic hand wash. 20% Chlorhexidine gluconate solution is used for pre-operative hand and skin preparation and for general skin disinfection. Chlorhexidine gluconate is also mixed with quaternary ammonium compounds such as cetrimide to get stronger and broader antimicrobial effects (e.g. Savlon). Chloroxylenols are less irritant and can be used for topical purposes and are more effective against gram positive bacteria than gram negative bacteria.

Hexachlorophene is chlorinated diphenyl and is much less irritant. It has marked effect over gram positive bacteria but poor effect over gram negative bacteria, mycobacteria, fungi and viruses. Triclosan is an organic phenyl ether with good activity against gram positive bacteria and effective to some extent against many gram negative bacteria including *Pseudomonas*. It also has fair activity on fungi and viruses.

Disadvantages: It is toxic, corrosive and skin irritant. Chlorhexidine is inactivated by anionic soaps. Chloroxylenol is inactivated by hard water.

Halogens

Mode of action

They are oxidising agents and cause damage by oxidation of essential sulphydryl groups of enzymes. Chlorine reacts with water to form hypochlorous acid, which is microbicidal.

Examples: Chlorine compounds (chlorine, bleach, hypochlorite) and iodine compounds (tincture iodine, iodophores).

Applications: Tincture of iodine (2% iodine in 70% alcohol) is an antiseptic. Iodine can be combined with neutral carrier polymers such as polyvinylpyrrolidone to prepare iodophores such as povidone-iodine. Iodophores permit slow release and reduce the irritation of the antiseptic. For hand washing iodophores are diluted in 50% alcohol. 10% Povidone Iodine is used undiluted in pre and postoperative skin disinfection. Chlorine gas is used to bleach water. Household bleach can be used to disinfect floors. Household bleach used in a stock dilution of 1:10. In higher concentrations chlorine is used to disinfect swimming pools. 0.5% sodium hypochlorite is used in serology and virology. Used at a dilution of 1:10 in decontamination of spillage of infectious material. Mercuric chloride is used as a disinfectant.

Disadvantages: They are rapidly inactivated in the presence of organic matter. Iodine is corrosive and staining. Bleach solution is corrosive and will corrode stainless steel surfaces.

Heavy Metals

Mode of action

Act by precipitation of proteins and oxidation of sulphydryl groups. They are bacteriostatic.

Examples: Mercuric chloride, silver nitrate, copper sulphate, organic mercury salts (e.g. mercurochrome, merthiolate).

Applications: 1% silver nitrate solution can be applied on eyes as treatment for opthalmia neonatorum (Crede's method). This procedure is no longer followed. Silver sulphadiazine is used topically to help to prevent colonisation and infection of burn tissues. Mercurials are active against viruses at dilution of 1:500 to 1:1000. Merthiolate at a concentration of 1:10000 is used in preservation of serum. Copper salts are used as a fungicide.

Disadvantages: Mercuric chloride is highly toxic, are readily inactivated by organic matter.

Surface Active Agents

Mode of actions

They have the property of concentrating at interfaces between lipid containing membrane of bacterial cell and surrounding aqueous medium. These compounds have long chain hydrocarbons that are fat soluble and charged ions that are water-soluble. Since they contain both of these, they concentrate on the surface of membranes. They disrupt membrane resulting in leakage of cell constituents.

Examples: These are soaps or detergents. Detergents can be anionic or cationic. Detergents containing negatively charged long chain hydrocarbon are called anionic detergents. These include soaps and bile salts. If the fat-soluble part is made to have a positive charge by combining with a quaternary nitrogen atom, it is called cationic detergents. Cationic detergents are known as quaternary ammonium compounds (or quat). Cetrimide and benzalkonium chloride act as cationic detergents.

Application: They are active against vegetative cells, Mycobacteria and enveloped viruses. They are widely used as disinfectants at dilution of 1–2% for domestic use and in hospitals.

Disadvantages: Their activity is reduced by hard water, anionic detergents and organic matter. Pseudomonas can metabolise cetrimide, using them as a carbon, nitrogen and energy source.

Dyes

Mode of action

Acridine dyes are bactericidal because of their interaction with bacterial nucleic acids.

Examples: Aniline dyes such as crystal violet, malachite green and brilliant green. Acridine dyes such as acriflavin and aminacrine. Acriflavine is a mixture of proflavine and euflavine. Only euflavine has effective antimicrobial properties. A related dye, ethidium bromide, is also germicidal. It intercalates between base pairs in DNA. They are more effective against gram positive bacteria than gram negative bacteria and are more bacteriostatic in action.

Applications: They may be used topically as antiseptics to treat mild burns. They are used as paint on the skin to treat bacterial skin infections. The dyes are used as selective agents in certain selective media.

Hydrogen Peroxide

Mode of action

It acts on the micro-organisms through its release of nascent oxygen. Hydrogen peroxide produces hydroxyl-free radical that damages proteins and DNA.

Application: It is used at 6% concentration to decontaminate the instruments, equipments such as ventilators. 3% hydrogen peroxide solution is used for skin disinfection and deodorising wounds and ulcers. Strong solutions are sporicidal.

Disadvantages: Decomposes in light, broken down by catalase, proteinaceous organic matter drastically reduces its activity.

Ethylene Oxide (EO)

Mode of action

It is an alkylating agent. It acts by alkylating sulphydryl-, amino-, carboxyl- and hydroxyl- groups.

Properties: It is a cyclic molecule, which is a colorless liquid at room temperature. It has a sweet ethereal odour, readily polymerises and is flammable.

Application: It is a highly effective chemisterilant, capable of killing spores rapidly. Since it is highly flammable, it is usually combined with CO_2 (10% CO_2 + 90% EO) or dichlorodifluoromethane. It requires presence of humidity. It has good penetration and is well absorbed by porous material. It is used to sterilise heat labile articles such as bedding, textiles, rubber, plastics, syringes, disposable petri dishes, complex apparatus like heart-lung machine, respiratory and dental equipments. Efficiency testing is done using *Bacillus subtilis* var niger.

Disadvantages: It is highly toxic, irritating to eyes, skin, highly flammable, mutagenic and carcinogenic.

Beta-Propiolactone (BPL)

Mode of action

It is an alkylating agent and acts through alkylation of carboxyl- and hydroxyl- groups.

Properties: It is a colourless liquid with pungent to slightly sweetish smell. It is a condensation product of ketane with formaldehyde.

Application: It is an effective sporicidal agent, and has broad-spectrum activity. 0.2% is used to sterilise biological products. It is more efficient in fumigation that formaldehyde. It is used to sterilise vaccines, tissue grafts, surgical instruments and enzymes.

Disadvantages: It has poor penetrating power and is a carcinogen.

Physio-chemical Method

Mode of action

A physio-chemical method adopts both physical and chemical method. Use of steamformaldehyde is a physio-chemical method of sterilisation, which takes into account action of steam as well as that of formaldehyde. Saturated steam at a pressure of 263 mm has a temperature of 70°C. The air is removed from the autoclave chamber and saturated steam at sub-atmospheric pressure is flushed in. Formaldehyde is then injected with steam in a series of pulses, each of 5–10 minutes. The articles are held at this holding temperature for 1 hr. Formaldehyde is then flushed by inflow of steam.

Disadvantages: Condensation of formaldehyde occurs and induction of large volume of formaldehyde wets the steam resulting in loss of latent heat.

Sterilisation control: using paper strips containing 106 spores of *G.stearothermophilus*.

TESTING OF DISINFECTANTS

A disinfectant must be tested to know the required effective dilution, the time taken to effect disinfection and to periodically monitor its activity. As disinfectants are known to lose their activity on standing as well as in the presence of organic matter, their activity must be periodically tested.

Different methods are:

1. Koch's method.
2. Rideal Walker method.
3. Chick Martin test.

4. Capacity use dilution test (Kelsey-Sykes test).

5. In use test.

Koch's method: Spores of *Bacillus anthracis* were dried on silk thread and were subjected to action of disinfectants. Later, it was washed and transferred to solid medium.

Rideal Walker method: This method relies on the estimation of phenol coefficient. Phenol coefficient of a disinfectant is calculated by dividing the dilution of test disinfectant by the dilution of phenol that disinfects under predetermined conditions. Both the phenol and the test disinfectant are diluted from 1/95 to 1/115 and their bactericidal activity is determined against *Salmonella typhi* suspension. Subcultures are performed from both the test and phenol at intervals of 2.5, 5, 7.5 and 10 minutes. The plates are incubated for 48–72 hr at 37°C. That dilution of disinfectant which disinfects the suspension in a given time is divided by that dilution of phenol which disinfects the suspension in same time gives its phenol coefficient.

Disadvantages of the Rideal-Walker test are: No organic matter is included; the micro-organism Salmonella typhi may not be appropriate; the time allowed for disinfection is short; it should be used to evaluate phenolic type disinfectants only.

Chick Martin test: This test also determines the phenol coefficient of the test disinfectant. Unlike in Rideal Walker method where the test is carried out in water, the disinfectants are made to act in the presence of yeast suspension (or 3% dried human feces). Time for subculture is fixed at 30 minutes and the organism used to test efficacy is *S. typhi* as well as *S. aureus*. The phenol coefficient is lower than that given by Rideal Walker method. The classical tests such as Rideal - Walker or Chick - Martin are not practicable.

Capacity use dilution test (Kelsey-Sykes test): Inoculum of four different test organisms, namely *Staphylococcus aureus*, *Escherichia coli*, *Pseudomonas aeruginosa* and *Proteus vulgaris* are added to the disinfectant in three successive. Dried yeast is included to simulate presence of organic matter. The method can be carried out under 'clean' or 'dirty' conditions. The dilutions of the disinfectant are made in hard water for clean conditions and in yeast suspension for dirty conditions. Test organism alone or with yeast is added at 0,10 and 20 minutes interval. The contact time of disinfectant and test organism is 8 min. The disinfectant is evaluated on its ability to kill micro-organisms or lack of it and the result is reported as a pass or a fail and not as a coefficient. The capacity test of Kelsey and Sykes gives a good guideline for the dilution of the preparation to be used. Disadvantage of this test is the fact that it is rather complicated.

In use test: The routine monitoring of disinfectant in use can be done by the 'in use' test of Maurer. This test is intended to estimate the number of living organism in a vessel of disinfectant in actual use. The disinfectant that is already in use is diluted 1 in 10 by mixing 1 ml of the disinfectant with 9 ml of sterile nutrient broth. Ten drops of the diluted disinfectant (each 0.02 ml) is placed on two nutrient agar plates. One plate is incubated at 37°C for 3 days while the other is held at room temperature for 7 days. The number of drops that yielded growth is counted after incubation. If there growth in more than five drops on either plate, it represents failure of disinfectant.

FILTER STERILISATION OF FERMENTATION MEDIA

Fermentation Medium contains different Ingredients in particular proportion.Different carbon source, Nitrogen source, minerals, additives, and antifoaming agents added having varied degree of heat sensitivity.Some components are heat labile for them sterilisation by heat is not advisable in such condition filter will be wise choice.In addition, Media for animal-cell culture cannot sterilised by steam because they contain heat-labile proteins, thus, filtration is the only method of choice.

Filter sterilisation of depends upon the content of Medium its size, type and moisture contents. Filters are of different types based on its applications.

Types of Filters

1 Filter with pore size smaller than the particle size:
   - They are absolute filters.
   - They are fixed pore filters.
   - If not physically damaged, are efficient to remove 100% micro-organisms.
2. Filter with pore size larger than the particle size:
   - They are known as depth filters.
   - They are non-fixed pore filters.
   - Composed of felts, woven yarns, asbestos pads and loosely packed fiber glass.

The terms absolute and depth can be misleading as they imply that absolute filtration only occurs at the surface of the filter, whereas absolute filters also have depth and filtration occurs within the filter as well as at the surface.

Methods of Suspended Solids Separation from a Fluid during Filtration

Inertial impaction

- Suspended particles in a fluid stream have momentum.
- The fluid in which the particles are suspended will flow through the filter by the route of least resistance.
- However, the particles, because of their momentum, tend to travel in straight lines and may therefore become impacted upon the fibres where they may then remain.
- Inertial impaction is more significant in the filtration of gases than in the filtration of liquids.

Diffusion

- Extremely small particles suspended in a fluid are subject to Brownian motion which is random movement due to collisions with fluid molecules.
- Thus, such small particles tend to deviate from the fluid flow pattern and may become impacted upon the filter fibres.
- Diffusion is more significant in the filtration of gases than in the filtration of liquids.

Electrostatic attraction

Charged particles may be attracted by opposite charges on the surface of the filtration medium.

Interception

- The fibres comprising a filter are interwoven to define openings of various sizes.
- Particles which are larger than the filter pores are removed by direct interception.
- Interception is equally important a mechanism in the filtration of gases and liquids.

- However, a significant number of particles which are smaller than the filter pores are also retained by interception.
- This may occur by several mechanisms - more than one particle may arrive at a pore simultaneously, an irregularly shaped particle may bridge a pore, once a particle has been trapped by a mechanism other than interception the pore may be partially occluded enabling the entrapment of smaller particles.
- Fixed pore or absolute filtration is the better system to use for media sterilisation.

Filter Sterilisation of Media

An ideal filtration system for the sterilisation of animal cell culture media must fulfill the following criteria:

- The filtered medium must be free of fungal, bacterial and mycoplasma contamination.
- There should be minimal adsorption of protein to the filter surface.
- The filtered medium should be free of viruses.
- The filtered medium should be free of endotoxins.
- Several filter manufacturers now supply absolute filtration systems for the sterilisation of animal cell culture medium.
- Such systems consist of membrane cartridges which are fitted into stainless steel, steam sterilisable modules.
- The membranes for media filtration are constructed from steam sterilisable hydrophilic material and are treated to produce a filtrate of particular quality.
- For example, if minimal protein adsorption is a major criterion then a specially coated filter membrane is used.
- It would be very difficult to construct a single filtration membrane which would fulfill all four criteria cited above thus a series of filters are used to achieve the desired result

Pall Process Filtration ltd. illustrates a system to produce sterile, mycoplasma free serum and consists of four filters arranged in sequence:

- The first filter is a positively charged polypropylene pre-filter with an absolute rating of 5 μm for the removal of coarse precipitates, clot-like material and other gross contaminants.
- The second filter is also positively charged polypropylene but with an absolute rating of 0.5 μm for bulk microbial removal, deformable gels, lipid-based materials and endotoxin reduction.
- The third filter is a single layered, nylon/polyester positively charged filter with a 0.1 μm absolute rating for further microbial and endotoxin removal and optimum protection of the final filter.
- The fourth filter is similar to the third and has the same rating, but is double layered and removes mycoplasmas, gives absolute sterility and final endotoxin control.
- Thus, the combination of four filters gives a sequential removal of decreasingly small particles and prolongs the life of the final filter.
- If it were necessary to remove viral contamination then a final 0.04 μm nylon/polyester filter would added.
- Similar systems may be used in downstream processing of animal cell products where the rating and properties of the filters would be optimised for the particular process

For the removal of cells and cell debris from an animal-cell fermentation broth:
- Two filter system is applied where the first filter is 1 μm and the second filter is 0.2 μm
- The pre-filter is a polypropylene 1.0 μm rated filter to remove the bulk of the cells and debris
- The second filter is an hydroxyl modified nylon/polyester 0.2 μm rated filter giving absolute cell removal with minimal protein adsorption.

FILTER STERILISATION OF AIR

Aerobic organisms require air for their growth. Some organisms require less quantity of air while other requires air in huge quantity. Air is composed of gases, suspended particles, and moisture. Thus, when we sterilise air we are removing all the particles and moisture from it. When air supplied, it should be free from organism otherwise; those organisms present in air will contaminate the fermenter. Thus, air must be sterile. It is generally sterilise by filter.

Filter sterilisation of depends upon the content of air like suspended matter its size, type and moisture content. Filters are of different types based on its applications

Types of Filters

1. Filter with pore size smaller than the particle size:
   - They are absolute filters.
   - They are fixed pore filters.
   - If not physically damaged, are efficient to remove 100% micro-organisms.
2. Filter with pore size larger than the particle size:
   - They are known as depth filters.
   - They are non-fixed pore filters.
   - Composed of felts, woven yarns, asbestos pads and loosely packed fiber glass.

The terms absolute and depth can be misleading as they imply that absolute filtration only occurs at the surface of the filter, whereas absolute filters also have depth and filtration occurs within the filter as well as at the surface

Methods of Suspended Solids Separation from a Fluid during Filtration

Inertial impaction

- Suspended particles in a fluid stream have momentum.
- The fluid in which the particles are suspended will flow through the filter by the route of least resistance.
- However, the particles, because of their momentum, tend to travel in straight lines and may therefore become impacted upon the fibres where they may then remain.
- Inertial impaction is more significant in the filtration of gases than in the filtration of liquids.

Diffusion

- Extremely small particles suspended in a fluid are subject to Brownian motion which is random movement due to collisions with fluid molecules.
- Diffusion is more significant in the filtration of gases than in the filtration of liquids.

- Thus, such small particles tend to deviate from the fluid flow pattern and may become impacted upon the filter fibres.

Electrostatic attraction

- Charged particles may be attracted by opposite charges on the surface of the filtration medium

Interception

- The fibres comprising a filter are interwoven to define openings of various sizes.
- Particles which are larger than the filter pores are removed by direct interception.
- However, a significant number of particles which are smaller than the filter pores are also retained by interception.
- This may occur by several mechanisms - more than one particle may arrive at a pore simultaneously, an irregularly shaped particle may bridge a pore, once a particle has been trapped by a mechanism other than interception the pore may be partially occluded enabling the entrapment of smaller particles.
- Interception is equally important a mechanism in the filtration of gases and liquids.

Sterilisation of Air

- Aerobic fermentations require the continuous addition of considerable quantities of sterile air.
- Although it is possible to sterilise air by heat treatment, the most commonly used sterilisation process is filtration.
- Fixed pore filters (which have an absolute rating) are very widely used in the fermentation industry and several manufacturers produce filtration systems for air sterilisation.
- These systems, like those for the sterilisation of liquids, consist of pleated membrane cartridges designed to be accommodated in stainless steel modules.
- The most common construction material used for the pleated membranes for air sterilisation is PTFE (PolyTetraFluoroEthylene), which is hydrophobic and is therefore resistant to wetting.
- Also, PTFE filters may be steam sterilised and are resistant to ammonia which may be injected into the air stream, prior to the filter, for pH control.
- As was seen for the filter sterilisation of liquids it is essential that a prefilter is incorporated upstream of the absolute filter.
- The prefilter traps large particles such as dust, oil and carbon (from the compressor) and pipe scale and rust (from the pipework).
- The use of a coalescing (combined, united) prefilter also ensures the removal of water from the air, entrained water is coalesced in the filter (air flow being from the inside of the filter to the outside) and is discharged via an automatic drain.

Sterilisation of Fermenter Exhausts Air

- Fixed pore membrane modules are also used for this application but the system must be able to cope with the sterilisation of water saturated air, at a relatively high temperature and carrying a large contamination level.

- In many traditional fermentations the exhaust gas from the fermenter was vented without sterilisation or vented through relatively inefficient depth filters.
- With the advent of the use of recombinant organisms and a greater awareness of safety and emission levels of allergic compounds the containment of exhaust air is more common (and in the case of recombinant organisms, compulsory).
  - Also, foam may overflow from the fermenter into the air exhaust line Thus, some form of pretreatment of the exhaust gas is necessary before it enters the absolute filter.
- This pretreatment may be a hydrophobic prefilter or a mechanical separator to remove water, aerosol particles and foam.
- The pretreated air is then fed to a 0.2 μm hydrophobic filter.
- Again, it is important to appreciate that the filtration system must be steam sterilisable.

Chapter 13

Development of Inocula for Industrial Fermentations

INTRODUCTION

Inoculum is a small amount of material containing bacteria, viruses, or other micro-organisms that is used to start a culture.

DEVELOPMENT OF INOCULA FOR YEAST PROCESSES

Industrial uses of yeasts are:

1. The brewing of beer.
2. The production of Baker's Yeast (biomass).
3. For the production of recombinant products from the yeast.

Brewing

Yeast used to inoculate a fresh batch of wort from previous fermentation or from propagator. It is common practice in the British brewing industry to use the yeast from the previous fermentation. The brewing terms used to describe this process and, 'crop' referring to the harvested yeast from the previous fermentation and 'pitch' meaning to inoculate. One of the major factors contributing to the continuation of this practice is the wort-based excise laws in the United Kingdom where duty charged on the sugar consumed rather than the alcohol produced.

Thus, dedicated yeast propagation systems are expensive to operate because duty charged on the sugar consumed by the yeast during growth. The problems with this technique are chances of contamination and degeneration of strains the most common problem with the degenerated cell is the change in the degree of flocculence and weakening of abilities of the yeast.

In breweries employing top fermentations in open fermenters, the above dangers minimised by collecting yeast to be used for future pitching from 'middle skimmings'. As the head of yeast develops the surface layer, (the most flocculent and highly contaminated yeasts) removed and discarded and the underlying cells (the 'middle skimmings') harvested and used for subsequent pitching.

Therefore, the 'middle skimmings' contain cells which have the desired flocculence and which have been protected from contamination by the surface layer of the yeast head. The pitching yeast may be treated to reduce the level of contaminating bacteria and remove protein and dead yeast cells by such treatments as reducing the pH of the slurry to 2.5 to 3, washing with water, washing with ammonium persulphate and treatment with antibiotics such as, polymixin, penicillin and neomycin. However, traditional open vessels are becoming rare and the bulk of beer brewed using cylindro-conical fermenters.

In these systems, the yeast flocculates and collects in the cone at the bottom of the fermenter where it is subject to the stresses of nutrient starvation, high ethanol concentration, low water activity, high carbon dioxide concentration and high pressure, which decreases the viability and physiological state of the yeast crop, would not be ideal for an inoculum. The situation is further complicate by the fact, that the harvested yeast is stored rapidly to about one degree before it is used as inoculum suspending in beer and storing in the absence of oxygen.

One of the key physiological features of yeast inoculum is the level of sterol in the cells. Sterols are required for synthesis of membrane but they are only produce in the presence of oxygen.

Thus, we have the irregularity of oxygen being required for sterol synthesis yet anaerobic conditions are required for ethanol production. This irregularity is resolved traditionally by aerating the wort before inoculation. The difficulties outlined above and the likelihood of strain degeneration and contamination mean that rarely used for more than five to ten consecutive fermentations that necessitates the periodical production of a pure inoculum.

Pure inocula prepared by a yeast propagation scheme utilising a 10 per cent of inoculum volume at each stage in the programme and employing conditions similar to those used during brewing.

Continuous aeration used during the propagation stage, which seems to have little effect on the beer produced in the subsequent fermentation.

Yeast inoculum produced in this way would also be sterol rich obviating the need for aerated wort. The simplest type of propagator is a single stage system resembling an unstirred aerated fermenter, which inoculated with a shake-flask culture developed from a single colony. Two-stage systems propagator operated semi-continuously. It consisted of two linked vessels one point five and one fifty cubic decimeters respectively. The smaller vessel filled with wort sterilised, cooled, aerated, and inoculated with a flask-grown culture. After growth for three to four days, the culture forced by air pressure into the second vessel, which, filled with, sterilised cooled wort and aerated. After mixing an aliquot of 1.5 dm^3 in second vessel, it is force back into the first vessel. In a further 3 to 4 days, the larger vessel contained sufficient biomass to pitch a thousand cubic decimeters fermenter and the first vessel contained sufficient inoculum for another second stage. However, although this procedure should produce a pure inoculum there is a danger of strain degeneration occurring in such a semi-continuous system.

Baker's Yeast

The commercial production of bakers' yeast involves the development of an inoculum through a large number of aerobic stages. Although the production stages, of the process, may not operate under strictly aseptic conditions, a pure culture is use for the initial inoculum thereby keeping contamination to a minimum in the early stages of growth. The development of inoculum for the production of bakers' yeast involve eight stages the first three being aseptic while the remaining stages were carried out in open vessels. The yeast pumped from one stage to the next or the seed cultures may be centrifuge and washed before transfer, which reduces the level of contamination. The yields obtained in the first five stages are relatively low because they are not fed-batch systems whereas the last three stages are fed-batch.

Outline of Production of Baker's Yeast

Baker's yeast is a major raw material in the baking process. Product requires good bake activity (rapid fermentation in doughs of high osmotic pressure) along with good shelf-life, after storage at low moisture levels. Bakers yeast is an example of a relatively low value product produced on a large scale.

Sucrose hydrolysed externally using sucrase secreted by the cell. The resultant monosaccharides diffuse into the yeast where it is degraded to pyruvate via glycolysis or converted to storage carbohydrates via anabolic pathways. A respiratory bottleneck exists in pyruvate transfer to the mitochondrion (TCA cycle). If rate of pyruvate production via glycolysis exceeds the rate of pyruvate transfer to the mitochondrion, a pool of excess pyruvate is maintained. Excess pyruvate leads to rapid conversion of the excess to ethanol via acetaldehyde. This is energy inefficient as a catabolic process—lower energy generation means lower biomass yields.

Storage carbohydrates very important, resuscitation of organism dependant on acceptable levels of storage carbohydrates in the yeast. Their presence and quantity totally dependant on the organisms growth rate. *S. cerevisiae* regulates budding using storage carbohydrates—this means that the organism can maintain a consistent doubling time irrespective of oscillating environmental conditions.

High growth rate—large fraction of cells budding, low levels of storage carbohydrates. Low growth rate—small fraction of cells budding, better levels of storage carbohydrates. Therefore, we need to control this metabolic phenomenon using process technology possibilities to consider for fermentation process. Aerobic batch culture—no control of metabolism. Large levels of fermentation (ethanol production). Continuous culturing—no down time, no cleaning time, continuous production of product, control of growth rate through dilution rate, accumulation of storage carbohydrates at low growth rates. Let's consider the product requirements. Product when inoculated into the dough does not require several doublings (2.75×10^8 cells ml^{-1} at the start, 3.00×10^8 cells ml^{-1} at the end)—just rapidly ferments based on the original innoculum size. This means that trace contamination is acceptable as it will not establish itself in the bread quickly enough. If trace contamination is acceptable, requirement for clean not sterile conditions. Clean conditions drastically reduce process overheads and downstream processing overheads. However, it is difficult to operate continuous culture under clean conditions (non-sterile) trace contaminants rapidly establish themselves in continuous culture, becoming a significant proportion of the biomass over long periods of time. Fed batch culture has all the advantages of batch and continuous culture. Controllability of continuous culture (D = F/V).

Batch nature of batch culture (periodical shutdown prevents the establishment of trace contaminants). Another feature of fed batch—at fixed F gradual slowing down in V—allows the accumulation of storage carbohydrates, trehalose and glycogen. Media formulation and preparation molasses (by-product of sugar manufacturing industry) is received by the plant at approximately 80° Brix (55 per cent sucrose).

The molasses is diluted to 40° Brix and clarified for fermentation. Molasses is not an adequate media: nitrogen (ammonia, ammonium sulphate and urea), phosphorous sources (orthophosphates or phosphoric acid), vitamin (biotin and thiamin) and mineral supplements (magnesium and others) are added.

Types of Baker's Yeast

Active dried yeast, a granulated form in which yeast is commercially sold. Baker's yeast is available in a number of different forms. Though each version has certain advantages over the others, the choice of which form to use is largely a question of the requirements of the recipe at hand and the training of the cook preparing it. With occasional allowances for liquid content and temperature, the different forms of commercial yeast are generally considered interchangeable.

1. Cream yeast is the closest form to the yeast slurries of the 19th century, being essentially a suspension of yeast cells in liquid, siphoned off from the growth medium. Its primary use is in industrial bakeries with special high-volume dispensing and mixing equipment, and it is not readily available to small bakeries or home cooks.

2. Compressed yeast is essentially cream yeast with most of the liquid removed. It is best known in the form of cake yeast, which is essentially a soft solid, beige in colour, but is also available in crumbled form for bulk usage. It is highly perishable; though formerly widely available for the consumer market, it has become less common in supermarkets in some countries due to its poor keeping properties, having been obsoleted in some such markets by active dry and instant yeast. It is still widely available for commercial use, and is somewhat more tolerant of low temperatures than other forms of commercial yeast; however, even there, instant yeast has made significant market inroads.

3. Active dry yeast is the form of yeast most commonly available to noncommercial bakers, as well as the yeast of choice for situations where long travel or uncontrolled storage conditions are likely. It consists of coarse oblong granules of yeast, with live yeast cells encapsulated in a thick jacket of dry, dead cells with some growth medium. Under most conditions, active dry yeast must be proofed or rehydrated first and, despite its better keeping qualities than other forms, is generally considered more sensitive than other forms to thermal shock when actually used in recipes. Active dry yeast also provides an alternative to butter and salt for seasoning pop corn.

4. Instant yeast appears similar to active dry yeast, but has smaller granules with substantially higher percentages of live cells. It is more perishable than active dry yeast, but also does not require rehydration, and can usually be added directly to all but the driest doughs. Instant yeast generally has a small amount of ascorbic acid added as a preservative. Some producers provide two or more forms of instant yeast in their product portfolio, for example, LeSaffre's 'SAF instant Gol' is designed specifically for doughs with high sugar contents.

5. Rapid-rise yeast is a variety of yeast (usually a form of instant yeast) designed to provide greater carbon dioxide output to allow faster rising at the expense of shortened fermentation times. There is considerable debate as to the value of such a product; while most baking experts believe it reduces the flavour potential of the finished product. Rapid-rise yeast is often marketed specifically for use in bread machines.

6. Flake yeast is dead yeast, sold primarily as a nutritional supplement. It has little to no leavening power.

For most commercial uses, yeast of any form is packaged in bulk (blocks or freezer bags for fresh yeast, vacuum-packed brick bags for dry or instant); however, yeast for home use is often packaged in premeasured doses, either small squares for compressed yeast or sealed packets for dry or instant. A single dose (reckoned for the average bread recipe of between 500 g and 1000 g of dough) is generally about 2.5 tsp or about 7 g, though comparatively lesser amounts are used when the yeast is used in a preferment.

Yeast Strains

Although genetic analyses and transformation can be performed with a number of taxonomically distinct varieties of yeast, extensive studies have been limited primarily to the many freely interbreeding species of the budding yeast *Saccharomyces* and to the fission yeast *Schizosaccharomyces pombe*. Although '*Saccharomyces cerevisiae*' is commonly used to designate many of the laboratory stocks of *Saccharomyces* used throughout the world, it should be pointed out that most of these strains originated from the interbred stocks of Winge, Lindegren and others who employed fermentation markers not only from *S. cerevisiae* but also from *S. bayanus*, *S. carlsbergensis*, *S. chevalieri*, *S. chodati*, *S. diastaticus*, etc. Nevertheless, it is still recommended that the interbreeding laboratory stocks of *Saccharomyces* be denoted as *S. cerevisiae*, in order to conveniently distinguish them from the more distantly related species of *Saccharomyces*.

Care should be taken in choosing strains for genetic and biochemical studies. Unfortunately there are no truly wild-type *Saccharomyces* strains that are commonly employed in genetic studies. Also, most domesticated strains of brewers' yeast and probably many strains of bakers' yeast and true wild-type strains of *S. cerevisiae* are not genetically compatible with laboratory stocks. It is often not appreciated that many 'normal' laboratory strains contain mutant characters. This condition arose because these laboratory strains were derived from pedigrees involving mutagenised strains, or strains that carry genetic markers. Many current genetic studies are carried out with one or another of the following strains or their derivatives, and these strains have different properties that can greatly influence experimental outcomes: S288C, W303, D273–10B, X2180, A364A, Σ1278B, AB972, SK1, and FL100. The haploid strain S288C (*MATα SUC2 mal mel gal2 CUP1 flo1 flo8-1 hap1*) is often used as a normal standard because the sequence of its genome has been determined, because many isogenic mutant derivatives are available, and because it gives rise to well-dispersed cells.

However, S288C contains a defective HAP1 gene, making it incompatible with studies of mitochondrial and related systems. Also, in contrast to Σ1278B, S288C does not form pseudohyae. While true wild-type and domesticated bakers' yeast give rise to less than 2 per cent *p*-colonies, many laboratory strains produce high frequencies of *p*-mutants. Another strain, D273-10B, has been extensively used as a typical normal yeast, especially for mitochondrial studies. One should examine the specific characters of interest before initiating a study with any strain. Also, there can be a high degree of inviability of the meiotic progeny from crosses among these 'normal' strains.

Many strains containing characterised auxotrophic, temperature-sensitive, and other markers can be obtained from the yeast genetics stock culture centre of the American type culture collection, including an almost complete set of deletion strains. Currently this set consists of 20382 strains representing deletants of nearly all non essential ORFs in different genetic backgrounds. Deletion strains are also available from EUROSCARF and research genetics.

Raw Materials

Ever since the 1920s, molasses has been the principal raw material in the production of baker's yeast. Both beet and cane molasses are used, separately or as a mixture, to supply fermentable supra as the major source of carbon and energy, together with minerals, trace elements, vitamins, and some organic nitrogen (amino acids). Additional nitrogen, in the form of ammonia, ammonium salts or urea and phosphorus (as phosphoric acid or phosphates) must always be supplied. Often some extra magnesium and/or zinc has to be added. When a mixture of beet molasses, with at least 20 per cent of cane molasses, is applied, no extra biotin supplement is needed; when pure cane molasses is used, pantothenate addition may be necessary. Although most molasses contain enough thiamine for optimum yeast growth, this vitamin is frequently added, since it stimulates the dough-leavening activity of the yeast.

One might conclude that molasses is not far from the ideal substrate for baker's yeast production. Certainly there is some truth in this conclusion. However, molasses may also contain harmful compounds deleterious to the growth yield and/or the quality of the yeast. These compounds include colloids and suspended solids, colouring substances, sulphurous acid, nitrates and nitrites, such substances as fungicides and sanitising agents (used at the sugar factory), short-chain fatty acids (especially butyric acid, which is highly toxic for yeast), hydroxymethylfurfural, and a vast number of compounds at the ppm-ppb level.

Carbon and energy sources

The main carbon and energy source present in molasses is sucrose. This disaccharide is hydrolysed to glucose and fructose by the enzyme invertase. Baker's yeast may express periplasmic as well as cytoplasmic invertase; these two enzymes are encoded by a single gene but are synthesised from different mRNAs. Synthesis of invertase, as well as such other enzymes as maltase, galactokinase, cytochromes and gluconeogenic and glyoxylate bypass enzymes, is prevented by the presence of glucose in the medium. This phenomenon is termed glucose repression. Analogous to the situation in *Escherichia coli*, the term 'catabolite repression' has been used, but in yeast no evidence has been obtained for a role of a metabolite (e.g. cAMP) as an important effector in this respect. It has been suggested that regulation of invertase synthesis by glucose takes place at the levels of transcription, translation and maturation of the enzyme prior to excretion.

Nitrogen sources

Since molasses does not contain sufficient amounts of assimilable nitrogen compounds to allow commercial yeast production, ammonia or urea is added during fermentation.

Vitamins

Baker's yeast requires biotin for growth, and compressed yeast contains about 0.75 to 2.5 ppm of this vitamin (dry weight basis). Cane molasses supplies ample amounts of biotin (0.5 to 0.8 ppm); beet molasses does not (0.01 to 0.02 ppm). Therefore, at least 20 per cent of cane molasses has to be blended with beet molasses in the preparation of the feed wort, or the fed has to be supplemented with synthetic biotin. For optimum growth it is also advisable to supplement the thiamin content of molasses with this vitamin. Thiamin is almost quantitatively taken up by baker's yeast during growth. Sufficient thiamin is usually added to the medium to obtain a content of 50 to 10 µg per g of final yeast solids because it improves the activity of compressed yeast in dough systems.

Minerals

For growth and good performance in fermentations, baker's yeast requires the addition of phosphates. The amounts added should give a final composition of the yeast of 2.5 to 3.5 per cent P_2O_5 for yeasts containing 7 to 9.5 per cent nitrogen (all based on dry weights). Phosphates are almost quantitatively taken up by the yeast during growth. The common sources of phosphorus are phosphoric acid, alkali phosphate salts, or ammonium phosphate. The latter can also serve as a source of nitrogen.

Fermentation Activators and Inhibitors

Many products have been reported to be activators of yeast growth, such as flour milling waste, sludge from aerobic digesters, etc. SO_2 inhibits yeast growth but concentrations up to 800 ppm in molasses can be well tolerated. *S. cerevisiae* adapts well to the presence of even higher concentrations of SO_2 as is known from the use of this species in the wine industry where fermentations are often carried out in the presence of 80 to 100 ppm of SO_2. Molasses contains variable amounts of nitrate which can be reduced to nitrite by bacterial action during the production of yeast.

Leavening agent

A leavening agent (also leavening or leaven) is any one of a number of substances used in doughs and batters that cause a foaming action which lightens and softens the finished product. The leavening agent—biological, chemical or even mechanical—reacts with moisture, heat, acidity, or other triggers to produce gas (usually carbon dioxide and sometimes ethanol) that becomes trapped as bubbles within the dough. When a dough or batter is mixed, the starch in the flour mixes with the water in the dough to form a matrix (often supported further by proteins like gluten or other polysaccharides like pentosans or xanthan gum), then gelatinises and 'sets', the holes left by the gas bubbles remain.

Biological leaveners

Micro-organisms that release carbon dioxide as part of their life cycle can be used to leaven products. Varieties of yeast are most often used, particularly *Saccharomyces* species (i.e. baker's yeast (*Saccharomyces cerevisiae*), though some recipes also rely on certain bacteria. Yeast leaves behind waste by-products (particularly ethanol and some autolysis products) that contribute to the distinctive flavour of yeast breads. In sourdough breads, the flavour is further enhanced by various lactic acid bacteria (lactobacilli) or acetic acid bacteria (acetobacilli).

Leavening with yeast is a process based on fermentation, biologically changing the chemistry of the dough or batter as the yeast works. Unlike chemical leavening, which usually activates as soon as the water combines the acid and base chemicals, yeast leavening requires proofing, which allows the yeast time to reproduce and consume carbohydrates in the flour.

Yeast can also be used to make alcoholic beverages like beer or wine. The resulting cast-off yeast, known as barm, can be used as a leavener and was probably ancestral to the use of modern pure-cultured yeast. While not as widely known, bacterial fermentation is sometimes used, occasionally providing a drastically changed flavour profile from a yeast fermentation; salt rising bread, which uses a culture of the *Clostridium perfringens* bacterium, is a well-known example.

Some typical biological leaveners are: (i) beer (unpasteurised—live yeast), (ii) buttermilk, (iii) ginger beer, (iv) kefir, (v) sourdough starter, (vi) yeast and (vii) yogurt.

Other carbon and energy sources

Any sugar-containing raw material or any starchy material that can be hydrolysed to fermentable sugars may serve as a carbon and energy source for the production of baker's yeast. These sugars are sucrose, maltose, glucose, fructose, and mannose. Lactose is not fermented by bakers' yeast, and galactose is fermented only very slowly. Such sugar-containing raw materials may be sugar cane juice or molasses, grape juice concentrates, date juice, wood hydrolysates, starch hydrolysates, or waste sulphite liquor. Up to the present time economics have dictated the use of molasses. Waste sulphite liquor is used to some extent in Finland. This liquor from paper pulp mills contains a mixture of hexoses and pentoses at very low concentrations. *S. cerevisiae* assimilates only the hexoses, and consequently very large volumes of liquor have to be passed through the fermentors.

Maximum Yield Versus Maximum Productivity

In general only a limited amount of seed yeast is available for commercial fermentation, which follows from the necessity to make optimum use of the fermentors, including those used for the production of seed.

Control

Demands for consistently high yeast quality and yield require an accurate control of commercial fermentations. Thus critical process variables such as temperature, aeration and pH are under closed-loop control; feed rates for molasses and ammonia frequently are controlled according to preset schedules (open-loop control). This may not be sufficient, however, since no corrections can be made when problems caused by, for example, the following conditions are encountered.

1. Changes in the composition of raw materials (especially molasses, a waste product of the sugar industry, is notorious in this respect).
2. Changes in the physiological state of seed yeast.

To overcome such problems, control strategies have been developed, the application of which leads to prevention of excessive alcohol formation.

Drying

Like commercial fermentation, the industrial practice of drying is based on technological as well as physiological knowledge. Dryers of four types—rotolouver, belt, fluidised-bed and spray dryers—have been reviewed. Application of spray drying for yeast cells is limited because the cell viability of the dried product obtained is very low even when a relatively low outlet air temperature (60°C) is used.

The heat of evaporation is a function of the actual dry matter content of the yeast. Up to 90 per cent dry matter, the heat of evaporation is constant and equals 10 kcal/mol, whereas above 90 per cent dry matter, the value increases linearly with dry matter content, reaching a value of 20 kcal/mol at 95 per cent dry matter. This is understandable, since when the yeast contains up to 90 per cent dry matter, free water is evaporated. To obtain a higher dry matter content, physically bound water must be removed. However, dry matter contents of commercial samples of dried yeast never exceed 96 per cent, since a higher dry matter content leads to irreversible damage of metabolic functions, probably due to removal of chemically bound water. During rehydration of dried yeast, the cells may lose up to 31 per cent of their dry matter, which includes proteins, peptides, amino acids, phosphate and vitamins. The leakage takes place only above 80 per cent dry matter content. This phenomenon may be explained by changes in membrane structures duo to liquid crystalline-gel transitions during the drying process. It is known that model membrane systems lose their barrier function when a gel phase is induced.

Improving Industrial Full-scale Production of Baker's Yeast by Optimising Aeration Control

Scientists have analysed the control of optimum dissolved oxygen of an industrial fed-batch procedure in which baker's yeast (*Saccharomyces cerevisiae*) is grown under aerobic conditions. Sugar oxidative metabolism was controlled by monitoring aeration, molasses flows, and yeast concentration in the propagator.

It has been found that proteins in yeast high yield, easy-to-use *K. lactis* expression at the later stage of the propagation, and keeping pH and temperature under controlled conditions. A large number of fed-batch growth experiments were performed in the tank for a period of 16 hr, for each of the three manufactured commercial products. For optimisation and control of cultivations, the growth and metabolite formation were quantified through measurement of specific growth and ethanol concentration. Data were adjusted to a model of multiple lineal regression, and correlations representing dissolved oxygen as a function of aeration, molasses, yeast concentration in the broth, temperature and pH were obtained. The actual influence of each variable was consistent with the mathematical model, further justified by significant levels of each variable, and optimum aeration profile during the yeast propagation was found. Baker's yeast is used extensively because of its ability to raise dough by fermenting mainly maltose and sucrose present in the dough to ethanol and carbon dioxide. It is also used in the leavening process because of its contribution to the aroma and flavour of bread. The conditions in dough differ from those in industrial baker's yeast production, since in the latter process, the environment is aerobic and the sugar concentration is low.

In a modern propagation plant, the question to be answered is how to control the process by ensuring optimal air supply and at the same time optimise the appropriate metabolic pathway that the yeast may encounter during its growth in the propagator. In particular, optimising processes under pre - and post-stationary phase conditions may produce a substantial economic. The production of baker's yeast involves the multistage propagation of the selected yeast strain using sugar as a carbon source. Baker's yeast is usually produced starting from a small quantity of yeast added to a liquid solution of essential nutrients (molasses, ammonia or ammonium salts, phosphate and vitamins) at a suitable temperature and pH. Once the cell population has grown enough, it is trans ferred into a larger bioreactor for a new growth stage; 4 or 5 stages are usually necessary to reach a satisfactory production quantity. The smaller bioreactors used for the initial stages operate under batch and anaerobic conditions, whereas in the larger bioreactors used for the later stages, aeration is provided and the fed-batch cultivation mode is adopted, i.e. the nutrients are fed to the culture medium at a variable rate.

The plant configuration and operative choices are the consequence of the effects that *Saccharomyces cerevisiae* metabolism produces on biomass yield and growth rate. During the aerobic growth of *S. cerevisiae*, both sugars and ethanol can be used as carbon and energy sources. Sugars can be metabolised via 2 different energy-producing pathways, oxidation or fermentation, depending on the sugar concentration in the medium.

Oxidative metabolism of glucose: Theoretically, glucose is entirely oxidised and provides a high level of energy for adenosine triphosphate (ATP) synthesis.

When the yeast production yield is maximised, 0.5 g of dry matter (biomass) is produced per gram of consumed glucose. Fermentative metabolism of glucose: When the glucose concentration is sufficiently high, yeasts ferment glucose and ethanol is produced.

The low level of energy produced is related to low yeast growth. This work analyses the control of optimum dissolved oxygen of an industrial fed-batch procedure in which baker's yeast (*Saccharomyces cerevisiae*) is grown under aerobic conditions. Sugar oxidative metabolism was controlled by monitoring aeration, molasses flows and yeast concentration in the propagator.

Indeed, at a high sugar concentration, oxidation is suppressed and fermentation takes place (the phenomenon often referred to as the Crabtree effect), oxidation predominates when sugar concentration is below 50–100 mg. On the other hand, under oxygen-limited growth conditions, the fermentative pathway leading to ethanol production predominates, even at a low sugar concentration.

Furthermore, an oxidative metabolism of ethanol may be produced. Without sugars, the ethanol produced during the initial fermentative metabolic pathway is reconsumed in the presence of molecular oxygen. Biomass yields on sugars are strongly related to the prevailing metabolic pathway, being maximal only when sugar is oxidised. At high growth rates, the biomass yield of baker's yeast (*S. cerevisiae*) decreases due to the production of ethanol. For this reason, it is standard industrial practice to use a fed-batch process whereby the growth rate is fixed at a level very close to the point of ethanol production. Optimally, growth should be maintained at this critical level, but in practice this is difficult because the critical growth rate is dependent upon strain and culture conditions.

The critical growth rate may vary from batch to batch and even during the experiment. In order to avoid the risk of decreasing the yield, an alternative approach is to use the overflow metabolite as an indicator of how close or far the actual growth rate is from the critical growth rate. Thus, if ethanol production is maintained constant, it is possible to fix the growth rate at a value slightly above the critical growth rate.

The effect of variables such as pH and temperature is well-known and their optimal set-points can easily be defined. On the contrary, yield and productiveness can largely be affected by the concentration of biomass, sugar, oxygen and ethanol formation, if any. The optimal conditions giving maximum yield and productiveness change along with time together with the biomass growth. Therefore, the feeding rate of the molasses is the most critical variable and the problem is to individuate the best feeding rate sequence. Furthermore, bioprocess control runs into a number of difficulties resulting from the nonlinear, non steady kinetic properties of the process dynamics as the micro-organisms multiply, adapt, and change their behaviour with time and with the environment, a lack of sensors providing direct measurements of the system state variables, such as biomass, substrates and metabolites. More often than not, sensors are not industrially available or used. The optimal process control must maximise both cell yield and productivity. The way to overcome this productivity and yield conflict is by accurately regulating the molasses feeding to ensure that the sugar concentration is tightly maintained in such a way that only oxidation occurs and the respiratory capacity of the cells is utilised to the maximum.

The carbohydrate feedstock is an important cost factor in baker's yeast production and, consequently, biomass yield on sugar is an important optimisation criterion. In order to maintain competitiveness, the fermentations must be highly consistent, with minimum variation in product quality, maximum yield on raw materials and minimum production of undesirable side products.

Many parameters impact the metabolic activities of micro-organisms and need to be controlled. Hence, many researchers have focused their attention on optimising fed-batch processes for the production of baker's yeast with different aims (productivity, quality of the yeast, or energy saving). The majority of them commonly developed their research work under laboratory conditions, seldom under pilot plant conditions, but never on a large industrial scale.

DEVELOPMENT OF INOCULA FOR BACTERIAL PROCESSES

The main objective of inoculum development for traditional bacterial fermentations is to decrease lag phase. A long lag phase is not only is wastage of time but also medium consumed in maintaining a viable culture prior to growth. The size of the inoculum and its physiological condition affect the length of the lag phase. Bacterial inocula should transfer when the cells are still metabolically active.

The age of the inoculum is particularly important in the growth of sporulating bacteria, for sporulation induced at the end of the logarithmic phase and the use of an inoculum containing a high percentage of spores would result in a long lag phase in a successive fermentation. The commercial production of proteases uses 5 per cent inoculum of thermophilic *Bacillus* in logarithm phase.

A two-stage inoculum development programme of *Bacillus subtilis* is used for the production of proteases. Inoculum for a seed fermenter was grown for 1 to 2 days on a solid or liquid medium and then transferred to a seed vessel where the organism was allowed to grow for a further ten generations before transfer to the production stage. The lag phase in plant fermenters eliminated by using inoculum medium of the same composition as used in the production fermenter and employing large inocula of actively growing seed cultures in the production of bacterial enzymes.

Inoculum development programme at pilot- scale for the production of vitamin B_{12} from *Pseudomonas denitrificans* shown below.

Stock culture
(Lyophilised with skim milk)
↓
Maintenance culture
(Agar slope incubated 4 days at 28°C)
↓
Seed culture - First stage
(2 dm³ flask containing 0.6 dm³ medium inoculated with culture from one slope; incubated with shaking for 48 hr at 28°C)
↓
Seed culture - Second stage
(40–80 dm³ fermenter containing 25–50 dm³ medium inoculated with 1–1.2% first stage seed culture. Incubated 25–30 hr at 32°C)
↓
Production culture
(500 dm³ fermenter with 300 dm³ medium inoculated with 5% second stage seed culture. Incubated at 32°C for 140–160 hr)

The acetic-acid bacteria used in the vinegar process are extremely sensitive to oxygen starvation therefore, it is essential to use an inoculum in an active physiological state. The cells at the end of fermentation are use as inoculum for the next batch by removing approximately 60% of the culture and restoring the original level with fresh medium. In this process, there are enough chances of strain degeneration and contaminant accumulation. However, strain stability is a major concern in inoculum development for fermentations employing recombinant bacteria.

Plasmid stability and productivity in *E. coli* biotin fermentation improved if stationary, rather than exponential phase, cells used as inoculum due to loss of plasmid in fermentation.

In the lactic-acid fermentation, lactic acid inhibits the production organism. Thus, production of lactic acid in the seed fermentation may result in generation of poor quality inoculum. High quality inoculum of *Lactococcus lactis* 10-1 on a laboratory scale obtained using electrodialysis, which reduced the lactate in the inoculum and reduced the length of the lag phase in the production fermentation.

DEVELOPMENT OF INOCULA FOR ANAEROBIC BACTERIAL PROCESSES

Clostridial Acetone-Butanol fermentation is anaerobic process.

However, the process was outcompeted by the petrochemical industry but there is still considerable interest in reestablishing the fermentation.

The inoculum development programme described by McNeil and Kristiansen given as below:

Heat-shocked spore suspension inoculated into 150 cm³ of potato glucose medium
↓
Stage 1 culture used as inoculum for 500 cm³ molasses medium
↓

Stage 2 culture used as inoculum for 9 dm^3 molasses medium

$\downarrow$

Stage 3 culture used as inoculum for 90000 dm^3 molasses medium

The stock culture is heat shocked to stimulate spore germination and to eliminate the weaker spores. The production stage inoculated with a very low volume. The use of such small inocula necessitates the achievement of as near perfect conditions as possible to prevent contamination and to avoid an abnormally long lag phase.

DEVELOPMENT OF INOCULA FOR MYCELIAL PROCESSES

The majority of the industrial fermentation processes carried out using mycelial (filamentous) organisms like fungi and Streptomycetes. Vegetative fungi and spores of fungi used as inoculum. The majority of industrially important fungi and Streptomycetes are capable of asexual sporulation so it is common practice to use a spore suspension as inoculum during an inoculum development programme. A major advantage of a spore inoculum is that it contains far more 'propagules' than a vegetative culture.

Three basic techniques developed to produce a high concentration of spores for use as an inoculum.

* Spores development (Sporulation) on solidified media.
* Spores development (Sporulation) on solid media.
* Spores development (Sporulation) in submerged culture medium.

First, we will see production of spores and then use of spore as inoculum.

Sporulation on Solidified Media

Most fungi and Streptomycetes will sporulate on suitable agar media but a large surface area must be employ to produce sufficient spores. Roll bottle technique given by Parker for the production of spores of *Penicillium chrysogenum* on solid media. In this technique three hundred cubic centimeters medium containing 3 per cent agar sterilised in one cubic decimeter cylindrical bottles, which then, cooled to forty-five degree and rotated on a roller mill so that the agar set as a cylindrical shell inside the bottle. These bottles inoculated with a spore suspension from a sub-master slope and incubated at twenty-four degree for six to seven days.

Sporulation on Solid Medium

Many filamentous organisms will sporulate freely on the surface of cereal grains from which the spores harvested. Substrates such as barley, hard wheat bran, ground maize, and rice are all suitable for the sporulation of a wide range of fungi. The sporulation of a given fungus affected by the amount of water added to the cereal before sterilisation and the relative humidity of the atmosphere, which should be as high as possible during sporulation. Fungi can produce relatively large number of spores on wheat bran or barley bran compared to solidified media like Nutrient agar and Sabouraud agar at particular temperature and humidity. Humidity is very important for the growth of fungi and production of spores about ninety to ninety-eight per cent of humidity is required.

Sporulation in Submerged Culture

Many fungi will sporulate in submerged culture provided a suitable medium is employed and suitable condition provided. This technique is more convenient than the use of solid or solidified media because it is easier to operate aseptically and it may apply on a large scale.

The technique first adopted by Foster and others. He induced submerged sporulation in *Penicillium notatum* by including two point five per cent calcium chloride in a defined nitrate-sucrose medium.

Medium components and other conditions favour the sporulation of fungi in submerged culture. According to Rhodes and others the conditions necessary for the submerged sporulation of the griseofulvin-producing fungus *Penicillium patulu,* and the nitrogen level had to be limited to between point zero five and point one per cent weight by volume and that good aeration had to maintain.

Most Actinomycetes do not sporulate in submerged culture due to this limitation they are more suitably cultivated using solid or solidified media for the production of spore inocula.

Uses of the Spore Inoculum

For the production of product at large scale, spore itself or vegetative cells developed from the spores used depending on the organism's fermentation quantity and processes.

Some fermentation process can proceed with both either spore or vegetative cell produced from the spore. In the clavulanic acid process the spore inoculum used to, inoculate the final seed stage. In the chlortetracycline process, a vegetative stage of fungi is use for the fermentation process.

Direct spore inoculation would avoid the cost of installation and operation of the seed tanks whereas the use of germinated spores would reduce the fermentation time of the final stage thus allowing a greater number of fermentations to carry out per year. However, labour costs for the production of the vegetative inoculum could be almost as high as for the final fermentation although some of these costs may recover.

Inoculum Development for Vegetative Fungi

Some fungi will not produce asexual spores and therefore such process must use an inoculum of vegetative mycelium. *Gibberella fujikuroi* is a fungus used for the commercial production of gibberellin. Cultures grow on long potato dextrose agar slants for one week at twenty-four degree.

Growth from three slants scraped off and transferred to a nine cubic decimeter. This medium aerated for seventy-five hours, at twenty-eight degree before transfer to a hundred cubic decimeter seed fermenter containing the same medium. The major problem in using vegetative mycelium as initial seed is the difficulty of obtaining a uniform standard inoculum. The procedure may improve by fragmenting the mycelium in a homogeniser such as a Waring blender prior to use as inoculum.

Effect of the Inoculum on the Morphology of Filamentous Organisms in Submerged Culture

When filamentous fungi grown in submerged culture they can grow as 'pellet' form consisting of compact discrete masses of hyphae or as the filamentous form in which the hyphae form a homogeneous suspension dispersed through the medium. When they form pellet they will not be able to grow properly due to nutrients and oxygen-limiting conditions inside the pellet while if they form filamentous forms than their distribution may not be proper in production medium to their filamentous growth. The information available on the morphology of Actinomycetes in submerged culture is very limited compared with that on fungi.

ASEPTIC OPERATION AND CONTAINMENT

An organism or many time more than one type of organisms are used to produce a particular type of fermentation product(s).

Following are the consequences which occurs, if the fermentation process is contaminated at any stage with an unwanted micro-organism(s):

- Due to competition for nutrients amongst actual product producing and foreign organisms, it may result in an overall loss of productivity.
- In case of a continuous fermentation process, contaminants may 'outgrow' the real product producing organism and as time passes, it may replace the interesting organisms from the fermentation media.
- The contaminants may contaminate the final products. For instance, if the biomass of the cells itself is a final products.
- Production of some unnecessary metabolites by the contaminants may create difficulties in extraction and purification of the products at final stage.
- During the process, the contaminant grow very fast using products produced by the desired organisms simultaneously and may time may it may degrade the desired products.
- If the media is contaminated with any lytic phase specific to the producing organisms, it may result in the lysis of organism and subsequently lead to the loss of yield.

How to Avoid the Entry of Contaminants?

- By using a sterile inoculum before starting a fermentation process.
- By sterilising the medium and other medium components to be added during the process properly.
- By sterilising the fermenter vessel and other ancillary equipment.
- Keeping aseptic conditions throughout the fermentation process.

Containment in the Fermentation Industries

- Protection against contamination can be achieved by aseptic operations during the process.
- Containment is a vigilance regarding prevention of escape of viable cells from a fermenter or any stage of downstream processing.
- East and others and Flickinger and Sansone, has initiated '*Containment guidelines*'.
- To grow a particular type of organism, an appropriate level of containment should be established.
- An entire process for potential hazards which are likely to happen when any organisms release accidentally, must be carefully assessed.
- Based on use of recombinant organisms that contains foreign DNA (genetically engineered) or not (non-genetically engineered), an appropriate assessment procedures must be established and based on hazard assessment, an organism can be categorised under a particular hazard group for which there is an appropriate level of containment.

Figure 13.1 shows the outline of the procedure accepted within the European industrial Community. When there is a non-genetically engineered organisms, based on assessment of the risk criteria, one should follow the specification given under hazardous group 1 to 4 given by Collins in 1992:

1. Information pertaining to the micro-organism to be employed.
2. The virulence power of the micro-organism (level of diseases it causes, i.e. mild or serious?).
3. The minimum number of micro-organisms needed to initiate an infection.
4. The routes through which the infection is likely to happen.

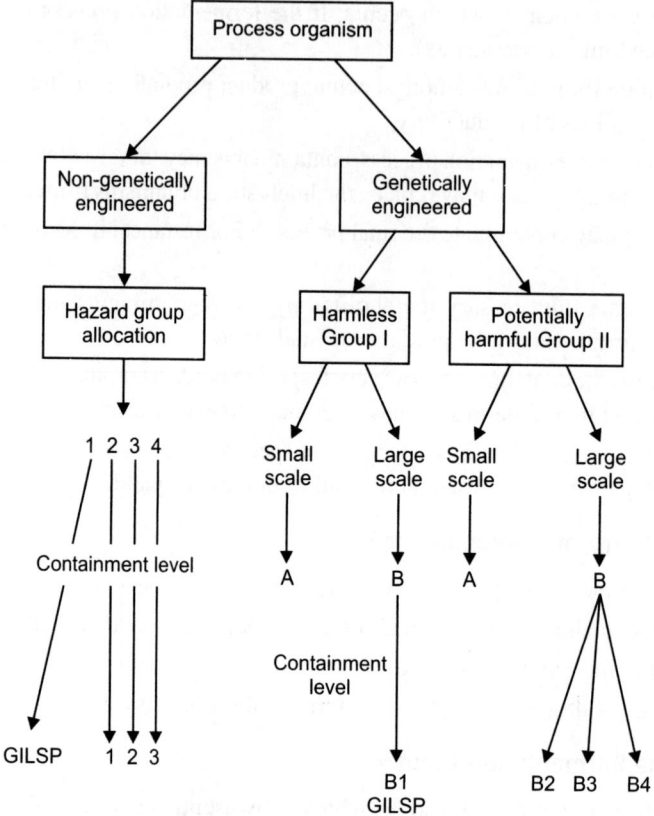

Fig. 13.1: Categorisation of a process micro-organism and designation of its appropriate level of containment at research or industrial sites within the European Federaion of Biotechnology (GILSP = Good Industrial Large Scale Practice).

6. Information related to prophylaxis and treatment.

5. The past history of the same type of incidence of infection in the community and information regarding locally existing latent reserves and vectors.

6. What is the amounts of organisms in how much volume of the media used in the fermentation process.

Winkler and Park in 1992, has described the details of assessment of risk at various stages and different ways to reduce it. Following are the containment requirements applied to any biotechnological operations once the organism has been categorised under a hazard group as described by Frommer and others in 1989.

General Containment

- Written instructions and code of practice.
- Biosafety manual.
- Good occupational hygiene.
- Good microbiological techniques.

- Biohazard sign.
- Restricted access.
- Accident reporting.
- Medical surveillance.

Primary Containment

Operation and equipment

- Work with viable micro-organisms should take place in closed systems, which minimise or prevent the release of cultivated micro-organisms.
- Treatment of exhaust air or gas from close system.
- Sampling from closed system.
- Addition of materials to close system and transfer of cultivated cells.
- Removal of material, products and effluents from closed system.
- Penetration of closed system by agitator shaft and measuring devices.
- Foam-out control.

Secondary Containment

Facilities

- Protective clothing appropriate to the risk category.
- Changing and washing facility.
- Disinfection facility.
- Emergency shower facility.
- Airlock and compulsory shower facilities.
- Effluents decontaminated.
- Controlled negative pressure.
- HEPA filters in air ducts.
- Tank for spilled fluids.
- Hermetically sealable area.

Hazard Group-1

When Hazard group 1 organisms are used on a large scale in any industrial process, the Good Industrial Large Scale Practice (GILSP) is the only requirement. As the prevention of escape of organisms during these processes is not a threat, processes under this category need to be operated aseptically only.

Hazard Group-4

- If the organism is belong to Hazard group 4 then the stringent requirements of level 3 must be followed before and during the fermentation.
- The systems for classifying organisms on the basis of hazard-assessment differs country as per geographical locations and country to country. Under these situations, the production and research experts must follow appropriate local official hazard lists and norms.

- When genetically engineered organisms are classified under either harmless (Group I) or potentially harmful (Group II) when they use in the in the process.
- The processes are also classified accordingly, like small scale or large scale as per the Health and Safety Executive document guidelines.
- The large scale processes can be categorised into two categories, i.e. IE or IIB.
- IE processes necessitate containment level B1 and are supposed to follow GILSP norms.
- IIB processes are further assessed to determine the most suitable containment level, ranging from B2 to B4.
- When non-genetically engineered organisms are used, the levels B2 to B4 correspond to levels 1 to 3.
- With the development of process engineering, availability of new micro-organisms and advancement in the sophisticated instrumentation, there are all chances of new legislations for these guidelines.
- Regardless of its genetic constitution, the organism is potentially harmful or harmless will be the key factor in future with the development and advancement in the fermentation processes.
- Though there are stringent norms for the use of genetically engineered and non-genetically engineered organisms which are categorised under potential hazardous group, many time differential official bodies when any industrial processes Problems can occur when different official bodies put the same organism in different hazard categories.
- To avoid this anomaly, the European Federation for Biotechnology (EFB) produced a consensus list of non-genetically engineered organism which were used frequently in the process in 1989.
- The EFB clarified that requirement of GILSP should met when micro-organisms used in industrial processes are in the lowest hazard group.
- As suggested by Schofield in 1992, to minimise the restrictions and reduces the cost of expensive equipment and other containment facilities, it is advisable to exploit organisms which cause minimum risk of hazards.

Overall Containment Categorisation

To accomplish the standards of the precise level of containment following aspects should be consider carefully:

- Procedures to be used.
- Staff training.
- The facilities in the laboratory and factory.
- Downstream processing.
- Effluents treatment.
- Work practice.
- Maintenance, etc.

It is imperative to ensure that all these aspects are of a sufficiently high standard to achieve the required levels of containment essential for a particular process by a government regulatory body. Unless and until, these criteria are not met as per the standard, one should not operate the process.

SECTION IV

Designing Aspects of Fermentator

Chapter 14

Fermentation Monitoring and Optimisation

INTRODUCTION

The term fermentation is derived from Latin fermentum to ferment, and it has been used to describe the metabolism of sugars by micro-organisms since ancient times. Today fermentation processes are used in the production of many different products and these processes can be divided into nine categories, according to the product (Table 14.1).

Table 14.1: Categories and typical application of fermentation processes.

Category	Typical processes
Production of microbial cells	Baker's yeast, single cell protein, lactic acid bacteria
Production of primary metabolites	Ethanol (both in beer and wine production and for technical use), lactic acid, citric acid, acetic acid, amino acids, vitamins
Production of secondary metabolites	Antibiotics (penicillins, cephalosporins, clavulanic acid, tetracycline), skikonin
Production of microbial enzymes	Lipases, proteases, amylases, glucose isomerase
Production of pharmaceutical proteins	tPA, human insulin, erythropoietin, growth hormone, hepatitis B vaccine, monoclonal antibodies
Production of polysaccharides	Xanthan gum, dextran
Production of DNA	Genes for vaccines or gene therapy
Biotransformations	Steroids, L-sorbose (from L-sorbitol)
Production of tissue	Bone marrow, skin

Development of fermentation processes can roughly be divided into four phases (Fig. 14.1). First the product is identified. In the case of a pharmaceutical this may be a result of random screening for different therapeutic effects by microbial metabolites (e.g. by high throughput screening of secondary metabolites from *Actinomycetes*) or it may be the result of a targeted identification of a novel product (e.g. a peptide hormone with known function). Outside the pharmaceutical sector the product may also be chosen after a random screening procedure (e.g. screening for a novel enzyme to be used in detergents)

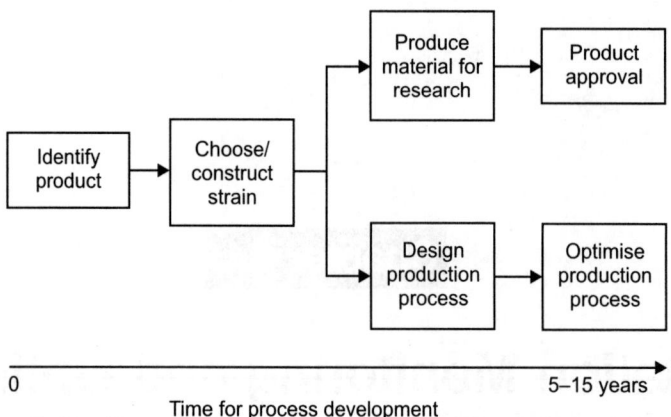

Fig. 14.1: Different phases in the development of a fermentation process.

or it may be chosen in a more rational fashion. The next step in the development phase is to choose a strain. In the past this choice was normally obvious after the product had been identified, for example, *Penicillium chrysogenum* was chosen for penicillin production because it was the first organism identified to produce penicillin.

With the introduction of recombinant DNA (rDNA) technology, it is now possible to choose almost any host for the production. Thus, a strain of *Escherichia coli* has been constructed that can produce ethanol at a high yield, and a recombinant strain of *P. chrysogenum* can now be used to produce 7-ADCA (a precursor used for synthesis of cephalosporins) directly by fermentation. The choice of strain does, however, often depend much on tradition within the company, and most of the fermentation industries have a set of favourite organisms that are used in the production of many different products. Thus the major players on the industrial enzyme market have chosen a few organisms (typically one to two strains of filamentous fungi and one to two species of bacteria) as production vehicles for a wide range of enzymes. With the production of a heterologous protein, that is, expression of a foreign gene in a given organism, it is also necessary to consider many other aspects (e.g. whether the protein is correctly folded and glycosylated).

After the strain has been constructed, one of the first aims is to produce sufficient materials for further research, and this is typically done in pilot-plant facilities. For a pharmaceutical compound sufficient material must be produced for clinical trials. For other products it may be necessary to carry out tests of the product and examine any possible toxic effects. In parallel to the continuing research in the application of the product, the production process is also designed. The final steps in development are product approval by the proper authorities and construction of the production facility (which in some cases may be through retrofitting of an existing plant). After the process has been developed, there is continuous optimisation of the process, and in some cases better strains are introduced. However, for pharmaceutical products new strains are, only introduced when there is a significant improvement in the process because it is costly to obtain approval from the authorities for redesign of the process.

CRITERIA FOR DESIGN AND OPTIMISATION

The criteria used for design and optimisation of fermentation processes depends on the product. Thus, the criteria used for a high-volume–low-value-added product are normally completely different from the criteria used for a low-volume–high-value added product. For products belonging to the first category (which includes most whole cell products, most primary metabolites, many secondary metabolites, most industrial enzymes, and most polysaccharides) the three most important design parameters are:

1. Yield of product on the substrate.
2. Productivity.
3. Final titer.

A high yield of product on the substrate is for these processes very important, since this gives a good utilisation of the raw materials, which often accounts for a significant part of the total costs. Thus, in penicillin production the costs of glucose alone may account for up to 15 per cent of the total production costs. Productivity is important because it ensures an efficient utilisation of the production capacity (i.e. the bioreactors). In an increasing market it is especially important to increase productivity, because this may prevent new capital investments. Final titer is of importance for the further treatment of the fermentation medium (e.g. purification of the product). Thus, if the product is present in a very low concentration at the end of the fermentation, it may be very expensive to perform purification of the product. In the production of novel pharmaceuticals, which typically belong to the category of low-volume–high-value added products, the three previously mentioned design parameters are normally not very important.

STRAIN CONSTRUCTION AND STRAIN IMPROVEMENT

Today high-performance production strains to be used in fermentation processes are often constructed using rDNA technology. However, the concept of metabolic pathway manipulation is by no means new. Thus, for decades better strains of *Saccharomyces* to be used for beer fermentation have been obtained through classical breeding and crossing of different strains, and in the production of penicillin productivity has increased by more than 500 times through repeated rounds of mutation and selection of new strains of *P. chrysogenum*. With the introduction of rDNA technology it has become possible to apply a more rational approach to strain improvement—namely by the introduction of targeted genetic changes resulting in strains with a phenotype that gives a better process.

This rational approach has been named cellular and metabolic engineering, and different definitions of the approach have been given. The term metabolic engineering was first introduced by Bailey, who defined it as 'improvement of cellular activities by manipulation of enzymatic transport and regulatory functions of the cell with the use of recombinant DNA technology'. His definition of metabolic engineering includes the following:

1. Inserting new pathways in micro-organisms with the aim of producing novel metabolites (e.g. production of polyketides by *Streptomyces*), with the aim of degrading toxic compounds (e.g. in bioremediation), or with the aim of constructing a novel biotransformation system.
2. Production of heterologous peptides, such as production of human insulin, erythropoitin, and tPA, or industrial enzymes, such as lipases and cellulases.
3. Improvement of pathway fluxes leading to higher yields of metabolites (e.g. increasing the flux toward antibiotics or primary metabolites) or yields of biomass (e.g. increasing the cell mass yield in baker's yeast production using industrial media containing mixed sugars).

What characterises metabolic engineering is the rational approach to performing genetic changes, and as with all other fields of engineering it consists of two steps: analysis and synthesis. As a consequence of the difficulties in performing detailed analysis of cellular metabolism these has mainly been focus on synthesis in the past, such as expression of new genes in various host cells, amplification of endogenous enzymes, and deletion of genes or modulation of enzyme activities. With modern experimental techniques it has, become possible, however, to perform detailed analysis of cellular function through both *in vivo* and *in vitro* measurements, and in the following section some of these techniques are described.

FERMENTER DESIGN

This section highlights the design of agitated vessels used for aerobic, single-cell (bacteria and yeast), and filamentous (bacteria and fungi) fermentations. It presents general design principles, combining basic concepts with practical matters including regulatory compliance, safety, maintenance, cleanability, ease of use, and cost, all of which are strongly interrelated. It focuses primarily on fermenters used to produce human and animal health products or their precursors, but the general approaches discussed are applicable to other types of products (e.g. industrial enzymes, food products), the primary differences are emphasised. Included in the presentation are fermenter vessels, agitation, aeration, heat transfer; sterilisation; cleaning, and piping systems. Control and data acquisition are discussed only to the extent that they influence the items already noted. It is assumed that the reader is familiar with the basics of fermentation, general fermenter construction, and nomenclature.

Designing Principles

The designing aspects comprises the following simple principles:

1. Each design case has unique requirements and calls for individualised application of the basic design concepts and practices discussed herein. The one-size-fits-all approach usually results in sleeves that drape into your food, arm holes that cut off circulation, pant legs that are made to trip over, or some combination of these features and others. The magic number approach—all problems have simple solutions based on codified numbers such as fixed geometric ratios—leads to the same place.

2. Successful design requires a systems approach. A fermenter is a system that is part of a process system, and the process is part of a plant system. All these systems interact with each other, with real people, with control systems, and with other systems related to regulatory compliance, safety, documentation (including protocols, SOPs, etc.), change control, maintenance, and so forth. Failure to take these interactions into account during fermenter design usually results in considerable pain, not only for those guilty of the omission, but also for innocents who were never asked for planning and/or design input, but who must live with the result. Note that appropriate documentation, protocols, SOPs, change control, and so on should be considered important elements of design and operation of any plant, licensed or not.

3. Compromise is always necessary. Nothing in any project is 100 per cent right, and nothing has to be. He or she who looks for 100 per cent of anything will hold up a project needlessly and will generate a lot of animus.

4. The time to be thinking about all of these points is prior to and during design, not when you are standing on the plant floor trying to validate a continuous mixer that has only one port and no instrumentation.

We discuss next the bases for safety and regulatory considerations, primarily to sensitise readers to these issues early on. They are all too often overlooked during early stages of design, when the focus is on satisfying process requirements (e.g. oxygen transfer). Unfortunately, this frequently leads to important constraints being ignored and hence to designs that require expensive and often cumbersome 'fixes'.

Safety and Regulatory Compliance

All fermenter designs are influenced by safety and regulatory factors, the extent depends on: (i) the nature of the product and its intended use, (ii) the process, and (iii) the nature and location of the facility.

For example, even fermentation products not regulated by the US. Food and Drug Administration (FDA) must adhere to safety requirements, which in some cases are very strict. Furthermore, no professional design can escape the effects of a long list of code requirements, for example, the ASME pressure vessel code. By the same token, not all products that are FDA regulated are equally affected by safety, regulatory, and code requirements. We must consider each case on the basis of its own unique requirements.

Containment: worker and community biosafety

A fermenter must be an integral part of a system designed to insure safety at three levels: product, worker, and community. Many of the methods used to protect the product also serve to protect the plant personnel and the community (e.g. use of dosed systems and HEPA filters); however, there are potential points of conflict. For example, some containment practices that call for completely welded hard piping to a contained drain line for any condensate that could be exposed to culture fluid could expose the product to drain line contaminants. One resolution of this problem is to use steam locks in all such lines. All such conflicts encountered to date have been resolved, but not always easily or relatively inexpensively.

Community safety

Protection must be provided against potential ill effects of fermentation products (e.g. cytotoxins, allergens), organisms (e.g. pathogens, whether recombinant or not), fermentation by-products (e.g. pharmacologically active precursors of the active product), or some combination. The practice of providing such protection is called containment, which is defined as insuring that deleterious fermentation components can't be transported to any area inside or outside the plant before they have been rendered harmless. To accomplish this, containment must be exercised at several levels. We consider here only direct or 'primary' containment of fermenters; however, one should not lose sight of the fact that the other levels must be considered during fermenter design, if for no other reason than to ensure that the fermenter design and operation will be consistent with the overall containment strategy.

Finally, it is important to note that containment is nothing new. Highly pathogenic organisms have long been used to produce therapeutics and biological warfare components; hence, there is a considerable body of experience dealing with the subject. There also is considerable guidance to be had from the nuclear industry; nevertheless, the natures, sizes, and large number of new commercial processes, along with new methods, materials, and so on, will force heightened awareness, scrutiny, review, and modernisation or change.

Physical safety

There are also many local construction codes that must be satisfied. Among these are various earthquake-resistant construction codes that are also applicable to containment considerations. And then there are the requirements of the final arbiters: the insurance companies.

Product safety

For example, regulatory scrutiny has been far stricter for fermenters used to produce active molecules directly than for those used to produce precursors; there has been considerable variation from product to product within a given class; and manufacture of human drugs has been regulated much more rigidly than manufacture of animal drugs. While understandable to some extent, such nonuniformity has caused considerable confusion. These gaps are beginning to be closed, and this may have considerable influence on the design of fermenters—particularly those used to produce precursors and animal drugs.

Requirements that flow from the need to have control are based on relatively simple ideas:

1. If an organism is subject to environmental, medium, or other conditions outside the range in which it is known to yield product meeting acceptable specifications and capable of being converted/ purified consistently to final product meeting approved product specifications, then we have no guarantee that it does not produce other products with which the recovery system cannot cope and that can escape detection by the analytical methods in place.

2. If a fermenter becomes contaminated with other microbes, said microbes could produce toxins that could be carried undetected to the final product. It is possible that this could occur without altering the behaviour of the process organism. It also is possible that products of the contaminant could cause the process organism to make toxins that could go undetected into the final product. Given these possibilities, plus the fact that it is not possible to prove that the contaminant will always be the same, evidence of contamination is evidence for lack of control.

3. If a fermenter is not cleaned properly, deleterious microbial or nonmicrobial products could remain to contaminate the next batch in such a way that impurities could be carried undetected to the final product. Similar statements can be made about other contaminants introduced via other routes as a result of poor cleaning practices.

Sterilisation/aseptic operation translates to: (i) destroying any microbial contaminants that may be present in any part of the equipment that might contact process fluids, and (ii) insuring that no microbes (other than the production organism) can enter after the equipment has been sterilised. The latter embodies the concept of the 'sterile barrier'. To these ends, the following points should considers:

1. The fermenter and all its ports and direct attachments must be sterilisable.

2. All piping that will contact process fluids (including additives) and/or provide paths into the system (e.g. the air exhaust line) must be sterilisable initially. Some (e.g. sampling lines, addition lines) must also be sterilisable at any time during a fermentation.

3. Inlet gases (e.g. air, oxygen) and all additives (e.g. medium components, acid, base, antifoam) must be sterilised before they contact any sterile process piping.

4. All penetrations (e.g. drive shaft, probe ports) must be sterilisable.

The preferred sterilisation method is automatic sterilisation-in-place (SIP) with steam. Clean operation requires, among other things, the following:

1. The system must be designed such that any surface that can contact a process stream can be cleaned consistently to a level that insures that the product will be free of soils resulting from a fermentation.

2. Nothing in the system that comes in contact with process fluid can introduce unacceptable and unidentifiable materials.

Design Basis and Other General Considerations

A design basis for fermentation equipment should derive from a facility/process design basis. The latter should include (among other items) the general nature of the facility (e.g. research and development vs. production, single product vs. multiuse), product(s) specifications, regulatory, containment, and other requirements, level of automatic operation, general processing scheme (e.g. batch), nature of individual process steps; productivity, concentrations, and so forth; cleaning requirements; special considerations (e.g. earthquake-proof construction), critical valving and instrumentation, staffing requirements and constraints; architectural and general floor plan constraints; and utilities. Most of these will have some influence on fermenter design—some in more subtle ways than others.

It is difficult to overstate the importance of the initial definition that will derive from the design basis. Obviously, a rational design basis for a fermenter must also be based on fermentation characteristics as well as operating cycle and productivity required (which should derive from the overall process design basis). From these will flow the sizes and number of vessels, and definitions of oxygen transfer, heat transfer, power, and bulk mixing requirements. The design must satisfy these but must also satisfy requirements for regulatory compliance, safety, cleaning, facile operation, and maintenance. All of these factors are highly interactive (e.g. design for oxygen transfer affects design for cleaning); hence, responsible design will almost always require several iterations to ensure the greatest probability of success and to minimise lost time caused by installation, operation, and various problems, the iterative nature of the design (and, unfortunately, construction) process makes most important the existence of the well-crafted, well-implemented, and well-documented change control process. Finally, as with any engineering project, failure to have a solid basis of design, a complete scope, accurate process flow diagrams and piping and instrumentation diagrams (P&IDs), and accurate process timing will almost certainly result in added cost, lost time, and worse.

Power

Power delivered to the broth is used for micromixing and gas dispersion, which are related to mass transfer, and macromixing, which provides overall homogeneity. Agitated vessel power is delivered via two mechanisms: direct mechanical power from the impellers and gas expansion. The bulk (>90 per cent) of the power comes from the impellers as long as the fluid motion is under their control, a condition that prevails so long as the impellers are not flooded. There are reasonably reliable correlations available to determine flooding conditions for Newtonian broths, however, flooding usually is not a problem under typical conditions used in most Newtonian broth fementation. Flooding is more likely in highly viscous, pseudoplastic broths typical of mycelial and polysaccharide fermentations because the viscosity near the impeller is much lower than in the rest of the broth. As a result (assuming air is introduced under the impeller), air tends to channel toward the impeller, thereby enshrouding it and decreasing the deliverable power. This can dramatically decrease overall mass transfer (and heat transfer) rates and quality of bulk mixing quality.

Multiimpeller systems: Most fermenters are equipped with more than one impeller. The reasons for this are to improve bulk mixing and power distribution and to avoid the need for very large impellers and/or very high agitator speeds. Impeller size is limited practically not only by vessel internals but also by the following:

1. Torque transmitted to the drive shaft: The larger the torque, the stronger and thicker the shaft must be. This also translates to higher torque and more expensive gear boxes.

2. The size of the vessel manway: This is particularly important if the impellers must be single piece for better cleaning/sterilisation characteristics.

Agitator speed is limited by the natural frequency of the agitation system. This is because severe and potentially dangerous vibrations will occur if the rotational speed of the agitator approaches the natural frequency.

Organism sensitivity to fluid mechanical forces: Finally, an additional constraint must be imposed if the organism is sensitive to fluid mechanical forces. This is seldom true of unicellular organisms. There are, however, some mycelial organisms that are sensitive. The extent of sensitivity and the nature of the forces that cause damage should be determined experimentally. One also should keep in mind that the character of fluid forces changes significantly with scale. For example, some turbulent forces that are negligible at small scale can be large and potentially destructive at large scale.

Bulk mixing

Good bulk mixing is needed to insure homogeneity, which is required for reliable data acquisition and control. There are several important factors that affect mixing quality.

Type(s) of impeller(s): Impeller types fall into three basic categories: radial flow (e.g. Rushton turbine), axial flow (e.g. marine propellor), and mixed axial and radial flow (e.g. Lightnin A-315). Radial flow impellers tend to deliver the high power required to enhance micromixing and mass transfer but do not promote top-to-bottom mixing; therefore, they are not the best choice for promoting homogeneity. Multiple radial flow impellers often are used in an attempt to compensate for poor bulk mixing. Pure axial flow impellers do not deliver much power but do tend to promote good top-to-bottom mixing and hence contribute significantly to bulk homogeneity. They do not, however, contribute much to mass transfer. Some mixed flow types appear to provide a good balance between bulk mixing and mass transfer requirements, particularly for non-Newtonian broths. It is recommend that they be considered seriously for such cases. Thus, considerable success can be obtained with combination systems, for example, those with turbines as the lower impellers and a hydrofoil on top.

Impeller size(s): In most cases, large-diameter impellers distribute power better and promote bulk mixing better than do small-diameter impellers.

Baffling: Baffles are placed in vessels to minimise fluid swirling and vortex formation. Baffling tends to increase transmittable power and to improve mixing (except for the dead spots, which tend to form behind the baffles). Elimination of significant vortexing is important for safety as well as for improved bulk mixing.

Gas flow: Gas flow has a complex effect on bulk mixing. The flow alone does tends to promote bulk mixing (as in bubble tanks), but it also tends to decrease the effect of the impellers, particularly at high values of gas linear velocity. Fortunately, this is not a major problem in most cases, but it can be for very large fermenters and for highly viscous non-Newtonian broths.

There are several problems that can interfere seriously with a fermentation. Gas holdup decreases the effective volume of a fermenter. Foaming and aerosol formation can constrain operation, cause major cleaning and asepsis problems, and in the extreme cause termination of a fermentation. All three are dependent on broth characteristics, power input, and gas flow rate. There is some information in the literature, but these points have not been given the attention that even begins to reflect their importance. In general, all one can say is that all three become bigger problems for a given fermentation as gas flow and power input increase. Heat transfer is required during fermentation to maintain constant temperature conditions, and at other times (i.e. sterilisation, induction) to increase or decrease broth temperature. In most cases, cooling is required during most of an active, aerobic fermentation.

Heat transfer rate: The rate at which heat can be transferred is dependent on: (i) the driving force for heat transfer, (ii) the area across which transfer must occur, and (iii) the resistance to heat transfer.

Sterilisation

Sterilisation and aseptic operation taken together (as they must be) have a much greater influence on fermenter design and operation than does any other requirement.

In theory: (i) sterilisation destroys or removes all foreign organisms in all process equipment (including piping, seals, etc.) that might come into contact with the process fluid, and (ii) aseptic operation insures that no contaminating organisms enter the fermenter after sterilisation. These ideals are not attainable in practice because (among other reasons): (i) there is a finite probability that a very low concentration of contaminating organisms will not be captured in a sample used to test for contamination, and (ii) there is a finite probability of false positives due to sample contamination and so forth. It is also important to note that there is a continuing debate concerning the definitions of pure culture and sterility.

The driving forces for sterilisation and aseptic operation range from minimising product losses to insuring strict compliance with regulatory requirements.

Among the many specific problems that contamination can cause are the following:

1. Production of a toxin that can't be removed by the purification system.

2. Production of an enzyme that degrades the product.

3. Decreased product yield due to use of substrate by contaminants.

4. Production of toxins that inhibit the producer strain.

5. Production of compounds (e.g. polysaccharides) that interfere with the operation of recovery and purification equipment.

Which specific problems will exist and where in the spectrum a particular case will lie depend primarily on the product, its economic value, whether it is regulated, and how it will be used. Given all these variables and the fact that absolute sterility is an unachievable abstraction, the extents to which one should go should be considered on a case-by-case basis. As a practical matter, sterilisation has to be interpreted as 'effective sterilisation', meaning that the design and procedures (including sampling and detection) are suitable for the specific case. For example, sterilisation of fermenters used to produce parenterals should be held to a much higher standard than sterilisation of fermenters used to produce amylases for starch hydrolysis.

For batch sterilising a fermentation medium as part of the fermenter design process to ensure that we will have adequate heating and cooling capacity. It should also be done as part of the process design to ensure adequate timing.

Heating and cooling phases usually contribute only a small fraction of the total sterilisation kill, but they do affect turnaround time significantly. Long heating and cooling times can also have other effects:

1. Cause damage to the fermentation medium to the extent that the fermentation can be compromised. The seriousness of this depends on some combination of economics and regulatory compliance.

2. Cause medium changes that have negative effects on recovery and purification without causing fermentation problems. This includes the possibility of introducing foreign materials that may pass undetected into the final product.

That cooling time is considerably longer than heating time. This results primarily from the lower temperature-driving forces during cooling. The problem is exacerbated considerably when the heat transfer coefficient is very low as is the case for viscous Non-Newtonian broths. Another method used for batch sterilisation is direct steam injection: live steam is injected directly into the medium as the main source of thermal energy. This decreases heating time as well as the overall steam requirement.

It also increases the medium volume by about 20 per cent (as a result of steam condensation), which can cause some problems, including the following:

1. Increased cooling time.

2. Medium dilution: This will be a significant problem if the initial medium cannot be made concentrated enough to account for the dilution. Some reasons include low solubility of medium components, increased viscosity, and increased reaction rates among medium components at elevated temperatures.

3. Introduction of impurities: This depends primarily on the quality of the injected steam. In some cases plant steam is acceptable if boiler cleaning agents do not cause problems. At the other extreme is the requirement to use clean steam (WFI quality). It should be noted with regard to this point that some steam will be injected directly even when jacket heating is used as the main energy source; therefore, one will always be in the position of having to evaluate the effects of contaminants carried by the steam.

For cases in which sterilisation times are too long or medium alterations cause too many problems, one might consider continuous sterilisation.

Piping system sterilisation: Heating and cooling times are not usually significant issues for sterilising piping, however, there can be some serious heat transfer problems related to piping length, diameter, and orientation.

Cleaning

Scrupulous cleaning is necessary to decrease nonbiological contamination and prevent cross-contamination of batches. Experience also has shown that reliable sterilisation is difficult or impossible to achieve in the absence of rigorous cleaning. Developing and designing reliable fermenter cleaning systems is not as straightforward as it might appear, and considerable controversy continues. Among the reasons for this is that while much is known about the basic science of cleaning in general, very little is known about the basic science of cleaning fermenters and little has been published. The approach taken is based primarily on experience derived from the dairy, food, and beverage industries. Such information is useful but is not directly applicable in general to pharmaceutical and biotech processes where soils are different and the cleaning requirements are far more stringent. What meager information there is concerning pharmaceutical and biotech soils has been obtained from experiments done on single soil components and/or studies done under conditions not truly representative of the process conditions. There have been no significant attempts to develop systematic analyses based on experiments done with complex mixtures typical of real fermentation soils under conditions found in real processes. As a result there are no reliable general methods, tools, or correlations on which to base the development of cleaning agents, protocols and CIP system design, decisions tend to be made based on arbitrary criteria (e.g. coupon bake on studies). In keeping with our general design philosophy, we think it important that such arbitrariness be avoided and that cleaning protocols be developed along with the fermentation. It is clear that this will help to ensure not only proper design but will also minimise cleaning validation studies and any questions concerning the presence of contaminants and/or cleaning residues in the commercial product that could not have been present in clinical trial material. The selection of fermenter cleaning agents and protocols should be based not only on the specific soil but also on the materials of construction, surface finishes, and so on. In addition, one should consider the issues of: (i) compatibility of each material with the cleaning agents and protocols, and (ii) potential materials interaction during cleaning. These are decisions that usually are made during the design phase, but really should be defined much earlier.

Mechanical Design

Much of the mechanical design follows from the results of calculations discussed already in this chapter; however, these must be tempered by other considerations that should be included in the process and/or facility design basis. Some of these (e.g. regulatory compliance, containment) have already been discussed and are revisited in greater detail in this section. Other items that should be included and will affect fermenter design include the following:

1. Extent of automation, nature of plant wide control system, etc.: Such items affect not only the instrumentation (beyond our scope) and similar factors, but also valving, piping, vessel ports, and so forth. Among some of the major issues here are identification of critical (as defined by cGMP requirements) valves and other components. It must be kept in mind that failure to identify critical instrumentation and control components frequently leads to very complex valving systems in which the failure of a single switch can bring everything to a grinding halt. Unfortunately, there are too many people who do not think this through (as a system) before final design begins.

2. Staff requirements and the anticipated nature of the operating staff: This can influence the complexity and physical layout of (among other things) the piping system(s).

3. Plant location: This will have some affect on overall design and component selection if for no other than service issues.

4. Available utilities: This can influence such things as the designs of the cooling and aeration systems.

5. Architectural constraints: These can have profound effects on the ways in which process requirements will be satisfied. Floor space and ceiling height constraints often require that a fermenter be designed in such a way that it violates some or most or even all of the rules of thumb mentioned in the previous section. This is not good but is better than building a vessel that does not fit into the plant.

6. Transportation constraints: The fermenter has to be moved from the fabricator's shop to the plant — at a cost not greater than the total project budget. This may require some thumb bending or breaking. In extreme cases it may be better to fabricate the fermenter on site; however, this can carry large penalties, particularly in licensed facilities.

7. Maintenance: Many design decisions can have major effects on the ease of maintenance. It is also often true that some of the design features that ease maintenance are more costly than those that do not. In most cases, the savings in capital expenditure will not come near paying for losses that will result later became of maintenance problems. This is particularly true in licensed facilities where regulators view good, facile maintenance as an integral, indispensible part of cGMP operation.

8. Definition of standards: Standards for welds, finishes, and so on should be standardised for all parts of the equipment that will contact process fluid. It does not, for example, make sense to call for a high-quality vessel finish (e.g. 320 grit, EP) and at the same time accept unpolished tubing and valves in inoculation and medium addition lines.

Finally, there are a few recurring themes that are encountered in fermenter design:

1. Building-in 'versatility': The types of versatility desired range from wanting a fermenter that can operate well over a very wide range of conditions but with a narrow range of organisms, to wanting a convertible bioreactor that can handle microbes, mammalian cells, and may be (some day) transgenic animals. Most people understand the value of versatility, but many do not understand the attendant problems and costs. As a general rule, convening laboratory glassware directly into large, stainless steel equivalents usually costs more than any rational person should be willing to pay. Also as a general rule, versatility should decrease as a process goes from the lab to the production floor. Versatility in the laboratory is almost a requirement because change is in the nature of laboratory work. Versatility on the plant floor, however, leads to more procedures, more paper work, more testing, and more confusion, particularly when equipment modifications are required to achieve the versatility; change is not in the basic nature of most plant work. Equipment capital savings can justify the added costs of all the preceding for many cases in which the equipment is designed to handle multiple, similar fermentations on a campaign basis and without significant equipment modification. This statement becomes less true as the fermentations become less similar, as more modifications are necessary, and as regulatory scrutiny increases. Bottom line: analyse very carefully any inclination to want versatility built into plant equipment (or even pilot equipment, in some cases).

2. Retrofitting existing equipment: The usual thinking is that capital and time savings can be had by refurbishing 'old faithful'. Just how true this is depends not only on the condition of the existing equipment but also on the intended application(s) of the reborn version. Comments similar to the ones made for versatility apply here. The chances of success decrease as one goes from a lab to licensed production facility. There are ample, expensive corpses to prove this point.

Vessel design

Materials of construction: The major choices that must be made are: (i) the type of metal to be used for the vessel and nozzles, and (ii) the type of elastomer to be used for static seals. The selections should be based on compatibility with the organism, compatibility with the product, corrosion resistance, cleanability (also related to finish), welding characteristics, and cost and durability. All of these should be determined during process development but seldom are. In almost all cases, the metal selected will be some grade of stainless steel (usually SS304, SS304L, SS316, or SS316L). The choice is usually associated with the nature and value of the product, although some of the other factors noted earlier may be considered. SS304 is usually good enough for lower-value, unlicensed products, whereas SS316L is the material of choice for high-value, licensed products. L-grade is selected when better corrosion resistance and good multipass welding characteristics are required; it adds about 15 per cent to the cost of the vessel.

Finally, one should be aware of that some O-ring/gasket-forming processes can leave very small quantities of metals in the seals. These may not be enough to cause fermentation or product problems *per se*, but they can be the cause of considerable corrosion.

Nozzle: Ports for additions and probes must be designed to be sterilisable, and cleanable. Among other things, this means they must have reliable seals and be free draining. The major debate here usually focuses on the choice between Ingold ports and those designed for sanitary clamp connection.

Baffles: Baffles are usually required in high-power systems to prevent swirling and vortexing, thereby increasing the power that can be delivered to the fluid. The usual practice is to use four baffles on 90°

centres welded directly to the wall. Each baffle should have a width equal to 10 per cent of the tank diameter and should have long slots cut out of the edge facing the wall so as to prevent solids build up and to make cleaning easier. Removable baffles are used in some cases; however, this practice, is discouraged particularly in cases where cleaning is a critical issue, simply because unsealed joints resulting from baffle removal make cleaning more difficult.

Jackets: Several types of jacket are used on fermenters; the choice of type is usually not critical for heat transfer purposes and is best left to the vessel fabricater. In some cases, however, a jacket type (e.g. half pipe) may be chosen to increase the vessel pressure rating.

The following have proven to be useful practices:

1. The jacket should extend from the probe ring (about 2 inches above the bottom tangent line) to the top tangent line and should be zoned. This allows additional active surface area to come in contact with the broth as the volume increase due to additions and to increasing gas hold-up, minimising cooling-loop pressure drop, and minimising medium bake-on.
2. Connections to the jacket should be via sanitary clamps or flanges, not by screwed fittings. This minimises possible damage to jacket welds when cooling lines are connected or disconnected.
3. Coolant should be filtered to avoid solids buildup and/or jacket fouling.
5. Jacket coving to accommodate view ports, decreases jacket area and effectiveness and increases cost considerably. It should be avoided.

Internal cooling surfaces: Coils are the most common internal cooling surfaces, although there are other types. They can easily double the available heat transfer area and tend to be more effective than jackets, however, they can have big disadvantages:

1. They make cleaning very difficult.
2. They can cause additional bulk mixing problems, particularly for very viscous non-Newtonian broths.
3. They will eventually leak nonsterile coolant into the broth.
4. They can add up to 25 per cent to the cost of the vessel.

If you absolutely, positively must use a coil, there are several points to remember:

1. Mount it in a way that will insure minimum stress during heat-up and cooldown.
2. Weld cladding over the butt welds used to join the coil pipe sections.
3. Leak test after construction and build in design features that will simplify leak testing on a regular basis thereafter.
4. Space coil turns at least 3 inches apart. Anything closer will insure major cleaning problems.

Spargers: There has been a lot of discussion concerning the pros and cons of ring and single-orifice spargers, and the details of design of each. We have seen both work well and have not found any evidence for the validity of claims concerning the importance of hole size (for example) for oxygen transfer *per se*. The primary focus should he on gas distribution (which will depend on other aspects of the agitation system) and on aseptic operation. We usually favour single-orifice spargers. There also have been advocates of porous (frits) spargers. The rationale presented has focused primarily on the small bubble size such spargers produce. There is some basis for these claims in cases where little mechanical energy is available for bubble break-up (e.g. in mammalian cell systems). There might also be some value in cases for which bubble coalescence is not a problem (far and few between in practical systems). They can be extremely difficult to clean, particularly for mycelial organisms.

Piping and valving

Design and construction details of the piping system depend on process, sterility, cleaning, and containment requirements, taken together.

Service lines for non-process-contacting fluids: All lines providing fluids that cannot contact process fluid surfaces (e.g. coolant lines, plant steam lines for heating only) fall into this category. Satisfactory service is provided by either copper or stainless steel piping along with a rational combination of welded and compression fittings. Durability, serviceability, cost, and corrosion resistance are major considerations. Ball valves are satisfactory on lines not turned on and off frequently (diaphragm valves tend to withstand a greater number of on-off cycles prior to failure).

Sterile piping systems: Before we discuss design of sterile piping systems, we reiterate the importance of a systems approach to integrating vessel and piping design. Independent design almost always results in a lot of aggravation, as well as higher cost and lost time.

General principles

There are several general principles that should be applied to all sterile piping:
1. Make all piping system components free draining. This requires special attention to the details of pipe pitches, valve orientations, and so forth.
2. Design all components and the piping layout so as to eliminate nooks and crannies where contamination (biological and nonbiological) can hide so as to escape cleaning and/or sterilisation. Frequently overlooked problems include such things as deadlegs and ridges formed by welding operations.
3. Make steam and condensate piping at least 3/8 inches diameter to insure proper steam flow and condensate draining.
4. Make all piping lengths as short as possible without interfering with good fabrication practices (GFP), operability, and maintenance.
5. Use check valves only when no other solution is possible (e.g. in overpressurised lines).
6. Do not use sight glasses in condensate drain lines except in the seal lubricant drain line.
7. Pay careful attention to piping orientation to avoid the possibility of air traps and inadequate heating.
8. Make sure that bottom drain valves are flush-mounted diaphragm valves capable of being steamed in the closed position. They must be free draining.
9. Avoid dip tubes unless there is no other way.

The extent of condensate problems during sterilisation can be remedied best by ensuring that whatever condensate does form can drain freely. This can be complicated by the fact that it is undesirable to have the sterile side of the filter connected directly to the drain line. Another approach is to steam heat the filter housing such that condensation cannot occur. This is effective but has the disadvantages of adding cost and decreasing filter life. Other factors that can affect condensate problems are the positioning and orientation of the housing. There are differing opinions concerning these; one is best advised to consider the advice the filter and the fermenter vendors for specific cases. Plugging of the exhaust filter by condensate and aerosols during fermentation require special consideration. Air leaving the fermenter will be essentially saturated with water vapour at fermentation temperature. The exhaust line, filter, and so on usually are colder than the fermenter, therefore, condensation is inevitable. The amount of condensation will depend on temperature differences, air flow rate, and the nature of the surfaces of the components in the exhaust line (the maximum potential is easily calculable).

Approaches that have been used to deal with these problems include:

1. Condensers.
2. Heat exchangers; used before the exhaust filter to avoid condensation from the fermenter off gases and after the pressure control valve to prevent reflux from the exterior exhaust line.
3. Steam-heated filter housings.
4. Heated exhaust lines.
5. Coalescers.
6. All of the above.

Agitation systems

Agitation can be done by direct mechanical coupling of the shaft to the drive or by magnetic coupling. The latter has been recommended for cases where high levels of containment are required. There also is the perception that magnetic drives are more suitable for maintaining more stringent levels of asepsis and cleanliness than is possible with direct drive. Neither claim has any substantive basis; indeed, there is good reason to believe that existing magnetic drives may present greater cleaning difficulties because of the manner in which the driven magnet must be mounted at the bottom of the vessel. In addition, power transfer by magnetic drive is quite low. Direct drive is the predominant current choice at almost any scale of operation. The major arguments for bottom drive are ease of maintenance, shorter shafts, less support structure, and lower overall height.

The arguments against bottom (and for top drive) focus primarily on the potential of catastrophic spills resulting from bottom seal failure, seal grinding as a result of broth particulates working into a bottom seal, and greater cleaning difficulties. If the seals are designed and maintained properly, none of these is a problem. There is no real difference in aseptic operability between top and bottom drives. The choice probably will continue to be driven primarily by personal preference.

There are very few cases in which double mechanical seals are not (or should not) be used in fermenters. The major debate focuses on seal orientation and the details of individual seal designs. There are basically two orientations used: inline and back-to-back. Inline design is recommended because it is found to operate more cleanly and require a simpler sterilising/lubricant system. Seal lubrication is usually provided by means of sterile steam condensate. It is extremely important that this condensate be free of particulates: their presence guarantees rapid seal failure, contaminated fermentations, and a hyperactive maintenance programme. It is also important to note that during sterilisation live steam flows through the seal housing.

There are some who insist on keeping the steam flowing throughout the fermentation: (Obviously, they have no faith in the seals.) The one thing this will guarantee is much more rapid wearing of the seals (perhaps supporting the lack of faith in the seals). One must also decide on the means for controlling lubricant flow rate. Most use a valve for this purpose.

It is suggest to use an orifice sized to deliver the proper flow. This avoids the cost and maintenance of a valve and insures fiddle-proof operation. It does, however, require the use of particulate-free condensate, but then so does proper operation of the seals. It is also recommend to include a sight glass in the lubricant drain line as well as a seal leak detector.

Other more complex detection systems might be considered for specific circumstances (e.g. severe containment requirements). Finally, preventing shaft vibration is another important factor in agitation system design. Shaft vibration is a safety hazard. It will also cause premature seal failure and other costly mechanical damage.

Cleaning systems

As noted earlier, specific requirements for cleaning depend on several factors including the nature of the fermentation broth. There is a major difference between most microbial broths and most mycelial broths. Single-cell microbial broths, with the exception of those containing a lot of undissolved particulates and high viscosity components (e.g. xanthan broths), tend to be free draining and readily amenable to cleaning based primarily on the physico-chemical action of the cleaning agents. Many fungi and mycelial bacteria, on the other hand, tend to cling to fermenter internals and may require mechanical action (e.g. high-velocity jets) in addition to cleaning agents. The following guidelines are applicable for most systems. These principles must be applied in light of the actual cleaning agents and protocols to be used:

1. Eliminate internals and nooks and crannies to the greatest extent possible in the vessel and throughout the piping system. This practice is consistent with design for aseptic operation.

2. Drill and position spray balls to insure complete coverage of all surfaces inside the vessel. This usually requires an empirical approach. Coverage can be tested by means of the riboflavin test. The reader is cautioned, however, that complete coverage is a necessary but not sufficient condition for cleaning. Also, please note the following: (i) spray balls designed for sanitary operation are self-draining and self-cleaning. They can be sterilised *in situ*. There is, however, considerable debate concerning whether they should be removed prior to fermentation and (ii) high-velocity, rotating devices that may be necessary when large clumps of sticky residue must be removed (e.g. as with a fungus) are not designed to be inherently self-cleaning, self-draining, or sterilisable.

3. Eliminate deadlegs in piping and deadspaces in valves, fittings, and other system components. This is also consistent with design for aseptic operation.

4. Specify the same materials and finishes for process and CIP piping as are specified for the fermenter. Obviously, these must be compatible with the cleansers and conditions used.

5. Avoid threaded joints: A completely welded system is best, but compression fittings can be satisfactory when cost is a major consideration, particularly for nonregulated products.

6. Make CIP piping as simple as possible. For example, make dual use of process and other piping as much as possible.

7. Avoid complex, expensive transfer panels wherever possible. Use swing elbows wherever practical.

8. Design and construct the system to facilitate validation and on-going testing for removal of contaminants, including the cleaning agents. This design should provide for swabbing, obtaining rinse samples, or whatever else the cleaning protocol requires.

There are many cases for which a portable CIP system is preferable to an integral system. For such cases, the fermenter and the portable unit should be designed to allow simple mechanical attachment of the necessary hoses between the two units and with utilities, and a straight forward means for interfacing the instrumentation, logging data, and controlling the systems of the two units.

To sum up, the classical CIP systems rely on continuous cleanser flow and the maintenance of a shallow puddle in the bottom of the fermenter. To achieve this they rely on special pumping devices such as eductors. Aside from the design, control, and other operating problems this causes, it has been observed in some facilities that the stable pool leads to the formation of 'cleaning rings' at the bottom of the fermenter. These rings may not be real problems (additional evidence is still required), but they do cause perception problems. It is suggest to use a pulsed-flow system to overcome not only the need for special pumping devices but also the 'cleaning ring' problem. Such pulsed systems have been found to accomplish both objectives in practice.

Chapter 15

Designing Parameters of Fermentor

INTRODUCTION

The function of the fermenter or bioreactor is to provide a suitable environment in which an organism can efficiently produce a target product—the target product might be cell biomass, metabolite and bioconversion product. It must be so designed that it is able to provide the optimum environments or conditions that will allow supporting the growth of the micro-organisms. The design and mode of operation of a fermenter mainly depends on the production organism, the optimal operating condition required for target product formation, product value and scale of production.

The choice of micro-organisms is diverse to be used in the fermentation studies. Bacteria, unicellular fungi, virus, algal cells have all been cultivated in fermenters. Now more and more attempts are tried to cultivate single plant and animal cells in fermenters. It is very important for us to know the physical and physiological characteristics of the type of cells which we use in the fermentation.

Before designing the vessel, the fermentation vessel must fulfill certain requirements that is needed that will ensure the fermentation process will occur efficiently. Some of the actuated parameters are: the agitation speed, the aeration rate, the heating intensity or cooling rate, and the nutrients feeding rate, acid or base valve. Precise environmental control is of considerable interest in fermentations since oscillations may lower the system efficiency, increase the plasmid instability and produce undesirable end products.

This chapter discusses various parameters of fermentation to enhance its productivity.

FERMENTATION TECHNOLOGY

Fermentation technology could be defined simply as the study of the fermentation process, techniques and its application. Fermentation should not be seen merely as a process that is entirely focused on the happenings occurring in the fermenter alone! There are many activities that occur upstream leading to the reactions that occur within the bioreactor or fermenter, despite the fermenter is regarded as the heart of the fermentation process. Fermentation technology is the whole field of study which involves studying, controlling and optimisation of the fermentation process right up from upstream activities, mid stream and downstream or post fermentation activities.

The study of fermentation technology requires essential inputs from various disciplines such as biochemistry, microbiology, genetics, chemical and bioprocess engineering and even a scatter of mathematics and physics.

Fermentation in Terms of Biochemistry and Physiology

Fermentation is now defined as a process of energy generation by various organisms especially micro-organisms. The fermentation process showed unique characteristics by which it generates energy in the absence of oxygen. The process of energy generation utilises the use of substrate level phosphorylation (SLP) which do not involved the use of electron transport chain and free oxygen as the terminal electron acceptor.

Engineering Definition of Fermentation

It is only up to recently with the rise of industrial microbiology and biotechnology that the definition of fermentation took a less specific meaning. Fermentation is defined more from the point of view of engineers. They see fermentation as the cultivation of high amount of micro-organisms and biotransformation being carried out in special vessels called fermenter or bioreactors. Their definitions make no attempt to differentiate whether the process is aerobic or anaerobic. Neither are they bothered whether it involves micro-organisms or single animal or plant cells. They view bioreactors as a vessel which is designed and built to support high concentration of cells.

Bioreactor

Bioreactor is also known as fermenter. A bioreactor is a specially designed vessel which is built to support the growth of high concentration of micro-organisms. It must be so designed that it is able to provide the optimum environments or conditions that will allow supporting the growth of the micro-organisms.

Bioreactors are commonly cylindrical vessels with hemispherical top and/or bottom, ranging in size from some liter to cube meters, and are often made of stainless steel and glass. The difference between a bioreactor and a typical composting system is that more parameters of the composting process can be measured and controlled in bioreactors. The sizes of the bioreactor can vary over several orders of magnitudes. The microbial cell (few mm^3), shake flask (100–1000 ml), laboratory fermenter (1–50 L), pilot scale (0.3–10 m^3) to plant scale (2–500 m^3) are all examples of bioreactors.

The design and mode of operation of a fermenter mainly depends on the production organism, the optimal operating condition required for target product formation, product value and scale of production. The design also takes into consideration the capital investment and running cost.

- Large volume and low value products like alcoholic beverages need simple fermenter and do not need aseptic condition.

- High value and low volume products require more elaborate system of operation and aseptic condition.

Bioreactors differ from conventional chemical reactors in that they support and control biological entities. As such, bioreactor systems must be designed to provide a higher degree of control over process upsets and contaminations, since the organisms are more sensitive and less stable than chemicals. biological organisms, by their nature, will mutate, which may alter the biochemistry of the bioreaction or the physical properties of the organism. Analogous to heterogeneous catalysis, deactivation or mortality occur and promoters or coenzymes influence the kinetics of the bioreaction. Although the majority of fundamental bioreactor engineering and design issues are similar, maintaining the desired biological activity and eliminating or minimising undesired activities often presents a greater challenge than traditional chemical reactors typically require. Other key differences between chemical reactors and bioreactors are selectivity and rate. In bioreactors, higher selectivity — that is, the measure of the system's capability for producing the preferred product (over other outcomes) — is of primary importance. In fact, selectivity is especially important in the production of relatively complex molecules such as antibiotics, steroids, vitamins, proteins and certain sugars and organic acids. Frequently, the activity and desired selectivity occur in a substantially smaller range of conditions than are present in conventional chemical reactors. Further, deactivation of the biomass often poses more severe consequences than a chemical upset.

The designing of a bioreactor also has to take into considerations the unique aspects of biological processes:

- The concentrations of starting materials (substrates) and products in the reaction mixture are frequently low; both the substrates and the products may inhibit the process. Cell growth, the structure of intracellular enzymes, and product formation depend on the nutritional needs of the cell (salts, oxygen) and on the maintenance of optimum biological conditions (temperature, concentration of reactants, and pH) within narrow limits.

- Certain substances inhibitors effectors, precursors, metabolic products influence the rate and the mechanism of the reactions and intracellular regulation.

- Micro-organisms can metabolise unconventional or even contaminated raw materials (cellulose, molasses, mineral oil, starch, wastewater, exhaust air, biogenic waste), a process which is frequently carried out in highly viscous, non-Newtonian media.

- In contrast to isolated enzymes or chemical catalysts, micro-organisms adapt the structure and activity of their enzymes to the process conditions, whereby selectivity and productivity can change. Mutations of the micro-organisms can occur under sub optimal biological conditions.

- Micro-organisms are frequently sensitive to strong shear stress and to thermal and chemical influences.

- Reactions generally occur in gas-liquid -solid systems, the liquid phase usually being aqueous.

- The microbial mass can increase as biochemical conversion progresses. Effects such as growth on the walls, flocculation, or autolysis of micro-organisms can occur during the reaction.

- Continuous bioreactors often exhibit complicated dynamic behaviour.

Requirements of bioreactors

Due to above mentioned demands made by biological systems on their environment, there is no universal bioreactor. However, the general requirements of the bioreactor are given below:

1. The vessel should be robust and strong enough to withstand the various treatments required such as exposure to high heat, pressure and strong chemicals and washings and cleanings.
2. The vessel should be able to be sterilised and to maintain stringent aseptic conditions over long periods of the actual fermentation process.
3. The vessel should be equipped with stirrers or mixers to ensure mass transfer processes occur efficiently.
4. It should have sensors to monitor and control the fermentation process.
5. It should be provided with inoculation point for aseptic transfer in inoculum.
6. Sampling valve for withdrawing a sample for different tests.
7. Baffles should be provided in case of stirred fermenter to prevent vertex formation.

8. It should be provided with facility for intermittent addition of an antifoam agent.

9. In case of aerobic submerged fermentation, the tank should be equipped with the aerating device.

10. Provision for controlling temperature and pH of fermentation medium.

11. Man hole should be provided at the top for access inside the fermenter for different purposes.

It is obvious that the design of the fermenter will involve co-operation between experts in microbiology, biochemistry, chemical engineering, mechanical engineering and costing.

Fermenter Design

The basic points of consideration while designing a fermenter:

- Productivity and yield.
- Fermenter operability and reliability.
- Product purification.
- Water management.
- Energy requirements.
- Waste treatment.

Few significant things of concern that should be taken into account while designing a fermenter:

- Design in features so that process control will be possible over reasonable ranges of process variables.
- Operation should be reliable.
- Operation should be contamination free.
- Traditional design is open cylindrical or rectangular vessels made from wood or stone.
- Most fermentation is now performed in close system to avoid contamination.
- Since the fermenter has to withstand repeated sterilisation and cleaning, it should be constructed from non-toxic, corrosion-resistant materials.
- Small fermentation vessels of a few liters capacity are constructed from glass and/or stainless steel.
- Pilot scale and many production vessels are normally made of stainless steel with polished internal surfaces.
- Very large fermenter is often constructed from mild steel lined with glass or plastic, in order to reduce the cost.
- If aseptic operation is required, all associated pipelines transporting air, inoculum and nutrients for the fermentation need to be sterilisable, usually by steam.
- Most vessel cleaning operations are now automated using spray jets, which are located within the vessels. They efficiently disperse cleaning fluids and this cleaning mechanism is referred to as cleaning-in-place (CIP).
- Associated pipe work must also be designed to reduce the risk of microbial contamination. There should be no horizontal pipes or unnecessary joints and dead stagnant spaces where material can accumulate; otherwise this may lead to ineffective sterilisation. Overlapping joints are unacceptable and flanged connections should be avoided as vibration and thermal expansion can result in loosening of the joints to allow ingress of microbial contaminants. Butt welded joints with polished inner surfaces are preferred.
- Normally, fermenters up to 1000 L capacity have an external jacket, and larger vessels have internal coils. Both provide a mechanism for vessel sterilisation and temperature control during the fermentation.

- Other features that must be incorporated are pressure gauges and safety pressure valves, which are required during sterilisation and operation. The safety valves prevent excess pressurisation, thus reducing potential safety risks. They are usually in the form of a metal foil disc held in a holder set into the wall of the fermenter. These discs burst at a specified pressure and present a much lower contamination risk than spring-loaded valves.

- For transfer of media pumps are used. However pumps should be avoided if aseptic operation is required, as they can be a major source of contamination. Centrifugal pumps may be used, but their seals are potential routes for contamination. These pumps generate high shear forces and are not suitable for pumping suspensions of shear sensitive cells. Other pumps used include magnetically coupled, jet and peristaltic pumps.

- Alternate methods of liquid transfer are gravity feeding or vessel pressurisation.

- In fermentations operating at high temperatures or containing volatile compounds, a sterilisable condenser may be required to prevent evaporation loss. For safety reasons, it is particularly important to contain any aerosols generated within the fermenter by filtersterilising the exhaust gases.

- Also, fermenters are often operated under positive pressure to prevent entry of contaminants.

CONSIDERATIONS THAT IMPROVE PRODUCTIVITY OF FERMENTER

Material of Construction

Laboratory scale fermenter

In fermentation with strict aseptic requirements it is important to select materials that can withstand repeated sterilisation cycles. On a small scale, it is possible to use glass and/or stainless steel. Glass is useful because it gives smooth surfaces, is non-toxic, corrosion proof and it is usually easy to examine the interior of vessel. The glass should be 100% borosilicate.

The following variants of the laboratory bioreactor can be made:

1. Glass bioreactor (without the jacket) with an upper stainless steel lid.
2. Glass bioreactor (with the jacket) with an upper stainless steel lid.
3. Glass bioreactor (without the jacket) with the upper and lower stainless steel lids.
4. Two-part bioreactor - glass/stainless steel. The stainless steel part has a jacket and ports for electrodes installation.
5. Stainless steel bioreactor with peepholes.

Vessels with two stainless steel plates cost approximately 50% more than those with just a top plate.

Pilot scale and large scale bioreactors

When all bioreactors are sterilised *in situ*, any materials use will have to assess on their ability to withstand pressure sterilisation and corrosion and their potential toxicity and cost. Pilot scale and large scale vessels are normally constructed of stainless steel or at least have a stainless steel cladding to limit corrosion. The American Iron and Steel Institute (AISI) states that steels containing less than 4% chromium are classified as steel alloys and those containing more than 4% are classified as stainless steel. Mild steel coated with glass or phenolic epoxy materials has occasionally been used. Wood, concrete and plastic have been used when contamination was not a problem in a process. Although stainless steel is often quoted as the only satisfactory material, it has been reported that mild-steel vessels were very satisfactory after 12 years use for penicillin fermentations and mild steel clad with stainless steel has been used for at least 25 years for acetone-butanol production. The corrosion resistance of stainless steel is thought to depend on the existence of a thin hydrous oxide film on the surface of metal. The composition of this film varies with different steel alloys and different manufacturing process treatment. The film is stabilised by chromium and is considered to be continuous, non-porous, insoluble and self healing. If damaged, the film will repair itself when exposed to air or an oxidising agent.

The minimum amount of chromium needed to resist corrosion will depend on the corroding agent in a particular environment, such as acid, alkalis, gases, soil, salt or fresh water. Increasing the chromium concentration enhances the resistance to corrosion, but only grades of steel containing at least 10 to 13% chromium develop the effective film. The inclusion of nickel in high per cent chromium steels enhances resistance and improves their engineering properties. The presence of molybdenum improves the resistance of stainless steels to solution of halogens salts and pitting by chloride ions in brine or sea water. Corrosion resistance can also be improved by tungsten, silicon and other elements. AISI grade 316 steels which contains 18% chromium, 10% nickel, 2–2.5% molybdenum are now commonly used for fermenter or bioreactor construction.

In citric acid fermentation where pH may be 1 to 2, it will be necessary to use a stainless steel with 3–4% molybdenum (AISI grade 317) to prevent leaching of heavy metals from the steel which would interfere with the fermentation. AISI grade 304, which contains 18.5% chromium and 10% nickel, is used extensively for brewing equipment. Now also Stainless steels (e.g. 1.4435, 1.4539, etc.) Hastelloy, Incolloy, Inconel, Monel, Titanium grades 1, 2, 7, 11 are used in construction of bioreactor. With plant and animal cell tissue culture, a low-carbon version (type 316L) is often used.

The thickness of the construction material will increase with scale. At 300,000 to 400,000 dm^3 capacity, 7-mm plate may be used for the side of the vessel and 10-mm plate for the top and bottom, which should be hemispherical to withstand pressure. It is also important to consider the ways in which a reliable aseptic seal is made between glass and glass, glass and metal, metal and metal joints such as between a fermenter vessel and a detachable top or base plate. With glass and metal a seal can be made with a compressible gasket, a lip seal or an 'O' ring. With metal to metal joints only 'O' ring is suitable.

Vessel Shape

Typical tanks are vertical cylinders with specialised top plates and bottom plates. In some cases, vessel design eliminates the need for a stirrer system especially in air lift fermenter. A tall, thin vessel is the best shape with aspect ratio (height to diameter ratio) around 10:1. Sometimes a conical section is used in the top part of the vessel to give the widest possible area for gas exchange.

Stainless steel top plates

The top plates are of an elliptical or spherical dish shape. The top plates can be either removable or welded. A removable top plate provides best accessibility, but adds to cost and complexity. Various ports and standard nozzles are provided on the stainless plate for actuators and probes. These include pH, thermocouple, and dissolved oxygen probes ports, defaming, acid and base ports, inoculum port, pipe for sparging process air, agitator shaft and spare ports.

Bottom plates

Tank bottom plates are also customised for specific applications. Almost most of the large vessels have a dish bottom, while the smaller vessels are often conical in shape or may have a smaller, sump type chamber located at the base of the main tank. These alternate bottom shapes aid in fluid management when the volume in the tank is low. In all cases, it is imperative that tank should be fully drainable to recover product and to aid in cleaning of the vessel. Often this is accomplished by using a tank bottom valve positioned to eliminate any 'dead section' that could arises from drain lines and to assure that all content will be removed from the tank upon draining.

If the bioreactor has a lower cover, then the following ports and elements should be placed and fastened there:

1. Discharge valve.
2. Sampling device.
3. Sparger.
4. Mixer's lower drive.
5. Heaters.

Height-to-diameter ratio (aspect ratio)

The height-to-diameter ratio is also a critical factor in vessel design. Although a symmetrical vessel maximises the volume per material used and results in a height-to-diameter ratio of one, most vessels are designed with higher ratio. The range of 2–3:1 is more appropriate and in some situation, where stratification of the tank content is not an issue or a mixer is used, will allow still higher ratio to be used in design. The vessels for microbiological work should have an aspect ratio of 2.5–3:1, while vessels for animal cell culture tend to have an aspect ratio closer to 1. The basic configuration of stirred tank bioreactors for mammalian cell culture is similar to that of microbial fermenter but the major difference is there in aspect ratio, which is usually smaller in mammalian cell culture bioreactor.

In stirred tank bioreactor (STR), height to diameter aspect ratio is 3:1 or 4:1 while in the case of CSTR, the aspect ratio is maintained more than 1, to ensure high residence time of gas phase, increase the transfer efficiency and to ensure less power input on introduction of gas, uniform power dissipation. In miniature bioreactor, aspect ratio is kept equal to large bioreactor in order to predict hydrostatic pressure and therefore oxygen solubility at the different scale of operation. The tower fermenter is an elongated non-mechanically stirred fermenter with aspect ratio of at least 6:1 for tubular section or 10:1 overall, through which there is unidirectional flow of gases. The tower fermenter used for citric acid production on a laboratory scale having height: diameter ratio of 16:1. Cylindro-conical vessels used for the brewing of lagers and beers having aspect ratio usually 3:1, with fermenter heights around 10 to 20 m.

Agitation

Agitator (impeller)

The agitator is required to achieve a number of mixing objective.

- Bulk fluid and gas-phase mixing.
- Air dispersion.
- Oxygen transfer.
- Heat transfer.
- Suspension of solid particles and maintain a uniform environment throughout the vessel contents.
- Enhancement of mass transfer between dispersed phases.

Bulk mixing and micro mixing both are influenced strongly by impeller type, broth rheology, and tank geometry and internals. Rushton disc turbines, vaned discs, open turbines of variable pitch and propeller. The disc turbine consists of a disc with a series of rectangular vanes set in vertical plane around the circumstances and vaned disc has a series of rectangular vanes attached vertically to the underside. Air from the sparger hits the underside of the disc and is displaced towards the vanes where the air bubbles are broken up into smaller bubbles. The vanes of variable pitch open turbine and the blade of marine impellers are attached directly to a boss on the agitator shaft. In this case air bubbles do not initially hit any surface before dispersion by the vanes or blades. The propeller and the open turbine flood when superficial velocity (Vs) exceeds 21 m h^{-1}, whereas the flat blade turbine can tolerate Vs of 120 m h^{-1} before being flooded, when two sets are used on the same shaft. Besides this, propeller is also less efficient in breaking up the bubbles and the flow it produces is axial rather than the radial. One of major drawback of Rushton disc turbine is that it provides very axial flow, resulting in poor overall top-to-bottom mixing. In addition, agitation intensity decrease with distance from the impeller, and this decrease can become more pronounced for viscous, pseudoplastic broths.

Various impellers use in bioreactors with their flow patterns

- Flat blade disk turbine.
- 45° Flat blade disk turbine.
- Curved blade disk turbine.
- Pitched blade turbine.
- Curved blade turbine.
- Large pitch blade impeller.
- Intermig.
- 3 segment blade impeller.
- Gate with turbine.
- Maxblend.
- Helical ribbon.

The axial flow hydrofoil impellers have become increasingly popular. These axial flow systems can pump liquid either down or up. They have been shown to give superior performance (compare to Rushton radial flow impellers) with respect to lower energy demands for the same level oxygen transfer. Further, they show reduced maximum shear rates, making them usable with sensitive cultures such as animal cell culture, while still being capable of giving excellent performance with viscous mycelial fermentation. Combination of axial flow and radial flow impeller systems are sometimes used.

One way to improve bulk mixing while maintaining good oxygen transfer rate is to use Rushtonturbine at lower position in tank and an axial flow impeller at the top. The designer can help to minimise top-to-bottom mixing problem by keeping the ratio of liquid height to tank diameter under 2, and by spacing the impeller properly. If the impellers are placed too closely, they interfere with each other, thereby decreasing OTR and mixing quality. If they are spaced too far apart, overall homogeneity suffers. Researchers have found that spacing between 1 and 1.5 impeller diameters gives good mixing in most practical cases, and oxygen transfer is not affected significantly by spacing within this range.

The effect of broth rheology on mixing are most pronounced for Non-Newtonian broths. If Rushton turbine used in a pseudoplastic broth, the shear rate drops off rapidly and the viscosity increases rapidly with distance from the turbine tip; therefore, there is good mixing only in the immediate vicinity of the impeller, and air tends to channel around the impeller and rise up to shaft. One way to overcome this problem is to use very large diameter turbines.

Agitation impellers

The vessel, 130 mm in diameter, had a working volume of 1.8 in a 3.0 litre nominal volume. The vessel is fitted with a thermocouple and a pH electrode. Good mixing and aeration in high viscosity broths may also achieved by a dual impeller combination, where the lower impeller acts as the gas dispenser and upper impeller acts primarily as a device for aiding circulation of vessel contents.

Agitation and aeration requirements for mammalian cell cultures are very different from those for microbial culture. OTR requirements are very much lower, but the cells are much more easily damaged by fluid mechanical forces generated by impellers or collapsing gas bubbles. In most cases, the impeller must provide enough mixing to keep cells or micro carriers suspended homogenously while creating as little fluid force as possible. A few, including marine propellers, have worked well under specific practical conditions up to several thousand liters, but most simply are not suitable for commercial application. The impeller, called the elephant ear is satisfactorily used in tissue culture vessels up to 500 L, provides adequate mixing and OTR, with little or no cell damage.

For mixing of mechanically sensitive mycelial micro-organisms, mixing systems are recommended, which generate dominating axial flows, thereby ensuring a more even mixing throughout the reactor's volume. Ekato Intermig mixing systems are among the most widespread ones. The approach to the mixing of more sensitive cells (tissue culture, animal cell, etc.) should be different, since; in this case, the mixing regime should have a laminar character.

Novel non rotational mixing

Almost all laboratory fermenter-bioreactors use circular rotation to agitate the culture medium. The major technical problem is that the axis of the stirrer (and the motors axis) rotates while the vessel is fixed. Thus, it is a physical necessity that a free space must exist between both, the moving axis and the immobile vessel, otherwise the rotation of the axis would not be possible.

This free space allows viruses and micro-organisms to get into the vessel. To limit the probability of contamination three ways are used:

The cheapest and less efficient solution is the use of so called lip-seals, which consist of elastic material with a central opening smaller than the axis diameter. This lip pushes onto the axis surface and should make the system tight. At the beginning, the closure can be satisfactory, but with time and especially at high rotation speed the lip is used up and the seal is no longer tight. Contaminating micro-organisms can penetrate into the vessel. Therefore, such a system is not recommended for long time cultures or continuous cultures.

The second solution is the so-called mechanical seal or axial face seal. In this mechanically more advanced joint the stirrer axis is connected to the head plate by two discs, which glide on each other under a given pressure. The problem of this solution is that the system is mechanically stable only for certain time and if medium salts dry out between these discs their destruction is fast and contamination inevitable. Hence, they must be changed even though they are quite expensive. Much larger seals of this type are used in large, industrial scale fermenter. However, because of the knowledge of the mentioned problems, they are used in double sets with sterile water in between to protect the culture if the packing breaks during a run.

Today's best solution with respect to contamination problems connected with the rotational stirring is the magnetic coupling. The stirrer axis is completely separated from the motor axis and from the outside environment of the vessel and the driving force is transmitted by two sets of magnets. Since the magnetic force diminishes strongly with distance between poles, the slot separating the rotating cup and the stationary one is very narrow. Frequently, medium deposits and dries out in this space which leads to problems. Because of the length of the axis and high transmitted force the magnetic coupling is technically quite complex and very expensive. For this reason, it is never proposed as standard equipment for laboratory fermenter. The client can sometimes buy it as an expensive option. In this way, the initial prices of many laboratory fermenters are kept lower despite of expensive consequences for the client at a later stage, when he is basically forced to buy the magnetic coupling option from the same producer.

LAMBDA has found a very simple, innovative solution for this mixing problem by selecting a non-rotational vertical up and down mixing solution. A simple elastic membrane allows the movement of the stirring axis and serves at the same time as a quality seal between the vessel and its central threaded cap. The membrane separates completely the interior of the vessel from the outside environment and this at low cost for the user.

Seals

Four basic types of seal assembly have been used: the stuffing box, the simple bush seal, the mechanical seal and the magnetic drive. Most modern fermenter stirrer mechanisms now incorporate mechanical seals instead of stuffing boxes and packed glands. Mechanical seals are most expensive but are more durable and less likely to be an entry point for organisms or contaminants or a leakage point for organisms or product which should be contained. Magnetic drives are also quite expensive and used in animal cell culture vessels.

Baffles

To augment mixing and gas dispersion, baffles are employed. They are normally incorporated into agitated vessels of all sizes to prevent vortex and to improve aeration efficiency. Baffles are metal strips roughly one-tenth of vessel diameter and attached radially to the wall of bioreactor. The agitation effect is only slightly increased with increase in width of baffles, but drops sharply with narrower baffles. Generally four to eight baffles are incorporated. Baffles should be installed in such a way that a gap exist between them and vessel wall, so that there is scouring acting around and behind the baffles thus minimising microbial growth on the baffles and fermenter walls. Extra cooling coils may be attached to baffles to improve the cooling capacity of the fermenter without affecting the geometry. With animal cell culture baffles causes shear damage, instead of baffles bottom drive axial impellers slightly off sight of centre is used.

Aeration System (Sparger)

Gas under pressure is supplied to the sparger (usually either a ring with holes or a tube with single orifice). It is defined as a device for introducing air into the liquid fermenter. Three basic types of sparger have been used and they are: the porous sparger, the orifice sparger (a perforated pipe), and the nozzle sparger (an open or partially closed pipe). A combined sparger-agitator may be used in laboratory fermenter.

Followings are adequate for good performance:

- The sparge holes in the ring should be in line with the inner edges of the impeller blades.

- The sparger holes should face downward to minimise medium retention in the sparger.
- Hole diameter should be chosen such that each hole is a critical orifice at maximum glass flow.
- The sparger inlet pipe should be placed so as to allow free draining back into the vessel.

Bubbles from a single orifice, bubble volume is directly proportional to orifice diameter and surface tension, and it is inversely proportional to the density of the liquid. Neither gas pressure, temperature, nor liquid viscosity is reported to have much effect on bubble size, but unfortunately no work has been done with systems exhibiting non-Newtonian viscosity (plastic flow) such as occurs in mold cultures. Bubble size is always larger than the pore size by a factor of 10 to 100 for porous spargers.

A well designed agitated gas dispersion system should generate a large interfacial area between the gas and the liquid phases, exhibit minimal influence of gassing on the power draw, and not be prone to flooding. These characteristics are shown to be strongly affected by the location of the sparger. Significant performance improvements, in terms of improved power draw and delayed onset of flooding on aeration, are achieved through the use of 'larger than impeller' ring spargers, positioned within the discharge stream from the impeller. There is little or no penalty in terms of the gas holdup generated. Results for the Rushton and upward and downward pumping pitched blade impellers are reported. The observed behaviour may be explained in terms of the loading regime of the impeller and the cavity forms observed behind the blades, which are governed by the interaction between the sparger (gas discharge) position and the flow patterns generated by each impeller.

Self-cleaning micro-sparger

Many fermentation processes had to be terminated because the air sparger loop was blocked by salt deposits. This problem has now been solved with the self-cleaning microsparger. When an air bubble forms and is released into the medium, a tiny portion of solution flows forth and back in the sparger orifice. This solution is partially dried up by the next bubble. When this happens many times then, especially in strongly mineralised media, a precipitate forms. This precipitate eventually completely closes the sparger openings. The deposit is sometimes so compact that it can be hardly removed. Some producers deliver special sparger end pieces to get rid of difficult cleaning procedures. A similar behaviour is also observed with micro-spargers.

Foam Control

The problem often encounters in fermentation is foaming. It is very important to control foaming. When foaming becomes excessive, there is a danger that filters become wet resulting in contamination, increasing pressure drop and decreasing gas flow. Foam can be controlled with mechanical foam breaker or the addition of surface active chemical agents, called anti foaming agents. Mechanical foam breaker available is 'turbosep', in which foam is directed over stationary turbine blades in a separator and the liquid is returned to fermenter. Foam is also controlled by addition of oils.

Control of foams by oil additions is of large economic importance to the fermentation industry. Excessive foaming causes loss of material and contamination, while excessive oil additions may decrease the product formation. Antifoam oils may be synthetic, such as silicones or polyglycols, or natural, such as lard oil or soyabean oil. Either will substantially change the physical structure of foam, principally by reducing surface elasticity.

Industrial antifoam systems usually operate automatically from level-sensing devices. Methods for metering of oil under aseptic conditions are: timed delivery through a solenoid, two solenoids with an expansion chamber between, a motor-driven hypodermic syringe, and certain industrial pumps.

Temperature Control

Normally in the design and construction of a fermenter there must be adequate provision for temperature control which will affect the design of the vessel body. Heat will be produced by microbial activity and mechanical agitation and if generated by these two processes is not ideal for the particular manufacturing process then heat may have to be added to or removed from, the system. On laboratory scale little heat is normally generated and extra heat has to be provided by placing fermenter in a thermostatically controlled bath, or by use of internal heating coils or by a heating jacket through which water is circulated or a silicone heating jacket. When certain size has been exceeded, the surface area covered by the jacket becomes too small to remove the heat produced by the fermentation. When this situation occurs internal coils must used and cold water is circulated to achieve correct temperature.

pH Control

Certain micro-organisms grow in particular pH only. In fermentation it is very essential to control pH in order to grow the desired micro-organisms for product formation. pH control sensors are used in fermenter for periodically checking of pH.

Valves and Steam Traps

Valves attached to fermenter are used for controlling the flow of liquids and gases in a variety of ways. A wide range of valves are available, but not all of them are suitable for use in fermenter construction. These are also having a significant role in the fermenter productivity.

The different valves available are: Gate valves, globe valves, piston valves, needle valves, plug valves, ball valves, butterfly valves, pinch valves, diaphragm valves, check valves, pressure control valves, safety valves and steam traps. Depending upon fermentation type and requirements these valves are chosen in designing bioreactor with good productivity.

Chapter 16

Bioreactor Design

INTRODUCTION

In any fermentation process the bioreactor plays a central role in determining the process efficiency. Even with recombinant products where stringent quality control implies that downstream processing is the major cost component, it is the bioreactor performance which determines product yields. Any vessel with facilities for aeration and agitation can be used as a bioreactor. Additionally, it must meet the requirements of aseptic operation and provide the cells with a controlled environment conducive to growth and product formation. Traditionally, the vessel of choice has been the stirred-tank bioreactor which consists of a vessel with a vertical rotating shaft with agitator blades (Fig. 16.1). Since the vessel must be sterilised, it must meet the requirements of pressure vessel design (since steam is used under pressure to sterilise the bioreactor). Thus, the material of construction is usually 4–5 mm thick stainless steel, which is also resistant to the acids that are typically produced during fermentation. The height to diameter ratio (H/D) varies from 1 (for small reactors) to 3 (for larger vessels). A high H/D ratio provides for a smaller footprint (space saving), ease of construction and better mixing since the agitator blades need to be proportionally larger as the vessel diameter increases. However, tall tabular reactor with a high H/D ratio often leads to oxygen starvation in the gas phase which affects oxygen transfer rates. The other factor which is critical for large reactors is heat transfer since the metabolic heat generated during growth has to be removed. Cooling coils are used to carry cold water or a mixture of glycerol and water if the temperature of the coolant has to be less than 0°C.

The design of the agitator blade plays a crucial role in oxygen transfer and mixing. Traditionally, the flat blade impeller, the so-called Rushton stirrer (Fig. 16.2) was used. This stirrer provides good radial mixing and does not get 'flooded' with air bubbles even with high air-flow rates. However, it provides poor bulk mixing and this becomes a problem for fungal fermentations. This is because fungal broths shows non-Newtonian behaviour where the viscosity changes with the shear force. Thus the air bubbles get channelled through the centre of the fermenter where the viscosity is low (due to high mixing) effectively starving the media which are close to the walls of oxygen. To prevent this, agitators with better bulk mixing characteristics have been designed, such as the Scaba agitator and the Prochem Maxflow agitator (Fig. 16.2).

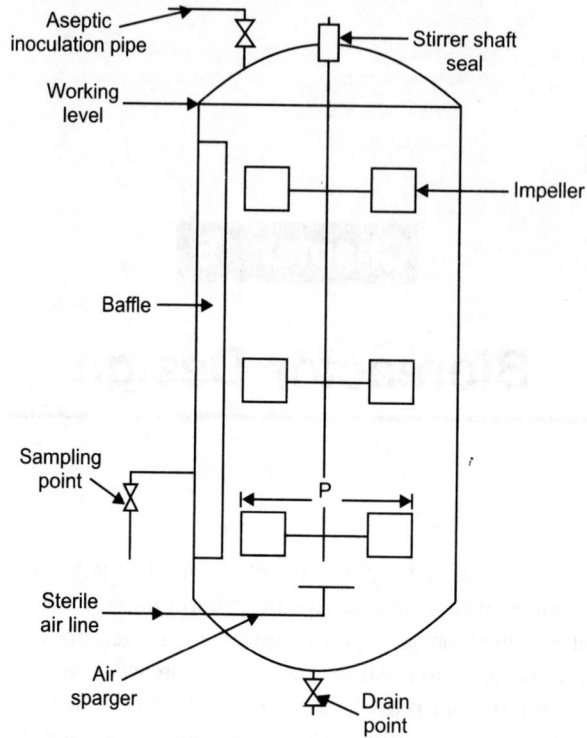

Fig. 16.1: Schematic of a stirred-tank bioreactor.

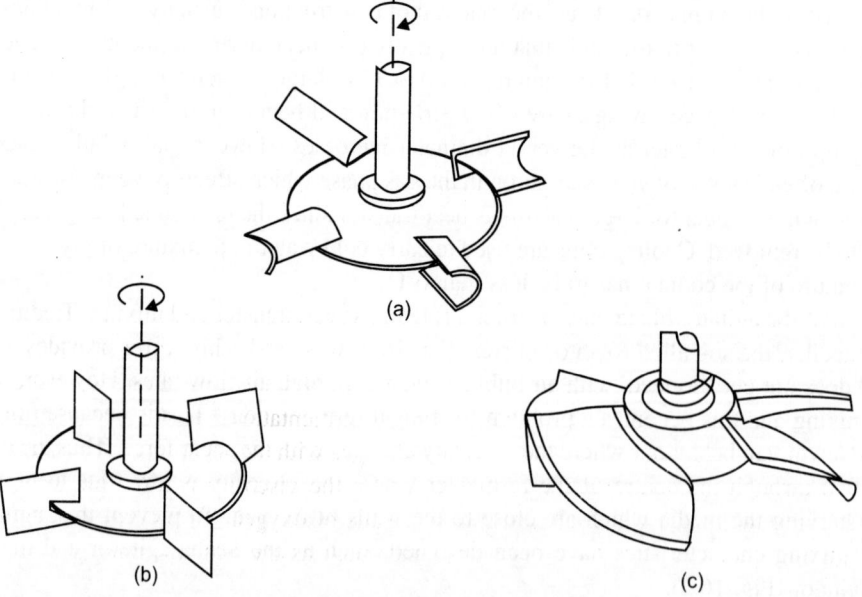

Fig. 16.2: Design of various agitators: (a) Scaba agitator, (b) Rushtor stirrer, and (c) Prochem Maxflow agitator.

When cells are shear sensitive the agitator is designed like a marine propeller which gives good axial mixing. Often the problems of heat and oxygen transfer cannot be addressed with the conventional stirred-tank design. Airlift reactors are then used to provide better oxygen transfer. These reactors are modifications of bubble column reactors which have been used in beer production.

In bubble columns the mixing is provided by the stream of bubbles entering the reactor at the bottom which helps to reduce construction and operating costs. In airlift reactors the liquid also rises with the bubbles through the 'riser', disengages with the gas phase and comes down through the downcomer. This 'downcomer' may be internal or external to the reactor. The circulation loop set up helps in improving heat and mass transfer rates. Other specialised bioreactors are required for specific needs. Thus, photobioreactors need to have a large specific surface area (i.e. surface area per unit volume) since the growth of cells is dependent on the incident sunlight. Thus, instead of a large tank, the cells are grown in tube banks (i.e. tubes arranged parallel to each other) and the material of construction is Plexiglas which is transparent to sunlight. The media (usually seawater) enters with a small inoculum, flows through the tubes like in a plug glow reactor and provides a high cell density at the outlet. Often a small fraction of the outlet cells is recycled to provide a continuous inoculum. Many novel bioreactors have been designed, like the pulsed column bioreactor which combines a pneumatic or mechanical pulsing of the reactor medium with a bubble column design. This helps in bubble break up thereby increasing the surface area and oxygen transfer rates. Since the shear forces are low, this set-up can be used for highly aerobic but shear sensitive organisms. However, most of the designs remain at the laboratory bench as they have not been taken up by industry which still prefers conventional and time-tested designs for large-scale operation.

CLASSIFICATION OF BIOCHEMICAL REACTOR

The biochemical reactor designs can be broadly classified into:

1. Submerged reactor—micro-organism remains submerged all the time inside the liquid. Gas-to-liquid mass transfer is achieved by dispersing the gas in the liquid through continuous input of energy.
2. Surface reactor—the culture adheres to solid surface and oxygen is supplied from the gas phase to the continuously wetted solid surface.

Submerged reactors are further classified depending upon the nature of the energy input, namely:

1. Mechanically stirred systems with agitators.
2. Forced convection of the liquid using recirculating pumps.
3. Operation by pumping compressed air.

Mechanically Stirred Tank Reactor

In a mechanically stirred tank reactor, air is introduced from the bottom through a sparging arrangement and the contents are agitated with mechanical stirrers. Different types of stirrer designs are available for creating a good mixing between the gas and liquid and for breaking the large gas bubbles into finer bubbles. An agitator has to create not only axial movement but also radial movement. As the shear near the tip of the agitator blades are high, shear sensitive micro-organisms break easily. Long thin organisms can break as they get entangled in the rotating shaft. Baffles are provided to prevent vortex formation and improve mixing. The reactor is provided with a jacket or coils for heating and cooling the liquid contents. Power consumption in large designs is an issue in this type of reactor. Gas sparging decreases the power requirement for the agitator motor.

When the gas is sparged in the eye of the impeller the blades break up the large bubbles into smaller ones. The larger bubbles rise in the fluid phase faster than smaller bubbles. In highly viscosity fluids the rise of the bubbles is very slow. If the impeller speed is high enough the gas bubbles, having diameters too small to escape from the liquid phase, will be continuously recirculated with the incoming fresh gas phase. As a result one can assume a complete mixing of the gas phase under well-agitated conditions. If the mixing mechanism is the same as the liquid phase, the mixing time for gaseous phase also can be approximated by the liquid phase mixing time.

The mixing pattern in a stirred vessel in the presence of air is different than that in the absence of it. At a higher impeller speed and low aeration rate, the pattern is more or less as that in an unaerated vessel. On increasing air flow rate, the flow pattern gets increasingly dominated by air flow. At a very high air flow rate, the air is not dispersed at all with a reversal of circulation flow at the centre. In this condition, the impeller is said to be flooded where energy dissipation by air is higher than the energy dissipation by the impeller.

Multicompartment Reactor or Cascade Reactor

A multicompartment reactor or a cascade reactor consists of a tall column, separated into several compartments and a single stirrer assembly with multiple blades agitates each of these compartments independently. Mixing and dispersion of gas into the liquid takes place in each chamber. Liquid flows down from the top compartment while gas rises upwards. The circulation pattern set up by multiple impellers depends on the type of the impellers, the speed of the agitator, geometry of the vessel and distance between impellers. For impellers spacing higher than the diameter of the vessel, the flow patterns creatred by each impeller are quite independent of each other. For spacing equal to the impeller diameter the flow patterns can either merge with each other or work against each other.

Typical reactors used in environmental applications are illustrated in Fig. 16.3. Table 16.1 contains a summary of typical applications for each type. The three basic reactors may find application as either suspended growth or biofilm reactors.

Table 16.1. Reactor types and their typical uses.

Reactor type	*Typical uses*
Basic reactors	
Batch	BOD test, high removal efficiency of individual waste-water constituents
Continuous-flow stirred-tank (CSTR)	Anaerobic digestion of sludges and concentrated wastes, aerated lagoon treatment of industrial wastes, stabilisation ponds for municipal and industrial wastes, part of activated sludge treatment of municipal and industrial waste-waters
Plug-flow (PFR)	Activated sludge treatment of municipal and industrial wastes, aerated lagoon treatment of industrial wastes, stabilisation ponds for municipal and industrial wastes, nitrification, high-efficiency removal of individual waste-water constituents
Biofilm reactors	
Packed bed	Aerobic and anaerobic treatment of municipal and industrial waste-waters, organic removal, nitrification, denitrification
Fluidised bed	Aerobic treatment of low BOD concentration waste-waters, toxic organic biodegradation, anaerobic treatment, denitrification
Rotating biological contactor (RBC)	Aerobic treatment of municipal and industrial waste-waters, organic removal, nitrification

(Cont'd...)

Reactor type	Typical uses
Reactor arrangements	
Recycle	General aerobic and anaerobic treatment of municipal and industrial waste-waters, especially medium to low concentration BOD, organic removal, nitrification
Series	BOD removal combined with nitrification or with nitrification and denitrification or combined with biological phosphorus removal, anaerobic staged treatment, stabilisation pond treatment, sequential anaerobic and aerobic treatment of waste-waters such as for removal of specific toxic organic chemicals
Parallel	Generally used for redundancy and reliability in plant operation, especially with high overall waste-water flow rates
Hybrid	Used for combined forms of treatment such as organic removal and nitrification, or organic removal, nitrification, and denitrification, or organic, nitrogen, and phosphorus removal, anaerobic treatment of industrial waste-waters
Sequencing batch	Useful for high-efficiency removal of individual constituents such as biodegradable but hazardous organics, combined removals of organics, nitrogen, and phosphorus, combination of aerobic and anaerobic processes with same micro-organisms

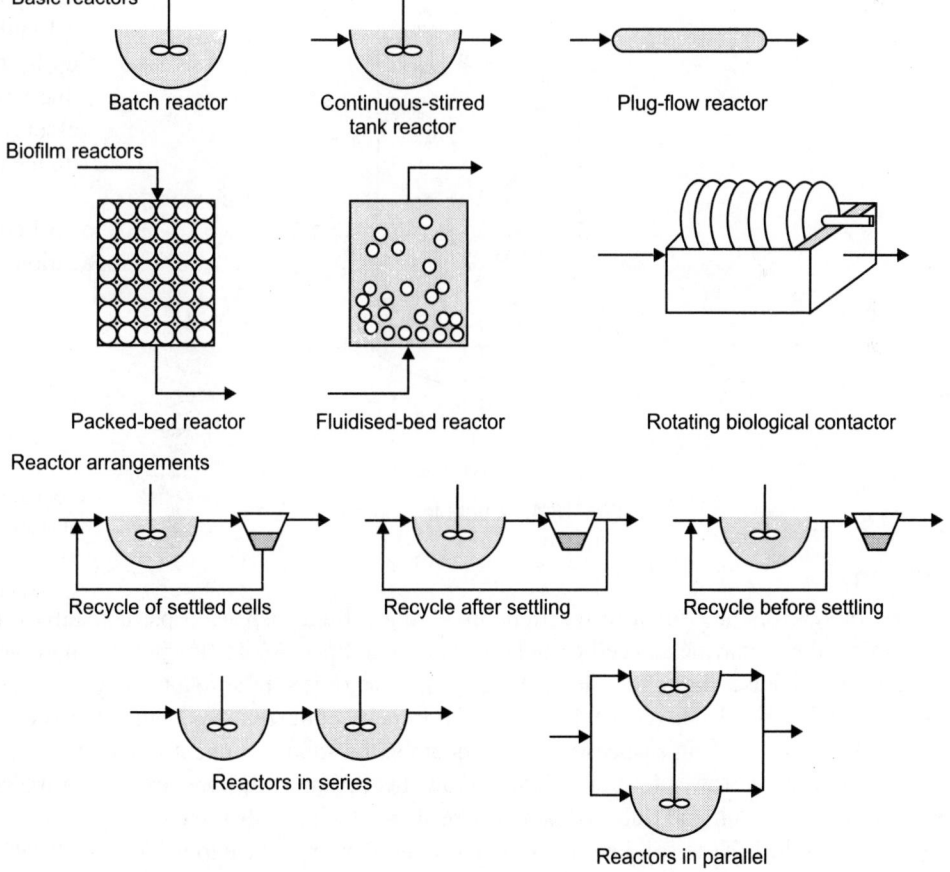

Fig. 16.3: Various reactor types and arrangements.

Bubble Columns

A bubble column bioreactor is shown in Fig. 16.4. Usually, the column is cylindrical with an aspect ratio of 4–6 (height-to-diameter). Gas is sparged at the base of the column through perforated pipes, perforated plates, or sintered glass or metal micro-porous spargers. O_2 transfer, mixing and other performance factors are influenced mainly by the gas flow rate and the rheological properties of the fluid. Internal devices such as horizontal perforated plates, vertical baffles and corrugated sheet packings may be placed in the vessel to improve mass transfer and modify the basic design. The column diameter does not affect its behaviour so long as the diameter exceeds 0.1 metre. One exception is the axial mixing performance. For a given gas flow rate, the mixing improves with increasing vessel diameter. Mass and heat transfer and the prevailing shear rate increase as gas flow rate is increased. In bubble columns the maximum aeration velocity does not usually exceed 0.1 ms⁻¹. The liquid flow rate does not influence the gas-liquid mass transfer coefficient so long as the superficial liquid velocity remains below 0.1 ms⁻¹.

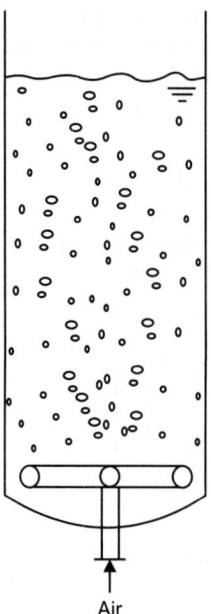

Air

Fig. 16.4: A bubble column.

Fluidised Beds

Fluidised bed bioreactors are suited to reactions involving a fluid-suspended particulate biocatalyst such as the immobilised enzyme and cell particles or microbial flocs. An up-flowing stream of liquid is used to suspend or 'fluidise' the solids. Geometrically, the reactor is similar to a bubble column except that the top section is expanded to reduce the superficial velocity of the fluidising liquid to a level below that needed to keep the solids in suspension. Consequently, the solids sediment in the expanded zone and drop back into the narrower reactor column below, hence, the solids are retained in the reactor whereas liquid flows out. A liquid fluidised bed may be sparged with air or some other gas to produce a gas-liquid-solid fluid bed. If the solid particles are too light, they may have to be artificially weighted, for example by embedding stainless steel balls in an otherwise light solid matrix. A high density of solids improves solid-liquid mass transfer by increasing the relative velocity between the phases. Denser solids are also easier to sediment but the density should not be too high relative to that of the liquid, or fluidisation will be difficult. Liquid fluidised beds tend to be fairly quiescent but introduction of a gas substantially enhances turbulence and agitation. Even with relatively light particles, the superficial liquid velocity needed to suspend the solids may be so high that the liquid leaves the reactor much too quickly, i.e. the solid-liquid contact time is insufficient for the reaction. In this case, the liquid may have to be recycled to ensure a sufficiently long cumulative contact time with the biocatalyst. The minimum fluidisation velocity. i.e. the superficial liquid velocity needed to just suspend the solids from a settled state — depends on several factors, including the density difference between the phases, the diameter of the particles, and the viscosity of the liquid.

Packed Bed Columns

A bed of solid particles, usually with confining walls, constitutes a packed bed. The biocatalyst is supported on, or within, the matrix of solids that may be porous or a homogeneous nonporous gel. The solids may be particles of compressible polymeric or more rigid material. A fluid containing nutrients flows continuously through the bed to provide the needs of the immobilised biocatalyst. Metabolites and products are released into the fluid and removed in the outflow. The flow may be upward or downward, but downflow under gravity is the norm. If the fluid flows up the bed, the maximum flow velocity is limited because the velocity cannot exceed the minimum fluidisation velocity or the bed will fluidise.

The depth of the bed is limited by several factors, including the density and the compressibility of the solids, the need to maintain a certain minimal level of a critical nutrient, such as O_2, through the entire depth, and the flow rate that is needed for a given pressure drop. For a given void volume (i.e. solids-free volume fraction of the bed) the gravity-driven flow rate through the bed declines as the depth of the bed increases. Nutrients and substrates are depleted as the fluid moves down the bed. Conversely, concentrations of metabolites and products increases. Thus, the environment of a packed bed is non-homogeneous but concentration variations along the depth can be decreased by increasing the flow rate. Gradients of pH may occur if the reaction consumes or produces H^+ or OH^-. Because of poor mixing, pH control by addition of acid and alkali is nearly impossible. Beds with greater void volume permit greater flow velocities through them but the concentration of the biocatalyst in a given bed volume declines as the voidage (void volume) is increased. If the packing, i.e. the biocatalyst-supporting solids, is compressible, its weight may compress the bed unless the packing height is kept low. Flow is difficult through a compressed bed because of a reduced voidage. Packed beds are used extensively as immobilised enzyme reactors. Such reactors are particularly attractive for product inhibited reactions: the product concentration varies from a low value at the inlet of the bed to a high value at the exit, thus, only a part of the biocatalyst is exposed to high inhibitory levels of the product.

BIOREACTOR DESIGN FEATURES

Irrespective of the specific bioreactor configuration used, the vessel must be provided with certain common features. The reactor vessel is provided with a vertical sight glass and side ports for pH, temperature, and dissolved O_2 sensors as minimum requirements. Retractable sensors that can be replaced during operation are increasingly used. Connections for acid and alkali (for pH control), antifoam agents, and inoculum are located above the liquid level in the reactor vessel. Air (or other gases, such as CO_2 or ammonia for pH control) is introduced through a sparger situated near the bottom of the vessel. The agitator shaft is provided with steam-sterilisable single or double mechanical seals. Double seals are preferred but they require lubrication with cooled, clean steam condensate. Alternatively, when torque limitations allow, magnetically coupled agitators may be used thereby eliminating the mechanical seals.

Aeration and agitation will inevitably produce foam that is controlled with a combination of chemical antifoam agents and mechanical foam breakers. Foam breakers are used exclusively when the presence of antifoam in the product is not acceptable or if the antifoam interferes with downstream processing operations such as membrane based separations or chromatography. The shaft of the high-speed mechanical foam breaker must also be sealed using double mechanical seals.

In most instances, the bioreactor is designed for a maximum allowable working pressure of 377–412 kPa (absolute). Although the sterilisation temperature generally does not exceed 121°C, the vessel is designed for a higher temperature, typically 150–180°C. The vessel is designed to withstand full vacuum, or it could collapse while cooling after sterilisation. The reactor can be sterilised in place using saturated clean steam at a minimum absolute pressure of 212 kPa. Over-pressure protection is provided by a rupture disc located on top of the bioreactor. Usually this is a graphite burst disc because it does not crack or develop pinholes without failing completely. The rupture disc is piped to a contained drain. Other items located on the head plate of the vessel are nozzles for media or feed addition and for sensors (e.g. the foam electrode), and instruments (e.g. the pressure gauge).

The vessel should have as few internals as practically possible and the design should take into account the needs of clean-in-place and sterilisation-in-place procedures. The vessel should be free of crevices and stagnant areas where pockets of liquids and solids may accumulate. Attention to design of such apparently minor items as the gasket grooves is important. Easy-to-clean channels with rounded edges are preferred. As far as possible, welded joints should be used in preference to couplings. Steam connections should allow for complete displacement of all air pockets in the vessel and associated pipework. Even the exterior of a bioprocess plant should be cleanly designed with smooth contours and minimum bare threads. The reactor vessel is invariably jacketed. In the absence of especial requirements, the jacket is designed to the same specifications as the vessel. The jacket is covered with chloride-free fibreglass insulation that is fully enclosed in a protective shroud. The jacket is provided with over-pressure protection through a relief valve located on the jacket or its associated piping. For a great majority of applications, austenitic stainless steels are the preferred material of construction for bioreactors. The bioreactor vessel is usually made in Type 316L stainless steel, while the less expensive Type 304 (or 304L) is used for the jacket, the insulation shroud and other surfaces not coming into direct contact with the fermentation broth. The L grades of stainless steel contain less than 0.03 per cent carbon, which reduces chromium carbide formation during welding and lowers the potential for later intergranular corrosion at the welds. The welds on internal parts should be ground flush with the internal surface and polished.

DESIGN FOR STERILE OPERATION

Most commercial fermentation processes are mono-cultures. To establish and maintain aseptic conditions are vital for the success of these processes. Hence, a bioreactor must be sterilised prior to inoculation and contamination during operation must be prevented. Contamination during culture is a common cause of process failure.

Sterilisation-in-place

A bioreactor intended for *in situ* sterilisation requires a complex arrangement of pipe work, valves, and filters to enable initial sterilisation and maintenance of sterility. A typical arrangement for *in situ* sterilisation is shown in Fig. 16.5. Because almost all biopharmaceutical production processes involve aeration, the

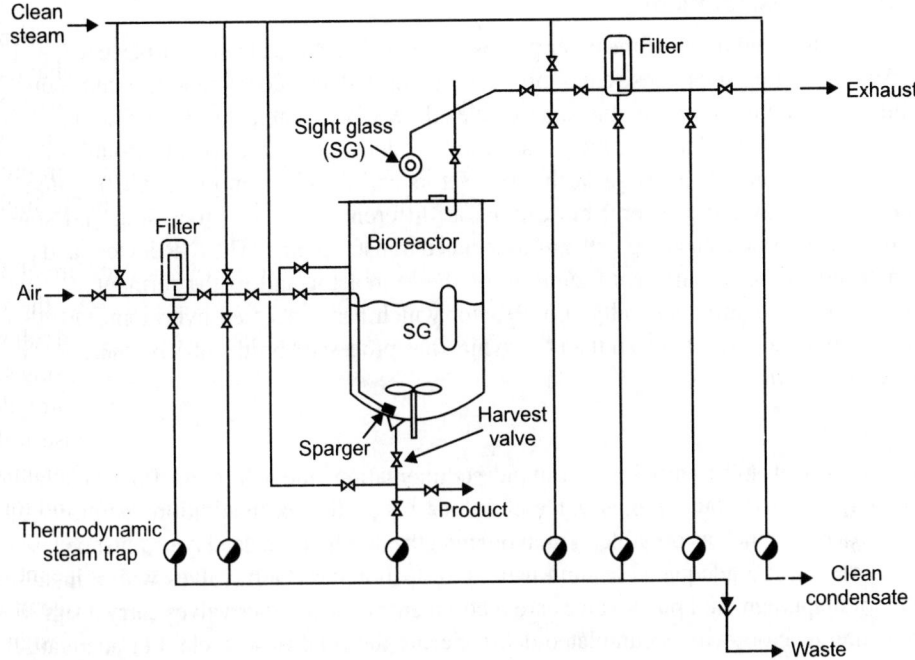

Fig. 16.5: A bioreactor with air inlet and exhaust groups arranged for in-place sterilisation with steam.

figure includes aeration and exhaust groups that must also be sterilised. The air inlet and exhaust lines have *in situ* steam-sterilisable gas filtres. Typically, hydrophobic membrane cartridge filters are used. These filters are rated for removing particles down to 0.45 μm or even 0.1 μm. Often the gas streams have two filter cartridges in series, with the first serving to protect the final filter.

A good system is designed so that the different sections can be sterilised independently of any of the others, thus sterilisation of any section during fermentation can be carried out if, and when, required.

Saturated clean steam (1.1–1.4 bar gauge) is used for sterilisation. The air inlet and exhaust groups are sterilised first, and then, in a second step, the bioreactor. The system is designed so that the filters, valves and the associated pipe-work reach sterilisation temperature (~121°C) very quickly (~1 minute), and are held at the temperature for the required time (25–30 minutes). Apart from the harvest valve, all other valves shown in Fig. 16.5 should either be diaphragm or pinch valves. The harvest valve is usually a piston valve with a metal bellows sealed stem. The valve closes flush with the internal surface of the bioreactor and there is an unobstructed flow path through the valve body. The valves may be operated manually, but pneumatic operation under automatic control is more efficient and reproducible.

The bioreactor is sterilised either filled with the medium or without. Empty sterilisation is the norm in cell culture applications where the media are invariably heat sensitive. In this case, filter sterilisation is used to sterilise the medium. Proper closing and opening sequence of the various valves is important for attaining sterility and preventing recontamination from the adjacent non-sterile areas. Once the sterilising steam supply is shut off, the bioreactor is immediately pressurised with sterile air through the air inlet filter so that any leakage from the outside to the sterile vessel is prevented. In bioreactors with stirrer or foam breaker shaft penetrations, the shaft seals require suitable piping and valves for steam sterilisation and maintenance of a sterile barrier fluid between the contents of the fermenter and the outside.

Clean-in-place Considerations

Industrial bioreactors and much of the other processing equipment are cleaned in-place using automated methods. Automation ensures consistency of cleaning and reduces down-time (i.e. unproductive time of a machine). Attaining an acceptable state of cleanliness is essential to prevent contamination and cross-contamination of biopharmaceuticals and food products. An effective and trouble-free cleaning capability requires attention to design of the bioreactors and the clean-in-place (CIP) systems. At any given time a plant may have several bioreactors at different stages of processing and some empty reactors which need to cleaned along with any associated transfer piping. The CIP devices and procedures must be matched to the specific configuration of the bioreactor and to the fermentation process to ensure satisfactory cleaning. Generally, a bioreactor which has processed hybridoma or other animal cell culture broth is far easier to clean than one which has processed broths of *Streptomyces* or mycelial fungi such as *Penicillium*.

Design aspects

To ensure removal of solid particles and avoid sedimentation, the minimum flow velocity through piping should be 1.5 ms^{-1}, but a higher value of 2.0 ms^{-1} is preferred. In addition, the piping should be free of dead spaces as much as possible, if unavoidable, the depth of the dead zone must be less than two pipe diameters to ensure adequate cleaning using CIP techniques. Only valves with a metal-bellows-sealed stem, or diaphragm and pinch valves are recommended as all other valves carry a significant risk of contaminating reactors with accumulated debris during the final rinse cycle. For adequate cleaning, the CIP solutions are sprayed into the reactor through one or more removable, static or dynamic spray balls, or dynamic spray nozzles (Fig. 16.6). In addition, the piping for air exhaust, which is upstream of the exhaust gas filter, and the air inlet piping, should also receive the cleaning solutions. For cleaning with jet spray, pressures of 308 to 377 kPa (absolute) are optimal. Permanently installed spray heads are not recommended for bioreactors because of potential difficulties with sterilisation. These devices must be inserted into the reactor through one of the ports on the head plate. Often, the spray heads are designed to spray the upper one-third of the tank and the falling liquid film irrigates the remaining surface.

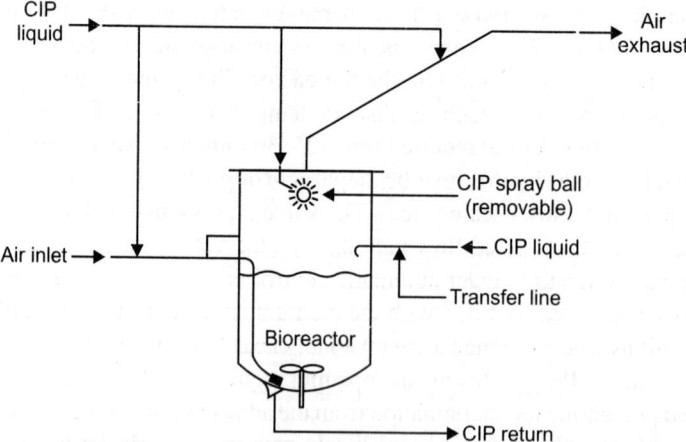

Fig. 16.6: Delivery of the clean-in-place (CIP) liquids to the bioreactor. The flow of CIP solutions is sequenced through the transfer line, the air inlet and exhaust groups and the spray ball of the bioreactor.

For bioreactors for parenteral (injectable) products and other biopharmaceuticals, potable quality deionised water is recommended for all pre-rinsing and detergent formulations. Pre-rinse should be on a once-through basis without recirculation. A five minute pre-rinse is usually sufficient for bacterial, yeast and animal cell culture reactors. Following pre-rinse, 1 per cent (w/v) NaOH at 75°–80°C should be circulated through the equipment so that all product contact surfaces are exposed to this solution for 15–20 minutes. The alkali should be discarded afterwards.

Dilution, contamination with soil and microbial spores that can survive for long periods and loss of quality definition of the starting material for the next cleaning, are some of the arguments against reuse of cleaning chemicals. A deionised or reverse osmosis water rinse at 25–35°C is used to remove all alkali from the system. Process equipment for products that are injected into the body must undergo a final wash with hot water-for-injection grade water. This ensures that all residual water complies with the requisite quality standards.

HEAT TRANSFER

The metabolic heat generated by the cells needs to be removed because it is important to maintain a constant temperature in the reactor. The overall heat balance on a reactor would include: the metabolic heat generated, the heat generated due to stirring and agitation, the heat lost due to evaporation, the sensible heat content of the input and output flows and the heat exchanged with the surroundings. In a large reactor only the first two terms are important, with the metabolic heat often contributing more than 80 per cent of the heat load on the system.

MASS TRANSFER

Mass transfer deals with the problem of diffusive transfer. Diffusion is controlled by concentration gradients. For the transfer of oxygen to the cells, oxygen has to first diffuse through the bubbles into the fermenter medium and then into the cells from the liquid medium. Let us now consider the situation where oxygen is transferred from bubbles to medium containing growing cells and simultaneously taken up from the medium by these cells for respiration.

SHEAR EFFECTS IN CULTURE

Shear rate is a measure of spatial variation in local velocities in a fluid. Cell damage in a moving fluid is sometimes associated with the magnitude of the prevailing shear rate. But the shear rate in the relatively turbulent environment of most bioreactors is neither easily defined nor easily measured. Moreover, the shear rate varies with location within the vessel. In bubble columns an average shear rate has been defined as a function of the superficial gas velocity.

In addition to turbulence within the fluid, other damage-causing phenomena in a bioreactor include inter-particle collisions, collisions with walls, other stationary surfaces, and the impeller, shear forces associated with bubble rupture at the surface of the fluid, phenomena linked with bubble coalescence and break-up, and bubble formation at the gas sparger. Effects of interfacial shear rate around rising bubbles and those due to bubble rupture at the surface can be minimised by adding non-ionic surfactants to the culture medium.

These surfactants reduce adherence of animal cells to bubbles, hence, fewer cells experience interfacial shear and rupture events at the surface of the liquid.

BIOREACTORS AND AIRLIFT REACTORS

The term airlift reactor (ALR) covers a wide range of gas liquid or gas–liquid–solid pneumatic contacting devices that are characterised by fluid circulation in a defined cyclic pattern through channels built specifically for this purpose. In ALRs, the content is pneumatically agitated by a stream of air or sometimes by other gases. In those cases, the name gas lift reactors has been used. In addition to agitation, the gas stream has the important function of facilitating exchange of material between the gas phase and the medium, oxygen is usually transferred to the liquid, and in some cases reaction products are removed through exchange with the gas phase. The main difference between ALRs and bubble columns (which are also pneumatically agitated) lies in the type of fluid flow, which depends on the geometry of the system. The bubble column is a simple vessel into which gas is injected, usually at the bottom, and random mixing is produced by the ascending bubbles. In the ALR, the major patterns of fluid circulation are determined by the design of the reactor, which has a channel for gas–liquid upflow—the riser—and a separate channel for the downflow (Fig. 16.7). The two channels are linked at the bottom and at the top to form a closed loop. The gas is usually injected near the bottom of the riser. The extent to which the gas disengages at the top, in the section termed the gas separator, is determined by the design of this section and the operating conditions. The fraction of the gas that does not disengage, but is entrapped by the descending liquid and taken into the downcomer, has a significant influence on the fluid dynamics in the reactor and hence on the overall reactor performance.

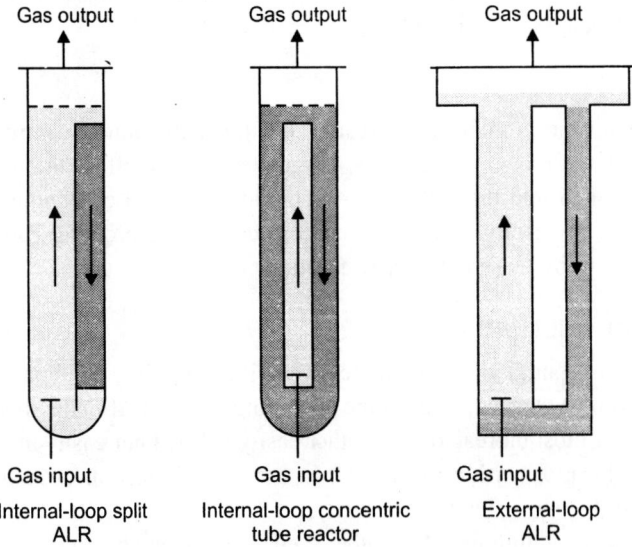

Fig. 16.7: Different types of ALRs.

Airlift Bioreactor Systems

An airlift bioreactor also known as a tower reactor can be described as a bubble column containing a draught tube. Many types of airlift bioreactors are currently in use today. The gas supply is given by means of a sparger at the bottom into the inner cylinder and moves into the annular space creating a swarm of bubbles that induce significant aeration. Air is fed through the sparger ring into the bottom of a central draught tube that controls the circulation of air and the medium. Air flows up the tube, forming bubbles, and exhaust gas disengages at the top of the column. The degassed liquid then flows downwards and the product is drained from the tank. The tube can be designed to serve as an internal heat exchanger, or a heat exchanger can be added to an internal circulation loop. Airlift reactors can be divided into two main types of reactors on the basis of their structure: (i) external-loop vessels, in which circulation takes place through separate and distinct conduits, and (ii) baffled (or internal-loop) vessels, in which baffles placed strategically in a single vessel create the channels required for the circulation. The designs of both types can be modified further, leading to variations in the fluid dynamics, in the extent of bubble disengagement from the fluid, and in the flow rates of the various phases.

All ALRs, regardless of the basic configuration (external loop or baffled vessel), comprise four distinct sections with different flow characteristics:

1. Riser: The gas is injected at the bottom of this section, and the flow of gas and liquid is predominantly upward.

2. Downcomer: This section, which is parallel to the riser, is connected to the riser at the bottom and at the top. The flow of gas and liquid is predominantly downward. The driving force for recirculation is the difference in mean density between the downcomer and the riser, this difference generates the pressure gradient necessary for liquid recirculation.

3. Base: In the vast majority of airlift designs, the bottom connection zone between the riser and downcomer is very simple. It is usually believed that the base does not significantly affect the overall behaviour of the reactor, but the design of this section can influence gas hold-up, liquid velocity, and solid phase flow.

4. Gas separator: This section at the top of the reactor connects the riser to the downcomer, facilitating liquid recirculation and gas disengagement. Designs that allow for a gas residence time in the separator that is substantially longer than the time required for the bubbles to disengage will minimise the fraction of gas recirculating through the downcomer (Fig. 16.8).

Momentum, mass transfer, and heat transfer will be different in each section, but the design of each section may influence the performance and characteristics of each of the other sections, since the four regions are interconnected.

Advantages of Airlift Bioreactors

For the growth of micro-organisms, ALRs are considered to be superior to traditional stirred-tank fermenters despite the fact that the conventional fermenters provide the major requirements for culturing micro-organisms: gas-medium interface for the supply of oxygen and the removal of waste gases, means of agitation to ensure proper nutrient distribution and to minimise damage resulting from addition of concentrated acid or base (for pH control), means of heat transfer (for temperature control), and a contamination-free environment. Therefore, the reason for the more successful growth reported in ALRs appears to lie in the difference in the fluid dynamics between ALRs and the more conventional fermenters. In conventional stirred tanks or bubble columns, the energy required for the movement of the fluids is introduced focally, at a single point in the reactor, via a stirrer or a sparger, respectively. Consequently, energy dissipation is very high in the immediate surroundings of the stirrer and decreases away from it toward the walls. Similarly, shear will be greatest near the stirrer, since the momentum is transferred directly to the fluid in that region, which, in turn, transfers this energy to the slower-moving, more distant elements of the fluid. This results in a wide variation of shear forces, for example, the maximum shear gradient in a stirred tank with a flat-blade turbine has been reported to be approximately 14 times

Gas separator configurations of internal-loop ALRs

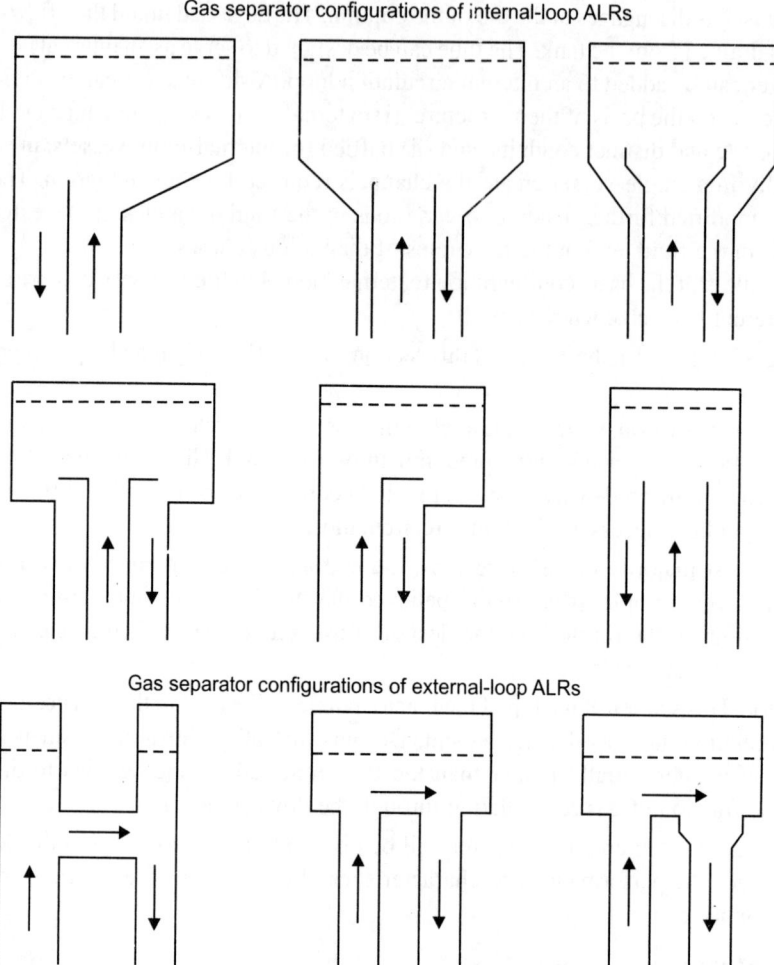

Gas separator configurations of external-loop ALRs

Fig. 16.8: Different types of gas separators.

the mean shear gradient. Cells in culture may thus be exposed to contrasting environments in a mechanically stirred vessel, either to minimal shear forces that may generate potentially undesirable gradients in temperature and in substrate, metabolite, and electrolyte concentrations or, alternatively, to highly turbulent zones, with no problems of heat or mass transfer, but with very high shear gradients that may endanger cell integrity or exert some influence on cell morphology and metabolism. Changes in the morphology of micro-organisms associated with high shear forces in the medium have frequently been observed. The nature of the relationship between such morphological changes and the rates of growth and metabolite production is still not properly understood, although it may be of great importance in the design and scale-up of bioreactors.

In ALRs, as in bubble columns, the gas is injected at a single point, but in ALRs the direct contribution of gas injection to the dynamics of the system is small, circulation of liquid and gas is facilitated by the difference in gas hold-up between the riser and the downcomer, which creates a pressure difference at the bottom of the equipment.

Mammalian and plant cells in culture are more susceptible than micro-organisms to the reactor conditions. Mammalian cells, which lack the rigid cell wall of micro-organisms, have a larger size (one order of magnitude) than micro-organisms and are very sensitive to mechanical stress. Plant cells have a rigid cellulose wall, but they are also much larger than micro-organisms (usually by about an order of magnitude) and are therefore also sensitive to reactor conditions. Another advantage of the ALR is the mechanical simplicity of the device. The absence of a shaft and of the associated sealing, which is always a weak element from the point of view of sterility, confers on the ALR an obvious advantage over agitated tanks. This consideration is especially important in processes involving slow-growing cultures, such as animal and plant cells, for which the risk of contamination is large.

Energy economy in the ALR may be improved by placing a second sparger in the upper part of the downcomer. If the liquid velocity is greater than the free rising velocity of the bubbles generated, the gas is carried down, resulting in a longer contact time between the bubble and the liquid. This diminishes the energy requirements, since part of the gas is injected against a lower hydrostatic pressure.

The advantages described above counterbalance the obvious disadvantage of ALRs, which is the requirement for a minimum liquid volume for proper operation. Indeed, the changes in liquid volume in these reactors are limited to the region of the gas separator, since the liquid height must always be sufficient to allow liquid recirculation in the reactor and must therefore be above the separation between the riser and the downcomer.

The main disadvantages are as follows:

1. Higher initial capital investment due to large-scale processes.
2. Greater air throughput and higher pressures needed, particularly for large-scale operations.
3. Low friction with an optimal hydraulic diameter for the riser and the downcomer.
4. Lower efficiency of gas compression.
5. Inherently impossible to maintain consistent levels of substrate, nutrients in oxygen with organisms circulating through the bioreactor and conditions changing.
6. Inefficient gas/liquid separation and foaming occurs.

Flow Configuration

Riser

In the riser, the gas and liquid flow upward, and the gas velocity is usually larger than that of the liquid. The only exception is homogeneous flow, in which case both phases flow at the same velocity. This can happen only with very small bubbles, in which case the free-rising velocity of the bubbles is negligible with respect to the liquid velocity. Although about a dozen different gas-liquid flow configurations have been developed, only two of them are of interest in ALRs:

1. Homogeneous bubbly flow regime, in which the bubbles are relatively small and uniform in diameter and turbulence is low.
2. Churn-turbulent regime, in which a wide range of bubble sizes coexist within a very turbulent liquid.

Downcomer

In the downcomer, the liquid flows downward and may carry bubbles down with it. For bubbles to be entrapped and flow downward, the liquid velocity must be greater than the free-rise velocity of the bubbles. At very low gas flow input, the liquid superficial velocity is low, practically all the bubbles disengage, and clear liquid circulates in the downcomer. As the gas input is increased, the liquid velocity becomes sufficiently high to entrap the smallest bubbles. Upon a further increase in liquid velocity larger bubbles are also entrapped. Under these conditions the presence of bubbles reduces the cross-section available for liquid flow, and the liquid velocity increases in this section. Bubbles are thus entrapped and carried downward, until the number of bubbles in the cross-section decreases, the liquid velocity diminishes, and the drag forces are not sufficient to overcome the buoyancy. This feedback loop in the downcomer causes stratification of the bubbles, which is evident as a front of static bubbles, from which smaller bubbles occasionally escape downward and larger bubbles, produced by coalescence, escape upward. The bubble front descends, as the gas input to the system is increased, until the bubbles eventually reach the bottom and recirculate to the riser. When this point is reached, the bubble distribution in the downcomer becomes much more uniform. This is the most desirable flow configuration in the downcomer, unless a single pass of gas is required. The correct choice of cross-sectional area ratio of the riser to the downcomer will determine the type of flow.

Gas separator

The gas separator is often overlooked in descriptions of experimental ALR devices, although it has considerable influence on the fluid dynamics of the reactors. The geometric design of the gas separator will determine the extent of disengagement of the bubbles entering from the riser. In the case of complete disengagement, clear liquid will be the only phase entering the downcomer. In the general case, a certain fraction of the gas will be entrapped and recirculated. Fresh gas may also be entrapped from the headspace if the fluid is very turbulent near the interface. The extent of this entrapment influences strongly gas hold-up and liquid velocity in the whole reactor. It is quite common to enlarge the separator section to reduce the liquid velocity and to facilitate better disengagement of spent bubbles. Experiments have been reported in which the liquid level in the gas separator was high enough to be represented as two mixed vessels in series. This point will be analysed further in the section devoted to mixing.

Gas Hold-up

Gas hold-up is the volumetric fraction of the gas in the total volume of a gas–liquid–solid dispersion:

The importance of the hold-up is twofold: (i) the value of the hold-up gives an indication of the potential for mass transfer, since for a given system a larger gas hold-up indicates a larger gas-liquid interfacial area, and (ii) the difference in hold-up between the riser and the downcomer generates the driving force for liquid circulation. It should be stressed, however, that when referring to gas hold-up as the driving force for liquid circulation, only the total volume of the gas is relevant. This is not the case for mass transfer phenomena, in this case, the interfacial area is of paramount importance, and therefore some information on bubble size distribution is required for a complete understanding of the process.

Because gas hold-up values vary within a reactor, average values, referring to the whole volume of the reactor, are usually reported. Values referring to a particular section, such as the riser or the downcomer, are much more valuable, since they provide a basis for determining liquid velocity and mixing. However, such values are less frequently reported.

A number of authors have measured the local hold-up profile along the riser of an external-loop ALR. In general, it was found that the hold-up increases with height. This finding concurs with the expected expansion of gas bubbles as regions of lower pressure are reached. Common sense indicates that this situation must be limited to a certain range, an increase in bubble size will enhance turbulence and result in an increase in bubble encounters, leading eventually to bubble coalescence. In the absence of gas recirculation, there is no effect on these variables. Moreover, this means that under conditions of no gas entrainment from the separator to the downcomer, it is possible to predict the riser gas hold-up as a function of the riser superficial gas velocity alone, which is of great importance for design purposes.

Liquid Velocity

The liquid velocity is one of the most important parameters in the design, of ALRs. It affects the gas hold-up in the riser and downcomer, the mixing time, the mean residence time of the gas phase, the interfacial area, and the mass and heat transfer coefficients. Circulation in ALRs is induced by the difference in hydrostatic pressure between the riser and the downcomer as a consequence of a difference in gas hold-up. Liquid velocity—like gas hold-up—is not an independent variable, because the gas flow rate is the only variable that can be manipulated.

Liquid velocity measurement

Several different methods can be used for measuring the liquid velocity. The most reliable ones are based on the use of tracers in the liquid. If a tracer is injected and two probes are installed in a section of the tube, the velocity of the liquid travelling the distance between probes can be taken directly from the recorded peaks, as the quotient of the distance between the two electrodes and the time required by the tracer to travel from the one to the other.

Liquid Mixing

For the design, modelling, and operation of ALRs, a thorough knowledge of mixing behaviour is necessary. This is of particular importance during the process of scale-up from laboratory-scale to industrial-scale reactors. The optimum growth rate of a micro-organism or the optimum production rate of a specific secondary metabolite usually relates to well-defined environmental conditions, such as pH range, temperature, substrate level, limiting factors, dissolved oxygen, and inhibitor concentration in a specific well-mixed laboratory-scale vessel. Because of the compromises made during scale-up, it is difficult to keep, at different scales of operation, the same hydrodynamic conditions established in the laboratory, mixing on an industrial scale may not be as good as mixing on a laboratory scale. In smaller-scale reactors it is easier to maintain the optimal conditions of pH, temperature, and substrate concentration required for maximum productivity of metabolites in a fermenter. Furthermore, in fermentation systems efficient mixing is required to keep the pH within the limited range, giving maximum growth rates or maximum production of the micro-organism during addition of acid or alkali for pH control.

Mixing time—or the degree of homogeneity—is also very important in fed-batch fermentation, where a required component, supplied either continuously or intermittently, inhibits the micro-organisms or must be kept within a particular concentration range. A large number of commercially important biological systems are operated in batch or fed-batch mode. In this operation mode, fast distribution of the incoming fluid is required, and the necessity for understanding the dynamics of mixing behaviour in these vessels is obvious. Even for batch systems, good control of the operating conditions, such as pH, temperature, and dissolved oxygen, require prior estimation of mixing so that the addition rates can be suitably adjusted. Deviation of the pH or temperature from the permitted range may cause a damage to the micro-organism, in addition to its effect on the growth and production rates. Moreover, a knowledge of the mixing characteristics is required for modelling and interpreting mass and heat transfer data.

A more comprehensive way of analysing mixing, applicable to continuous systems, is a study of the residence time distribution (RTD). Although ALRs are usually operated in a batch-wise manner, at least in the laboratory, advantage is taken of the fact that the liquid circulates on a definite path to characterise the mixing in the reactor. Hence, a single-pass RTD through the whole reactor or through a specific section is usually measured. Based on the observed RTD, several models have been proposed. These models have the advantage of reducing the information of the RTD to a small number of parameters, which can later be used in design and scale-up. One way of obtaining information on the mixing characteristics of each of the regions of the ALR is the simultaneous measurement of the response in the ALR at several points, so that after one single pulse injection the response of each section in the loop can be obtained. This method of measurement has the advantage that multiple measurements are made for the same tracer injection experiment. This enables us to check the consistency of the liquid velocity results obtained, since independent measurements can obtained in the same run.

Mixing in the Gas Phase

For all practical purposes, the gas in the riser of an ALR exhibits plug flow behaviour. Only for extremely high or hindered liquid circulation will the axial dispersion of the bubbles have some effect on the gas RTD. In the downcomer, the gas flow is almost plug-flow when the bubble recirculation is fully developed. But at the stage at which a stationary phase of suspended bubbles appears at the top of the downcomer, appreciable dispersion will occur. This zone has a large degree of mixing due to coalescence and consequent rise of larger bubbles amid smaller ones, with repeated events of break-up and coalescence. However, this type of operation has no relevance to practical applications. It is an operation mode to be avoided at all costs. Indeed, no data on mixing under these conditions have been reported. The main question related to the mixing of the gas phase is, in fact, related to gas recirculation. When a particular gas flow has developed in the downcomer, part of the gas is being recirculated. A pulse of gas tracer at the inlet would produce, as a response, a series of pulses, separated one from the other by the gas circulation time. In practice, not many of these pulses would be detected, due to dilution and disengagement of the tracer in each pass through the separator.

Mass Transfer Rate Measurements

Methods for the determination of $k_L a$ in a reactor can be grossly classified as steady-state and nonsteady-state methods. In the steady-state methods, the rates of oxygen uptake in steady-state operation are evaluated, either by measurement of inlet and outlet rates of oxygen or by direct analysis of a compound that reacts with the oxygen, as in the case of the sulphite method. One of the problems associated with these procedures is that the changes in oxygen concentration in the gas streams are usually small, and the errors of measurement thus have a substantial influence. When a chemical is added to the system, there is the question of whether the addition has provoked changes in the physico-chemical properties, which thus become different from the properties of the original system.

Transient methods may be applied to follow the response of the dissolved oxygen concentration in the system after a step-change of oxygen concentration in the inlet gas stream. These methods have the advantage that addition of an alien material is not required and that a single concentration is measured. The correct use of this method has been analysed in depth by Linek.

One important point to take into account is the dynamics of the oxygen electrode. The lag in the response of the oxygen electrode makes it necessary to discern between the electrode response and the real oxygen concentration, especially when close to a sharp change in concentration. A correct analysis should also include the model of the dynamic behaviour of the electrode. In order to simplify these procedures, approximations based on truncation of parts of the response curve have been proposed.

These methods are based on truncating the first part of the electrode response obtained in a transient experiment. Once the error included in the value of $k_L a$ is set, the extent of truncation is fixed, allowing simplification of the analysis of the remainder of the curve. It should be kept in mind that this simplification implies the loss of part of the information, and due care should be given to statistical analysis of the results. Variations of the method have been proposed to minimise disturbances in the system by introduction of step variations of agitation or pressure. In this way, the method can be applied to bioreactors during real operation of the system.

One problem that may appear in the measurement of mass transfer rates, especially when viscous liquids are used, is related to the presence of very small bubbles that are depleted of oxygen very rapidly but do not disengage in the gas separator, thus constituting an inert volume of gas in the reactor, Kawase and Moo-Young analysed the use of transient absorption of CO_2 for the determination of $k_L a$ and concluded that the error due to small bubble retention was much smaller than that in the case of O_2.

Bubble Size and Interfacial Area

As said earlier, the interfacial area per unit volume is an important component of the volumetric mass transfer coefficient. In fact, it is the part of $k_L a$ that is most susceptible to changes in operation variables and fluid properties. The mass transfer coefficient k_L varies only within a limited range, but the interfacial area is the main component responsible for the changes in mass transfer rate due to variations in turbulence, initial bubble size, and liquid properties.

Data Correlations for Mass Transfer Rate

There are two ways of correlating experimental data from ALRs. First, the hydrodynamic point of view suggests that the movement of the fluid in the reactor determines its overall behaviour, the gas superficial velocity is therefore the more appropriate independent variable. Second, the thermodynamic point of view is based on a consideration of energy balance, a more global approach to the system. This will lead to correlation of the phenomena in the system as a function of the energy input. Indeed, it is easier to compare mass transfer coefficients in ALRs with those in conventional reactors when the data are presented as a function of the total power input (both mechanical and pneumatic) per unit volume of the medium.

Heat Transfer

Because of the relatively low reaction rates of processes involving micro-organisms and cells, it may—in a very general way—be said that heat-effect problems related to local variations of temperature are not common in bioreactors. Even in the case in which polymeric products are released into the medium and very high viscosity is reached, heat transfer is not the controlling step, because such viscous media will hinder the mass transfer, and heat generation will consequently be limited. In such cases the main point of focus is thus, mass, rather than heat, transfer. Reactions catalysed by immobilised enzymes, however, may require different considerations, because of higher reaction rates.

THREE-PHASE AIRLIFT REACTORS

The special qualities of the ALR stem, as stated before, from its fluid dynamic characteristics. One such characteristic is the directionality of the liquid flow. Independently of superimposed fluctuations, a clear net flow is present in the reactor, with exception of the gas separator in internal-loop designs. Therefore, it is to be expected that the fluidisation capacity of the ALR will be markedly superior to that of a bubble column. Several studies have been conducted on the suspension of solids in ALRs, particularly on the use of this type of device for catalytic processes in the chemical industry, where the solid support is usually heavy. In this regard, a very important point is the minimum gas superficial velocity that leads to complete solid fluidisation. Hysteresis has been observed in some cases, once total fluidisation has been attained, the superficial velocity can be reduced to values lower than that required to reach this state. This is due to the high pressure drop related to passage of liquid through a bed of solids, before fluidisation, as compared to the drag forces required to maintain the solid in suspension after all the solids are suspended. Contradictory data on the effect of the suspended solids on the reactor performance have been reported. Fan claimed that the overall gas hold-up increased due to the presence of the solids, whereas Koide showed the opposite effect on the gas hold-up and reported a small decrease in $k_L a$ as well. It is possible that these discrepancies are due to the use of different solids. One of the properties of solids that is often overlooked is wettability. Small bubbles may adhere on wettable solids, leading to a change in the apparent density of the particle and thus changing their solid circulation velocity.

In the case of suspended solids that take an active part in the process, the mass transfer rate from the liquid to the solid may become the limiting step. The dependence of particle size on the mass transfer to the suspended solids has been studied by several authors.

AIRLIFT REACTOR—SELECTION AND DESIGN

Scale-up of Airlift Bioreactors

The problems encountered in the scale-up of bioreactors can be concentrated into two groups. The first includes the cases in which a high power input per unit volume is used on a laboratory scale, but cannot be maintained on an industrial scale due to economic or mechanical limitations. This is not the case for plant or animal cultures, for which a very high specific power input cannot be used because of cell fragility. The second group of problems can be generalised as lack of knowledge about the hydrodynamics of large-volume vessels.

The methods available for scale-up of bioreactors have been reviewed by Oosterhuis, Kossen and Oosterhuis, Sweere and, Sola and Godia, among others. Because, in general, design from first principles cannot be undertaken because of the lack of basic knowledge about the hydrodynamics of the bioreactors, one possible solution is the semifundamental method, which comprises using approximate simple models for fluid dynamics and integrating them with basic known kinetics and heat and mass transfer rates.

It does not happen very often that all this information is available with a degree of certainty that allows safe design of a large system. Thus, the designer must usually resort to dimensional analysis. This method requires a knowledge of all the variables affecting the process, which can be obtained from a qualitative, but realistic, model.

A simplified version of this method is to limit the number of variables to one or two and to follow rules of thumb, which, depending on the specific case, may be constant P/V, constant $k_L a$, etc. The literature shows, however, many cases of inconsistency for this method. For example, design of a scaled-up bioreactor in which the oxygen transfer rate is kept constant can lead to a better performance than expected, as in the example reported by Taguchi for glucoamylase production by *Endomyces* sp., or worse than expected, as in the case of protease production by *Streptomyces* sp. The method of regime analysis and scale down, proposed by Kossen and Oosterhuis, combines two tools to overcome the problem posed by the complexity of biochemical reactors. Their method is based on considering the regime of the full-scale process as the objective and planning the strategy of process development from this point on. This method is therefore applicable only to conventional and well-studied bioreactors that are to be used in a new bioprocess. Regime analysis is based on the consideration that, generally, biochemical processes involve a series of steps, some being mass or heat transfer by convection, some being diffusive mechanisms (activated or not), and others being chemical reaction steps. In the latter case, a mass-transfer mechanism is superimposed, since molecules must encounter one another in order to react, and usually a heat effect will accompany the reaction. Depending on whether these steps take place in parallel or in series and on the relaxation time for each step, the rate of the total process is often given as the rate of one single step. But the equilibrium between all the individual rates can be (and usually is) upset by a change in scale. This is to be expected, because a change in scale will not bring a change in the physico-chemical or kinetic parameters (scale-insensitive variables), but will affect the overall convective mass and heat rates (scale-sensitive variables).

A new equilibrium will be established, and the interplay of all the parameters of the system may lead to a regime in which a different step becomes the step-controlling the process rate. The method of Kossen and Oosterhuis starts with an analysis of the operation of the large-scale system. Once the regime is clarified, a small-scale system is designed in such a way that it simulates the operation regime of the larger one. Optimisation studies can be done on the smaller model, and conclusions will then be extrapolated to the full-scale process.

Design Improvements

In the design of ALR, several modifications aiming at the improvement of some of the characteristics of the equipment have been proposed. One of the earliest modifications was the two-staged ALR, proposed by Orazem and Erickson. Their design was inspired by the improvement observed by them in the performance obtained with multistage sieve trays over single-stage bubble columns. They claimed that a substantially higher mass transfer coefficient was obtained, as was a better performance in terms of oxygen transferred per unit of energy invested. As mentioned above, it has been shown repeatedly that the mass transfer rate in a bubble column is usually higher than that in the conventional ALR. It, therefore, makes sense to try to bring into the ALR some of the characteristics of bubble columns in a controlled fashion. This was done by Bando, who tested a perforated draft tube in a concentric-tube ALR. The perforations in the draft tube facilitated communication between the less-well aerated liquid in the downcomer and the better aerated riser. The reported improvements in mass transfer rates were undoubtedly obtained at the cost of a reduction of circulation velocity.

One of the advantages of the directionality of flow in ALRs is the improved fluidisation capacity. The strengthening of this advantage was the aim of another modification, the helical flow promoter, proposed by Gluz and Merchuk. The helical flow promoter causes the fluid to flow down in the downcomer in a helical pattern. The device comprises several fins or baffles, which have the effect of modifying the flow paths, instead of going in straight lines along the axis, the flow paths move along an helix. The baffles may be installed in a small section at almost any place along the riser or the downcomer, and the effect is perceived throughout the reactor. One of the best positions for the helical flow promoter is the top of the riser. A helical flow is then generated in the downcomer to produce a swirl at the bottom and a corkscrew-like path in the riser. This has a strong potential for the culture of photosynthetic micro-organisms. The helical movement causes secondary flow, which leads to enhanced radial mixing, and therefore more homogeneous distribution of light and heat among liquid elements and suspended particles. With the helical flow promoter, it is thus more likely that all the elements of the fluid get the same exposure to light and heat exchange.

One of the most important characteristics of the helical flow promoter is the enhanced capacity for fluidising solid particles. This is due to the swirls that develop at the bottom of the reactor. Thus, this modification is especially suited to processes operating with cells immobilised on a solid. The minimal gas flow rate for complete fluidisation of solids in an ALR may be reduced drastically by the use of a helical flow promoter. In addition, the mass transfer rate to suspended solids may be enhanced up to 50 per cent, because of the higher relative velocity between the particles and solids.

Thus, ALRs are popular in modern bioprocess research and development. These reactors are particularly suitable for biological processes in which a high mass transfer rate is required, but excessive power input may lead to damage of the cells due to shear effects. ALRs also have very appealing characteristics for bioprocesses for low-value products in which efficiency of energy utilisation may become the key point for design. Such is the case for ALRs for waste-water treatment. The ALR is particularly effective for solid fluidisation, which is important in many biological processes in which the biocatalyst is available in the form of pellets or is immobilised on a solid support.

The distinctive characteristics of ALRs are conferred by the fluid dynamics of the liquid–gas or liquid–gas–solid mixtures in it. These characteristics are expressed as gas hold-up, liquid velocity, and mixing in each of the zones of the ALR. It is important to recognise the differences in the fluid dynamic characteristics of these zones: the riser, the gas separator, and the downcomer. Only a correct understanding of behaviour and interconnection of these regions can make possible the correct design of a new reactor or the scale-up of a laboratory device up to pilot or industrial size.

The purpose of scale-up is to conserve and repeat on a larger scale the fluid dynamics of the reactor. Therefore, one of the most important factors in the design and scale-up of reactors is the influence of the geometric characteristics of the system on the flow of the different phases present.

The variables affecting the performance of the reactor are geometric design, operation variables, and fluid properties. Several correlations are available in the technical literature for the prediction of the fluid dynamic characteristics of the reactor (gas hold-up, liquid velocity, mixing time of axial dispersion coefficient) and of the mass and heat transfer coefficients. Although attempts have been made to study all the aspects described, no single research group has managed to cover all the variables over a wide range. It is, therefore, extremely important that the engineer confronting scale-up or *de novo* design of an ALR analyses in depth the range of validity of the correlations used for the calculations.

Chapter 17

Aeration and Agitation

INTRODUCTION

The main function of aeration is to supply enough oxygen to the microbes in submerge culture technique for proper metabolism, while agitation provides proper mixing of the nutrient so that each and every organisms get proper nutrients. Each fermentation process requires unique type of aeration and agitation system.

The parts of the fermenter involved in aeration and agitation are:

- The agitator (impeller).
- The aeration system (sparger).

Agitator (impeller): The main aim of the agitator is to provide homogenous environment all over the fermenter. It is also used for mixing of different phases, oxygen and heat transport.

Aeration system (sparger): A sparger is a tool used for introducing air into the fermentation medium.

Three basic types of sparger:

1. The porous sparger.
2. The orifice sparger (a perforated pipe).
3. The nozzle sparger (an open or partially closed pipe).

Porous sparger: The porous sparger is mainly used for laboratory scale non agitated fermenter. It is made up of sintered glass, ceramics or metal.

Orifice sparger: In small stirred fermenters the perforated pipes were arranged below the impeller in the form of crosses or rings (ring sparger), approximately three-quarters of the impeller diameter.

Nozzle sparger: Most modern mechanically stirred fermenter designs from laboratory to industrial scale have a single open or partially closed pipe as a sparger to provide the stream of air bubbles. Ideally the pipe should be positioned centrally below the impeller and as far away as possible from it to ensure that the impeller is not flooded by the air stream. The single-nozzle sparger causes a lower pressure loss than any other sparger and normally does not get blocked.

NEED FOR AERATION AND AGITATION

The majorities of fermentation processes are aerobic and, therefore, require the provision of oxygen. If the stoichiometry of respiration is considered, then the oxidation of glucose may be represented as:

$$C_6H_{12}O_6 + 6O_2 = 6H_2O + 6CO_2$$

Thus, 192 grams of oxygen are required for the complete oxidation of 180 grams of glucose. However, both components must be in solution before they are available to a micro-organism and oxygen is approximately 6000 times less soluble in water than is glucose (a fermentation medium saturated with oxygen contains approximately 7.6 mg dm^{-3} of oxygen at 30°C). Thus, it is not possible to provide a microbial culture with all the oxygen it will need for the complete oxidation of the glucose (or any other carbon source) in one addition.

Therefore, a microbial culture must be supplied with oxygen during growth at a rate sufficient to satisfy the organisms' demand. The aeration and agitation of the fermentation medium, provides necessary oxygen to the industrial fermentation process. However, the productivity of many fermentations is limited by oxygen availability and, therefore, it is important to consider the factors which affect a fermenter's efficiency in supplying microbial cells with oxygen.

OXYGEN REQUIREMENTS OF INDUSTRIAL FERMENTATIONS

Although a consideration of the stoichiometry of respiration gives an appreciation of the problem of oxygen supply, it gives no indication of an organism's true oxygen demand as it does not take into account the carbon that is converted into biomass and products.

It has been studied that a culture's demand for oxygen is very much dependent on the source of carbon in the medium. Thus, the more reduced the carbon source the greater will be the oxygen demand. However, it is inadequate to base the provision of oxygen for fermentation simply on an estimation of overall demand, because the metabolism of the culture is affected by the concentration of dissolved oxygen in the broth.

It may be seen that the specific oxygen uptake rate increases with increase in the dissolved oxygen concentration up to a certain point (referred to as C_{crit}) above which no further increase in oxygen uptake rate occurs.

Thus, maximum biomass production may be achieved by satisfying the organism's maximum specific oxygen demand by maintaining the dissolved oxygen concentration greater than the critical level. If the dissolved oxygen concentration were to fall below the critical level then the cells may be metabolically disturbed. However, it must be remembered that it is frequently the objective of the fermentation technologist to produce a product of the micro-organism rather than the organism itself and that metabolic disturbance of the cell by oxygen starvation may be advantageous to the formation of certain products.

Equally, provision of a dissolved oxygen concentration greater than the critical level may have no influence on biomass production, but may stimulate product formation. Thus, the aeration conditions necessary for the optimum production of a product may be different from those favouring biomass productions.

The oxygen demand of fermentation largely depends on the concentration of the biomass and its respiratory activity, which is related to the growth rate. By limiting the initial concentration of the medium, the biomass in the vessel may be kept at a reasonable level and by supplying some nutrient component as a feed, the rate of growth and hence the respiratory rate, may be controlled.

Oxygen Supply

Oxygen is normally supplied to microbial culture in the form of air, this being the cheapest available source of the gas.

The method for provision of a culture with a supply of air varies with the scale of the process:

1. Laboratory scale: Cultures may be aerated by means of the shake-flask method. Flask are shaken on a platform contained in a controlled environment chamber.
2. Pilot and industrial scale: Air is provided to the cultures by specific types of fermenter (Bubble fermenter).

Bartholomew and others represented the transfer of oxygen from air to the cell, during fermentation, as occurring in a number of steps:

- The transfer of oxygen from an air bubble into solution.
- The transfer of the dissolved oxygen through the fermentation medium to the microbial cell.
- The uptake of the dissolved oxygen by the cell.

The rate of oxygen transfer from air bubble to the liquid phase may be given by the equation:

$$dC_L / dt = K_L a (C^* - C_L)$$

where, C_L is the concentration of dissolved oxygen in the fermentation broth (mmole dm^{-3}), t is time (hr), dC_L/dt is the change in oxygen concentration over a time period, i.e. the oxygen transfer rate (mmole O_2 dm^{-3} h^{-1}), K_L is the mass transfer coefficient (cm h^{-1}), a is the gas/liquid interface area per liquid volume (cm^2 cm^{-3}), C* is the saturated dissolved oxygen concentration (mmoles dm^{-3}).

K_L may be considered as the sum of the reciprocals of the resistances to the transfer of oxygen from gas to liquid and (C^*-C_L) may be considered as the 'driving force' across the resistances. It is extremely difficult to measure both K_L and 'a' in a fermentation and, therefore, the two terms are generally combined in the term $K_L a$, the volumetric mass-transfer coefficient, the units of which are reciprocal time (h^{-1}).

The volumetric mass-transfer coefficient ($K_L a$) is used as a measure of the aeration capacity of a fermenter. The aeration capacity of the system will be more if $K_L a$, is higher.

The $K_L a$ value will depend upon the design and operating conditions of the fermenter and will be affected by such variables as aeration rate, agitation rate and impeller design.

These variables affect 'K_L' by reducing the resistances to transfer and affect 'a' by changing the number, size and residence time of air bubbles.

Determination of $K_L a$ Values

The sulphite oxidation technique:

$$Na_2SO_3 + 0.5O_2 = Na_2SO_4$$

The rate of reaction is such that as oxygen enters solution it is immediately consumed in the oxidation of sulphite, so that the sulphite oxidation rate is equivalent to the oxygen-transfer rate.

Gassing-out techniques:

1. The static method of gassing out: The technique was first described by Wise, the concentration of oxygen in the solution is decreased by passing nitrogen gas into the liquid, this will remove all the oxygen from the solution. The aeration and agitation of deoxygenated liquid increase the dissolved oxygen which is monitored using some form of dissolved oxygen probe. This technique has the advantage over the sulphite oxidation method in that it is very rapid (normally taking up 15 minutes)

and may utilise the fermentation medium, to which may be added dead cells or mycelium at a concentration equal to that produced during the fermentation.

2. The dynamic method of gassing out: The procedure involves stopping the supply of air to the fermentation which results in a linear decline in the dissolved oxygen concentration due to the respiration of the culture.

   The aeration and agitation of deoxygenated liquid increase the dissolved oxygen which is monitored using some form of dissolved oxygen probe.

   The dynamic gassing out method has the advantage over the previous methods of determining the K_La during an actual fermentation and may be used to determine K_La values at different stages in the process. The technique is also rapid and only requires the use of a dissolved oxygen probe, of the membrane type.

The oxygen-balance technique: The oxygen balance technique is used for the determination of transportation of amount of oxygen into the fermentation medium in a given period of time. It is also used for the measurement of K_La of a fermenter.

The procedure involves measuring the following parameters:

- The amount of medium in the fermenter (dm^3).
- The rate of flow of air (incoming and outgoing air), dm^3 min^{-1}.
- The total pressure at inlet and outlet (atm).
- The temperature of the gases of the inlet and outlet, (K).
- The partial pressure of oxygen of the inlet and outlet.

The oxygen balance technique appears to be the simplest method for the assessment of K_La and has the benefit of measuring aeration efficiency during fermentation. The sulphite oxidation and static gassing out techniques have the disadvantage of being carried out using either a salt solution or an uninoculated, sterile fermentation medium.

The factors affecting K_La values in fermentation vessels:

- The air flow rate employed.
- The degree of agitation.
- The rheology properties of the medium.
- The presence of antifoam agents.

The effect of air-flow rate on K_La: The rate of air flow in fermentation media in agitated and non-agitated fermenter will be different.

Mechanically agitated reactors: The air-flow rate of 0.5–1.5 volumes of air per volume of medium per minute is to be maintained constant on scale-up. If the impeller is unable to disperse the incoming air then extremely low oxygen transfer rates may be achieved. Thus proper flow rate should be maintained by agitator.

Non-mechanically agitated reactors: Bubble columns and air-lift reactors are not mechanically agitated. Mixing and aeration is dependent on the air passage.

Bubble columns: Bubble column reactor cannot be used for highly viscous medium. Pattern of gas bubbles in a bubble column reactor is dependent on the gas superficial velocity. Gas velocity should be 1–4 cm per second for uniform bubbles throughout medium which will provide proper mixing. If gas velocity is higher or lower than uniform bubbles will not be produced, thus when bubbles coalesce produces differences in fluid density which will disturb air flow rate.

Air lift reactors: In this fermenter, medium circulation is also accomplished with bubble formation. K_La obtained in air-lift reactor will be less than bubble fermenter due to shorter contact time between bubble and medium.

Degree of agitation

- Agitation is playing a vital role in the oxygen transfer rate in agitated fermenter.
- Agitation increases the area available for oxygen transfer by dispersing air into the medium.
- It increase the contact time for bubbles in the medium.
- It prevents coalesces of air bubbles.
- It decreases thickness of liquid film at gas-liquid interface.

Medium rheology (medium flow characteristics)

Mostly the product of fermentation process is not interfering with medium flow rate or viscosity. But certain bacterial strain producing polysaccharide which can increase the viscosity and hence affect the medium rheology. Thus bacterial polysaccharide will decrease the oxygen transfer rate and bulk mixing.

Effect of Foam and Antifoams on Oxygen Transfer

Antifoam agents collapse the foam and thus increase the oxygen transfer rate of the fermentation medium. Thus K_La value decreases due to use of antifoam agent. If inadequate space is provided above the liquid level for foam control, then abundant amount of antifoam must be used to prevent loss of broth from the vessel. Thus, it is more productive to operate a fermenter at a lower working volume.

The high degree of aeration and agitation required in a fermentation frequently gives rise to the undesirable phenomenon of foam formation. In extreme circumstances the foam may overflow from the fermenter via the air outlet or sample line resulting in the loss of medium and product, as well as increasing the risk of contamination. The presence of foam may also have an adverse effect on the oxygen-transfer rate. It was pointed out that Waldhof and vortex-type fermenters were particularly affected due to the bubbles becoming entrapped in the continuously recirculating foam, resulting in high bubble residence times and therefore, oxygen-depleted bubbles. The presence of foam in a conventional agitated, baffled fermenter may also increase the residence time of bubbles and therefore result in their being depleted of oxygen. Furthermore, the presence of foam in the region of the impeller may prevent adequate mixing of the fermentation broth. Thus, it is desirable to break down a foam before it causes any process difficulties and this may be achieved by the use of mechanical foam breakers or chemical antifoams. However, mechanical foam control consumes considerable energy and is not completely reliable so that chemical antifoams are preferred.

All antifoams are surfactants and may, themselves, be expected to have some effect on oxygen transfer. The predominant effect observed by most workers is that antifoams tend to decrease the oxygen-transfer rate. Antifoams cause the collapse of bubbles in foam but they may favour the coalescence of bubbles within the liquid phase, resulting in larger bubbles with reduced surface area to volume ratios and hence a reduced rate of oxygen transfer. Thus, a balance must be struck between the necessity for foam control and the deleterious effects of the controlling agent. Foam formation has a particular influence on the liquid height in the fermenter at which it is practical to operate. If inadequate space is provided above the liquid level for foam control, then copious amounts of antifoam must be used to prevent loss of broth from the vessel.

Balance between Oxygen Supply and Demand

Both the demand for oxygen by a micro-organism and the supply to the organism by the fermenter have been considered in this chapter. This section attempts to bring these two aspects together and considers how processes may be designed such that the oxygen uptake rate of the culture does not exceed the oxygen transfer rate of the fermenter. It will also be recalled that the dissolved oxygen concentration during the fermentation should not fall below the critical dissolved oxygen concentration (C_{crit}) or the dissolved oxygen concentration which gives optimum product formation. Thus, it is necessary that the oxygen-transfer rate of the fermenter matches the oxygen uptake rate of the culture whilst maintaining the dissolved oxygen above a particular concentration. To balance supply and demand conditions of fermenter it should be adjusted to match the supply. This may be achieved by:

1. Controlling biomass concentration.
2. Controlling the specific oxygen uptake rate.
3. A combination of 1 and 2.

SCALE-UP AND SCALE-DOWN

Scale-up means increasing the scale of a fermentation, for example from the laboratory scale to the pilot plant scale or from the pilot plant scale to the production scale. Increase in scale means an increase in volume and the problems of process scale-up are due to the different ways in which process parameters are affected by the size of the unit. It is the task of the fermentation technologist to increase the scale of a fermentation without a decrease in yield or if a yield reduction occurs, to identify the factor which gives rise to the decrease and to rectify it. The major factors involved in scale-up are:

1. Inoculum development: An increase in scale may mean the extra stages have to be incorporated into the inoculum development programme.
2. Sterilisation: Sterilisation is a scale dependent factor because the number of contaminating micro-organisms in a fermenter must be reduced to the same absolute number regardless of scale. Thus, when the scale of a process is increased the sterilisation regime must be adjusted accordingly, which may result in a change in the quality of the medium after sterilisation.
3. Environmental parameters: The increase in scale may result in a changed environment for the organism. These environmental parameters may be summarised as follows:
   (a) Nutrient availability.
   (b) pH.
   (c) Temperature.
   (d) Dissolved oxygen concentration.
   (e) Shear conditions. ·
   (f) Dissolved carbon dioxide concentration.
   (g) Foam production.

All the above parameters are affected by agitation and aeration, either in terms of bulk mixing or the provision of oxygen. Points a, b, c and e are related to bulk mixing whist d, e, f and g are related to air-flow and oxygen transfer. Thus, agitation and aeration tends to dominate the scale-up literature. However, it should always be remembered that inoculum development and sterilisation difficulties may be the reason for a decrease in yield when a process is scaled up and that achieving the correct aeration/ agitation regime is not the only problem to be addressed.

Scale-up of Aeration/Agitation Regimes in Stirred Tank Reactors

From the list of environmental parameters affected by aeration and agitation it will be appreciated that it is extremely unlikely that the conditions of the small-scale fermentation will be replicated precisely on the large scale.

Thus, the most important criteria for a particular fermentation must be established and the scale-up based on reproducing those characteristics. The scale-up window represents the boundaries imposed by the environmental parameters and cost on the aeration/agitation regime. Thus, if dissolved oxygen concentration is perceived as the overriding environmental condition then power consumption per unit volume and volumetric air-flow rate per unit volume should be maintained constant on scale-up. However, as a result, the other parameters will not be the same in the larger scale and therefore, neither will the environmental factors which they influence. The most important environmental domains affected by aeration and agitation for the majority of fermentations are oxygen concentration and shear.

Scale-up of Air-lift Reactors

Bubble columns and air-lift vessels tend to be scaled-up on the basis of geometric similarity and constant gas velocity. The other problem in the scale-up of air-lift systems is that the organism is exposed to extremes of oxygen levels in the riser and downcomer and the effects of these conditions should be investigated on the laboratory scale.

Scale-down Methods

Scale-down is the situation where laboratory or pilot-scale experiments are conducted under conditions which mimic the industrial-scale conditions. This approach is important in both the development of a new product and the improvement of an existing full-scale fermentation. Frequently, conditions achievable on a laboratory scale are impractical on an industrial scale, which means that if inappropriate conditions have been used in the laboratory unrealistic yield objectives may be set for the scaled-up process. The aspects to consider in the design of laboratory or pilot-plant experiments in the context of scale-down may be summarised as follows.

Medium design

Media relevant to the industrial situation should be used in development experiments.

Medium sterilisation

If the medium is to be batch sterilised on the large scale its exposure time at a high temperature will be much greater than that experienced in the laboratory or pilot plant. Thus, the sterilisation times on the smaller scales should be increased to mimic the industrial situation. Alternatively, medium sterilised in the production fermenter may be used in the laboratory and pilot plant. This highlights the advantage of continuous sterilisation where little loss of medium quality occurs. Furthermore, the same continuous steriliser may be used for both full-scale and pilot scale vessels.

Inoculation procedures

Due to a range of circumstances, it may not always be possible to inoculate every production fermentation with inoculum in optimum condition. The scale-down approach can be used to predict the consequences of such events by mimicking these situations in the laboratory, for example by storing inoculum or using inocula of different ages.

Number of generations

An industrial scale fermentation requires a greater number of generations than does a laboratory one; this may place more severe stability criteria on the process strain than may have been appreciated on the small scale. The industrial situation may be modelled in the laboratory by using serial sub-culture to ensure that the strain is sufficiently stable. This approach is particularly pertinent in the development of recombinant fermentations.

Mixing

As indicated in the previous section it is almost inevitable that the degree of mixing will decrease with an increase in scale. Thus, it is possible to model inadequate mixing in the laboratory by subjecting the organism to pulse medium feeds or fluctuating process conditions such as oxygen concentration, pH and temperature. Such scaled-down experiments then allow predictions to be made about the suitability of new strains for industrial exploitation.

Oxygen transfer rate

Far higher oxygen transfer rates can be achieved in laboratory fermenters than in industrial-scale ones. Thus, unrealistic demands may be made of a fermentation plant if the development work has been done at very high oxygen-transfer rates. Therefore, the laboratory and pilot fermenters should reflect the oxygen transfer rates achievable in the full-scale fermenters. The adoption of these simple approaches to small scale experimentation can prevent many scale-up problems before they even occur.

SECTION V

Biosensors and Instrumentation and Control Systems

Chapter 18

Biosensors and Nanobiosensors: Design and Applications

INTRODUCTION

Biosensors are the device in which there is a coupling of biological sensing element with a detector system using a transducer. In comparison with any other currently available diagnostic device, biosensors are much higher in performance in terms of sensitivity and selectivity both. Biosensors have found potential applications in the industrial processing and monitoring, environmental pollution control, also in agricultural and food industries. Important features for commercialisation of the biosensors are selectivity, sensitivity, stability, reproducibility and low cost. A biosensor is an analytical device which converts a biological response into an electrical signal (Fig. 18.1). The term 'biosensor' is often used to cover sensor devices used in order to determine the concentration of substances and other parameters of biological interest even where they do not utilise a biological system directly. Biosensors function by coupling a biological sensing element with a detector system using a transducer.

The first scientifically proposed as well as successfully commercialised biosensors were electro-chemical sensors for multiple analytes. The following statement is also defined for the biosensor, 'A chemical sensing device in which a biologically derived recognition is coupled to a transducer, to allow the quantitative development of some complex biochemical parameter.' The schematic diagram shown in Fig. 18.2 for the biosensor is mainly divided into three sections: (i) sensor: a sensitive biological element (biological material (e.g. tissue, micro-organisms, organelles, cell receptors, enzymes, antibodies, nucleic acids, etc.), (ii) transducer: it is the detector element (works in a physico-chemical way, optical, piezoelectric, electrochemical, etc.) that transforms the signal resulting from the interaction of the analyte with the biological responsible for the display of the results in a user-friendly way, (iii) third section is the associated electronics, which comprises of signal conditioning circuit (amplifier), processor and a display unit.

PRINCIPLE OF A BIOSENSOR

The desired biological material (usually a specific enzyme) is immobilised by conventional methods (physical or membrane entrapment, non-covalent or covalent binding). This immobilised biological

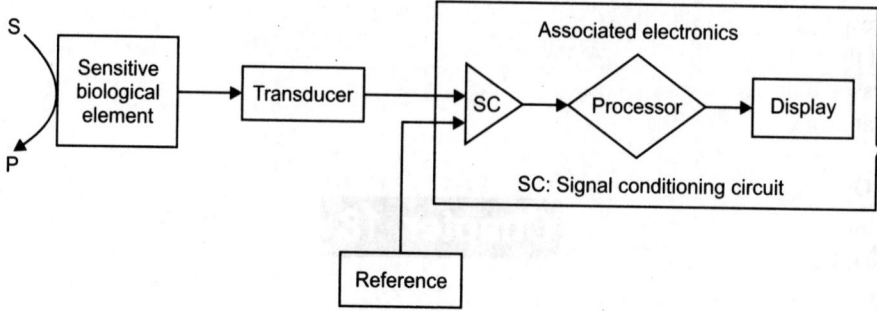

Fig. 18.1: Schematic diagram showing main components of a biosensor.

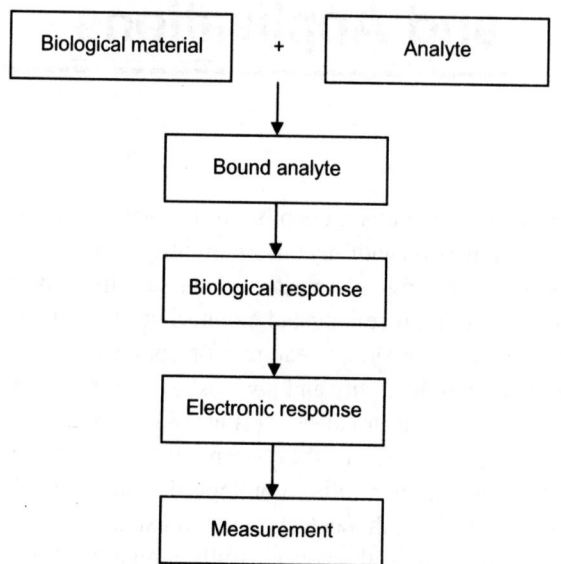

Fig. 18.2: Measurement flow for a biosensor.

material is in intimate contact with the transducer. The analyte binds to the biological material to form a bound analyte which in turn produces the electronic response that can be measured. In some instances, the analyte is converted to a product which may be associated with the release of heat, gas (oxygen), electrons or hydrogen ions. The transducer can convert the product linked changes into electrical signals which can be amplified and measured.

Working of a Biosensor

The electrical signal from the transducer is often low and superimposed upon a relatively high and noisy (i.e. containing a high frequency signal component of an apparently random nature, due to electrical interference or generated within the electronic components of the transducer) baseline. The signal processing normally involves subtracting a 'reference' baseline signal, derived from a similar transducer without any biocatalyst membrane, from the sample signal, amplifying the resultant signal difference and electronically filtering (smoothing) out the unwanted signal noise. The relatively slow nature of the biosensor response considerably eases the problem of electrical noise filtration. The analogue signal produced at this stage may be output directly but is usually converted to a digital signal and passed to a microprocessor stage where the data is processed, manipulated to desired units and output to a display device or data store.

TECHNOLOGY USED FOR TRANSDUCER IN BIOSENSORS

The technology used for transducer can be any one of the four types listed below and depend upon the biological sensor used. In biosensors, suitable transducers are designed, keeping in view the following:

1. Specific desired interaction between the analyte and the biological elements.
2. The intended use of the biosensors and the.
3. Manufacturing cost of the device.

Biosensing Method

The critical aspect of the biosensor is matching the appropriate biological and electronic components to produce a relevant signal during analysis. Isolation of the biological component is very much essential to ensure that only the molecule of interest is bound or immobilised on the electronic component or the transducer. Attachment of the biological component to the electronic component is vital for the success of these devices. The stability of the biological component is also quite critical, since it is being used outside its normal biological environment.

TYPE OF BIOSENSORS

Biosensors can be categorised according to the basic principles of signal transduction and biorecognition elements. In the general scheme of a biosensor, the biorecognition element responds to the target compound and the transducer converts the biological response to a detectable signal, which can be measured electrochemically, optically, acoustically, mechanically, calorimetrically, or electronically, and then correlated with the analyte concentration. Biological elements include enzymes, antibodies, micro-organisms, biological tissue, and organelles. When the binding of the sensing element and the analyte is the detected event, the instrument is described as an affinity sensor. When the interaction between the biological element and the analyte is accompanied or followed by a chemical change in which the concentration of one of the substrates or products is measured the instrument is described as a metabolism sensor. Finally, when the signal is produced after binding the analyte without chemically changing it but by converting an auxiliary substrate, the biosensor is called a catalytic sensor. The method of transduction depends on the type of physico-chemical change resulting from the sensing event. Often, an important ancillary part of a biosensor is a membrane that covers the biological sensing element and has the main functions of selective permeation and diffusion control of analyte, protection against mechanical stresses, and support for the biological element.

The research field of biosensors started with the introduction of the first generation glucose oxidase (GO_x) biosensor in 1962 by Clark and Lyons. Since then, biosensors have been intensively studied and extensively utilised in various applications, ranging from public health and environmental monitoring to homeland security and food safety. Though a lot of research activity has been involved in developing biosensors for various purposes the time has come to bring this technology to the forefront and make it commercially available. Efforts and funds need to be mobilised to manufacture biosensors on a large scale so as to benefit and be of use to the general public. With exposure to the commercial market the applications of this technology would be greatly enhanced. A few such applications could be detection of virulence of a vaccine just before it is injected so as to prevent accidental acquisition of a disease, bandages detecting a septic wound, deadly viruses in the environment or from the patient sample (rapid and early detection) and so forth in the medical field. Real time monitoring of dairy products and breweries might help foster a cleaner and hygienic environment and experiment with different tastes imparted by specific micro-organisms in specific concentrations giving rise to new products. A farfetched and plausible use of this technology could be in space exploration where if present the concentration of the living organisms would be very low and might lead to answering many of the long standing questions regarding the presence of life in space.

MICRO-BIOSENSORS

The major progress in microsystem technologies for creating small, integrated and reliable microtransducers devices in combination with biological sensing elements has revolutionised the field of biosensors during the last decade. Such micro-biosensor systems raised the expectation to get a comprehensive insight into dynamic cellular metabolic events and subsequently a complete understanding of the metabolism of human biology. Currently, cancer can be detected by monitoring the concentration of certain antigens present in the bloodstream or other bodily fluids, or through tissue examinations. Correspondingly, diabetes is monitored by determining the glucose concentrations in the blood over time. However, despite their widespread clinical use, these techniques have a number of potential limitations. For example, a number of diagnostic devices have slow response times and are burdensome to patients. Furthermore, these assays are expensive and cost the health care industry billions of dollars every year. Therefore, there is a need to develop more efficient and reliable sensing and detection technologies.

Biosensing Techniques

Biosensors can be classified either by the type of biological signalling mechanism they utilise or by the type of signal transduction they employ. Transduction can be accomplished via a great variety of methods. Most forms of transduction can be categorised in one of three main classes. These classes are: (i) electrochemical detection methods, (ii) optical detection methods and (iii) mass detection methods. However, new types of transducers are constantly being developed for use in biosensors. Each of these three main classes contains many different subclasses, creating a nearly infinite number of possible transduction methods or combination of methods. Here, we will discuss each of these three detection mechanisms.

Electrochemical Biosensors

The first scientifically proposed as well as successfully commercialised biosensors were those based on electrochemical sensors for multiple analytes. At present, there are many proposed and already commercialised devices based on the electrochemical principle including those for pathogens and toxins. This stems from a number of attributes of electrochemistry including the high sensitivity of electrochemical transducers, their compatibility with modern miniaturisation/microfabrication technologies, minimal power requirements, economical cost, and independence of sample turbidity and colour.

The basic principle for this class of biosensors is that chemical reactions between immobilised biomolecule and target analyte produce or consume ions or electrons, which affects measurable electrical properties of the solution, such an electric current or potential. The electrochemical signal produced is then used to relate quantitatively to the amount of analyte present in a sample solution. Potentiometry, amperometry, voltammetry, and, more recently, electrochemical impedance spectroscopic measurements are among the electrochemical detection techniques often used in conjunction with immunoassay systems and immunosensors, leading to their respective categories according to the type of signal measured. The fundamental principles of each of these techniques are presented below, followed by discussions based on some recent work that have specifically addressed problems encountered in these areas.

Potentiometric Biosensors

These biosensors are based on ion-selective electrodes (ISE) and ion-sensitive field effect transistors (ISFET). The primary outputting signal is possibly due to ions accumulated at the ion-selective membrane interface. Current flowing through the electrode is equal to or near zero. The electrode follows the presence of the monitored ion resulting from the enzyme reaction. For example, glucose oxidase can be immobilised on a surface of the pH electrode. Glucose has only minimal influence on pH in the working medium, however, the enzymatically formed gluconate causes acidification. A biorecognition element is immobilised on the outer surface or captured inside the membrane. In the past the pH glass electrode was used as a physico-chemical transducer. Nowadays, semiconductor based physico-chemical transducers are more common. ISFETs and LAPS (light addressable potentiometric sensor) based systems especially are convenient for biosensor construction. The ISFET principle is based on a local potential generated by surface ions from a solution. This potential modulates the current flow across a silicon semiconductor. The transistor gate surface in ISFET is covered by a selective membrane, for pH detection this could be made from compounds such as silicon nitride (Si_3N_4), alumina (Al_2O_3), zirconium oxide (ZrO_2) and tantalum oxide (Ta_2O_5). The LAPS principle is based on semiconductor activation by a light-emitting diode (LED). The sensor is made from an n-type silicon typically coated with 30 nm of silicon oxide, 100 nm of silicon nitride, and indium-tin oxide. The LAPS measures a voltage change as a function of medium pH in the LED activated zone. This opens the way for multiposition sensing and construction of an array of biorecognition zones. A good example of a potentiometric immunosensor involves the detection of enzyme-labelled immunocomplexes formed at the surface of a polypyrrole coated screen-printed electrode.

Amperometric Biosensors

In amperometry, the current produced by the oxidation or reduction of an electroactive analyte species at an electrode surface is monitored under controlled potential conditions. The magnitude of the current is then related to the quantity of analyte present. Clark oxygen electrodes perhaps represent the basis for the simplest forms of amperometric biosensors, where a current is produced in proportion to the oxygen concentration. This is measured by the reduction of oxygen at a platinum working electrode in reference to a Ag/AgCl reference electrode at a given potential. Typically, the current is measured at a constant potential and this is referred to as amperometry. If a current is measured during controlled variations of the potential, this is referred to as voltammetry. Furthermore, the peak value of the current measured over a linear potential range is directly proportional to the bulk concentration of the analyte, i.e. the electroactive species. However, not all protein analytes are intrinsically capable to serve as redox partners in electrochemical reactions, a suitable label must be introduced to promote the electrochemical reaction of the analyte at the working electrode. Despite the disadvantage of this often indirect sensing system, it is claimed that amperometric devices maintain a sensitivity superior to potentiometric devices. An example of an amperometric device is the aforementioned glucose biosensor, which is based on the amperometric detection of hydrogen peroxide. A very tangible application of amperometry is used in combination with immunosensing techniques to measure levels of the human chorionic gonadotropin β-subunit (β-HCG) in advanced pregnancy testing.

Voltammetric Biosensors

Voltammetry belongs to a category of electro-analytical methods, through which information about an analyte is obtained by varying a potential and then measuring the resulting current. It is, therefore, an amperometric technique. Since there are many ways to vary a potential, there are also many forms of voltammetry, such as: polarography (DC Voltage), linear sweep, differential staircase, normal pulse, reverse pulse, differential pulse and more. Cyclic voltammetry is one of the most widely used forms and it is useful to obtain information about the redox potential and electrochemical reaction rates (e.g. the chemical rate constant) of analyte solutions. More recently, interdigitated array (IDA) microelectrodes have gained popularity as an alternative transducer in electrochemical immunoassays. In general, a simple design of an IDA consists of a pair of interdigitated microelectrode 'fingers'. When an IDA is used as a sensing electrode in a voltammetric experiment, the two interdigitated electrodes are usually held at different potentials to achieve 'redox' cycling of the electroactive species to be detected. A major advantage of this redox cycling is that it improves the signal-to-noise ratio by enhancing the Faradaic current relative to the background current, resulting in lower detection limits and improved sensitivity. These features opened up many opportunities in which IDA electrodes were applied as electrochemical detectors in analytical chemistry and biosensor systems.

OPTICAL BIOSENSORS

Optical detection biosensors are the most diverse class of biosensors because they can be used for many different types of spectroscopy, such as absorption, fluorescence, phosphorescence, Raman, SERS, refraction, and dispersion spectrometry. In addition, these spectroscopic methods can all measure different properties, such as energy, polarisation, amplitude, decay time, and/or phase. Amplitude is the most commonly measured as it can easily be correlated to the concentration of the analyte of interest. In optical biosensors, the optical fibres allow detection of analytes on the basis of absorption, fluorescence or light scattering. Since they are non-electrical, optical biosensors have the advantages of lending themselves to *in vivo* applications and allowing multiple analytes to be detected by using different monitoring wavelengths. The versatility of fibre optics probes is due to their capacity to transmit signals that reports on changes in wavelength, wave propagation, time, intensity, distribution of the spectrum, or polarity of the light. In general, acquisition of the signal from these devices is accomplished through flexible cables, which can transmit light to the biological component. Optical methods are readily multiplexed, samples can be interrogated with many wavelengths simultaneously without interfering with one another. A large variety of optical methods have been used in biosensors, however, those devices based on fluorescence spectroscopy, surface plasmon resonance, interferometry and spectroscopy of guided modes in optical waveguide structures (grating coupler and resonant mirror) are the most common. However, other emerging optical sensing technologies have been under investigation, such as optical ring resonators and photonic crystals.

Fluorescence-based Biosensors

Fluorescence is a widely used optical method for biosensing due to its selectivity and sensitivity. A fluorescence-based device monitors the frequency change of electromagnetic radiation emission stimulated by previous absorption of radiation and subsequent generation of an excited state that only exits for a very short time. Single molecules could be repeatedly excited and detected to produce a bright signal easily measured even at single-cell level. There are three types of fluorescence biosensing. The first is direct sensing when a specific molecule is detected before and after a change or reaction takes place. The second form is indirect biosensing when a dye is added that will optically transduce the presence of a specific target molecule. The use of green fluorescent protein (GFP) is a powerful fluorescent tag that has enabled investigators to study the location, structure and dynamics of molecular events within living cells. However, binding interactions between an activated signalling molecule and its target could be difficult to detect due to the difficulty of seeing this localised interaction over background fluorescence. A third type of fluorescence biosensing, called fluorescence energy transfer (FRET), can be used and it generates a unique fluorescence signal. In a typical fluorescence measurement, the fluorophore is excited by a specific wavelength of light and emits light at a different wavelength. However, when two fluorophores are paired in such a way that the emission wavelength of one overlaps with the excitation wavelength of the other, the excitation of one of them will stimulate fluorescence of the complementary pairing one (if they reside within about few Angstroms from each other). FRET has tremendous utility because the unique fluorescence signal generated under these circumstances can be used to visualise and quantify the position and concentration of interacting fluorophores.

Two major strategies have been used to develop FRET biosensors: (i) two chain probes in which the fluorophores are on two different molecules resulting in intermolecular FRET when the two molecules come into proximity, or (ii) single chain probes in which different regions of a single molecule are tagged and undergo FRET due to intramolecular, conformational changes.

Surface Plasmon Resonance Biosensors

Surface plasmon resonance (SPR) biosensor was first demonstrated for biosensing in 1983 by Liedberg and others. Since then it has been extensively explored and has gradually become a very powerful label-free tool to study the interactions between the target and biorecognition molecules. The principle, development, and applications of SPR biosensors have been well described in several excellent review papers. The success of surface plasmon resonance (SPR) biosensors is a wide range of fields from fundamental biological studies to clinical diagnosis applications. In SPR biosensing, the adsorption of a targeted analyte by a surface bioreceptor is measured by tracking the change in the conditions of the resonance coupling of incident light to the propagating surface plasmon wave (SPW). The SPW is a charge density oscillation that occurs at the interface of two media with dielectric constants of opposite signs, such as a metal and a dielectric. The existence of this surface plasmon wave is dictated by the electromagnetic (EM) properties of the metal, typically gold or silver, and the dielectric interface (sample medium). The resonance coupling appears as a dip in the reflectivity of the light spectrum, which is traditionally tracked by measuring the wavelength, the incident angle or the intensity of the reflected light. The coupling of the light to the SPW requires, for electromagnetic reasons, a high-index prism or a periodic grating surface. The sensitivity of the SPR lies in the strong electromagnetic enhancement of the SPW. Commercial SPR biosensors are generally capable of detecting 1 pg/mm^2 of absorbed analytes. This sensitivity is strongly dependent on many parameters, but is particularly dependent on surface functionalisation. Sensor detection limit (DL) is another important parameter to characterise the sensor performance. The DL can be deduced by taking into account the noise in the transduction signal, σ, i.e. the minimum resolvable signal: DL = σ/S, where S is the sensitivity. Improvement in the DL can be accomplished by increasing the sensitivity or reducing the noise level. Sensitivity can be enhanced by increasing the light-matter interaction. Today, the key challenge in the SPR biosensor development lies not primarily in the integration of the various components of the biosensor (sampling handling, control electronics, etc.) but on providing robust integrated SPR biosensors that are as or more sensitive ($<pg/mm^2$) than their current counterparts, such as interferometer, optical ring resonator, and optical fibre based biosensors.

Optical Fibre Based Biosensors

Among optical-based biosensors, optical fibers are new, tiny, flexible platforms that are being used with increasing frequency as biosensor transducers. Optical fibers are able to make quick and sensitive responses, and can be employed as an intrinsic or extrinsic biosensor. Optical fibers are a convenient material for optical sensor design because they can be inexpensive and provide easy and efficient signal delivery. Fibre Bragg gratings (FBGs), while developed as a tool for the telecommunications industry, have flourished as a versatile sensor with a wide breadth of applications. They are currently among the most popular of all fibre-based optical sensors for analysing load, strain, temperature, vibration, and RI.

Optical fibers transmit light on the basis of the principle of total internal reflection (TIR). Fibre optic biosensors are based on the transmission of light along silica glass fibre, or plastic optical fibre to the site of analysis. Optical fibre biosensors can be used in combination with different types of spectroscopic technique, e.g. absorption, fluorescence, phosphorescence, surface plasmon resonance (SPR), etc. Optical biosensors based on the use of fibre optics can be classified into two different categories: intrinsic sensors, where interaction with the analyte occurs within an element of the optical fibre, and extrinsic sensors, in which the optical fibre is used to couple light, usually to and from the region where the light beam is influenced. In practice, fibre optics can be coupled with all optical techniques, thus increasing their versatility.

The simplest optical biosensors use absorbance measurements to determine any changes in the concentration of analytes that absorb a given wavelength of light. The system works by transmitting light through an optical fibre to the sample, the amount of light absorbed by the analyte is detected through the same fibre or a second fibre. The biological material is immobilised at the distal end of the optical fibers and either produces or extracts the analyte that absorbs the light.

ACOUSTIC BIOSENSORS

Electroacoustic devices used in biosensors are based on the detection of a change of mass density, elastic, viscoelastic, electric, or dielectric properties of a membrane made of chemically interactive materials in contact with a piezoelectric material. Bulk acoustic wave (BAW) and surface acoustic wave (SAW) propagation transducers are commonly used. In the first, a crystal resonator, usually quartz, is connected to an amplifier to form an oscillator whose resonant frequency is a function of the properties of two membranes attached to it. The latter is based on the propagation of SAWs along a layer of a substrate covered by the membrane whose properties affect the propagation loss and phase velocity of the wave. SAWs are produced and measured by metal interdigital transducers deposited on the piezoelectric substrate. Even though SAW-based biosensor systems have been the focus of academic and industrial research for a number of years, most of these approaches only feature laboratory setups that are suitable for proof-of-principle evaluation and first experimental tests. For real commercial success, two crucial issues need to be solved: an appropriate production process is required, as is an applicable handling process for future SAW based biosensors. Most contributions to the scientific community relating to SAW based sensor technology do not suggest overall system designs but rather basic approaches limited to the sensor element itself. Apart from the sensor, there are a number of additional issues which must be addressed when considering a market-compatible overall system. However, surface acoustic wave biosensors are inexpensive devices to manufacture, and require inexpensive electronics to run and to disseminate data. They offer a flexible approach to point-of-care, realtime diagnostics and their small size allows flexibility in the samples to be analysed, the devices can be incorporated into airway tubing to capture proteins and other molecules in breath condensate. Therefore SAW-based biosensor technology is a promising approach which may eventually be able to compete against established but much more complex optical- based biosensing techniques, like surface plasmon resonance (SPR).

Quartz Crystal Microbalance

The Quartz Crystal Microbalance (QCM) has been the most used acoustic device for sensor applications since 1959, when Sauerbrey established the relation between the change in the resonance frequency and the surface mass density deposited on the sensor face. The classical QCM sensor typically consists of an oscillator circuit containing a thin AT-cut quartz disc with circular electrodes on both sides of the quartz. Due to the piezoelectric properties of the quartz material, an alternating voltage between these electrodes leads to a mechanical oscillations of the crystal. These oscillations are generally very stable due to the high quality of the quartz. If a mass is adsorbed or placed onto the quartz crystal surface, the frequency of oscillation changes in proportion to the amount of mass. Therefore, these devices can be used as high sensitivity microbalances intended to measure mass changes in the nanogram range by coating the crystal with a material which is selective towards the species of interest. The quartz crystal acts as a signal transducer, converting mass changes due to the hybridisation process into frequency changes. One of the main advantages of this device is the ability to control a QCM's selectivity by applying different coatings, which makes this sensor type extremely versatile. Despite of the extensive use of QCM technology, some challenges such as the improvement of the sensitivity and the limit of detection in high fundamental frequency QCM, remain unsolved, recently, an electrodeless QCM biosensor for 170 MHz fundamental frequency, with a sensitivity of 67 Hz cm^{-2} ng^{-1}, has been reported, this shows that the classical QCM technique still remains as a promising technique.

APPLICATION AND EXAMPLES OF MICRO-BIOSENSORS

In Vivo Biosensor

The field of biosensors may be viewed as comprising essentially two broad categories of instrumentation: (i) sophisticated laboratory machines capable of rapid, accurate and convenient measurement of complex biological interactions, (ii) easy-to-use, portable devices for use by nonspecialists for *in situ* or home analysis. Although biosensor development made a huge progress in recent years, their application in clinical diagnosis is not very common, except for glucose biosensors which represent about 90 % of the global biosensor market. Interferences with undesired molecules during measurements with real samples and also high selectivity and accuracy are still serious issue.

This is very important, since treatment is often dependent on individual levels of clinical markers. The emergence of semi-synthetic and synthetic receptors is yielding more robust, versatile and widely applicable sensors, while nanomaterials are facilitating highly sensitive and convenient transduction of the resulting binding and catalytic events. Escalating healthcare costs together with consumer demand is likely to generate a new generation of inexpensive wearable, integrated and less-invasive sensors amenable to mass production to support the maintenance of wellbeing, care of the elderly, pharmaceutical development and testing, and distributed diagnostics.

Electrochemical biosensors currently dominate the field, but are focused mainly on metabolite monitoring, while bioaffinity monitoring is carried out principally using optical techniques. However, both transducers find utility across the whole field, along with piezoelectric, thermometric, and micro-mechanical transducers.

Glucose Biosensor

Blood glucose measurement for the management of diabetes comprises approximately 90% of the world market for biosensors. Millions of diabetics test their blood glucose levels daily, making glucose the most commonly tested analyte. Such huge market size makes diabetes a model disease for developing new biosensing concepts. Glucose concentration is also one of the most monitored indicators in many endocrine metabolic disorders. The glucose biosensor is the most widely used example of an electrochemical biosensor which is based on a screen-printed amperometric disposable electrode.

The first developed glucose enzyme electrode relied on a thin layer of glucose oxidase (GO_x) entrapped over an oxygen electrode via a semi-permeable dialysis membrane. Measurements were made based on the monitoring of the oxygen consumed by the enzyme-catalysed reaction:

$$\text{Glucose} + O_2 \rightarrow \text{Gluconic acid} + H_2O_2 \qquad \ldots(18.1)$$

The entire field of biosensors can trace its origin to this original glucose enzyme electrode. A wide range of amperometric enzyme electrodes, differing in electrode design or material, immobilisation approach, or membrane composition, has since been described. In 1973, Guilbault and Lubrano described an enzyme electrode for the measurement of blood glucose based on amperometric (anodic) monitoring of the hydrogen peroxide product:

$$H_2O_2 \rightarrow O_2 + 2H^+ + 2e^- \qquad \ldots(18.2)$$

The resulting biosensor offered good accuracy and precision in connection with 100 µl blood samples. The most suitable concept for glucose determination rely on the use of the natural oxygen co-substrate and generation and detection of hydrogen peroxide (equations 18.1 and 18.2). The biocatalytic reaction involves reduction of the flavin group (FAD) in the enzyme by reaction with glucose to give the reduced form of the enzyme ($FADH_2$) followed by re-oxidation of the flavin by molecular oxygen to regenerate the oxidised form of the enzyme GO_x(FAD). The resulting electric current from the re-oxidation on the working electrode at applied constant potential is proportional to the glucose concentration:

$$GO_x(\text{FAD}) + \text{Glucose} \rightarrow GO_x(\text{FADH}_2) + \text{Gluconolactone } GO_x(\text{FADH}_2) + O_2 + GO_x(\text{FAD}) + H_2O_2$$

This model is now utilised in the most commercially successful glucose biosensors utilising glucose oxidase or glucose dehydrogenase. Varieties of enzymes were used for biosensor construction, for example oxidoreductase enzymes were used for lactate, malate, and ascorbate. This type of biosensor has been used widely throughout the world for glucose testing in the home bringing diagnosis to on site analysis.

Array-type Biosensors

The rapid detection and monitoring of toxins in clinical fluids require new approaches in order to expedite appropriate countermeasures. For example, many toxins are secreted by bacteria during the course of infection and should be detected in low quantities (ngmL^{-1}) in urine or blood products following intoxication. In recent years, the fabrication of biosensors able to distinguish multiple analytes in a single sample has become an increasingly well recognised research goal. In such bioassays, it is imperative to implement the sensor in an array format. Instruments for reading microarray systems and microplate readers are a few examples. The detectors for such systems need to measure the analyte quantity at different locations, which is typically carried out sequentially by a single detector across the array (scanning) or to dedicating an individual detector to each site. Most array-biosensors require a single measurement (usually when the assay reaches its biochemical equilibrium) per detection site (pixel). Others require the capturing of the reaction kinetics, necessitating multiple measurements per pixel. For instance, fluorescence detection in microarray systems is extensively used for a variety of applications. These include but are not limited to nucleic acid/protein detection and quantification, DNA sequencing, blotting, and real-time PCR analysis. The single frequency excitation induces photon emission (at a different frequency) from the fluorescent label in the sample material. Filtration of the emission spectrum and detection follow to form the fluorescence image of the sample. There are two types of fluorescence detection systems for array-based platforms. In one system, the excitation source is scanned across the array and a single-pixel very high-performance detector such as a photomultiplier tube is used for detection. The other approach is to use a homogeneous excitation light source for the entire assay at once and measure fluorescence emission at different pixels through a 2D array of detectors e.g. a CCD camera. In these implementations, the resolution of the CCD camera and the number of photosensitive pixels determine the possible array size.

Array-type biosensors offer many advantages over the conventional analytical methods, the most significant of which are: (i) a variety of analytes can be investigated simultaneously in the same sample, (ii) the required sample quantities are minimal, (iii) low consumption of scarce reagents, (iv) high miniaturisation and (v) high sample throughput.

These advantages become very evident if we consider the workflow during a typical drug screening process. At the beginning there is the need for a selection of few eligible compounds out of a large variety of molecules for a given purpose. Since the most limiting factor is the ratio between the number of data points per day and the cost per data point, this first mass-selection is best done at a molecular level where DNA-chips and protein microarrays find their most common application.

The biochip is used to simultaneously analyse a panel of related tests in a single sample, producing a patient profile. The patient profile can be used in disease screening, diagnosis, monitoring disease progression or monitoring treatment. Performing multiple analyses simultaneously, described as multiplexing, allows a significant reduction in processing time and the amount of patient sample required. Biochip Array Technology is a novel application of a familiar methodology, using sandwich, competitive and antibody-capture immunoassays. The difference from conventional immunoassays is that the capture ligands are covalently attached to the surface of the biochip in an ordered array rather than in solution.

Microbial Biosensor

Micro-organisms have been integrated with a variety of transducers such as amperometric, potentiometric, conductimetric, luminescence and fluorescence to construct biosensor devices. Since microbial biosensor response, operational stability and long-term use are, to some extent, a function of the immobilisation strategy used, immobilisation technology plays a very important role and the choice of immobilisation technique is critical. Micro-organisms can be immobilised on transducer or support matrices by chemical or physical methods. Based on the sensing technique, recent reported microbial biosensors can be classified into two major groups: electrochemical microbial biosensors and optical microbial biosensors. Amperometric microbial biosensors have been extensively exploited for the determination of biochemical oxygen demand (BOD) for the measurement of biodegradable organic pollutants in aqueous samples. Amperometric microbial biosensors also provide rapid and sensitive tools in health and fermentation applications. The detection of glucose, which is of great interest in the diagnosis of diabetes and quality control of fermentation and food, accounts for about 90% of the entire biosensor market. Several microbial biosensors for glucose detection have been fabricated based on the oxygen consumption of the respiratory activity in the microbes. Generally, the bacteria, which can uptake glucose, also possess the enzyme activity to metabolise other carbohydrates, such as galactose, catechol, mannose, and xylose. Selective biosensors can still be developed for different sugars as long as the bacteria are adapted to the specific analyte in advance through the selective cultivation. The qualitative and quantitative detection of alcohols with high sensitivity, selectivity, and accuracy is required in many fields. Several micro-organisms which can metabolise ethanol with the consumption of oxygen, such as *G. oxydans*, *Pichia angusta*, and *Candida tropicalis* have been applied to the construction of ethanol whole-cell biosensors. Potentiometric microbial biosensors detect the amount of analytes by measuring the potential difference between the working electrode and the reference electrode separated by a selective membrane. Recently, a potentiometric biosensor based on the pH electrode modified by permeabilised *P. aeruginosa* was developed for selective and rapid detection of cephalosporin group of antibiotics. The hydrolysis of cephalosporin, due to the enzyme activity of the microbial layer, was accompanied by the production of protons near the pH electrode. The response came from the change of electric potential difference between the working electrode and the reference electrode. Another potentiometric biosensor for the identification of β-lactam residues in milk was also reported.

Microbial biosensors have been under extensive investigation over decades. Particularly, some electrochemical and optical microbial biosensors developed for environmental applications have been commercialised. For example, commercial on-line BOD microbial biosensors were available from Biosensores SL Moncofar, Spain and Isco GmbH, Gross Umstadt, Germany. The Green Screen Environmental Monitoring (EM) with a yeast cellular sensing element was designed for the simultaneous detection of genotoxicity and cytotoxicity by Gentronix Ltd., Manchester, UK. In addition, amperometric and bioluminescent whole-cell toxicity biosensors have been developed by Euroclone Ltd., West Yorkshire, UK and Remedios Ltd., Aberdeen, UK. Nevertheless, commercial microbial biosensors are just tips of the iceberg compared to the great amount of academic research on them. The intrinsic disadvantages (slow response, low sensitivity, and poor selectivity) using micro-organisms as the biosensing element limit the widespread interest of microbial biosensors on the market. Microbial biosensors typically suffer from the poor selectivity because of the non-specific cellular response to substrates. With the development of biotechnology and the availability of genome sequence for more micro-organisms, we can genetically engineer microbes with specific metabolic pathways up-regulated or downregulated, thus providing enhanced selectivity to specific targets. Another way to improve the selectivity of microbial biosensors is to develop microbial sensor arrays. The introduction of analyte to the microbial sensor arrays will generate a finger-printed response pattern. By combining with artificial neural network analysis, the target compound can be identified.

NANOBIOSENSORS: BASIC CONCEPTS AND APPLICATIONS

Advances in nanotechnology have led to the development of nanoscale biosensors that have exquisite sensitivity and versatility. The ultimate goal of nanobiosensors is to detect any biochemical and biophysical signal associated with a specific disease at the level of a single molecule or cell. They can be integrated into other technologies such as lab-on-a-chip to facilitate molecular diagnostics. Their applications include detection of micro-organisms in various samples, monitoring of metabolites in body fluids and detection of tissue pathology such as cancer. Their portability makes them ideal for pathogenesis of cancer applications but they can be used in the laboratory setting as well.

The ability to detect disease-associated biomolecules, such as disease-specific metabolites, nucleic acids, proteins, pathogens, and cells such as circulating tumour cells, is essential not only for disease diagnosis in the clinical setting but also for biomedical research involving drug discovery and development. Nanotechnology, with its enhanced sensitivity and reduced instrumentation size, will rapidly improve our current biodiagnostic capacity with respect to specificity, speed, and cost. Reduction in sensor size provides great versatility for incorporation into multiplexed, transportable, portable, wearable, and even implantable medical devices. The integration of nanoscale ultrasensitive biosensors with other medical instruments will open the door to emerging medical fields, including point-of-care diagnostics and ubiquitous healthcare systems. The biomedical application of nanobiosensors is wide, moreover, the future impact of nanobiosensor systems for point-of-care diagnostics will be unmatched. This technology will revolutionise conventional medical practices by enabling early diagnosis of chronic debilitating diseases, ultrasensitive detection of pathogens, and long-term monitoring of patients using biocompatible integrated medical instrumentation.

There are different strategies for creating next generations of nanobiosensor devices: (i) the use of a completely new class of nanomaterial for sensing purposes, (ii) new immobilisation strategies, and (iii) the new nanotechnological approaches. In the section, current state-of-the art principles of nanobiosensor systems are discussed along with future perspectives.

Nanomaterials for New Biosensing Principles

One of the first new nanomaterials to impact on amperometric biosensors was the carbon nanotube (CNT), which was blended into a number of formulations to improve current densities and overall performance of enzyme electrodes and enzyme-labelled immunosensors. Amperometric enzyme electrodes benefitted from enhanced reactivity of NADH and hydrogen peroxide at CNT-modified electrodes and aligned CNT 'forests' appeared to facilitate direct electron transfer with the redox centers of enzymes. The most widely used nanomaterial in industry overall to date, however, is the silver nanoparticle. These have also been harnessed as a simple electrochemical label in a highly sensitive amperometric immunoassay intended for distributed diagnostics and as an inexpensive solution for immunoassays performed in developing countries. In this electrochemical sandwich immunoassay, silver nanoparticles are used as a robust label, which can be solubilised after the binding reaction has occurred, using thiocyanate, to form a silver chelate. This benign chemistry replaces earlier versions using aggressive chemical oxidants such as nitric acid. Once solubilised, the silver concentration can be very sensitively determined using stripping voltammetry on a single-use screen-printed carbon electrode. The silver colloid aggregates due to the presence of thiocyanate and the negatively charged aggregates are attracted to the positive potential of the carbon electrode during the pre-treatment. Once in direct contact with the electrode surface, the silver is oxidised at 0.6 V to form soluble silver ions, which are immediately complexed by the thiocyanate and detected by the ensuing anodic stripping voltammetry. Hence, the analyte concentration yields a signal which is directly proportional to the anodic stripping voltammetry peak of silver. In one example, the cardiac marker myoglobin, was measured down to 3 ng mL^{-1}, which was comparable with the conventional enzyme-linked immunosorbent assay (ELISA). Samples volumes of less than 50 mL could be handled and the assay worked in turbid solutions without the need for sample clean up.

A variety of other nanoparticle-based strategies have been described in the literature for electrochemical affinity assays. Most recently, nanostructured materials have been used to deliver label-free electrochemical immunoassays. Gooding's group in Australia described a direct electrochemical immunosensor for detection of veterinary drug residues in undiluted milk. They used a displacement assay for with a mixed layer of oligo(phenylethynylene) molecular wire, to facilitate electrochemical communication, and oligo(ethylenelycol) to control the interaction of proteins and electroactive interferences with the electrode surface. More recently, Turner and others reported on the use of a highly conductive N-doped graphene sheet-modified electrode, which exhibited significantly increased electron transfer and sensitivity towards the breast cancer marker CA 15-3. This label-free immunosensor delivered a low detection limit of 0.012 U mL^{-1} and worked well over a broad linear range of 0.1–20 U mL^{-1}.

Despite using new nanotechnologies for biosensors the application of nanomaterials to bioanalytics in array-type assays or *in vivo* monitoring is currently a replacement of organic dyes, radioactive or metal labels and contrast agents by metal, oxide or luminescent nanocrystals. Such methods have to be used to investigate metabolic pathways on cellular levels where conventional device-based nanobiosensors have no chance to measure. Using such new labelling nanomaterials the bioanalytical and imaging methods remain mostly unchanged, whereas the tagged or labelled biomolecule is replaced by a bionano-system. The conjugation between biomolecule and nanocrystal is crucial for every bionano-system as it determines the overall biological properties of the conjugate.

Immobilisation Strategies at the Nanoscale

Since the development of the first biosensor, biosensors technology has experienced a considerable growth in terms of applicability and complexity of devices. In the last decade this growth has been accelerated due the utilisation of electrodes -modified nanostructured materials in order to increase the power detection of specific molecules. Other important feature can be associated with the development of new methodologies for biomolecules immobilisation. This includes the utilisation of several biological molecules such as enzymes, nucleotides, antigens, DNA, amino acids and many others for biosensing. Moreover, the utilisation of these biological molecules in conjunction with nanostructured materials opens the possibility to develop several types of biosensors such as nanostructured and miniaturised devices and implantable biosensors for real time monitoring. The interface between the nanostructure and the biomolecule requires significant attention as it dictates the biosensor performance and sensitivity. Based on the physical and chemical properties of both the nanostructure and the biomolecule, a number of immobilisation methods have been proposed. The key problem during the immobilisation is how to fully maintain the biomolecule's conformation and activity. Non-specific biomolecule adsorption onto the nanostructure is the initial stage of the degradation mechanisms that will ultimately compromise the functionality of the biosensor. The different methods of conjugation between nanostructures and biomolecules can be divided into three categories.

The first category includes methods where biomolecules are bound non-covalently to nanoparticles. Therefore, nanoparticles are first derivatised with a chemisorbed monolayer or the capping agent from synthesis to have hydrophobic surfaces. In a second step, these hydrophobic nanoparticles are precipitated and redissolved in water within tensidic micelles. In principle, this method works with all common micelle building agents such as phospholipids and sodium dodecylsulfate. In a final step, biomolecules are coupled covalently to functional groups at the outer sphere of the micelles. A major advantage of this method is that the whole process, from non-polar/polar solvent transfer to the coupling, is relatively easy to perform. The bond between nanoparticles and biomolecules is based on hydrophobic interactions within the micelles. Therefore, the conjugate disintegrates relatively easily.

The second category contains methods in which biomolecules are chemisorbed onto nanoparticles by means of a 'linker'. This can be realised in two variations: first, the biomolecules contain surface active groups such as, e.g. thiols, and are directly chemisorbed onto the nanoparticles. Second, a bifunctional molecule is chemisorbed onto the nanoparticles and biomolecules are coupled to these molecules in a second step, similar to the micelles from the first category. Chemisorption of thiols onto gold surfaces is well known and as long as the adsorption energy is less than '40 kJ mol^{-1}, this bond has mostly covalent character. However, from a practical point of view, on a longer time-scale these conjugates can disintegrate by desorption, what could become critical for long-term experiments in the range of days.

The third category of coupling methods includes methods in which biomolecules are bound covalently to modified nanoparticles. Therefore, the nanoparticles have to be derivatised with a cross-linked surface shell, which contains binding sites for biomolecules. This cross-linked surface shell could consist of functionalised polymers or inorganic networks like silica. Second, the biomolecules have to be coupled to these surface shells. Such conjugates are very stable due to covalent bonds. The major disadvantage of these methods is the sometimes difficult and costly preparation. Compared to the other categories these methods are recommended when long-term stability of the conjugate is necessary.

In conclusion, it can be stated that different methods of producing bionanosystems with different advantages and disadvantages are available. The major problem with all of them is that the biomolecule is turned into a colloid by attaching it to a nanocrystal. Because colloids have very different 'solubility' from biomolecules, there is always a tendency for coagulation within biological media.

EXAMPLES OF NANOBIOSENSORS

Nanowire Biosensors

Nanowire biosensors are a class of nanobiosensors, of which the major sensing components are made of nanowires coated by biological molecules such as DNA molecules, polypeptides, fibrin proteins, and filamentous bacteriophages. A bionanowire is a one-dimensional fibrillike nanostructure, with the diameter constrained to tens of nanometers or less and unconstrained length. Since their surface properties are easily modified, nanowires can be decorated with virtually any potential chemical or biological molecular recognition unit, making the wires themselves analyte independent. The nanomaterials transduce the chemical binding event on their surface into a change in conductance of the nanowire in an extremely sensitive, real time and quantitative fashion. One dimensional nanowires, nanotubes, nanobelts and nanosprings have become the focus of intensive research in biosensing due to their unique properties and their potential for fabrication into high density nanoscale devices. The nanowires can be used for both efficient transport of electrons and optical excitation, and these two factors make them critical to the function and integration of nanoscale devices. In fact, they are the smallest dimension structures that can be used for efficient transport of electrons and are thus critical to the function and integration of these nanoscale devices. Because of their high surface-to-volume ratio and tunable electron transport properties due to quantum confinement effect, their electrical properties are strongly influenced by minor perturbations. One of the excellent candidates for development of enzyme/protein-based biosensors is the CNT, due to its unique electric, electrocatalytic, and mechanical properties. Wang and co-workers employed CNT/Nafion-based electrodes for immobilising glucoseoxidase (GOx) enzyme for sensitive detection of glucose. This CNT/Nafion composite was prepared by dispersing solubilised CNTs in Nafion solution onto an electrode surface. The CNT-based biosensor offers substantially greater signals, especially at low potential, reflecting the electrocatalytic activity of CNTs. Such low potential operation of the CNT-based biosensor results in a wide linear range and a fast response time. Boron-doped silicon nanowires (SiNWs) have been used to create highly sensitive, real-time electrically based sensors for biological and chemical species. Biotin-modified SiNWs were used to detect streptavidin down to at least a picomolar concentration range. The small size and capability of these semiconductor nanowires for sensitive, label-free, real-time detection of a wide range of chemical and biological species could be exploited in array-based screening and *in vivo* diagnostics. Furthermore, the highly ordered nanowires array combined with multiple biorecognition holds the promise of developing multiplexed nanobiosensors. They will be very useful for high-throughput diagnosis and screening. Due to their small size and robustness, the nanowires are a good candidate material for fabricating nanoscale biosensors for in-body biosensing and for making remote-controlled nanbiosensors for environmental monitoring. In general, nanobiosensors based on nanowires such as CNTs show great promise for future applications in health-care testing, disease diagnostics and environmental monitoring.

Cantilever Biosensors

The best example of the use of nanotransducers is cantilever based biosensors which utilise a micromechanically produced cantilever in a similar manner as for the production of (atomic force microscopy) AFM probes. A several 100 nm thick cantilever is bent due to biosensing interaction on the surface, which can be optically sensed by a laser. The sensitivity can be tuned down to single molecule interaction analysis. The high sensitivity of microcantilever sensors has proven to be a powerful platform for detecting molecular interactions in a label-free, time resolved manner. By an asymmetrical chemisorption of molecules (i.e. on one side of the microcantilever), the sensors can detect processes in 'static' mode by measuring the bending of a microcantilever due to stress formation during the adsorption process, or in 'dynamic' mode where the resonant frequency of an oscillating microcantilever shifts due to mass adsorption on its surface. The versatility of the microcantilever technique as a chemical/biological sensor has been demonstrated for vapours, ions, DNA, proteins, antibiotics and pathogenic micro-organisms. The mechanical sensitivity of the static mode technique stems from changes in surface stress caused by molecular interactions with the surface (change in the electronic charge distribution of the substrate's surface atoms) and by lateral interactions within the molecular layer (electrostatic forces, structural changes and steric competition).

This sensitivity to structural changes in static mode operation has shown to be particularly suited for measuring binding processes based on conformational changes of molecules attached to the microcantilever's surface such as proteins, DNA or lipid bilayers. Recently, Bumbu and others applied the static mode technique to study the behaviour of poly(methyl methacrylate) brushes that had been polymerised from the silicon surface of a microcantilever sensor, i.e. using a 'grafting from' approach. While this allowed the authors to study the *in situ* swelling and collapse of poly(methyl methacrylate) brushes, the kinetics of brush formation could not be monitored in real-time. The driving impetus behind this work is to apply microcantilever sensors operated in static mode to study in real-time: (i) the kinetic aspects of surface PEGylation, and (ii) conformational changes in the PEG layer over a timescale of tens of minutes *in situ*.

In the last decade, several research groups observed that microcantilevers can transduce a number of different signal domains, e.g. mass, temperature, heat, electromagnetic field, stress, into a mechanical deformation: either a bending or a change in the resonance frequency, with a resolution which is orders of magnitude higher than that achievable with macroscopic structures. Over the last years, the number of applications of these sensors has shown a fast growth in diverse fields, such as genomic or proteomic, because of the biosensor flexibility, the low sample consumption, and the non pretreated samples required.

Ion-channel Based Sensing

Biological ion channels are water-filled subnanosised pores formed by protein molecules in the membranes of all living cells. Ion channels play a crucial role in living organisms by selectively regulating the flow of ions into and out of a cell thereby controlling the cell's electrical and biochemical activities. New generations of nanobisensors have been developed in which the conductance of a population of molecular ion channels is switched by the recognition event. The approach mimics biological sensory functions and can be used with most types of receptor, including antibodies and nucleotides. The technique is very flexible and even in its simplest form it is sensitive to picomolar concentrations of proteins. The sensor is essentially an impedance element whose dimensions can readily be reduced to become an integral component of a microelectronic circuit. It may be used in a wide range of applications and in complex media, including blood. These uses might include cell typing, the detection of large proteins, viruses, antibodies, DNA, electrolytes, drugs, pesticides and other low-molecular-weight compounds.

The active elements of the ion-channel switch comprise a gold electrode to which is tethered a lipid membrane containing gramicidin ion channels linked to antibodies. The molecular structure of the tethered membrane results in an ionic reservoir being formed between the gold electrode and the membrane. The ionic reservoir can be accessed electrically through connection to the gold electrode. In the presence of an applied potential, ions flow between the reservoir and the external solution when the channels are conductive. The ion current is switched off when mobile channels diffusing within the outer half of the membrane become cross-linked to antibodies immobilised at the membrane surface. This prevents them forming dimers with channels immobilised within the inner half of the membrane. The number of dimers is measured from the electrical conduction of the membrane. The switch has a high gain, a single channel facilitates the flux of up to a million ions per second. A quantitative model of the biosensor has been verified experimentally. This structure is assembled using a combination of sulphur–gold chemistry and physisorption. The membrane consists of lipids and channels, some immobilised on the gold surface and some diffusing laterally within the plane of the membrane. Thus with a low density of channels and a high density of immobilised antibodies, each channel can access up to 10^3 more capture antibodies than if the gating mechanism were triggered by a directing binding of analyte to the channels. The speed and sensitivity of the biosensor response may be adjusted in direct proportion to the number of binding sites accessible to each mobile channel. This allows for quantitative detection of analyte from sub-picomolar to micromolar concentrations in less than 10 minutes.

Several companies/research groups have developed biosensors based on synthetic lipid monolayers and bilayers. For example, OhmX Corporation is currently developing a reagentless biosensor system using self-assembled monolayers tethered to a gold surface for the electronic detection of biomarkers in clinical samples. Stochastic signal analysis has been employed by Bayley's group at Oxford and has made substantial contributions in advancement of ion channel biosensors.

The fabrication of the ion-channel switch (ICS) biosensor has several interesting properties that make it an appealing case study. The ICS biosensor incorporates a self-assembled monolayer providing enhanced stability. The tethered bilayer permits 2-D diffusion of gramicidin channels that provides a remarkable gating mechanism. Since gramicidin has a terminal ethanolamine group that permits a range of chemistries, the biosensor may be prepared for use with a wide range of receptors to detect many different analytes. The ICS sensing mechanism does not require washing (unlike an ELISA assay), provides large transduction amplification (millions of ions for every channel dimerisation), and a high detection sensitivity since a single channel can diffuse and identify analyte molecules bound to many capture sites. The ICS biosensor also provides an objective electrical readout that is intrinsically digital.

The digital output permits the use of sophisticated statistical signal processing algorithms to estimate the type and concentration of analyte.

NANOBIOSENSORS IN NANOMEDICINE

Nanomedicine involves cell-by-cell regenerative medicine, either repairing cells one at a time or triggering apoptotic pathways in cells that are not repairable. Multilayered nanoparticle systems are being constructed for the targeted delivery of gene therapy to single cells. Cleavable shells containing targeting, biosensing, and gene therapeutic molecules are being constructed to direct nanoparticles to desired intracellular targets. Therapeutic gene sequences are controlled by biosensor-activated control switches to provide the proper amount of gene therapy on a single cell basis. The central idea is to set up gene therapy 'nanofactories' inside single living cells. Molecular biosensors linked to these genes control their expression. Gene delivery is started in response to a biosensor detected problem, gene delivery is halted when the cell response indicates that more gene therapy is not needed. Cell targeting of nanoparticles, both nanocrystals and nanocapsules, has been tested by a combination of fluorescent tracking dyes, fluorescence microscopy and flow cytometry. Intracellular targeting has been tested by confocal microscopy. Successful gene delivery has been visualised by use of GFP reporter sequences. DNA tethering techniques were used to increase the level of expression of these genes. Integrated nanomedical systems are being designed, constructed, and tested *in vitro*, *ex vivo*, and in small animals. While still in its infancy, nanomedicine represents a paradigm shift in thinking – from destruction of injured cells by surgery, radiation, chemotherapy to cell-by-cell repair within an organ and destruction of non-repairable cells by natural apoptosis.

Thus, simple, easy-to-use measurement devices for a diverse range of biologically relevant analytes have an intuitive appeal as portable or pocket-sized analysers, and this has driven the diverse range of applications reported in the literature. However, both historical precedent and a critical analysis of potential markets leads to an indisputable conclusion that healthcare is and will continue to be the most important area for the application of biosensors. The maintenance of health is one of the most laudable technological objectives challenging science and technology and diagnosis is an essential prerequisite for treatment and prevention of disease. Moreover, related applications of biosensors, such as the maintenance of food safety and environmental monitoring can be aligned with this central objective. The developing world has a desperate need for robust diagnostics that can be deployed in the field by both healthcare professionals and volunteers. Infectious diseases account for around a quarter of worldwide deaths, although they are projected to decline as a percentage of total deaths over the coming 20 years, as other cause become more prevalent. In developing countries we are faced with diseases of poverty such as HIV/AIDS and tuberculosis, where the former kills 2 million people each year and the latter still affects around a third of the world's population and accounts for an estimated 1.6 million deaths, according to the WHO, although the incidence has been falling globally at a rate of 2.2% in recent years. In addition there are 2.5 million deaths from diarrheal infections and almost 800000 from malaria. Of the estimated 80 million global deaths in 2015, 36 million (63%) were due to non-communicable diseases.

Technology needs to offer more economic solutions and distributed diagnostics enabled by biosensors and enhanced by consumer products available over-the-counter are a key part of the solution. This is also commercially attractive, with in vitro diagnostics already worth an estimated US$40 billion per year. While glucose biosensors for diabetes have had the most profound effect on disease management to date, biosensors for other metabolites promise utility for other noncommunicable diseases such as kidney disease, which is increasingly being recognised as an emerging problem in a rapidly ageing population. Multifarious affinity biosensors have been described to detect cardiac disease markers such as creatine kinase and troponin, while cancer markers and single cell cancer detection have attracted considerable recent literature.

Thus this chapter has illustrated that biosensors have achieved considerable success both in the commercial and academic arenas and that the need for new, easy-to-use, home and decentralised diagnostics is greater than ever. The enormous success of the glucose sensor serves as a model for future possibilities and should not over shadow the multifarious other applications that this versatile technology can address. Emerging science, driving new sensors to deliver the molecular information that underpins all this, includes the development of semisynthetic ligands that can deliver the exquisite sensitivity and specificity of biological systems without the inherent instability and redundancy associated with natural molecules. Currently affibodies, peptide arrays and molecularly imprinted polymers are particularly promising research directions in this respect. Chances of success are enhanced by the potential utility of some of these materials for novel therapeutic, antimicrobial and drug release strategies, since these complimentary areas will drive investment in these approaches.

Chapter 19

Instrumentation and Control Systems

INTRODUCTION

The widespread use of advanced control and process automation for biochemical applications has been lagging as compared with industries such as refining and petrochemicals whose feedstocks are relatively easy to characterise and whose chemistry is well understood and whose measurements are relatively straight forward. Biological processes are extraordinarily complex and subject to considerable variability. The reaction kinetics cannot be completely determined in advance in a fermentation process because of variations in the biological properties of the inoculant. Therefore, information regarding the activity of the process must be gathered as the fermentation progresses. Directly measuring all the necessary variables which characterise and govern the competing biochemical reactions, even under optimum laboratory conditions, is not yet achievable. Developing mathematical models which can be utilisad to infer the biological processes underway from the measurements available, although useful, is still not sufficiently accurate. Add to this the constraints and compromises imposed by the manufacturing process and the task of accurately predicting and controlling the behaviour of biological production processes is formidable indeed. This chapter discusses some of the more innovative measurement and control instrumentation and systems available as well as to review the more traditional measurement, control and information analysis technologies currently in use.

MEASUREMENT TECHNOLOGY

Measurements are the key to understanding and therefore controlling any process. As it relates to biochemical engineering, measurement technology can be separated into three broad categories. These are biological, such as cell growth rate, florescence, and protein synthesis rate, chemical, such as glucose concentration, dissolved oxygen, pH and offgas concentrations of CO_2, O_2, N_2, ethanol, ammonia and various other organic substances, and physical, such as temperature, level, pressure, flow rate and mass. The most prevalent are the physical sensors while the most promising for the field of biotechnology are the biological sensors. It is necessary to prevent foreign organisms from contaminating the process. In-line measurement devices must conform to the International Sanitary Standards specifying the exterior surface and materials of construction for the 'wetted parts.' Instruments must also be able to withstand steam sterilisation which is needed periodically to prevent bacterial buildup. Devices located in process lines should be fitted with sanitary connections to facilitate their removal during extensive clean-in-place and sterilise-in-place operations, Sample ports, used for the removal of a small portion of the contents from the bioreactor for analysis in a laboratory, must be equipped with sterilisation systems to ensure organisms are not inadvertently introduced during the removal of a sample.

BIOSENSORS

Biosensors are literally the fusing of biological substrates onto electric circuits. These have long been envisioned as the next generation of analytical sensors measuring specific biomolecular interactions. The basic principle is first to immobilise one of the interacting molecules, the ligand, onto an inert substrate such as a dextran matrix which is bonded (covalently bound) to a metal surface such as gold or platinum. This reaction must then be converted into a measurable signal typically by taking advantage of some transducing phenomenon.

Four popular transducing techniques are:

1. Potentiometric or amperometric, where a chemical or biological reaction produces a potential difference or current flow across a pair of electrodes.

2. Enzyme thermistors, where the thermal effect of the chemical or biological reaction is transduced into an electrical resistance change.

3. Optoelectronic, where a chemical or biological reaction evokes a change in light transmission.

4. Electrochemically sensitive transistors whose signal depends upon the chemical reactions underway.

One example is the research to produce a biomedical device which can be implanted into a diabetic to control the flow of insulin by monitoring the glucose level in the blood via an electrochemical reaction. One implantable glucose sensor, designed by Leland Clark of the Childrens Hospital Research Center in Cleveland, utilises a microprobe where the outside wall is constructed of glucose-permeable membrane such as cuprophan. Inside, an enzyme which breaks the glucose down to hydrogen peroxide is affixed to an inert substrate. The hydorgen peroxide then passes through an inner membrane, constructed of amaterial such as cellulose acetate, where it reacts with platinum producing a current which is used to monitor the glucose concentration. A commercial example of a biosensor, introduced by Pharmacia Biosensor, is utilising a photoelectric principle called *sufluceplusmon resonance* (SPR) for detection of changes in concentration of macromolecular reactants. This principle relates the energy transferred from photons bombarding a thin gold film at the resonant angle of incidence to electrons in the surface of the gold. This loss of energy results in a loss of reflected light at the resonant angle.

The resonant angle is affected by changes in the mass concentration in the vicinity of the metal's surface which is directly correlated to the binding and dissociation of interacting molecules.

Pharmacia claims its BIAcore system can provide information on the affinity, specificity, kinetics, multiple binding patterns, and cooperativity of a biochemical interaction on line without the need of washing, sample dilution or labelling of a secondary interactant. Their scientists have mapped the epitope specificity patterns of thirty monoclonal antibodies (Mabs) against recombinant core HIV-1 core protein.

CELL MASS MEASUREMENT

The on-line direct measurement of cell mass concentration by using optical density principles promises to dramatically improve the knowledge of the metabolic processes underway within a bioreactor. This measurement is most effective on spherical cells such as *E. Coli*. The measurement technology is packaging a sterilisable stainless steel probe which is inserted directly into the bioreactor itself via a flange or quick-disconnect mounting. By comparing the mass over time, cell growth rate can be determined. This measurement can be used in conjunction with metabolic models which employ such physiological parameters as oxygen uptake rate (OUR), carbon dioxide evolution rate (CER) and respiratory quotient (RQ) along with direct measurements such as dissolved oxygen concentration, pH, temperature, and offgas analysis to more precisely control nutrient addition, aeration rate and agitation. Harvest time can be directly determined as can shifts in metabolic pathways possibly indicating the production of an undesirable by-product. Cell mass concentrations of up to 100 grams per liter are directly measured using the optical density probe. In this probe, light of a specific wavelength, created by laser diode or passing normal light through a sapphire crystal, enters a sample chamber containing a representative sample of the bioreactor broth and then passes to optical detection electronics.

The density is determined by measuring the amount of light absorbed, compensating for backscatter. Commercial versions such as those manufactured by Cerex, Wedgewood, and Monitec are packaged as stainless steel probes that can be mounted directly into bioreactors ten liters or greater, and offer features such as sample debubblers to eliminate interference from entrained air. Another technique used to determine cell density is spectrophotometric titration which is a laboratory procedure which employs the same basic principles as the probes discussed above. This requires a sample to be withdrawn from the broth during reaction and therefore exposes the batch to contamination.

CHEMICAL COMPOSITION

The most widely used method for determining chemical composition is chromatography. Several categories have been developed depending upon the species being separated. These include gas chromatography and several varieties of liquid chromatography including low pressure (gel permeation) and high pressure liquid chromatography and thin layer chromatography. The basic principle behind these is the separation of the constituents travelling through a porous, sorptive material such as a silica gel. The degree of retardation of each molecular species is based on its particular affinity for the sorbent. Proper selection of the sorbent is the most critical factor in determining separation. Other environmental factors such as temperature and pressure also play a key role. The chemical basis for separation may include adsorption, covalent bonding or pore size of the material. Gas chromatography is used for gases and for liquids with relatively low boiling points. Since many of the constituents in a biochemical reaction are of considerable molecular weight, high pressure liquid chromatography is the most commonly used. Specialisad apparatus is needed for performing this analysis since chromatograph pressures can range as high as 10,000 psi. Thin layer chromatography requires no pressure but instead relies on the capillary action of a solvent through a paper-like sheet of sorbent. Each constituent travels a different distance and the constituents are thus separated. Analysis is done manually, typically using various colouring or fluorescing reagents. Gel permeation chromatography utilises a sorbent bed and depends on gravity to provide the driving force but usually requires a considerable time to effect a separation. All of these analyses are typically performed in a laboratory, therefore they require the removal of samples. As the reaction is conducted in a sterile environment, special precautions and sample removal procedures must be utilisad to prevent contaminating the contents of the reactor.

DISSOLVED OXYGEN

Dissolved oxygen is one of the most important indicators in a fermentation or bioreactor process. It determines the potential for growth. The measurement of dissolved oxygen is made by a sterilisable probe inserted directly into the aqueous solution of the reactor. Two principles of operation are used for this measurement: the first is an electrochemical reaction while the second employs an amperometric (polarographic) principle.

The electrochemical approaches a sterilisable stainless steel probe with a cell face constructed of a material which will enable oxygen to permeate across it and enter the electrochemical chamber which contains two electrodes of dissimilar reactants (forming the anode and cathode) immersed in a basic aqueous solution. The entering oxygen initiates an oxidation reduction reaction which in turn produces an EMF which is amplified into a signal representing the concentration of oxygen in the solution.

In the amperometric (polarographic) approach, oxygen again permeates a diffusion barrier and encounters an electrochemical cell immersed in basic aqueous solution. A potential difference of approximately 1.3 V is maintained between the anode and cathode. As the oxygen encounters the cathode, an electrochemical reaction occurs:

$$O_2 + 2H_2O + 4e^- + 4OH^- \text{ (at cathode)}$$

The hydroxyl ion then travels to the anode where it completes the electrochemical reaction process:

$$4OH^- + O_2 + 2H_2O + 4e^- \text{ (at anode)}$$

The concentration of oxygen is directly proportional to the amount of current passed through the cell.

EXHAUST GAS ANALYSIS

Much can be learned from the exchange of gases in the metabolic process such as O_2, CO_2, N_2, NH_2, and ethanol. Infact, most of the predictive analysis is based upon such calculations as oxygen uptake rate, carbon dioxide exchange rate or respiratory quotient. This information is best obtained by a component material balance across the reactor. A key factor in determining this is the analysis of the bioreactor offgas and the best method for measuring this is with a mass spectrometer because of its high resolution. Two methods of operation are utilisad. These are magnetic deflection and quadrapole. The quadrapole has become the primary commercial system because of its enhanced sensitivity and its ability to filter out all gases but the one being analysed.

Magnetic deflection mass spectrometers inject a gaseous sample into an inlet port, bombard the sample with an electron beam to ionise the particles and pass the sample through a magnetic separator. The charged particles are deflected by the magnet in accordance with its mass-to-energy (or charge) ratio- the greater this ratio, the less the deflection. Detectors are located on the opposing wall of the chamber and are located to correspond to the trajectory of specific components. As the ionisad particles strike the detectors, they generate a voltage proportional to their charge. This information is used to determine the percent concentration of each of the gasses.

The quadrapole mass spectrometer also employs an electron beam to ionise the particles using the quadrapole instead of a magnet to deflect the path of the particles and filter out all but the specific component to be analysed. The quadrapole is a set of four similar and parallel rods with opposite rods electrically connected. A radio frequency and DC charge of equal potential, but opposite charge, is applied to each set of the rods. By varying the absolute potential applied to the rods, it is possible to eliminate all ions except those of a specific mass-to-energy ratio. Those ions which successfully travel the length of the rods strike a Faraday plate which releases electrons to the ions thereby generating a measurable change in EMF. For a given component the strength of the signal can be compared to references to determine the concentration. The quadrapole, when used in conjunction with a gas chromatograph to separate the components, can measure a wide range of gases, typically from 50 to

1000 atomic mass units (amu). As mass spectrometers are relatively expensive, the exhaust gas of three or more bioreactors is typically directed to a single analyser. This is possible because the offgas analysis is done outside the bioreactors themselves. However, the multiplexing of the streams results in added complexity with regard to sample handling and routing, particularly if concerns of cross contamination need be addressed. The contamination issue is usually handled by placing ultrafilters in the exhaust lines. Care, however, must be taken to ensure that these filters don't plug resulting in excessive back pressure. Periodic measurement calibration utilising reference standards must be sent to the spectrometer to check its calibration.

MEASUREMENT OF pH

Metabolic processes are typically highly susceptible to even slight changes in pH, and therefore, proper control of this parameter is critical. Precise manipulation of pH can determine the relative yield of the desired species over competing by-products. Deviations of as little as 0.2 to 0.3 may adversely affect a batch in some cases. Like the cell mass probe and dissolved oxygen probes described earlier, the pH probe is packaged in a sterilisible inert casing with permeable electrode facings for direct insertion into the bioreactor. The measurement principle is the oxidation reduction potential of the hydrogen ion and the electrode materials are selected for that purpose.

WATER PURITY

Water purity is often very important in biochemical processes. One of the best methods to detect the presence of salts or other electrolytic materials is to measure its resistivity. Conductivity or resistivity probes are capable of measuring conductivities as high as 20,000 microsiemens per centimeter and resistivities as high as 20 megoluns per centimeter

TEMPERATURE

Precise temperature control and profiling are key factors in promoting biomass growth and controlling yield. Temperature is one of the more traditional measurements in bioreactors so there is quite a variety of techniques. Filled thermal systems, are among the more traditional temperature measuring devices. Their operating principle is to take advantage of the coefficient of thermal expansion of a sealed fluid to transduce temperature into pressure or movement. This has the advantage of requiring essentially no power and therefore is very popular in mechanical or pneumatic control loops. Although the trend in control is toward digital electronic, pneumatic and mechanical systems are still very popular in areas where solvent or other combustible gases may be present and therefore represent a potential safety hazard. The primary constraint in these types of systems is that the receiver (indicator, recorder, controller) must be in close proximity to the sensor.

Thermocouple assemblies, are a popular measurement choice in electronic systems or in pneumatics where the sensor must be remote. The thermoelectric principle, referred to as the *Seebeck Effect,* is that two dissimilar metals, when formed into a closed circuit, generate an electromotive force when the junction points of the metals are at different temperatures. This conversion of thermal energy to electric energy generates an electric current. Therefore, if the temperature of one juncture point (the cold junction) is known, the temperature of the hot juncture point is determined by the current flow through the circuit. Depending upon the alloys chosen, thermocouples can measure a wide temperature range (-200 to $+350°C$ for copper, constantan) and are quite fast acting assuming the assembly doesn't contribute too much lag in its absorbance and dissipation of heat. Its primary disadvantages are its lack of sensitivity (copper, constantan generates only 40.5 microvolts per °C) and requirement for a precise cold junction temperature reading.

Resistance temperature detectors, RTD's, are more sensitive than thermocouples especially when measuring small temperature ranges. As a result, they are preferred for accurate and precise measurements. The principle behind these devices is based on the use of materials, such as platinum or nickel, whose resistance to current flow changes with temperature. These materials are used as one leg in a Wheatstone bridge circuit with the other legs being known precision resistors. A voltage is applied across the bridge and the voltage drop midway through each path of the circuit is compared. The potential difference at the midway point is directly related to the ratio of each set of resistances in series. Since three of these are known, the resistance of the RTD can be calculated and the temperature inferred. If the RTD is remote from the bridge circuit, the resistance of lead wires can affect the measurement. Therefore, for highly precise measurements, compensating circuits are included which require increasing the wiring for this measuring device from two to as many as four leads. Thermistors are a special class of RTD's and are constructed from semiconductor material. Their primary advantage is their greater sensitivity to changes in temperature, therefore making them a more precise measuring method. Their disadvantage is their nonlinear response to temperature changes. This form of RTD is gaining popularity for narrow range applications, particularly in laboratory environments.

PRESSURE

Pressure is an important controlled variable. The measurement is obtained by exposing a diaphragm surface or seal to the process via a flange or threaded tap through the vessel wall. The signal is translated through a filled capillary to a measurement capsule which will transduce the signal to one measurable by an electronic circuit by one of several methods. One method is to employ a piezoelectric phenomenon whereby the pressure exerted on an asymmetric crystal creates an elastic deformity which in turn causes the flow of an electric charge. A second technology is variable resistance whereby flexure on a semiconductive wafer affects its resistivity which is measured in a similar fashion to RTD's. Several types of pressure measurements can be taken. These include absolute pressure, where one side of the capsule is exposed to 0 psia in a sealed chamber. Gauge pressure is measured with one side of the capsule vented to atmosphere. Vapour pressure transmitter seals one side of the capsule, filling it with the chemical composition of the vapour to be measured. The vapour pressure in the sealed chamber is compared with the process pressure (at the same temperature). If equal, the compositions are inferred to be equal. This technique is used primarily for binary mixtures as multicomponent compositions have too many degrees of freedom.

MASS

Weigh cells or load cells are typically used to measure the mass of the contents of a vessel. These are electromechanical devices which convert force or weight into an electrical signal. The technique is to construct a Wheatstone bridge similar to that used in the RTD circuit with one resistor being a rheostat which changes resistance based on load.

Three configurations are popular. These are the column, where the cell is interposed between one leg of the vessel and the ground (Fig. 19.1) and is typically used for weights exceeding 2500 kg. The second is the cantilever design, where the weight is applied to a bending bar and is used for weights under 250 kg. The third is the shear design, where the weight is applied to the center of a dual strain gage arrangement.

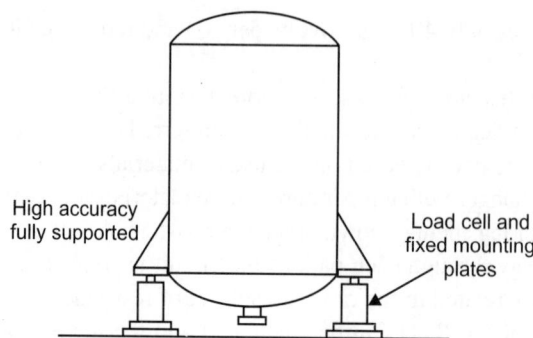

Fig. 19.1: Schematic of the installation of a load cell.

MASS FLOW RATE

A Coriolis meter utilises a measurement technology which is capable of directly measuring mass flow (instead of inferring mass flow from volumetric flow and density). The Coriolis effect is the subtle correction to the path of moving objects to compensate for the rotation of the earth. This phenomenon is used by the mass flow meter to create a vibration whose frequency is proportional to the mass of the fluid flowing through the meter. This is accomplished via the geometry of the meter, specifically the bends to which the fluid is subjected IC it trmdc thrnlloh the meter

VOLUMETRIC FLOW RATE

Quite a number of technologies are available for measuring volumetric flow rates. These include differential pressure transmitters, vortex meters and magnetic flow meters. Each has its advantages and disadvantages. The differential pressure transmitter is the most popular and has been in use the longest. Its measurement principle is quite simple. Create a restriction in the line with an orifice plate and measure the pressure drop across the restriction. The measurement takes advantage of the physical relationship between pressure drop and flow. That is, the fluid velocity is proportional to the square root of the pressure drop, and in turbulent flow, the volumetric flow rate is essentially the velocity of the fluid multiplied by the cross-sectional area of the pipe.

Inaccuracies with regard to transmitting the pressures between the sensor and transducer occur at very low flow rates, therefore closely coupled units have been designed for this purpose. Using this approach and small bore orifice plates, extremely low flows can be measured. A 0.38 millimeter diameter bore can accurately measure flows in the 0.02 liters per minute range for liquids and 0.03 cubic meters per hour for gases. Jewelled orifice plates can have a bore as small as 0.05 millimeters in diameter. The primary disadvantages of the differential pressure producing flow measurements are the permanent pressure drop caused by the restriction in the line, sediment buildup behind the orifice plate (which could be a source of bacterial buildup) and loss of accuracy over time as the edge of the plate is worn by passing fluid and sediment. This type of transmitter typically has a limited range (turndown)-usually a 4 to 1 ratio between its maximum and minimum accurate flow rates.

Vortex meters utilise a precision constructed bar or bluff through the diameter of the flow path to create a disruption in flow which manifests itself as eddy currents or vortices being generated, starting at the downstream side of the bar. The frequency at which the vortices are created are directly proportional to velocity of the fluid. Although these devices contain a line obstruction, the turbulence created by the vortices make the bluff self cleaning and they are available for sanitary applications. Also, their linear nature makes them a wide-range device with a ratio of as much as 20:1 between the maximum and minimum flow rate. Line sizes as small as 1″ are available which are capable of reading flow rates as low as 0.135 liters per minute. Magnetic flow meters take advantage of the electrolytes in an aqueous solution to induce a magnetic field in the coils surrounding the meters flow tube. The faster the flow rate, the greater the induced field. Interestingly, the ionic strength of the electrolytes has only negligible effect on the induced field so long as it is above the threshold value of 2 microsiemens per centimeter. Because these meters create no obstructions to the flow path they are the preferred meter for sanitary applications.

BROTH LEVEL

As the broth in a fermenter or bioreactor becomes more viscous and is subjected to agitation from sparging (the introduction of tiny sterilisad air bubbles at the bottom of the liquid) and from mixing by the impeller, it has a tendency to foam. This can be a serious problem as the level may rise to the point where it enters the exhaust gas lines clogging the ultrafilters and possibly jeopardising the sterile environment within the reactor. Various antifoam strategies can be employed to correct this situation, however, detection of the condition is first required.

Capacitance probes are one means to accomplish this. The basic principle is to measure the charge between two conductive surfaces maintained at different voltage potentials and separated by a dielectric material. The construction of the probe provides an electrode in the center surrounded by an insulator, air, and a conductive shell. The length of the probe is from the top of the reactor to the lowest level measuring point. As the level in the reactor rises the broth displaces the air between the capacitance plates and thereby changes the dielectric constant between the plates to the level of the broth. The result is a change in the charge on the plate. If the vessel wall can act as a plate (is sufficiently conductive), the preferred approach would be to use an unshielded probe (inner electrode with insulator) to prevent erroneous readings resulting from fouling of the probe. Because of the uncertain dielectric character of the broth, this measurement should only be used as a gross approximation of level for instituting antifoaming strategies. Several other forms of level measurement technologies are available. One is the float and cable system, where the buoyancy of the float determines the air-broth interface boundary and the length of the cable determines the level. The density of the broth may render this measurement questionable. A second is hydrostatic tank gauging, where level is inferred from pressure. Again, density, particularly if two phases exist (aqueous and foam), may render this approach questionable. A third is sonic, which computes the distance from the device to the broth surface based on the time it takes for the sound wave initiating from the device to reflect off the surface of the air-liquid boundary and return. Several other ingenious variations of these basic approaches are commercially available as well.

REGULATORY CONTROL

Automatic regulatory control systems have been in use in the process industries for over fifty years. Utilising simple feedback principles, measurements were driven toward their setpoints by manipulating a controlled variable such as flow rate through actuators like throttling control valves. Through successive refinements in first mechanical, then pneumatic, then electronic and finally digital electronic systems, control theory and practice has progressed to a highly sophisticated state.

Single Stage Control

The fundamental building block has been the proportional plus integral plus derivative (PID) controller whereby the proportional term would adjust the manipulated variable to correct for a deviation between measurement and target or setpoint, the integral term would continue the action of the proportional term over time until the measurement reached the setpoint and the derivative term would compensate for lags in the action in the measurement in responding to actions of the manipulated variable.

Judicious application of this control strategy on essentially linear single variable control systems which don't exhibit a prolonged delay (dead time) between action by the manipulated variable and measured response by the controlled variable has proven quite effective. Fortunately most single loop control systems exhibit this behaviour.

In highly nonlinear applications such as pH control, or in situations where the dynamics of the process change over time as occurs in many chemical reactions, adjustments to the tuning coefficients are needed to adequately control the modified process dynamics. When the process under control exhibits significant dead time, the problem is considerably more difficult. One approach is to use a simple model-based predictor corrector algorithm such as the Smith predictor which is interposed between the manipulated and controlled variable in parallel with a conventional controller and conditions the measurement signal to the controller based on time conditioned changes to the manipulated variable made by the controller. This works exceedingly well if properly tuned, but is sensitive to changes in process dynamics. Another scheme, introduced by Shinsky recently, utilises a standard PID controller with a dead time function added to the external reset feedback portion of the loop. This appears to be less sensitive to changes in process conditions.

MULTIVARIABLE CONTROL

Characterising a process as a set of nonlinear time dependent equations and then developing a strategy which manipulates sets of outputs based on changes to the inputs is another approach gaining momentum in other industries such as petroleum refining. One approach is called Dynamic Matrix Control (DMC) which first automates the process of determining the coefficients for the set of nonlinear equations based on sets of controlled and manipulated variables declared. The method perturbs each of the manipulated variables and determines the corresponding response of the controlled variables. Once the model is constructed, the information is represented in a relative gain matrix to predict the control actions necessary to correct for changing process conditions. Once the DMC is correctly tuned, including dynamic compensations, a predictor corrector algorithm is applied to compensate to changes in the process dynamics over time. This technique has been applied quite successfully to reaction processes in the petroleum industry including fluid catalytic cracking units and catalytic reformers.

Batch Control

Batch is a general term given to a diverse set of time dependent control. State variable control, such as the opening and closing of a solenoid or the starting and stopping of a motor, including the use of any timing circuits which may be used for alarming in the event the action doesn't achieve its specified results in the allotted time. The interlocking, sequencing or coordinating of systems of devices to ensure their proper and coordinated operation. Examples include interlocking a discharge pump to the opening of the discharge valve and the alignment of pumps and valves to transfer materials from one vessel to another. This may include actions such as the resetting and starting of totalisers to ensure the proper amount of material was successfully transferred.

The modification of selected process variables in accordance with a prespecified time-variable profile. Two examples are the changing of the reactor temperature over time to conform with a specified profile or the timed periodic addition of nutrient into the bioreactor. Conducting event driven actions such as adding antifoam upon the detection of excess foam or invoking an emergency shut down routine if an exothermic reaction goes beyond controllable limits. Performing a sequence of operations in a coordinated manner to produce the desired changes to the contents of a process unit. This would typically include combinations of the above mentioned activities on various sets of equipment associated with the unit.

ARTIFICIAL INTELLIGENCE

A considerable amount of attention is being given to the use of various forms of artificial intelligence for the control of bioreactor systems. Two forms of systems are currently being explored. These are expert systems and neural networks. Expert systems combine stored knowledge and rules about a process with inference engines (forward and backward chaining algorithms) to choose a best or most reasonable approach among a large number of choices when no correct answer can be deduced and in some situations the information may appear to be contradictory.

Neural networks are also being seriously explored for certain classes of optimisation applications. These employ parallel solution techniques which are patterned after the way the human brain functions. Statistical routines and back propagation algorithms are used to force closure on a set of cross linked circuits. Weighting functions are applied at each of the intersections.

The primary advantage for using neural networks is that no model of the problem is required (some tuning of the weighting functions may facilitate 'learning', however). The user merely furnishes the system with cause and effect data which the programme uses to learn the relationships and thereby model the process from the data. Given an objective function, it can assist in the selection of changes to the causes (manipulated variables) to achieve the optimum results or effects (controlled variables).

DISTRIBUTED CONTROL SYSTEMS

Distributed control systems are organisad into five subsystems. Process interface, which is responsible for the collection of process data from measurement instruments and the issuing of signals to actuating devices such as pumps, motors and valves. Process control, which is responsible for translating the information collected from the process interface subsystem and determining the signals to be sent to the process interface subsystem based on preprogrammed algorithms and rules set in its memory.

Process operations, which is responsible for communicating with operations personnel at all levels including operator displays, alarms, trends of process variables and activities, summary reports, and operational instructions and guidelines. It also tracks process operations and product batch lots.

Applications engines, which are the repository for all of the programmes and packages for the system from control, display and report configuration tools to programme language compilers and programme libraries to specialisad packages such as database managers, spreadsheets and optimisation or expert system packages to repositories for archived process information.

SECTION VI

Enzymes and their Importance in Bioprocesses

Chapter 20

Characteristics of Enzymes

INTRODUCTION

Enzymes are proteins that speed up chemical reactions. Without enzymes most chemical reactions would still occur, but they would happen much to slowly to sustain life. Because the body is essentially a 'chemical processing plant,' enzymes are crucial in every aspect to physiology. All enzymes have certain characteristics in common, which enable them to be effective.

Enzymes are unaffected by the reactions that they speed up: This characteristic greatly increases the efficiency of enzymes, because they can be reused over and over again.

Enzymes are highly specific: Each enzyme generally works with only particular kinds of molecules, called the substrate.

Enzymes can speed up the same chemial reaction going in opposite directions: An enzymes may ordinarily break a molecule into two pieces, but will put it back together again if it is provided only with the pieces.

GENERAL CHARACTERISTICS OF ENZYMES

- The catalytic behaviour of proteins acting as enzymes is one of the most important functions that they perform in living cells.
  - › Without catalysts, most cellular reactions would take place too slowly to support life.
  - › With the exception of some RNA molecules, all enzymes are globular proteins.
  - › Enzymes are extremely efficient catalysts, and some can increase reaction rates by 10–20 times that of the uncatalysed reactions.
- Enzymes are well suited to their roles in three major ways: They have enormous catalytic power, they are highly specific in the reactions they catalyse, and their activity as catalysts can be regulated.

Catalytic Efficiency

- Enzyme-catalysed reaction accomplish many important organic reactions, such as ester hydrolysis, alcohol oxidation, amide formation, etc.

> Enzymes cause these reactions to proceed under mild pH and temperature conditions, unlike the way they are done in a test tube.

- Catalysts increase the rate of chemical reactions without being used up in the process.
  > Although catalysts participate in the reaction, they are not permanently changed, and may be used over and over.
  > Enzymes act like many other catalysts by lowering the activation energy of a reaction, allowing it to achieve equilibrium more rapidly.
  > Enzymes can accomplish in seconds what might take hours or weeks under laboratory conditions.
- The removal of carbon dioxide out of the body is sped up by the enzyme *carbonic anhydrase*, which combines CO_2 with water to form carbonic acid much more quickly than would be possible without the enzyme (36 million molecules per minute).

$$CO_2 + H_2O \xrightarrow[\text{anhydrase}]{\text{Carbonic}} H_2CO_3$$

Specificity

- Enzymes are often very specific in the type of reaction they catalyse, and even the particular substance that will be involved in the reaction.
  > Strong acids catalyse the hydrolysis of any amide or ester, and the dehydration of any alcohol. The enzyme *urease* catalyses the hydrolysis of a single amide, urea.

$$H_2N - \overset{\overset{\text{O}}{\|}}{C} - NH_2 + H_2O \xrightarrow{\text{Urease}} CO_2 + 2NH_3$$

- An enzyme with absolute specificity catalyses the reaction of one and only one substance.
- An enzyme with relative specificity catalyses the reaction of structurally related substances (lipases hydrolyse lipids, proteases split up proteins, and phosphatases hydrolyse phosphate esters).
- An enzyme with stereochemical specificity catalyses the reaction of only one of two possible enantiomers (D-amino acid oxidase catalyses the reaction of D-amino acids, but not L-amino acids).

Regulation

- The catalytic behaviour of enzymes can be regulated.
- A relatively small number of all of the possible reactions which could occur in a cell actually take place, because of the enzymes which are present.
- The cell controls the rates of these reactions and the amount of any given product formed by regulating the action of the enzymes.

Enzymes are Catalysts

A catalyst is a chemical that increases the rate of a chemical reaction without itself being changed by the reaction. The fact that they aren't changed by participating in a reaction distinguishes catalysts from substrates, which are the reactants on which catalysts work. Enzymes catalyse biochemical reactions. They are similar to other chemical catalysts in many ways:

1. Enzymes and chemical catalysts both affect the rate but not the equilibrium constant of a chemical reaction. Reactions proceed downhill energetically, in accord with the Second Law of Thermo-

dynamics. Catalysts merely reduce the time that a thermodynamically favoured reaction requires to reach equilibrium. Remember that the Second Law of Thermodynamics tells whether a reaction can occur but not how fast it occurs.

2. Enzymes and chemical catalysts increase the rate of a chemical reaction in both directions, forward and reverse. This principle of catalysis follows from the fact that catalysts can't change the equilibrium of a reaction. Because a reaction at equilibrium occurs at the same rate both directions, a catalyst that speeds up the forward but not the reverse reaction necessarily alters the equilibrium of the reaction.

3. Enzymes and chemical catalysts bind their substrates, not permanently, but transiently—for a brief time. There is no action at a distance involved. The portion of an enzyme that binds substrate and carries out the actual catalysis is termed the active site.

Enzymes differ from ordinary chemical catalysts in several important respects:

1. Enzymes are specific: Chemical catalysts can react with a variety of substrates. For example, hydroxide ions can catalyse the formation of double bonds and also the hydrolysis of esters. Usually enzymes catalyse only a single type of reaction, and often they work only on one or a few substrate compounds.

2. Enzymes work under mild conditions: Chemical catalysts often require high temperature and/or pressure to function. For example, nitrogen can be reduced to ammonia industrially by the Haber process, catalysed by iron at $500°C$ and at 300 atmospheres pressure of N_2. In contrast, the same reaction is carried out enzymatically at $25°C$. and less than 1 atmosphere pressure of N_2. These gentle conditions of temperature, pressure, and pH characterise enzymatic catalysis, especially within cells.

3. Enzymes are stereospecific: Chemical catalysis of a reaction usually leads to a mixture of stereo-isomers. For example, the addition of acid-catalysed water to a double bond leads to an equimolar (50:50) mixture of D and L isomers where the water is added. In contrast, catalysis of water addition by enzymes results in complete formation of either the D or L isomer, but not both.

4. Enzymes are macromolecules: The macromolecules are composed of protein, or in a few cases, RNA. Most chemical catalysts are either surfaces, for example, metals like platinum, or else small ions, such as hydroxide ions.

5. Enzymes are often regulated: The regulation occurs either by the concentration of substrates, by binding small molecules or other proteins, or by covalent modification of the enzymes amino acid side chains. Thus, an enzyme's effectiveness can be altered without changing the concentration of the enzyme; on the other hand, the effectiveness of a chemical catalyst is generally determined by its overall concentration.

Enzymes as Peptides

Peptides are naturally occurring biological molecules. They are short chains of amino acid monomers linked by peptide (amide) bonds. The covalent chemical bonds are formed when the carboxyl group of one amino acid reacts with the amino group of another. The shortest peptides are dipeptides, consisting of 2 amino acids joined by a single peptide bond, followed by tripeptides, tetrapeptides, etc. A polypeptide is a long, continuous, and unbranched peptide chain. Hence, peptides fall under the broad chemical classes of biological oligomers and polymers, alongside nucleic acids, oligo- and polysaccharides, etc. Peptides are distinguished from proteins on the basis of size, and as an arbitrary benchmark can be understood to contain approximately 50 or fewer amino acids. Proteins consist of one or more polypeptides arranged in a biologically functional way, often bound to ligands such as coenzymes and cofactors, or to another protein or other macromolecule (DNA, RNA, etc.), or to complex macromolecular assemblies. Finally, while aspects of the techniques that apply to peptides versus polypeptides and proteins differ (i.e. in the specifics of electrophoresis, chromatography, etc.), the size boundaries that distinguish peptides from polypeptides and proteins are not absolute: long peptides such as amyloid beta have been referred to as proteins, and smaller proteins like insulin have been considered peptides. Amino acids that have been incorporated into peptides are termed 'residues' due to the release of either a hydrogen ion from the amine end or a hydroxyl ion from the carboxyl end, or both, as a water molecule is released during formation of each amide bond. All peptides except cyclic peptides have an *N*-terminal and C-terminal residue at the end of the peptide.

ENZYME NOMENCLATURE AND CLASSIFICATION

Enzyme Nomenclature — EC System

- Some of the earliest enzymes to be discovered were given names ending in–*in* to indicate that they were proteins (e.g. the digestive enzymes pepsin, trypsin, chymotrypsin).

- Because of the large number of enzymes that are now known, a systematic nomenclature called the Enzyme Commission (EC) system is used to name them.

- Enzymes are grouped into six major classes on the basis of the reaction which they catalyse. Each enzyme has an unambiguous (and often long) systematic name that specifies the substrate of the enzyme (the substance acted on), the functional group acted on, and the type of reaction catalysed. All EC names end in –*ase*. EC classification of enzymes is shown in Table 20.1.

Table 20.1: EC classification of enzymes.

EC code	Group name	Type of reaction catalysed
EC 1	Oxidoreductases	Oxidation-reduction reactions
EC 2	Transferases	Transfer of functional groups
EC 3	Hydrolases	Hydrolysis reactions
EC 4	Lyases	Addition to double bonds or the reverse of that reaction
EC 5	Isomerases	Isomerisation reactions
EC 6	Ligases	Formation of bonds with ATP cleavage

Examples

Enzymes are also assigned common names derived by adding -*ase* to the name of the substrate or to a combination of substrate name and type of reaction:

$$H_2N-\overset{\overset{\displaystyle O}{\|}}{C}-NH_2 \ + \ H_2O \xrightarrow{\text{Enzyme}} CO_2 \ + \ 2NH_3$$

IEC name: Urea amidohydrolase (EC 3.5.1.5)
Substrate: Urea
Functional group: Amide
Type of reaction: Hydrolysis
Common name: Urea + ase = Urease

$$CH_3CH_2OH \ + \ NAD^+ \ \xrightarrow{\text{Enzyme}} \ CH_3CHO \ + \ NADH \ + \ H^+$$

IEC name: Aalcohol dehydrogenase (EC 1.1.1.1)
Substrate: Alcohol (ethyl alcohol)
Type of reaction: Dehydrogenation (removal of hydrogen)
Common name: Alcohol dehydrogenation + ase = Alcohol dehydrogenase

Examples: enzyme nomenclature

- Predict the substrates for the following enzymes:
  - › Maltase.
  - › Peptidase.
  - › Glucose 6-phosphate isomerase.

Examples: enzyme substrates

The following general enzyme names and the reactions that they catalyse are shown in Table 20.2.

Table 20.2: Enzymes and reactions catalysed.

Enzyme	Reaction catalysed
Decarboxylase	Formation of ester linkages
Phosphatase	Removal of carboxyl groups from compounds
Peptidase	Hydrolysis of peptide linkages
Esterase	Hydrolysis of phosphate ester linkages

ENZYME COFACTORS

Enzyme cofactors are discussed below:

- Conjugated proteins function only in the presence of specific nonprotein molecules or metal ions called prosthetic groups.
  - › If the nonprotein component is tightly bound, and forms an integral part of the enzyme structure, it is a true prosthetic group.
  - › If the nonprotein component is weakly bound, and easily separated from the rest of the protein, it is called a cofactor.
- When the cofactor is an organic substance, it is a coenzyme. The cofactor may also be an inorganic ion (usually a metal cation, such as Mg^{2+}, Zn^{2+}, or Fe^{2+}).
- The protein portion is called an apoenzyme:

  Apoenzyme + Cofactor (coenzyme or inorganic ion) → Active enzyme

- Many organic coenzymes are derived from vitamins.

MECHANISM OF ENZYME ACTION

Mechanism of enzyme action is discussed below:

- Enzymes differ widely in structure and specificity, but a general theory that accounts for their catalytic behaviour is widely accepted.

- The enzyme and its substrates interact only over a small region of the surface of the enzyme, called the active site.
    - › When the substrate binds to the active site via some combination of intermolecular forces, an enzyme-substrate (ES) complex is formed.
    - › Once the complex forms, the conversion of the substrate (S) to product (P) takes place:

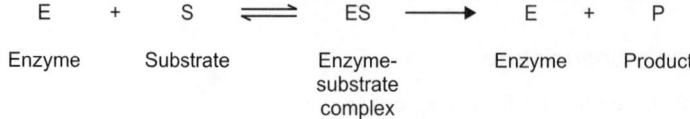

$$E \quad + \quad S \quad \rightleftharpoons \quad ES \quad \longrightarrow \quad E \quad + \quad P$$

| Enzyme | Substrate | Enzyme-substrate complex | Enzyme | Product |

Enzyme Action

$$\text{Sucrase} + \text{Sucrose} \rightleftharpoons \left\{ \begin{array}{c} \text{Sucrase-} \\ \text{sucrose} \\ \text{complex} \end{array} \right\} \longrightarrow \text{Sucrase} + \text{Glucose} + \text{Fructose}$$

| Enzyme | Substrate | | Enzyme | Products |

- The chemical transformation of the substrate occurs at the active site, aided by functional groups on the enzyme that participate in the making and breaking of chemical bonds.
- After the conversion is complete, the product is released from the active site, leaving the enzyme free to react with another substrate molecule.

Lock-and-key theory

- The lock-and-key theory explains the high specificity of enzyme activity.
- Enzyme surfaces accommodate substrates having specific shapes and sizes, so only specific substances 'fit' in an active site to form an ES complex (Fig. 20.1).
- A limitation of this theory is that it requires enzymes conformations to be rigid. Research suggests that instead enzymes are at least somewhat flexible.

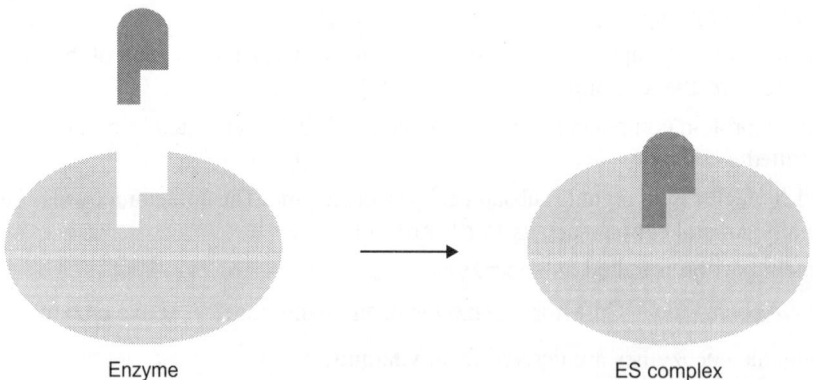

Enzyme   ES complex

Fig. 20.1: Enzyme surfaces accommodate substrates.

Induced-fit Theory

- A modification of the lock-and-key theory called the induced-fit theory proposes that enzymes have flexible conformations that may adapt to incoming substrates (Fig. 20.2).

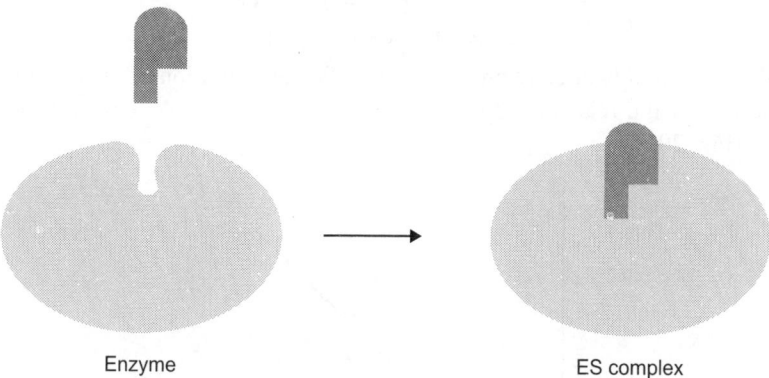

Enzyme   ES complex

Fig. 20.2: Enzymes showing flexible conformations.

- The active site adopts a shape that is complementary to the substrate only after the substrate is bound.

ENZYME ACTIVITY

Enzyme activity is discussed below:

- Enzyme activity refers to the catalytic ability of an enzyme to increase the rate of a reaction.
- The turnover number is the number of molecules of substrate acted on by one molecule of the enzyme per minute.
  - › Carbonic anhydrase is one of the highest at 36 million molecules per minute.
  - › More common numbers are closer to 1000 molecules per minute.
- Enzyme assays are experiments that are performed to measure enzyme activity.
- Assays for blood enzymes are routinely performed in clinical labs.
- Assays are often done by monitoring the rate at which a characteristic colour of a product forms or the colour of a substrate decreases. For reactions involving H^+ ions, the rate of change in pH over time can be used.

Enzyme International Units

- Enzyme activity levels are reported in terms of enzyme international units (IU), which defines enzyme activity as the amount of enzyme that will convert a specified amount of substrate to a product within a certain time.
  - › One standard IU is the quantity of enzyme that catalyses the conversion of 1 micromole (1 mol) of substrate per minute under specified conditions.
  - › Unlike the turnover number, IUs measure how much enzyme is present. An enzyme preparation having an IU of 40 is forty times more concentrated than the standard solution.

Factors Affecting Enzyme Activity

Enzyme concentration

- The concentration of an enzyme, [E], is typically low compared to that of the substrate. Increasing [E] also increases the rate of the reaction:

$$E \quad + \quad S \; \rightleftharpoons \quad ES$$
Increased [E] gives more [ES]

- The rate of the reaction is directly proportional to the concentration of the enzyme (doubling [E] doubles the rate of the reaction), thus, a graph of reaction rate vs. enzyme concentration is a straight line (Fig. 20.3).

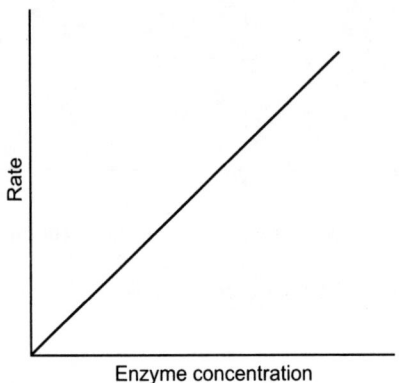

Fig. 20.3: Enzyme rate and concentration.

Substrate concentration

- The concentration of substrate, [S], also affects the rate of the reaction.
- Increasing [S] increases the rate of the reaction, but eventually, the rate reaches a maximum (V_{max}), and remains constant after that.
- The maximum rate is reach when the enzyme is saturated with substrate, and cannot react any faster under those conditions (Fig. 20.4).

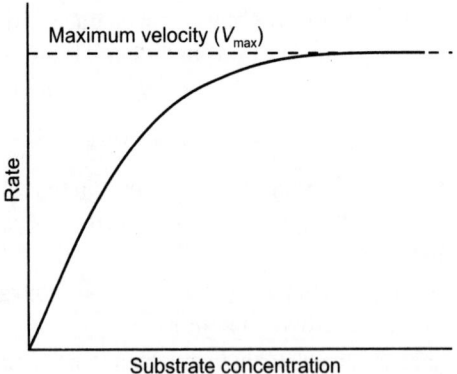

Fig. 20.4: Maximum rate and substrate concentration.

Temperature

- Like all reactions, the rate of enzyme-catalysed reactions increases with temperature.
- Because enzymes are proteins, beyond a certain temperature, the enzyme denatures.

- Every enzyme-catalysed reaction has an optimum temperature at which the enzyme activity is highest, usually from 25–40°C, above or below that value, the rate is lower as shown in Fig. 20.5.

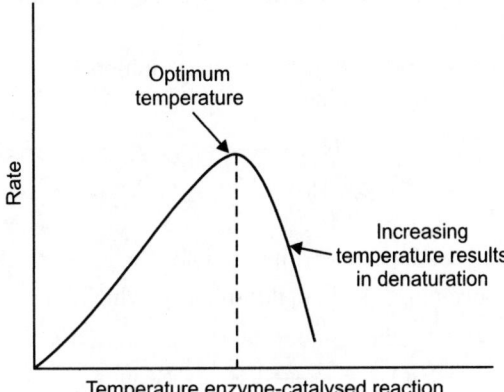

Fig. 20.5: Enzyme catalysed reaction and optimum temperature.

Effect of pH

- Raising or lowering the pH influences the acidic and basic side chains in enzymes. Many enzymes are also denatured by pH extremes (e.g. pickling in acetic acid [vinegar] preserves food by deactivating bacterial enzymes.)
- Many enzymes have an optimum pH, where activity is highest, near a pH of 7, but some operate better at low pH (e.g. pepsin in the stomach) as shown in Fig. 20.6.

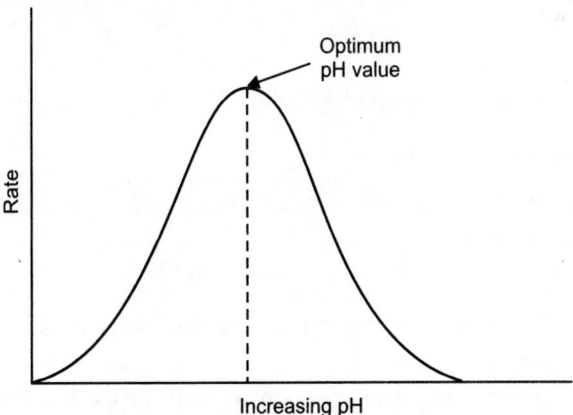

Fig. 20.6: Optimum pH value and rate of enzyme.

Enzyme inhibition – irreversible inhibition

- An enzyme inhibitor is a substance that decreases the rate of an enzyme-catalysed reaction.
  - › Many poisons and medicines inhibit one or more enzymes and thereby decrease the rate of the reactions they carry out.

> Some substances normally found in cells inhibit specific enzymes, providing a means for internal regulation of cell metabolism.

- Irreversible inhibition occurs when an inhibitor forms a covalent bond with a specific functional group of an enzyme, thereby inactivating it.
  > Cyanide ion, CN^-, is a rapidly-acting, highly toxic inhibitor, which interferes with the iron-containing enzyme cytochrome oxidase:

$$Cyt - Fe^{3+} + CN^- \longrightarrow Cyt - Fe - CN^{2+}$$

| Cytochrome oxidase | Cyanide ion | Stable complex |

- Cell respiration stops, and death occurs within minutes.
- An antidote for cyanide poisoning is sodium thiosulphate, which converts cyanide into thiocyanate, which does not bind to cytochrome:

$$CN^- + S_2O_3^{2-} \longrightarrow SCN^- + SO_3^{2-}$$

| Cyanide ion | Thiosulphate ion | Thiocyanate ion | Sulphate ion |

- Heavy metal poisoning results when mercury or lead ions bind to —SH groups on enzymes. Heavy metals can also cause protein denaturation. Pb and Hg can cause permanent neurological damage.
- Heavy-metal poisoning is treated by administering chelating agents, which bind tightly to metal ions, allowing them to be excreted in the urine.

$$CaEDTA^{2-} + Pb^{2+} \rightarrow PbEDTA^{2-} + Ca^{2+}$$

Antibiotics

- Antibiotics are enzyme inhibitors that act on life processes that are essential to certain strains of bacteria.
  > Sulpha drugs.
  > Penicillins—interfere with transpeptidase, which bacteria use in the construction of cell walls.

Enzyme Inhibition—Reversible Inhibition

- A reversible inhibitor binds reversibly to an enzyme, establishing an equilibrium between the bound and unbound inhibitor:

$$E + I \rightleftharpoons EI$$

  > Once the inhibitor combines with the enzyme, the active site is blocked, and no further catalysis takes place.
  > The inhibitor can be removed from the enzyme by shifting the equilibrium.
  > There are two types of reversible inhibitors: competitive and noncompetitive.

Enzyme Inhibition—Competitive Inhibitors

- Competitive inhibitors bind to the enzyme's active site and compete with the normal substrate molecules. They often have structures that are similar to those of the normal substrate (Fig. 20.7).
- Sulpha drugs such as sulphanilamide are similar in structure to *p*-aminobenzoic acid (PABA), which bacteria need to build folic acid in order to grow. Sulphanilamide blocks PABA from fitting into

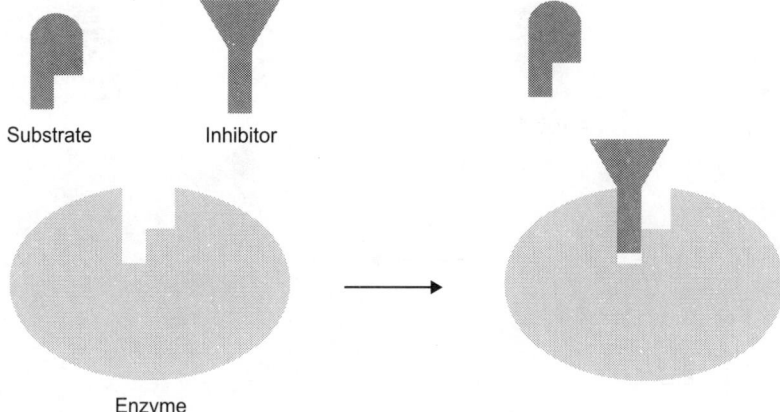

Fig. 20.7: Enzyme showing substrate and inhibitor.

the active site of the enzyme which builds folic acid, causing the bacteria to eventually die. Since humans obtain folic acid from their diet rather than by manufacturing it, it is not harmful to the patient.

p-aminobenzoic acid   Folic acid   Sulphanilamide

- In competitive inhibition, there are two equilibria taking place:

Increasing [S] ⟶

Equilibrium 1:   E + S ⇌ ES

◀── Decreasing [E]

Equilibrium 2:   E + I ⇌ EI

- Competitive inhibition can be reversed by either increasing the concentration of the substrate, or decreasing the concentration of the enzyme, in accordance with Le Châtelier's principle.

Enzyme Inhibition—Noncompetitive Inhibitors

- Noncompetitive inhibitors bind reversibly to the enzyme at a site *other than the active site*, changing the 3D shape of the enzyme (Fig. 20.8) and the active site, so that the normal substrate no longer fits correctly.

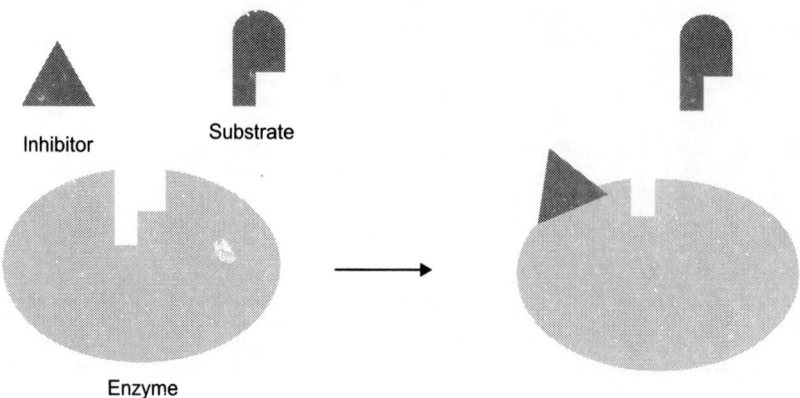

Fig. 20.8: 3D shape of enzyme.

> Noncompetitive inhibitors do not look like the enzyme substrates.
> Increasing the substrate concentration does not affect noncompetitive inhibition because it can't bind to the site occupied by the inhibitor.

Thus, enzyme inhibitors also occur naturally and are involved in the regulation of metabolism. For example, enzymes in a metabolic pathway can be inhibited by downstream products. This type of negative feedback slows the production line when products begin to build up and is an important way to maintain homeostasis in a cell. Other cellular enzyme inhibitors are proteins that specifically bind to and inhibit an enzyme target.

This can help control enzymes that may be damaging to a cell, like proteases or nucleases. Summary of enzyme inhibitors is shown in Fig. 20.9.

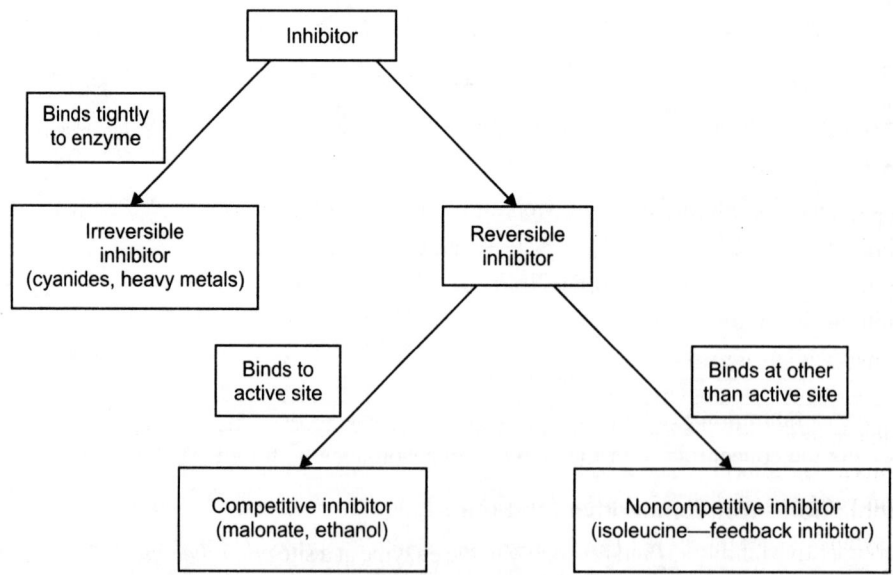

Fig. 20.9: Flow diagram showing summary of enzyme inhibitors.

REGULATION OF ENZYME ACTIVITY

Enzyme Regulation

- Enzymes work together to facilitate all the biochemical reactions needed for a living organism. To respond to changing conditions and cellular needs, enzyme activity requires very sensitive controls:
  › Activation of zymogens.
  › Allosteric regulation.
  › Genetic control.

Activation of Zymogens

- Zymogens or proenzymes are inactive precursors of an enzyme:
  › Some enzymes in their active form would degrade the internal structures of the cell.
  › These enzymes are synthesised and stored as inactive precursors, and when the enzyme is needed, the zymogen is released and activated where it is needed.
  › Activation usually requires the cleavage of one or more peptide bonds.
  › The digestive enzymes pepsin, trypsin, and chymotrypsin, as well as enzymes involved in blood clotting, are activated this way.

  Examples of Zymogens are shown in Table 20.3.

Table 20.3: Examples of Zymogens.

Zymogen	Active enzyme	Function
Chymotrypsinogen	Chymotrypsin	Digestion of proteins
Pepsinogen	Pepsin	Digestion of proteins
Procarboxypeptidase	Carboxypeptidase	Digestion of proteins
Proelastase	Elastase	Digestion of proteins
Prothrombin	Thrombin	Blood clotting
Trypsinogen	Trypsin	Digestion of proteins

Allosteric Regulation

- Compounds that alter enzymes by changing the 3D conformation of the enzyme are called modulators.
- They may increase the activity (activators) or decrease the activity (inhibitors). (Noncompetitive inhibitors are examples of this activity.)
- Enzymes with quaternary structures with binding sites for modulators are called allosteric enzymes.
- These variable-rate enzymes are often located at key control points in cell processes.
- Feedback inhibition occurs when the end product of a sequence of enzyme-catalysed reactions inhibits an earlier step in the process. This allows the concentration of the product to be maintained within very narrow limits.
- The synthesis of isoleucine from threonine is an example of allosteric regulation.
  › Threonine deaminase, which acts in the first step of the conversion pathway, is inhibited by the isoleucine product.

> When isoleucine builds up, it binds to the allosteric site on threonine deaminase, changing its conformation so that threonine binds poorly. This slows the reaction down so that the isoleucine concentration starts to fall.

> When the isoleucine concentration gets too low, the enzyme becomes more active again, and more isoleucine is synthesised.

Genetic Control

- The synthesis of all proteins and enzymes is under genetic control by nucleic acids. Increasing the number of enzymes molecules present through genetic mechanisms is one way to increase production of needed products.

- Enzyme induction occurs when enzymes are synthesised in response to cell need.

- This kind of genetic control allows an organism to adapt to environmental changes. The coupling of genetic control and allosteric regulation allows for very tight control of cellular processes.

- β-galactosidase is an enzyme in the bacterium *Escherichia coli* that catalyses the hydrolysis of lactose to D-galactose and D-glucose.

  > In the absence of lactose in the growth medium, there are very few β-galactosidase molecules.

  > In the presence of a lactose-containing medium, thousands of molecules of enzyme are produced.

  > If lactose is removed, the production of the enzyme once again decreases.

ENZYME KINETICS

Enzyme kinetics is the study of the chemical reactions that are catalysed by enzymes. In enzyme kinetics, the reaction rate is measured and the effects of varying the conditions of the reaction are investigated. Studying an enzyme's kinetics in this way can reveal the catalytic mechanism of this enzyme, its role in metabolism, how its activity is controlled, and how a drug or an agonist might inhibit the enzyme. Enzymes are usually protein molecules that manipulate other molecules—the enzymes' substrates. These target molecules bind to an enzyme's active site and are transformed into products through a series of steps known as the enzymatic mechanism. These mechanisms can be divided into single-substrate and multiple-substrate mechanisms. Kinetic studies on enzymes that only bind one substrate, such as triosephosphate isomerase, aim to measure the affinity with which the enzyme binds this substrate and the turnover rate. Some other examples of enzymes are phosphofructokinase and hexokinase, both of which are important for cellular respiration (glycolysis).

When enzymes bind multiple substrates, such as dihydrofolate reductase, enzyme kinetics can also show the sequence in which these substrates bind and the sequence in which products are released. An example of enzymes that bind a single substrate and release multiple products are proteases, which cleave one protein substrate into two polypeptide products. Others join two substrates together, such as DNA polymerase linking a nucleotide to DNA. Although these mechanisms are often a complex series of steps, there is typically one rate-determining step that determines the overall kinetics. This rate-determining step may be a chemical reaction or a conformational change of the enzyme or substrates, such as those involved in the release of product(s) from the enzyme. Knowledge of the enzyme's structure is helpful in interpreting kinetic data. For example, the structure can suggest how substrates and products bind during catalysis; what changes occur during the reaction; and even the role of particular amino acid residues in the mechanism. Some enzymes change shape significantly during the mechanism, in such cases, it is helpful to determine the enzyme structure with and without bound substrate analogues that do not undergo the enzymatic reaction. Not all biological catalysts are protein enzymes; RNA-based catalysts such as ribozymes and ribosomes are essential to many cellular functions, such as RNA splicing and translation. The main difference between ribozymes and enzymes is that RNA catalysts are composed of nucleotides, whereas enzymes are composed of amino acids. Ribozymes also perform a more limited set of reactions, although their reaction mechanisms and kinetics can be analysed and classified by the same methods.

REACTIONS CATALYSED BY ENZYMES

The reaction catalysed by an enzyme uses exactly the same reactants and produces exactly the same products as the uncatalysed reaction. Like other catalysts, enzymes do not alter the position of equilibrium between substrates and products. However, unlike uncatalysed chemical reactions, enzyme-catalysed reactions display saturation kinetics. For a given enzyme concentration and for relatively low substrate concentrations, the reaction rate increases linearly with substrate concentration; the enzyme molecules are largely free to catalyse the reaction, and increasing substrate concentration means an increasing rate at which the enzyme and substrate molecules encounter one another. However, at relatively high substrate concentrations, the reaction rate asymptotically approaches the theoretical maximum; the enzyme active sites are almost all occupied and the reaction rate is determined by the intrinsic turnover rate of the enzyme. The substrate concentration midway between these two limiting cases is denoted by K_M. Low and high substrate enzymes are shown in Fig. 20.10.

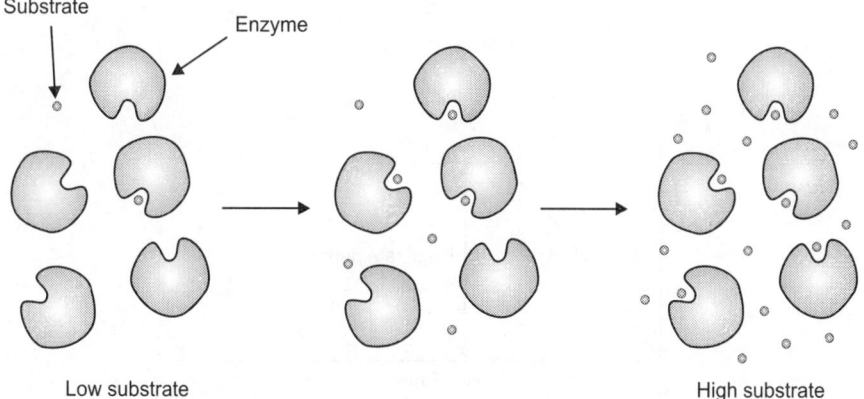

Fig. 20.10: Low substrate and high substrate enzymes.

The two most important kinetic properties of an enzyme are how quickly the enzyme becomes saturated with a particular substrate, and the maximum rate it can achieve. Knowing these properties suggests what an enzyme might do in the cell and can show how the enzyme will respond to changes in these conditions.

ENZYME ASSAYS

Enzyme assays are laboratory procedures that measure the rate of enzyme reactions. Because enzymes are not consumed by the reactions they catalyse, enzyme assays usually follow changes in the concentration of either substrates or products to measure the rate of reaction. There are many methods of measurement.

Spectro-photometric assays observe change in the absorbance of light between products and reactants; radiometric assays involve the incorporation or release of radioactivity to measure the amount of product made over time. Spectrophotometric assays are most convenient since they allow the rate of the reaction to be measured continuously. Although radiometric assays require the removal and counting of samples (i.e. they are discontinuous assays) they are usually extremely sensitive and can measure very low levels of enzyme activity. An analogous approach is to use mass spectrometry to monitor the incorporation or release of stable isotopes as substrate is converted into product. The most sensitive enzyme assays use lasers focused through a microscope to observe changes in single enzyme molecules as they catalyse their reactions. These measurements either use changes in the fluorescence of cofactors during an enzyme's reaction mechanism, or of fluorescent dyes added onto specific sites of the protein to report movements that occur during catalysis.

These studies are providing a new view of the kinetics and dynamics of single enzymes, as opposed to traditional enzyme kinetics, which observes the average behaviour of populations of millions of enzyme molecules. An example progress curve for an enzyme assay is shown in Fig. 20.11. The enzyme produces product at an initial rate that is approximately linear for a short period after the start of the reaction. As the reaction proceeds and substrate is consumed, the rate continuously slows (so long as substrate is not still at saturating levels). To measure the initial (and maximal) rate, enzyme assays are typically carried out while the reaction has progressed only a few per cent towards total completion.

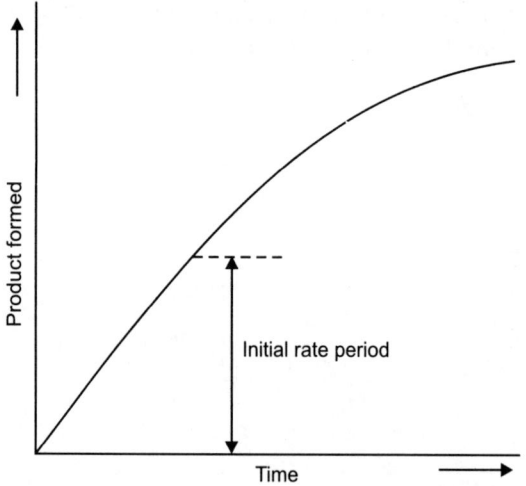

Fig. 20.11: Curve for an enzyme reaction.

The length of the initial rate period depends on the assay conditions and can range from milliseconds to hours. However, equipment for rapidly mixing liquids allows fast kinetic measurements on initial rates of less than one second. These very rapid assays are essential for measuring pre-steady-state kinetics, which are discussed below.

Most enzyme kinetics studies concentrate on this initial, approximately linear part of enzyme reactions. However, it is also possible to measure the complete reaction curve and fit this data to a non-linear rate equation. This way of measuring enzyme reactions is called progress-curve analysis. This approach is useful as an alternative to rapid kinetics when the initial rate is too fast to measure accurately.

SINGLE-SUBSTRATE REACTIONS

Enzymes with single-substrate mechanisms include isomerases such as triosephos-phateisomerase or bisphosphoglycerate mutase, intramolecular lyases such as adenylate cyclase and the hammerhead ribozyme, an RNA lyase. However, some enzymes that only have a single substrate do not fall into this category of mechanisms. Catalase is an example of this, as the enzyme reacts with a first molecule of hydrogen peroxide substrate, becomes oxidised and is then reduced by a second molecule of substrate. Although a single substrate is involved, the existence of a modified enzyme intermediate means that the mechanism of catalase is actually a ping–pong mechanism, a type of mechanism that is discussed in the Multi-substrate reactions section below.

Michaelis–Menten Kinetics

As enzyme-catalysed reactions are saturable, their rate of catalysis does not show a linear response to increasing substrate. If the initial rate of the reaction is measured over a range of substrate concentrations (denoted as $[S]$), the reaction rate (v) increases as $[S]$ increases, as shown in Fig. 20.12. However, as $[S]$ gets higher, the enzyme becomes saturated with substrate and the rate reaches V_{max}, the enzyme's maximum rate.

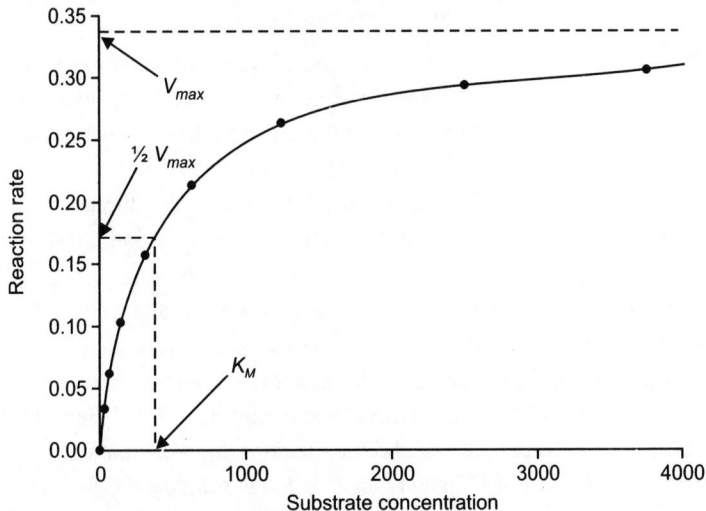

Fig. 20.12: Saturation curve for an enzyme showing the relation between the concentration of substrate and rate.

The Michaelis–Menten kinetic model of a single-substrate reaction is shown in Fig. 20.13. There is an initial bimolecular reaction between the enzyme E and substrate S to form the enzyme–substrate complex ES. Although the enzymatic mechanism for the unimolecular reaction $ES \xrightarrow{k_{cat}} E + P$ can be quite complex, there is typically one rate-determining enzymatic step that allows this reaction to be modelled as a single catalytic step with an apparent unimolecular rate constant k_{cat}. If the reaction path proceeds over one or several intermediates, k_{cat} will be a function of several elementary rate constants, whereas in the simplest case of a single elementary reaction (e.g. no intermediates) it will be identical to the elementary unimolecular rate constant k_2. The apparent unimolecular rate constant k_{cat} is also called turnover number and denotes the maximum number of enzymatic reactions catalysed per second.

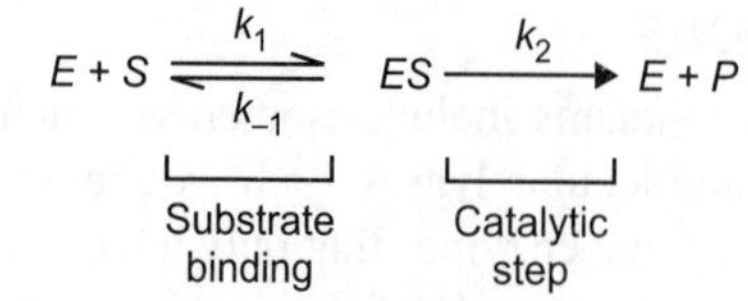

$$E + S \underset{k_{-1}}{\overset{k_1}{\rightleftharpoons}} ES \xrightarrow{k_2} E + P$$

$$\underbrace{\qquad}_{\substack{\text{Substrate} \\ \text{binding}}} \quad \underbrace{\qquad}_{\substack{\text{Catalytic} \\ \text{step}}}$$

Fig. 20.13: Single-substrate mechanism for an enzyme reaction. k_1, k_{-1} and k_2 are the rate constants for the individual steps.

The Michaelis–Menten equation describes how the (initial) reaction rate v_0 depends on the position of the substrate-binding equilibrium and the rate constant k_2.

$$v_0 = \frac{V_{max}[S]}{K_M + [S]}$$

with the constants

$$K_M = \frac{k_2 + k_{-1}}{k_1} \approx K_D$$

$$V_{max} = k_{cat}[E]_{tot}$$

This Michaelis–Menten equation is the basis for most single-substrate enzyme kinetics. Two crucial assumptions underlie this equation (apart from the general assumption about the mechanism only involving no intermediate or product inhibition, and there is no allostericity or cooperativity). The first assumption is the so-called quasi-steady-state assumption (or pseudo-steady-state hypothesis), namely that the concentration of the substrate-bound enzyme (and hence also the unbound enzyme) changes much more slowly than those of the product and substrate and thus the change over time of the complex can be set to zero $d[ES]/dt = 0$.

The second assumption is that the total enzyme concentration does not change over time, thus $[E]_{tot} = [E] + [ES] =$ constant. A complete derivation is discussed below.

The Michaelis constant K_M is experimentally defined as the concentration at which the rate of the enzyme reaction is half V_{max}, which can be verified by substituting $[S] = K_m$ into the Michaelis–Menten equation and can also be seen graphically. If the rate-determining enzymatic step is slow compared to substrate dissociation ($k_2 \ll k_{-1}$), the Michaelis constant K_M is roughly the dissociation constant K_D of the ES complex.

If $[S]$ is small compared to K_M then the term $[S]/(K_M + [S]) \approx [S]/K_M$ and also very little ES complex is formed, thus $[E]_0 \approx [E]$. Therefore, the rate of product formation is:

$$v_0 \approx \frac{k_{cat}}{K_M}[E][S] \qquad \text{if } [S] \ll K_M$$

Thus the product formation rate depends on the enzyme concentration as well as on the substrate concentration, the equation resembles a bimolecular reaction with a corresponding pseudo-second order rate constant k_2/K_M. This constant is a measure of catalytic efficiency. The most efficient enzymes reach a k_2/K_M in the range of 10^8–10^{10} M^{-1} s^{-1}. These enzymes are so efficient they effectively catalyse a reaction each time they encounter a substrate molecule and have thus reached an upper theoretical limit for efficiency (diffusion limit), these enzymes have often been termed perfect enzymes.

Direct Use of the Michaelis–Menten Equation for Time Course Kinetic Analysis

The observed velocities predicted by the Michaelis–Menten equation can be used to directly model the time course disappearance of substrate and the production of product through incorporation of the Michaelis–Menten equation into the equation for first order chemical kinetics. This can only be achieved however if one recognises the problem associated with the use of Euler's number in the description of first order chemical kinetics, i.e. e^{-k} is a split constant that introduces a systematic error into calculations and can be rewritten as a single constant which represents the remaining substrate after each time period.

$$[S] = [S]_0 (1-k)^t$$

$$[S] = [S]_0 (1 - v/[S]_0)^t$$

$$[S] = [S]_0 [1 - (V_{max} [S]_0/(K_M + [S]_0)/[S]_0)]^t$$

Stuart Beal and Santiago Schnell and Claudio Mendoza derived a closed form solution for the time course kinetics analysis of the Michaelis-Menten mechanism. The solution, known as the Schnell-Mendoza equation, has the form:

$$\frac{[S]}{K_M} = W[F(t)]$$

where, W is the Lambert-W function. and where $F(t)$ is:

$$F(t) = \frac{[S]_0}{K_M} \exp\left(\frac{[S]_0}{K_M} - \frac{V_{max}}{K_M} t \right)$$

This equation is encompassed by the equation below, obtained by Berberan-Santos, which is also valid when the initial substrate concentration is close to that of enzyme:

$$\frac{[S]}{K_M} = W[F(t)] - \frac{V_{max}}{k_{cat} K_M} \frac{W[F(t)]}{1 + W[F(t)]}$$

where, W is again the Lambert-W function.

Linear Plots of the Michaelis–Menten Equation

The plot of v versus $[S]$ above is not linear, although initially linear at low $[S]$, it bends over to saturate at high $[S]$. Before the modern era of nonlinear curve-fitting on computers, this nonlinearity could make it difficult to estimate K_M and V_{max} accurately. Therefore, several researchers developed linearisations of the Michaelis–Menten equation, such as the Lineweaver–Burk plot, the Eadie–Hofstee diagram and the Hanes–Woolf plot. All of these linear representations can be useful for visualising data, but none should be used to determine kinetic parameters, as computer software is readily available that allows for more accurate determination by nonlinear regression methods. The Lineweaver–Burk plot or double reciprocal plot is a common way of illustrating kinetic data. This is produced by taking the reciprocal of both sides of the Michaelis–Menten equation, this is a linear form of the Michaelis–Menten equation and produces a straight line with the equation $y = mx + c$ with a y-intercept equivalent to $1/V_{max}$ and an x-intercept of the graph representing $-1/K_M$.

$$\frac{1}{v} = \frac{K_M}{V_{max}[S]} + \frac{1}{V_{max}}$$

Naturally, no experimental values can be taken at negative $1/[S]$; the lower limiting value $1/[S] = 0$ (the y-intercept) corresponds to an infinite substrate concentration, where $1/v = 1/V_{max}$ as shown in the Fig. 20.14, thus, the x-intercept is an extrapolation of the experimental data taken at positive concentrations. More generally, the Lineweaver–Burk plot skews the importance of measurements taken at low substrate concentrations and, thus, can yield inaccurate estimates of V_{max} and K_M. A more accurate linear plotting method is the Eadie-Hofstee plot. In this case, v is plotted against $v/[S]$. In the third common linear representation, the Hanes-Woolf plot, $[S]/v$ is plotted against $[S]$. In general, data normalisation can help diminish the amount of experimental work and can increase the reliability of the output, and is suitable for both graphical and numerical analysis.

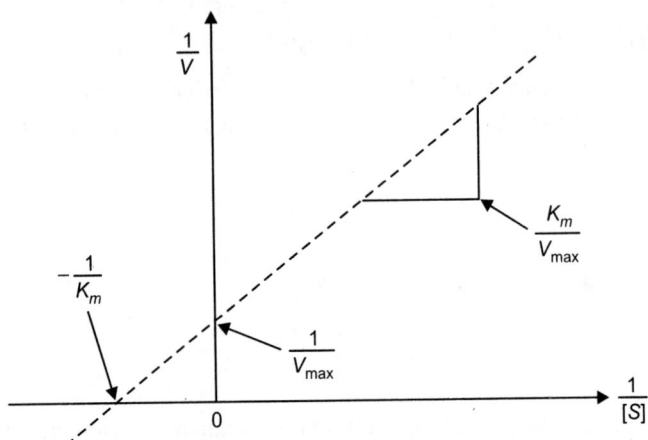

Fig. 20.14: Lineweaver–Burk or double-reciprocal plot of kinetic data, showing the significance of the axis intercepts and gradient.

Practical Significance of Kinetic Constants

The study of enzyme kinetics is important for two basic reasons. Firstly, it helps explain how enzymes work, and secondly, it helps predict how enzymes behave in living organisms. The kinetic constants defined above, K_M and V_{max}, are critical to attempts to understand how enzymes work together to control metabolism.

Making these predictions is not trivial, even for simple systems. For example, oxaloacetate is formed by malate dehydrogenase within the mitochondrion. Oxaloacetate can then be consumed by citrate synthase, phosphoenolpyruvate carboxykinase or aspartate aminotransferase, feeding into the citric acid cycle, gluconeogenesis or aspartic acid biosynthesis, respectively. Being able to predict how much oxaloacetate goes into which pathway requires knowledge of the concentration of oxaloacetate as well as the concentration and kinetics of each of these enzymes. This aim of predicting the behaviour of metabolic pathways reaches its most complex expression in the synthesis of huge amounts of kinetic and gene expression data into mathematical models of entire organisms. Alternatively, one useful simplification of the metabolic modelling problem is to ignore the underlying enzyme kinetics and only rely on information about the reaction network's stoichiometry, a technique called flux balance analysis.

Michaelis–Menten Kinetics with Intermediate

Now considering the less simple case:

$$E + S \underset{K_{-1}}{\overset{k_1}{\rightleftharpoons}} ES \overset{K_2}{\longrightarrow} EI \overset{k_3}{\longrightarrow} E + P$$

where a complex with the enzyme and an intermediate exists and the intermediate is converted into product in a second step. In this case we have a very similar equation

$$v_0 = k_{cat} \frac{[S][E]_0}{K'_M + [S]}$$

but the constants are different

$$K'_M = \frac{k_3}{k_2 + k_3} K_M = \frac{k_3}{k_2 + k_3} \cdot \frac{k_2 + k_{-1}}{k_1}$$

$$K_{cat} = \frac{k_3 k_2}{k_2 + k_3}$$

We see that for the limiting case $k_3 \gg k_2$, thus when the last step from EI to $E + P$ is much faster than the previous step, we get again the original equation. Mathematically we have then $K'_M \approx K_M$ and $k_{cat} \approx k_2$.

MULTI-SUBSTRATE REACTIONS

Multi-substrate reactions follow complex rate equations that describe how the substrates bind and in what sequence. The analysis of these reactions is much simpler if the concentration of substrate A is kept constant and substrate B varied. Under these conditions, the enzyme behaves just like a single-substrate enzyme and a plot of v by $[S]$ gives apparent K_M and V_{max} constants for substrate B. If a set of these measurements is performed at different fixed concentrations of A, these data can be used to work out what the mechanism of the reaction is. For an enzyme that takes two substrates A and B and turns them into two products P and Q, there are two types of mechanism: ternary complex and ping–pong.

Ternary-complex Mechanisms

In these enzymes, both substrates bind to the enzyme at the same time to produce an EAB ternary complex. The order of binding can either be random (in a random mechanism) or substrates have to bind in a particular sequence (in an ordered mechanism). When a set of v by $[S]$ curves (fixed A, varying B) from an enzyme with a ternary-complex mechanism are plotted in a Lineweaver–Burk plot, the set of lines produced will intersect.

Enzymes with ternary-complex mechanisms include glutathione S-transferase, dihydrofolate reductase and DNA polymerase. The animations of the ternary-complex mechanisms of the enzymes dihydrofolate reductase[β] and DNA polymerase[γ] are shown in Fig. 20.15.

Ping–Pong Mechanisms

As shown in Fig. 20.16, enzymes with a ping-pong mechanism can exist in two states, E and a chemically modified form of the enzyme E^*; this modified enzyme is known as an intermediate. In such mechanisms, substrate A binds, changes the enzyme to E^* by, for example, transferring a chemical group to the active site, and is then released. Only after the first substrate is released can substrate B bind and react with the modified enzyme, regenerating the unmodified E form. When a set of v by $[S]$ curves (fixed A, varying B) from an enzyme with a ping–pong mechanism are plotted in a Lineweaver–Burk plot, a set of parallel lines will be produced. This is called a secondary plot.

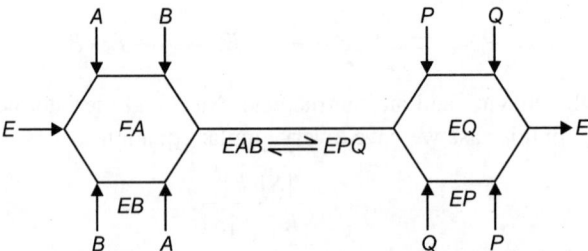

Fig. 20.15: Random-order ternary-complex mechanism for an enzyme reaction. The reaction path is shown as a line and enzyme intermediates containing substrates *A* and *B* or products *P* and *Q* are written below the line.

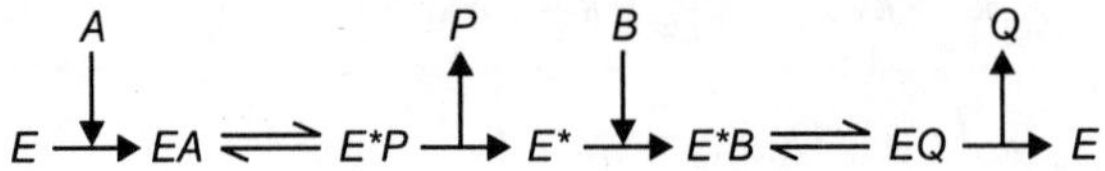

$$E \xrightarrow{\;A\;} EA \rightleftharpoons E^*P \xrightarrow{\;P\;} E^* \xrightarrow{\;B\;} E^*B \rightleftharpoons EQ \xrightarrow{\;Q\;} E$$

Fig. 20.16: Ping–pong mechanism for an enzyme reaction. Intermediates contain substrates *A* and *B* or products *P* and *Q*.

Enzymes with ping–pong mechanisms include some oxidoreductases such as thioredoxin peroxidase, transferases such as acylneuraminate cytidylyltransferase and serine proteases such as trypsin and chymotrypsin. Serine proteases are a very common and diverse family of enzymes, including digestive enzymes (trypsin, chymotrypsin, and elastase), several enzymes of the blood clotting cascade and many others. In these serine proteases, the *E** intermediate is an acyl-enzyme species formed by the attack of an active site serine residue on a peptide bond in a protein substrate.

NON-MICHAELIS–MENTEN KINETICS

Some enzymes produce a sigmoid *v* by [*S*] plot, which often indicates cooperative binding of substrate to the active site. This means that the binding of one substrate molecule affects the binding of subsequent substrate molecules. This behaviour is most common in multimeric enzymes with several interacting active sites. Here, the mechanism of cooperation is similar to that of haemoglobin, with binding of substrate to one active site altering the affinity of the other active sites for substrate molecules (Fig. 20.17).

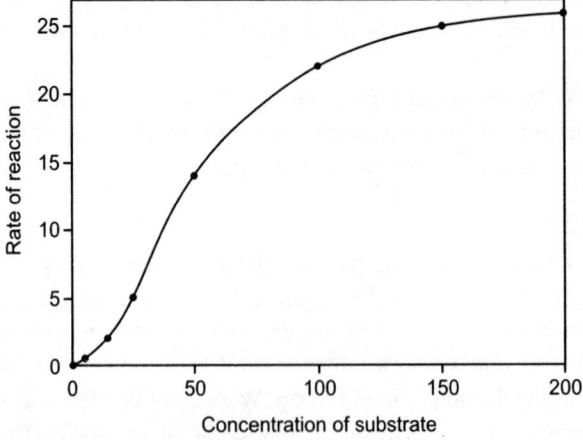

Fig. 20.17: Saturation curve for an enzyme reaction showing sigmoid kinetics.

Positive cooperativity occurs when binding of the first substrate molecule increases the affinity of the other active sites for substrate. Negative cooperativity occurs when binding of the first substrate decreases the affinity of the enzyme for other substrate molecules. Allosteric enzymes include mammalian tyrosyl tRNA-synthetase, which shows negative cooperativity, and bacterial aspartate transcarbamoylase and phospho-fructokinase, which show positive cooperativity.

Cooperativity is surprisingly common and can help regulate the responses of enzymes to changes in the concentrations of their substrates. Positive cooperativity makes enzymes much more sensitive to [S] and their activities can show large changes over a narrow range of substrate concentration. Conversely, negative cooperativity makes enzymes insensitive to small changes in [S]. The Hill equation is often used to describe the degree of cooperativity quantitatively in non-Michaelis–Menten kinetics. The derived Hill coefficient n measures how much the binding of substrate to one active site affects the binding of substrate to the other active sites. A Hill coefficient of <1 indicates negative cooperativity and a coefficient of >1 indicates positive cooperativity.

Pre-steady-state Kinetics

In the first moment after an enzyme is mixed with substrate, no product has been formed and no intermediates exist. The study of the next few milliseconds of the reaction is called Pre-steady-state kinetics also referred to as Burst kinetics as shown in Fig. 20.18. Pre-steady-state kinetics is therefore concerned with the formation and consumption of enzyme–substrate intermediates (such as ES or $E*$) until their steady-state concentrations are reached.

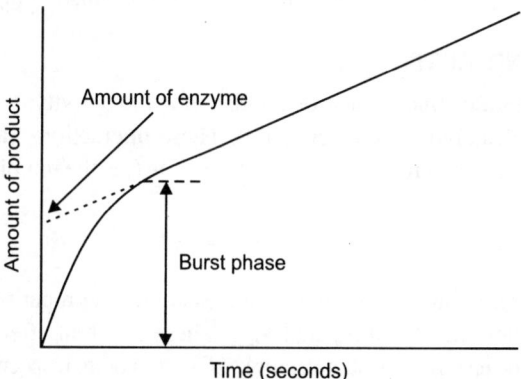

Fig. 20.18: Pre-steady state progress curve, showing the burst phase of an enzyme reaction.

This approach was first applied to the hydrolysis reaction catalysed by chymotrypsin. Often, the detection of an intermediate is a vital piece of evidence in investigations of what mechanism an enzyme follows. For example, in the ping–pong mechanisms that are shown above, rapid kinetic measurements can follow the release of product P and measure the formation of the modified enzyme intermediate $E*$. In the case of chymotrypsin, this intermediate is formed by an attack on the substrate by the nucleophilic serine in the active site and the formation of the acyl-enzyme intermediate.

The enzyme produces $E*$ rapidly in the first few seconds of the reaction. The rate then slows as steady state is reached. This rapid burst phase of the reaction measures a single turnover of the enzyme. Consequently, the amount of product released in this burst, shown as the intercept on the y-axis of the graph, also gives the amount of functional enzyme which is present in the assay.

Chemical Mechanism

An important goal of measuring enzyme kinetics is to determine the chemical mechanism of an enzyme reaction, i.e. the sequence of chemical steps that transform substrate into product. The kinetic approaches discussed above will show at what rates intermediates are formed and inter-converted, but they cannot identify exactly what these intermediates are. Kinetic measurements taken under various solution conditions or on slightly modified enzymes or substrates often shed light on this chemical mechanism, as they reveal the rate-determining step or intermediates in the reaction. For example, the breaking of a covalent bond to a hydrogen atom is a common rate-determining step. Which of the possible hydrogen transfers is rate determining can be shown by measuring the kinetic effects of substituting each hydrogen by deuterium, its stable isotope. The rate will change when the critical hydrogen is replaced, due to a primary kinetic isotope effect, which occurs because bonds to deuterium are harder to break than bonds to hydrogen. It is also possible to measure similar effects with other isotope substitutions, such as $^{13}C/^{12}C$ and $^{18}O/^{16}O$, but these effects are more subtle.

Isotopes can also be used to reveal the fate of various parts of the substrate molecules in the final products. For example, it is sometimes difficult to discern the origin of an oxygen atom in the final product; since it may have come from water or from part of the substrate. This may be determined by systematically substituting oxygen's stable isotope ^{18}O into the various molecules that participate in the reaction and checking for the isotope in the product. The chemical mechanism can also be elucidated by examining the kinetics and isotope effects under different pH conditions, by altering the metal ions or other bound cofactors, by site-directed mutagenesis of conserved amino acid residues, or by studying the behaviour of the enzyme in the presence of analogues of the substrate(s).

ENZYME INHIBITION AND ACTIVATION

Enzyme inhibitors are molecules that reduce or abolish enzyme activity, while enzyme activators are molecules that increase the catalytic rate of enzymes. These interactions can be either reversible (i.e. removal of the inhibitor restores enzyme activity) or irreversible (i.e. the inhibitor permanently inactivates the enzyme).

Reversible Inhibitors

Traditionally reversible enzyme inhibitors have been classified as competitive, uncompetitive, or non-competitive, according to their effects on K_m and V_{max}. These different effects result from the inhibitor binding to the enzyme E, to the enzyme–substrate complex ES, or to both, respectively as shown in Fig. 20.19.

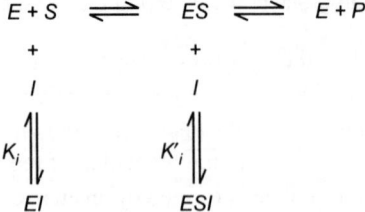

Fig. 20.19: Kinetic scheme for reversible enzyme inhibitors.

The division of these classes arises from a problem in their derivation and results in the need to use two different binding constants for one binding event. The binding of an inhibitor and its effect on the enzymatic activity are two distinctly different things, another problem the traditional equations fail to acknowledge. In noncompetitive inhibition the binding of the inhibitor results in 100% inhibition of the enzyme only, and fails to consider the possibility of anything in between. The common form of the inhibitory term also obscures the relationship between the inhibitor binding to the enzyme and its relationship to any other binding term be it the Michaelis–Menten equation or a dose response curve associated with ligand receptor binding.

Thus, it can be written as:

$$\frac{V_{max}}{1+\dfrac{[I]}{K_i}}$$

$$\frac{V_{max}}{\dfrac{[I]+K_i}{K_i}}$$

Adding zero to the bottom $([I] - [I])$

$$\frac{V_{max}}{\dfrac{[I]+K_i}{[I]+K_i -[I]}}$$

Dividing by $[I] + K_i$

$$\frac{V_{max}}{\dfrac{1}{1-\dfrac{[I]}{[I]+K_i}}}$$

$$V_{max} - V_{max}\frac{[I]}{[I]+K_i}$$

This notation demonstrates that similar to the Michaelis–Menten equation, where the rate of reaction depends on the per cent of the enzyme population interacting with substrate. Fraction of the enzyme population bound by substrate

$$\frac{[S]}{[S]+K_m}$$

fraction of the enzyme population bound by inhibitor

$$\frac{[I]}{[I]+K_i}$$

the effect of the inhibitor is a result of the per cent of the enzyme population interacting with inhibitor. The only problem with this equation in its present form is that it assumes absolute inhibition of the enzyme with inhibitor binding, when in fact there can be a wide range of effects anywhere from 100% inhibition of substrate turn over to just >0%. To account for this the equation can be easily modified to allow for different degrees of inhibition by including a delta V_{max} term.

$$V_{\max} - \Delta V_{\max} \frac{[I]}{[I] + K_i}$$

$$V_{\max} 1 - (V_{\max} 1 - V_{\max} 2) \frac{[I]}{[I] + K_i}$$

This term can then define the residual enzymatic activity present when the inhibitor is interacting with individual enzymes in the population. However the inclusion of this term has the added value of allowing for the possibility of activation if the secondary $V_{\max}$ term turns out to be higher than the initial term.

To account for the possibly of activation as well the notation can then be rewritten replacing the inhibitor 'I' with a modifier term denoted here as 'X'.

$$V_{\max} 1 - (V_{\max} 1 - V_{\max} 2) \frac{[X]}{[X] + K_x}$$

While this terminology results in a simplified way of dealing with kinetic effects relating to the maximum velocity of the Michaelis–Menten equation, it highlights potential problems with the term used to describe effects relating to the K_m. The K_m relating to the affinity of the enzyme for the substrate should in most cases relate to potential changes in the binding site of the enzyme which would directly result from enzyme inhibitor interactions. As such a term similar to the one proposed above to modulate $V_{\max}$ should be appropriate in most situations:

$$K_m 1 - (K_m 1 - K_m 2) \frac{[X]}{[X] + K_x}$$

Irreversible Inhibitors

Enzyme inhibitors can also irreversibly inactivate enzymes, usually by covalently modifying active site residues. These reactions, which may be called suicide substrates, follow exponential decay functions and are usually saturable. Below saturation, they follow first order kinetics with respect to inhibitor.

MECHANISMS OF CATALYSIS

The favoured model for the enzyme–substrate interaction is the induced fit model. This model proposes that the initial interaction between enzyme and substrate is relatively weak, but that these weak interactions rapidly induce conformational changes in the enzyme that strengthen binding. These conformational changes also bring catalytic residues in the active site close to the chemical bonds in the substrate that will be altered in the reaction. Conformational changes can be measured using circular dichroism or dual polarisation interferometry. After binding takes place, one or more mechanisms of catalysis lower the energy of the reaction's (Fig. 20.20) transition state by providing an alternative chemical pathway for the reaction. Mechanisms of catalysis include catalysis by bond strain; by proximity and orientation; by active-site proton donors or acceptors; covalent catalysis and quantum tunnelling. Enzyme kinetics cannot prove which modes of catalysis are used by an enzyme.

However, some kinetic data can suggest possibilities to be examined by other techniques. For example, a ping–pong mechanism with burst-phase pre-steady-state kinetics would suggest covalent catalysis might be important in this enzyme's mechanism. Alternatively, the observation of a strong pH effect on $V_{\max}$ but not K_m might indicate that a residue in the active site needs to be in a particular ionisation state for catalysis to occur.

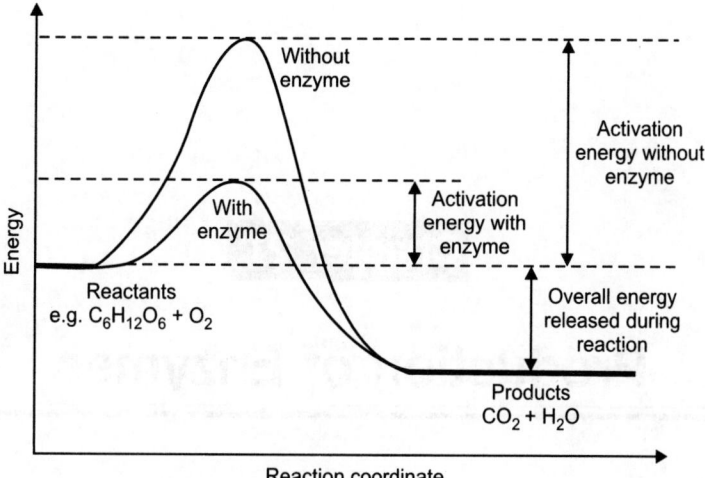

Fig. 20.20: The energy variation as a function of reaction coordinate shows the stabilisation of the transition state by an enzyme.

Chapter 21

Production of Enzymes

INTRODUCTION

Enzymes are protein catalysts that increase the velocity of a chemical reaction and are not consumed during the reaction they catalyse. All enzymes contain a protein backbone. In some enzymes this is the only component in the structure. However there are additional non-protein moieties usually present which may or may not participate in the catalytic activity of the enzyme. Covalently attached carbohydrate groups are commonly encountered structural features which often have no direct bearing on the catalytic activity, although they may well effect an enzyme's stability and solubility. Other factors often found are metal ions (cofactors) and low molecular weight organic molecules (coenzymes).

These may be loosely or tightly bound by noncovalent or covalent forces. They are often important constituents contributing to both the activity and stability of the enzymes. This requirement for cofactors and coenzymes must be recognised if the enzymes are to be used efficiently and is particularly relevant in continuous processes where there may be a tendency for them to become separated from an enzyme's protein moiety.

BIOLOGICAL CATALYSTS

This section discusses shared properties with chemical catalysts and differences between enzymes and chemical catalysts.

1. Shared properties with chemical catalysts:
   (a) Enzymes are neither consumed nor produced during the course of a reaction.
   (b) Enzymes do not cause reactions to take place, but they greatly enhance the rate of reactions that would proceed much slower in their absence. They alter the rate but not the equilibrium constants of reactions that they catalyse.
2. Differences between enzymes and chemical catalysts:
   (a) Enzymes are proteins.
   (b) Enzymes are highly specific and produce only the expected products from the given reactants, or substrates (i.e. there are no side reactions).

(c) Enzymes may show a high specificity toward one substrate or exhibit a broad specificity, using more than one substrate.

(d) Enzymes usually function within a moderate pH and temperature range.

Functions of Enzymes

Enzymes allow many chemical reactions to occur within the homeostasis constraints of a living system. Enzymes function as organic catalysts. A catalyst is a chemical involved in, but not changed by, a chemical reaction. Many enzymes function by lowering the activation energy of reactions. By bringing the reactants closer together, chemical bonds may be weakened and reactions will proceed faster than without the catalyst. The use of enzymes can lower the activation energy of a reaction. Enzymes can act rapidly, as in the case of carbonic anhydrase (enzymes typically end in the -ase suffix), which causes the chemicals to react 10^7 times faster than without the enzyme present. Carbonic anhydrase speeds up the transfer of carbon dioxide from cells to the blood. There are over 2000 known enzymes, each of which is involved with one specific chemical reaction. Enzymes are substrate specific. The enzyme peptidase (which breaks peptide bonds in proteins) will not work on starch (which is broken down by human-produced amylase in the mouth).

Enzymes are proteins. The functioning of the enzyme is determined by the shape of the protein. The arrangement of molecules on the enzyme produces an area known as the active site within which the specific substrates will 'fit'. It recognises, confines and orients the substrate in a particular direction.

Functions of Enzymes in Cells

Enzymes are proteins that do the everyday work within a cell. Their basic function is to speed up the process and efficiency of a reaction without themselves being consumed in the process. Enzymes are responsible for moving large parts of a cell's internal structure, such as pulling chromosomes apart when a cell divides. Enzymes make the energy molecules that are constantly needed for the cell to survive, and they break down molecules, recycle the old parts and make new molecules that allow the cell to grow.

Catalysts for change: Enzymes are catalysts, meaning they speed up the rate at which reactants interact to form products in a chemical reaction, while not being consumed in the reaction. They physically combine chemical reactants in a way that lowers the energy required for bonds to break and new bonds to form, making the formation of a product much faster. They lower what is called the activation energy of the reaction, or the amount of energy required for a hybrid of the reactants and products to form. The hybrid then becomes the product. Without enzymes, these chemical reactions would proceed at a rate that is hundreds to thousands of times slower.

Making energy: Living organisms store the energy required for daily life in the form of chemical energy. Adenosine triphosphate, or ATP, is the main form of chemical energy. ATP is a charged battery that can be discharged to release energy that powers the movement of enzymes. Enzymes are also required to make ATP, however. The main enzyme that produces ATP is called ATP Synthase, which is part of the electron transport chain in the mitochondria of cells. For every molecule of glucose that is broken down for energy, ATP Synthase makes about 32 to 34 ATP molecules.

Molecular motors: Enzymes are the protein machines that perform the day-to-day functions within cells. They deliver packages from one part of the cell to another. They pull chromosomes apart when the cell undergoes mitosis. They pull cilia, which are like the oars of a cell, to help cells move or to help cells move mucus up our airway into our throat. Common motor proteins are myosins, kinesins and dyneins. These families of motor proteins catalyse the breakage of ATP into ADP (adenosine diphoshphate) to get the energy they need to do their grunt work.

Breaking and building: The cells that comprise organisms obtain energy by breaking down organic carbon compounds such as sugar, protein and fat. Breaking these molecules down into smaller parts is called catabolism, while building new molecules from these recycled smaller parts is called anabolism. Enzymes perform these functions at every step of the way. Energy sources such as glucose, a simple sugar, store a lot of energy. But the cell cannot access that energy to make ATP unless it is able to break the bonds within the glucose molecule.

CLASSIFICATION OF ENZYMES

Enzymes are divided into six major classes with several sub classes (Table 21.1 and Fig. 21.1).

Table 21.1: Six major classes of enzymes and examples of their sub classes.

Classification	*Distinguishing feature*
1. Oxidoreductases	$A_{red} + B_{ox} \rightarrow A_{ox} + B_{red}$
Oxidases	Use oxygen as an electron acceptor but do not incorporate it into the substrate
Dehydrogenases	Use molecules other than oxygen (e.g. NAD^+) as an electron acceptor
Oxygenases	Directly incorporate oxygen into the substrate
Peroxidases	Use H_2O_2 as an electron acceptor
2. Transferases	$A–B + C \rightarrow A + B– C$
Methyltransferases	Transfer one-carbon units between substrates
Aminotransferases	Transfer NH_2 from amino acids to keto acids
Kinases	Transfer $PO_3\sim$ from ATP to a substrate
Phosphorylases	Transfer $PO_{3\_}$ from inorganic phosphate (P) to a substrate
3. Hydrolases	$A–B + H_2O \rightarrow A–H + B–OH$
Phosphatases	Remove $PO_3\sim$ from a substrate
Phosphodiesterases	Cleave phosphodiester bonds such as those in nucleic acids
Proteases	Cleave amide bonds such as those in proteins
4. Lyases	$A(XH)–B \rightarrow A–X + B–H$
Decarboxylases	Produce CO_2 via elimination reactions
Aldolases	Produce aldehydes via elimination reactions
Synthases	Link two molecules without involvement of ATP
5. Isomerases	Isomers have the same molecular formula but differ in their structural formula
Racemases	Interconvert L and D stereoisomers
Mutases	Transfer groups between atoms within a molecule
6. Ligases	$A + B + ATP \rightarrow A–B + ADP + Pi$
Carboxylases	Use CO_2 as a substrate
Synthetases	Link two molecules via an ATP-dependent reaction

EC 1 oxidoreductases: Catalyse the transfer of hydrogen or oxygen atoms or electrons from one substrate to another, also called oxidases, dehydrogenases, or reductases. Note that since these are 'redox' reactions, an electron donor/acceptor is also required to complete the reaction.

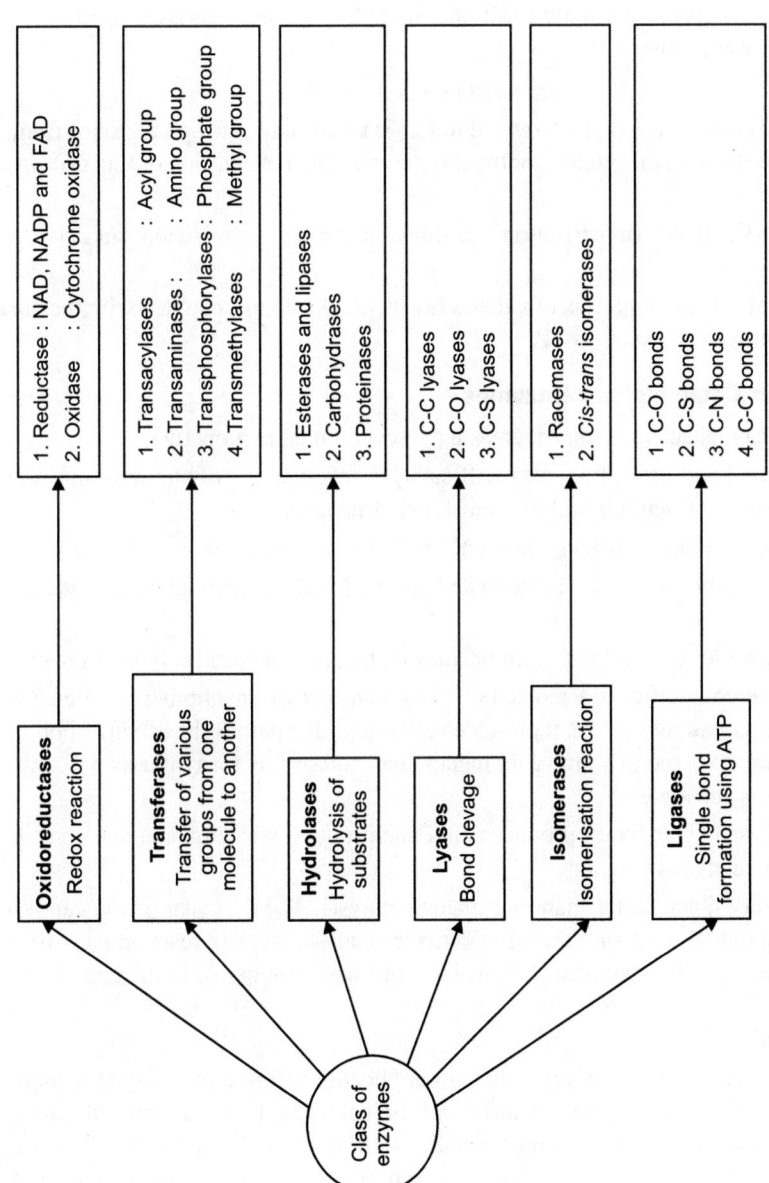

Figure 21.1: Classification of enzyme and examples of related enzyme classes.

EC 2 transferases: Catalyse group transfer reactions, excluding oxido-reductases (which transfer hydrogen or oxygen and are EC 1). These are of the general form:

$$A–X + B \leftrightarrow BX + A$$

EC 3 hydrolases: Catalyse hydrolytic reactions. Includes lipases, esterases, nitrilases, peptidases/ proteases. These are of the general form:

$$A–X + H_2O \leftrightarrow X–OH + HA$$

EC 4 lyases: Catalyse non-hydrolytic (covered in EC 3) removal of functional groups from substrates, often creating a double bond in the product, or the reverse reaction, i.e. addition of function groups across a double bond.

EC 5 isomerases: Catalyses isomerisation reactions, including racemisations and *cis-tran* isomerisations.

EC 6 ligases: Catalyses the synthesis of various (mostly C-X) bonds, coupled with the breakdown of energy-containing substrates, usually ATP.

Difference between Catalysts and Enzymes

When compared to man made catalysts, enzymes have several unique properties.

1. They are extremely specific. Enzymes will catalyse reactions involving only one substrate or a very small group of structurally related substrates. Enzymes show:
   (a) Absolute specificity - working upon only a single substrate.
   (b) Group specificity - working upon a related group of molecules containing a specific functional group.
   (c) Linkage specificity - working on molecules that contain a specific type of chemical bond.
2. Enzymes are stereospecific. If a molecule exists as a pair of enantiomers, the enzyme will use only one of the pair as substrate and produce only one of the pair as the product. For example, the enzymes that are involved in amino acid metabolism and/or protein synthesis will only utilise the L-amino acids as substrates.
3. Reactions catalysed by enzymes produces only one product. Wasteful side reactions do not occur during enzyme catalysed reactions.
4. Enzymes are very much faster than man made catalysts. The best man made catalysts increase reaction rates about 10^7 fold, on average man made catalysts increase reaction rates 10^2 to 10^4 fold. Enzymes can enhance the rate from 10^{17} to 10^{20} fold when compared to the uncatalysed reaction.

IMMOBILISED ENZYMES

Immobilised enzymes are widely used for variety of applications. Based on the type of application, the method of immobilisation and support material can be selected. The immobilised enzymes can be separated from the reaction mixture and reused and also immobilised in order to prevent the enzyme from being exposed to harsh conditions, high temperature, surfactants and oxidising agents, etc. The immobilised enzymes are also widely used in food industry, pharmaceutical industry, bioremediation, detergent industry, textile industry, etc. Enzyme immobilisation improves the operational stability and is also due to the increased enzyme loading which causes the controlled diffusion. Several hundreds of enzymes are immobilised and used for various large scale industries. Immobilisation technique reduces the effluent treatment costs.

An enzyme derived from an organism or cell culture that catalyses metabolic reaction in living organisms and/or substrate conversions in various chemical reactions. The enzymes are the potential catalyst which works in mild temperature, pressure, pH, substrate specificity under suitable reaction conditions and for the production of desired products without any intermediate products as contaminations for these advantages enzyme are used in variety of application such as cosmetics, paper industry, textile industry, food industry, pharmaceutical industry, laundry and in detergents, etc. The biotechnological method of producing enzyme is expensive, hence new methods have been implemented to reduce the cost.

The enzymes have various other limitations such as low stability, highly sensitive to the process conditions and these problems can be overcome by the immobilisation techniques. Immobilised enzymes are being used since 1916, when Nelson and Griffin discovered that invertase when absorbed to charcoal has the ability to hydrolyse the sucrose. The possibility of immobilised enzyme for its reuse and stability was identified by Grubhofer and Schelth, who reported the covalent immobilisation of several enzymes. The repeated assay can be done with the immobilised enzyme which reduces the cost of assay and the reuse of enzyme process is also very simple and it can be attained through ultrafiltration technique. Irreversible enzyme immobilisation includes covalent binding and entrapment.

Reversible enzyme immobilisation includes adsorption, ionic binding, affinity binding and metal binding. Immobilisation of enzyme by enzymatic process is recently identified by researchers in order to avoid harsh immobilisation.

Enzyme Immobilisation Methods

Covalent binding

Covalent binding is a conventional method for immobilisation, it can be achieved by direct attachment with the enzyme and the material through the covalent linkage. The covalent linkage is strong and stable and the support material of enzymes includes polyacrylamide, porous glass, agarose and porous silica. Covalent method of immobilisation is mainly used when a reaction process does not require enzyme in the product, this is the criteria to choose covalent immobilisation method. This covalent binding of the enzyme with the support material involves two main steps such as, the activation of the support material by the addition of the reactive compound and the second one is the modification of the polymer backbone to activate the matrix (Fig. 21.2).

The activation step produces the electrophilic group on the support material, so that the support material couples/reacts with the strong nucleophiles on the proteins. For example glutaraldehyde is the activation method, in this reaction the amine group reacts with the activated matrix. The covalent binding is normally formed between the functional group in the support matrix and the enzyme surface that contains the amino acid residues. The amino acid residues involved in the covalent binding are the sulphydryl group of cysteine, hydroxyl group of serine and threonine. The attachment between the enzyme and the support material can be achieved either through direct linkage or through the spacer arm. The potentiality of using the spacer arm is that it provides the greater degree of the mobility to the enzymes hence the enzymes show the higher activity when compared to the direct attachment.

Advantages of covalent coupling

1. The strength of binding is very strong, so, leakage of enzyme from the support is absent or very little.
2. This is a simple, mild and often successful method of wide applicability

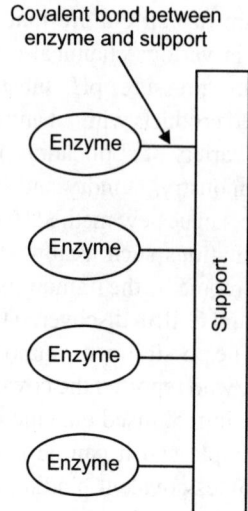

Fig. 21.2: Covalent binding.

Disadvantages of covalent coupling

1. Enzymes are chemically modified and so many are denatured during immobilisation.
2. Only small amounts of enzymes may be immobilised (about 0.02 grams per gram of matrix).

Entrapment

Enzymes are occluded in the synthetic or natural polymeric networks, it is a permeable membrane which allows the substrates and the products to pass, but it retains the enzyme inside the network, the entrapment can be achieved by the gel, fibre entrapping and microencapsulation. The advantage of entrapment of enzyme immobilisation is fast, cheap and mild conditions required for reaction process. The disadvantage is that limitation in mass transfer. The support matrix protects the enzymes from microbial contamination, proteins and enzymes in the micro environment. Microencapsulation method is that the enzyme molecules are capsulated within spherical semipermeable membranes with a selective controlled permeability. This method provides the large surface area between polymeric material and the enzyme. The drawback of this method is inactivation of enzyme during encapsulation.

Advantages and disadvantages of entrapment

Advantages of entrapment: Loss of enzyme activity upon immobilisation is minimised.

Disadvantages of entrapment: (i) The enzyme can leak into the surrounding medium, (ii) another problem is the mass transfer resistance to substrates and products and (iii) substrate cannot diffuse deep into the gel matrix.

Adsorption

This is a simple method of preparing an immobilised enzymes and the materials used for adsorption are activated charcoal, alumina, ion exchange resins, this method is cheap and easy for use and the disadvantage is a weak binding force between the carrier and the enzyme. This method comes under carrier bound immobilisation and the process of immobilisation is reversible. Adsorption (Fig. 21.3) is the easiest and

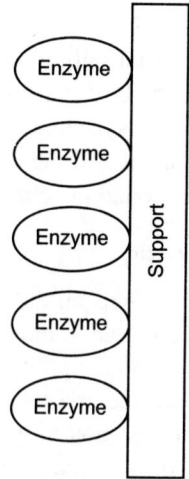

Fig. 21.3: Adsorption.

oldest immobilisation techniques. The interaction between the enzyme and the surface of the matrix through weak forces by salt linkage, hydrogen bonds, hydrophobic bonds, ionic bonds and van der waals forces. Based on the charges of the matrix and the protein arrangements the strongly bound, but not distorted enzyme will be formed. The advantage of enzyme adsorption is minimum activation step and as a result of minimum activation, no reagents required. It is cheap and easy way of immobilisation.

Advantages of adsorption

1. Little or no confirmation change of the enzyme.
2. Simple and cheap.
3. No reagents are required.
4. Wide applicability and capable of high enzyme loading.

Disadvantages of adsorption

1. Desorption of the enzyme protein resulting from changes in temperature, pH and ionic strength.
2. Slow method.

Ionic Binding

The bonding involved between the enzyme and the support material is salt linkages. In non-covalent immobilisation the process will be reversed by changing the temperature polarity and ionic strength conditions. This principle is similar to protein-ligand interactions principles used in chromatography.

Affinity Binding

The immobilisation of enzyme linked to the matrix through the specific interactions. The two methods are being followed in affinity immobilisation. The first method is the activation of the support material which contains the coupled affinity ligand, so that the enzyme will be added. The advantage of this method is the enzyme is not exposed to any harsh chemicals conditions. The second method, the enzyme modified to another molecule which has the ability to bind towards a matrix.

Metal Linked Immobilisation

In metal linked enzyme immobilisation, the metal salts are precipitated over the surface of the carriers and it has the potential to bind with the nucleophilic groups on the matrix. The precipitation of the metal ion on the carrier can be achieved by heating. This method is simple and the activity of the immobilised enzymes is relatively high (30–80%). The carrier and the enzyme can be separated by decreasing the pH, hence it is a reversible process. The matrix and the enzyme can be regenerated, by the process.

PROPERTIES OF IMMOBILISED ENZYMES

The stability of the immobilised enzymes is based on the temperature and time. The activity of the enzyme is retained throughout series of cycles. Due to immobilisation, the properties of enzymes will be altered such as catalytic activity with respect to the support matrix. The change in the enzyme properties in the immobilised enzyme is due to the enzyme and the substrate reacts in the micro-environment which is different from the enzyme substrate reaction in the bulk solution environment. The change is also due to the change in the three dimensional conformation of the protein when linked with the support matrix. These conformational changes are to a lesser extent and these changes occur in the limited enzyme systems. Enzyme immobilisation improves the operational stability and is also due to the increased enzyme loading which causes the controlled diffusion. The stabilisation as a result of, number of bonds formed between the enzyme and the support matrix. When the immobilised enzyme acts on the macro-molecular substrates, active site of the enzymes does not able to access with the substrates, hence the enzyme loses its activity.

APPLICATION OF THE IMMOBILISED ENZYMES

Food Industry

In food industry, the purified enzymes are used but during the purification the enzymes will denature. Hence the immobilisation technique makes the enzymes stable. The immobilised enzymes are used for the production of syrups. Immobilised beta-galactosidase used for lactose hydrolysis in whey for the production of bakers yeast. The enzyme is linked to porous silica matrix through covalent linkage. This method is not preferably used due to its cost and the other technique developed by Valio in 1980, the enzyme galactosidase was linked to resin (food grade) through cross-linking. This method was used for the various purposes such as confectionaries and icecreams.

Biodiesel Production

Biodiesel is monoalkyl esters of long chain fatty acids. Biodiesel is produced through triglycerides (vegetable oil, animal fat) with esterification of alcohol (methanol, ethanol) in the presence of the catalyst.

Wastewater Treatment

The increasing consumption of fresh water and water bodies are mixed up with polluted industrial wastewater and the wastewater treatments are necessary at present. The sources of dye effluents are textile industry, paper industry, leather industry and the effluents are rich in dye colourants. These effluents are threat to the environment and even in low concentration it is carcinogenic. Now-a-days enzymes are used to degrade the dye stuffs. The enzymes used in the wastewater treatments are preoxidases, laccase, azo reductases. These enzymes due to harsh conditions like extreme temperature, low or high pH and high ionic strength may lose its activity, to overcome this problem immobilised enzymes are used. The Horseradish peroxidases are entrapped in calcium alginate beads.

The immobilised laccase enzyme has the ability to degrade dyes anthracinoid dye, Lancet blue and Ponceau red. Adsorption method is widely used because of its easy regeneration. During the covalent method of immobilisation the conformational change in the enzyme occurs which will affect the activity of the enzyme. In single enzyme nanoparticle (SEN), the enzyme is protected by a nanometer thick substance as it provides the large surface area. SEN has the ability to retain its activity during the extreme conditions. SEN is also used in the removal of heavy metals from the wastewater.

Lipase has the ability to hydrolyse oil and fats to long chain fatty acid and glycerol. The immobilised lipase is of high interest for the hydrolysis of oils and fats for treating the wastewater from the food industry. The drawback of the conventional treatment methods is slow biodegradability, oil and fats are absorbed on the surface of sludge. Researchers are now focusing on the treatment with immobilised lipase. Lipase immobilised on the sol gel/calcium alginate with the size of 82 μm, immobilised lipase. Immobilised lipase operated for 100 days in continuous sludge without any problem, does not produce foam in the reactor.

Textile Industry

The enzymes derived from microbial origin are of great interest in textile industry. The enzymes such as cellulase, amylase, liccase, pectinase, cutinase, etc. are used for various textile applications such as scouring, biopolishing, desizing, denim finishing, treating wools, etc. Among these enzymes cellulase has been widely used from the older period to till now. The textile industries now turned to enzyme process instead of using harsh chemical which affects the pollution and cause damage to the fabrics. The processing of fabrics with enzymes requires high temperatures.

Hence, enzyme immobilisation for this process is able to withstand at extreme and able to maintains its activity for more than 5–6 cycles. Endoglucanase is a component of cellulase enzyme, endoglucanase is microencapsulated with Arabic gum is a natural polymer with the biodegradable property is used as a matrix for encapsulation of endoglucanase. Encapsulation of endoglucanase prevented it to retain its activity in the presence of detergents.

Detergent Industry

The detergent industry also employs enzymes for removal of stains. The enzymes used in detergent industry are protease which is used to remove the stains of blood, egg, grass and human sweat. Amylase used to remove the starch based stains like potatoes, gravies, chocolate. Lipase used to remove the stains of oil and fats and also used to remove the stains in cuffs and collars. Cellulase is used for cotton based fabrics in order to improve softening, colour brightening and to remove soil stains. Now-a-days Biotech cleaning agents are widely used in the detergent industries. When compared to synthetic detergents the biobased detergents have good cleaning property.

Biomedical Industry

Immobilised enzymes are used for diagnosis and treatment of diseases in the medical field. The inborn metabolic deficiency can be overcome by replacing the encapsulated enzymes (i.e. enzymes encapsulated by erythrocytes) instead of waste metabolites which act as a carrier for the exogenous enzyme drugs and the enzymes are biocompatable in nature, hence there is no immune response. The enzyme encapsulation through the electroporation is a easiest way of immobilisation in the biomedical field and it is a reversible process for which enzyme can be regenerated. The enzymes when combined with the biomaterials provides biological and functional systems. The biomaterials are used in tissue engineering application for repair of the defect. The advantage of the enzyme immobilisation in biomedical is that the free enzymes are consumed by the cells and not active for prolonged use, hence the immobilised enzymes remains stable, to stimulate the growth and to repair the defect. The cancer therapy is delivery of enzymes to the oncogenic sites have been improved with new methods. The nanoparticles and nanospheres are often used as enzyme carriers for the delivery of therapeutic agents.

NANOPARTICLES AS IMMOBILISATION MATRIX

Nanoparticles act as very efficient support materials for enzyme immobilisation, because of their ideal characteristics for balancing the key factors that determine biocatalysts efficiency, including specific surface area, mass transfer resistance and effective enzyme loading. Diffusion problem is more relevant when we are dealing with the macro-molecular substrates, for such systems the nano-particles are the ideal candidates. Moreover, the enzyme bound nanoparticles show Brownian movement, when dispersed in aqueous solutions showing that the enzymatic activities are comparatively better than that of the unbound enzyme. In addition, magnetic nanoparticles possess additional advantage, can be separated easily using an external magnetic field. Studies have shown that immobilisation of enzymes to the nanoparticles can reduce protein unfolding and can improve stability and performance.

Enzymatic immobilisation on Au and Ag nanoparticles have been studied using either as whole cells or isolated enzymes, which include lysozyme, glucose oxidase, aminopeptidase, as well as alcohol dehydrogenase. Cruz and others reported the Immobilisation of enzymes S. Carlsberg and Candida antarctica lipase B (CALB) on fumed silica nanoparticles for applications in nonaqueous media and they observed catalytic activities were remarkably high. Won and others immobilised acetylcholinesterase onto magnetic glasses based on iron oxide/silica, for paraoxon sensing. Ganesana and others performed the immobilisation of acetylcholinesterase on nickel nanoparticles and obtained a highly sensitive detection method for organophosphate pesticides.

Uygun and others employed magnetic poly (2-hydroxyethyl methacrylate-N-methacryloyl-(l)-phenylalanine) to immobilise α-amylase. They reported a substrate affinity increases upon enzyme immobilisation and showed that a specific activity of 85% was maintained after 10 reuses. Khoshnevisan and others immobilised cellulase on magnetic nanoparticles obtaining a smaller activity than for the free enzyme, but at 80°C the immobilised enzyme showed slightly greater activity. Lee and others used amino-functionalised silica-coated magnetic nanoparticles to immobilise trypsin. They applied this system to a pressure-assisted digestion for proteome analysis. It was observed for each of the experiments in which the magnetic nanoparticles were employed an increased number of protein identification in comparison with the experiment with free trypsin. Recently Smith and others have reported the immobilisation of enzymes (Peroxidase, cellulase, trypsin and alpha amylase) on TiO_2 nanoparticles. The immobilised enzymes show higher activity than free enzymes. It also showed enhanced thermal stability compared to its soluble counterpart at higher temperature. All the advantages of immobilised enzymes on micron-sized particles are inherited when nanomaterials are used as solid supports. Broadly there are four main approaches to link a protein or enzyme to the nanoparticles.

Electrostatic adsorption: The most widely used linkage approach consists of electrostatic adsorption. This is the simplest approach and is already used routinely as an electron dense marker in histology. The interaction between the nanoparticle and protein may be modulated by the pH or charge screening by controlling the ionic strength of the medium.

Covalent attachment to the surface modified nanoparticles: Another general method for nanoparticle–protein conjugation is covalently linking a protein to the nanoparticle ligand. This approach has been greatly advanced by extreme control over the surface chemistry of the nanoparticles. For example, a variety of organic functional groups can be introduced to the surface using mild conditions. The popular labelling chemistry utilises the covalent binding of primary amines with sulpho-NHS esters or R-COOH groups via reaction with EDC. Nanoparticles labelled with NHS esters can react to form covalent bonds with the primary amine of lysine on a protein. In addition, nanoparticles coated with maleimide groups can react with the thiol of cysteine on a protein. Oxide nanoparticles (TiO iron oxide, Copper oxide, silver and gold oxide) can be easily modified by Silanisation yielding a modified surface exhibiting amino groups, which can be used as adsorbent or as coupling sites for linking various proteins.

Conjugation using specific affinity of protein: Nanoparticle–protein conjugation can also be achieved by using specific labelling strategies.

Direct conjugation to the nanoparticles surface: A direct reaction of a chemical group on the protein without the use of a linker is usually desired if the particle is used as a biosensor electron transfer is used. For Au and Ag nanoparticles, this can be achieved by the Au-thiol or Ag- thiol chemistry where a protein with a cysteine covalently bonds to an Au or Ag nanoparticle. The conjugation requires incubation of the protein and the nanoparticle together as the Au–S or Ag-S bond is strongly favoured. Similarly, for sulphur containing nanoparticles such as ZnS/CdSe, cysteine can directly form a disulphide bridge with the surface S atom. Some important new consequences arise when the size of the carrier approaches nanodimensions. Mostly, these all work out in the favour of using nanosized materials.

Present studies shows that different types of nanomaterial are available for enzyme immobilisation for examples carbon nanotubes (CNTs), nanoparticles, magnetic nanoparticles, mesoporous media, nanofibres, nanocomposites, nanorods and sol–gel materials containing nanometer-size particles and single-enzyme nanoparticles. However, the major problem for their application is their high cost and complex supports preparation. Strategies or protocols for synthesising the nanoparticles should be developed which are low cost, ecofriendly and can be used for large scale synthesis.

Thus, immobilisation process has been used for enhancing enzyme activity and stability in aqueous and non-aqueous media. Selecting and designing the support matrix are important in enzyme immobilisation. Recently, the use of nanoparticles has emerged as a versatile tool for generating excellent supports for enzyme stabilisation due to their small size and large surface area. It has been observed that the stability and activity of enzymes increases when immobilised on such materials. The nanomaterials are key components in the future market of high technology. Nanoparticles strongly influence the mechanical properties of the material like stiffness and elasticity and provide biocompatible environments for enzyme immobilisation.

Advantages of Enzyme Immobilisation

1. Multiple or repetitive use of a single batch of enzymes.
2. Immobilised enzymes are usually more stable.
3. Ability to stop the reaction rapidly by removing the enzyme from the reaction solution.
4. Product is not contaminated with the enzyme.
5. Easy separation of the enzyme from the product.
6. Allows development of a multienzyme reaction system.
7. Reduces effluent disposal problems.

Disadvantages of Enzyme Immobilisation

1. It gives rise to an additional bearing on cost.
2. It invariably affects the stability and activity of enzymes.
3. The technique may not prove to be of any advantage when one of the substrate is found to be insoluble.
4. Certain immobilisation protocols offer serious problems with respect to the diffusion of the substrate to have an access to the enzyme.

PRODUCTION OF INDUSTRIAL ENZYMES

Enzymes can be used as chemicals to determine the concentration of substrates, measure the catalytic activity of enzymes present in biological samples and serve as labels in immunoassays to determine the concentrations of enzymatically inert substances. For instance, enzymes are routinely used in determination of glucose (glucose oxidase, horse-radish peroxidase), urea (urease, glutamate dehydrogenase), triglycerides (lipase, carboxylesterase, glycerol kinase, etc.), in clinical diagnosis. Carbohydrates, organics acids, alcohols and other food ingredients are routinely determined in food analysis using enzymes.

Fermentation is a method of generating enzymes for industrial purposes. Fermentation involves the use of micro-organisms, like bacteria and yeast to produce the enzymes. There are two methods of fermentation used to produce enzymes. These are submerged fermentation and solid-state fermentation. Submerged fermentation involves the production of enzymes by micro-organisms in a liquid nutrient media. Solid-state fermentation is the cultivation of micro-organisms and hence enzymes on a solid substrate. Carbon containing compounds in or on the substrate are broken down by the micro-organisms, which produce the enzymes either intracellular or extracellular. The enzymes are recovered by methods such as centrifugation, for extracellularly produced enzymes and lysing of cells for intracellular enzymes.

LARGE SCALE FERMENTATION OF ENZYMES

Use of an aerobic submerged culture in a stirred tank reactor is the typical industrial process for enzyme production involving a micro-organism that produces an industrial enzyme. Figure 21.4 shows a flow diagram of a typical production process.

Organism and Enzyme Production

The basic mechanisms of enzyme synthesis, including transcription, translation and post-translational processing, seem to be highly conservative. However, several differences exist between various classes of organisms, as well as some fundamental differences between prokaryotic and eukaryotic organisms. The enzymes themselves differ enormously in molecular mass, number of polypeptide chains, isoelectric point and degree of glycosylation. In addition, a variety of enzyme-producing species exist. Although all the differences may influence the characteristics of synthetic patterns, the basic mechanisms underlying enzyme synthesis are similar enough to allow a general treatment of the microbiological production process. However, the differences in production kinetics among various species are large enough to make individual optimisation programmes necessary. Different organisms may also differ in their suitability for fermentation, such process characteristics as viscosity or recoverability, legal clearance of the organism and available knowledge about the selected organism, must be considered. As cellular enzyme expression is heavily influenced by regulatory mechanism, enzyme synthesis rates range from no synthesis to maximum synthesis allowed by the synthetic apparatus, as in a normal control loop.

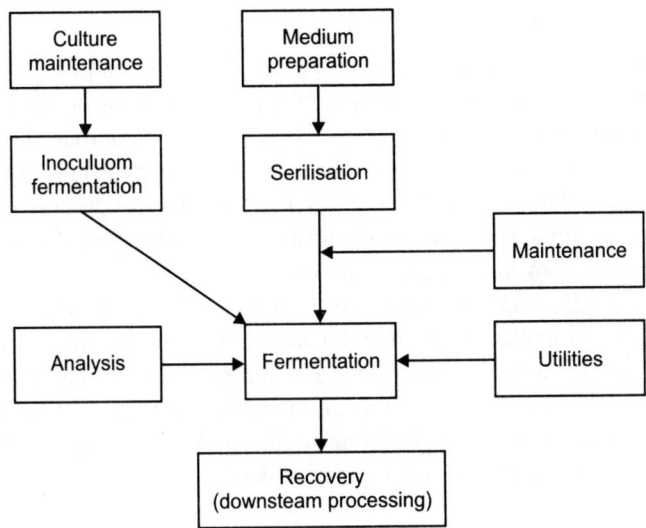

Fig. 21.4: Flow diagram of a fermentation process.

The complexity of mechanisms ranges from relatively simple and well-understood induction and repression systems, to very complex global regulatory networks. Process development must deal with the complexity of the enzyme synthesis system either by changing the structural characteristics, including structural elements of the regulatory systems, or by selecting optimal environmental conditions.

Media of Enzyme Production

Substrates for industrial applications are usually included in complex media, which must be supplemented with special compounds, such as a nitrogen source, various nutrient salts, or certain trace elements.

The main sources for microbial processes consist of sugars such as sucrose or hexoses. They represent the source for both carbon and energy. Typical feedstock materials are molasses, unrefined sugar, sulphite liquor from cellulose production plants, hydrolysates of wood and starch, or fruit juices, such as the grape juice used in wine making. In many cases starches from various cereals (barley, corn, rice, rye and wheat) or tubers (potatoes) are used as inexpensive raw materials. Starch-containing substrates from wheat are mostly used in raw form, i.e. in the form of grist and therefore contain a number of additional substances.

Nitrogen, phosphorus and potassium are important nutrients to maintain growth. They are added in the form of inorganic compounds such as phosphates, ammonium compounds, or potassium chloride. Soya meal, fish meal, cotton seed, low-quality protein materials such as casein or its hydrolysates, millet, stillage and especially corn steep liquor are also used as low-price nutrients. In addition, these chemically complicated mixtures contain trace elements and growth promoters. A number of trace elements are provided in sufficient quantity by the tap water that is used for the preparation of the medium. In general, the raw materials are dissolved or suspended in water and the resulting medium is heated, filtered and sterilised. The complex composition of the media used in industry causes considerable problems. For downstream processing (harvest, concentration and purification of product) or for analytical assays during the process, additional pre-treatment of the raw material is needed to avoid unfavourable side effects.

Sterilisation

Most bioprocesses for the production of enzymes are based on pure cultures. Because contamination would prevent proper functioning of the process, bioprocessing must be carried out under aseptic conditions. Solid substrates, such as grist used for amylase production or treated soil used in mushroom cultivation, are kept in rooms at elevated temperatures for an appropriate amount of time. Liquid media are sterilised *in situ*, e.g. in the reaction vessel or in separate containers, usually by external heating. In some cases, media are prepared in a concentrated form and steam is injected directly into the solution, with the resulting condensate making up the final volume.

The temperature for sterilisation is normally above 100°C for an appropriate period. Flow sterilisation through heat exchangers also applies temperatures far above 100°C, but residence times are shorter than in batch sterilisation. In general, the duration of the heat treatment does not only depend on the material to be processed but also on the pH of the medium and the initial number of viable germs or spores. In aerobic processes the culture must be supplied with sterile air. Air is usually filtered through glass wool filters, sintered materials, or membranes of appropriate design.

Inoculation

After inoculation with the micro-organisms, the process should start immediately and the reaction should proceed fast. The amount of active cell culture added is therefore critical and depends on the size of the batch. Scaling-up from the original starter culture to the inoculation broth is done in several steps in large-scale industrial processes. The starter culture is kept deep frozen (–70 to –90°C) for preservation. In fungal inoculates proper wetting of the spores is achieved by adding small amounts of surfactants to the broth. If inoculation by spare suspensions is not optimal, mycelial pellets can be used for start-up. Bacterial spores must be activated by thermal treatment before they can be used for inoculation. During the exponential growth phase of the bioprocess cells can be harvested for following inoculations.

Fermentation

For enzyme production, economy of scale leads to the use of fermenters with a volume of 20–200 m^3. The higher energy yield from aerobic combustion results in the use of aerobic processes which require continuous transfer of poorly soluble oxygen into the culture broth.

Process design is the complicated task of choosing the optimal conditions for maximal process outcome. The number of interdependent factors is high and the available physiological knowledge is seldom complete. The relationship between synthesis rate and growth rate could be very complex and depends on the presence of inductors or the absence of repressors. The use of genetically developed strains may considerably ease process design. The designed process is usually first tested on a pilot-plant scale and optimised in a number of fermentation runs, it is then scaled up to production size. The total synthesis rate depends not only on growth rate but also on biomass concentration.

Two types of cultivation can be distinguished: In batch culture, the growth rate cannot be controlled by dosed feeding, because all substrates are added at the beginning. Batch processes are rarely used today. In fed-batch processes, a low initial biomass concentration should be chosen to maintain the desired growth rate for a certain period, without exceeding the transport capacity of the equipment. In addition, fed-batch processes can be designed in a way that the enzyme concentration at harvesting is higher than in batch or continuous cultures and the productivity of fed-batch processes can be increased several-fold compared to batch processes. Fed-batch processes probably constitute the most frequently used process type.

Continuous cultures are ideally suited for high productivity because the excess of biomass is continuously withdrawn and both synthesis rate and biomass concentration will be optimal. Therefore, in principle, continuous culture is preferred for biotechnological production. However, enzyme concentrations are lower than those reached in fed-batch cultures. In addition, the use of continuous cultures is limited by technical reasons such as the higher contamination risk and the problem of strain degeneration.

DOWNSTREAM PROCESSING

Downstream processing is a very important step in biotechnology because costs for collection, concentration and purification of the final product are substantial. High product concentrations in the supernatant or inside the cells and efficient purification are therefore important aspects in the overall economy of enzyme manufacture. The degree of purity of commercial enzymes ranges from raw enzymes to highly purified forms and depends on the application. Raw materials for the isolation of enzymes are animal organs, plant material and micro-organisms.

Often enzymes may be purified several hundred-fold but the yield of the enzyme may be very poor, frequently below 10% of the activity of the original material. In contrast, industrial enzymes will be purified as little as possible, only other enzymes and material likely to interfere with the process which the enzyme is to catalyse, will be removed. Unnecessary purification will be avoided as each additional stage is costly in terms of equipment, manpower and loss of enzyme activity. As a result, some commercial enzyme preparations consist essentially of concentrated fermentation broth, plus additives to stabilise the enzyme's activity.

However, the content of the required enzyme should be as high as possible (e.g. 10% w/w of the protein) in order to ease the downstream processing task. This may be achieved by developing the fermentation conditions or, often more dramatically, by genetic engineering. It may well be economically viable, e.g. to spend some time cloning extra copies of the required gene together with a powerful promoter back into the producing organism in order to get over-producers. Downstream processing involves isolation and purification steps and ends up in the formulation of the enzyme preparation.

Enzymes are universally present in living organisms, each cell synthesises a large number of different enzymes to maintain its metabolic reactions. The choice of procedures for enzyme purification depends on their location. On the one hand, isolation of intracellular enzymes often involves the separation of complex biological mixtures. On the other hand, extracellular enzymes are generally released into the medium with only a few other components. Enzymes are very complex proteins and their high degree of specificity as catalysts is manifest only in their native state. The native conformation is attained under specific conditions of pH, temperature and ionic strength. Hence, only mild and specific methods can be used for enzyme isolation. Figure 21.5 shows the sequence of steps involved in the recovery of enzymes.

Preparation of Biological Starting Materials

Animal organs: Animal organs must be transported and stored at low temperature to retain enzymatic activity. Frozen organs can be minced with machines generally used in the meat industry and the enzymes can be extracted with a buffer solution. Besides mechanical grinding, enzymatic digestion can also be employed.

Plant material: Plant material can be ground with various crushers or grinders and the desired enzymes can be extracted with buffer solutions. The cells can also be disrupted by previous treatment with lytic enzymes.

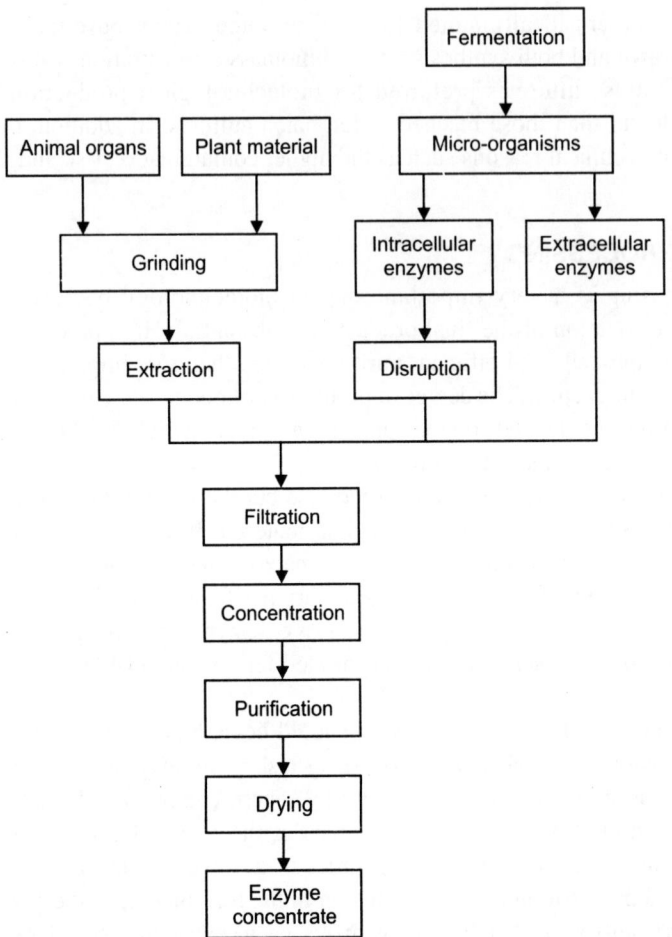

Fig. 21.5: Flowchart of the downstream processing of enzymes.

Micro-organisms: Enzymes might accumulate inside the cells or be released into the medium. Most enzymes used commercially are extracellular enzymes and the first step in their isolation is the separation of the cells from the solution. For intracellular enzymes, which are being isolated today in increasing amounts, the first step involves grinding to rupture the cells. A number of methods for the disruption of cells are known, corresponding to the different types of cells and the problems involved in isolating intracellular enzymes. However, only a few of these methods are used on an industrial scale.

Cell disruption by mechanical methods

High-pressure homogenisation is the most common method of cell disruption. The cell suspension is pressed through a valve and hits an impact ring. The cells are ruptured by shearing forces and simultaneous decompression. The wet grinding of cells in a high-speed bead mill is another effective method of cell disruption. Glass balls with a diameter of 0.2–1 mm are used to break the cells. The efficiency of this method depends on the geometry of the stirrer system.

Cell disruption by non-mechanical methods

Cells may frequently be disrupted by chemical, thermal, or enzymatic lysis. The drying of micro-organisms and the preparation of acetone powders are standard procedures in which the structure of the cell wall is altered to permit subsequent extraction of the cell contents.

Separation of Solid Matter

After cell disruption, the next step is separation of extracellular or intracellular enzymes from cells or cellular fragments, respectively. This operation is rather difficult because of the small size of bacterial cells and the slight difference between the density of the cells and that of the fermentation medium. Continuous filtration is used in industry. Large cells, e.g. yeast cells, can be removed by decantation. Today, efficient centrifuges have been developed to separate cells and cellular fragments in a continuous process. Residual plant and organ matter can be separated with simpler centrifuges or filters.

Filtration

Pressure filters: A filter press (plate filter, chamber filter) is used to filtrate small volumes or to remove precipitates formed during purification. The capacity to retain solid matter is limited and the method is rather work-intensive. However, these filters are highly suitable for the fine filtration of enzyme solutions.

Vacuum filters: Vacuum filtration is generally the method of choice because biological materials are easily compressible. A rotary vacuum filter is used in the continuous filtration of large volumes. The suspension is usually mixed with a filter aid (e.g. kieselguhr) before being applied to the filter.

Cross-flow filtration: In recent years, a new method of filtration, cross-flow filtration, has been devised. In conventional methods, the suspension flows perpendicular to the filtering material. In cross-flow filtration, the input stream flows parallel to the filter area, thus preventing the accumulation of filter cake and an increased resistance to filtration. Cross-flow filtration can be conveniently used in recombinant DNA techniques to separate organisms in a closed system.

Centrifugation

The sedimentation rate of a bacterial cell with a diameter of 0.5 mm is less than 1 mm/h. An economical separation can be achieved only by sedimentation in a centrifugal field. The range of applications of centrifuges depends on the particle size and the content of the solids. Decanters (scroll-type centrifuges) work with low centrifugal forces and are used in the separation of large cells or protein precipitates. Solid matter is discharged continuously by a screw conveyer moving at a differential rotational speed. Tubular bowl centrifuges are built for very high centrifugal forces and can be used to sediment very small particles. However, these centrifuges cannot be operated in a continuous process. Moreover, solid matter must be removed by hand after the centrifuge has come to a stop. A further disadvantage is the appearance of aerosols.

Separators (disk stack centrifuges) can be used in the continuous removal of solid matter from suspensions. Solids are discharged by a hydraulically operated discharge port (intermittent discharge) or by an arrangement of nozzles (continuous discharge). Bacteria and cellular fragments can be separated by a combination of high centrifugal forces. Disk stack centrifuges that can be sterilised with steam are used for recombinant DNA techniques in a closed system.

Extraction

An elegant method used to isolate intracellular enzymes is liquid-liquid extraction in an aqueous two-phase system. This method is based on the incomplete mixing of different polymers, e.g. dextran and poly(ethylene glycol), or a polymer and a salt in an aqueous solution. The first extraction step separates cellular fragments. Subsequent purification can be accomplished by extraction or, if high purity is required, by other methods. The extractability can be improved by using affinity ligands or modified chromatography gels, e.g. phenyl-sepharose.

Flocculation and flotation

The flocculation of bacterial cells to form larger particles can be achieved by the addition of mineral colloids, salts, or organic polymers. The neutralisation of charges on the cell surface and the formation of bridges between individual cells lead to agglomeration. Agglomerates can then be removed by filtration or centrifugation.

Concentration

The enzyme concentration in starting material is often very low. The volume of material to be processed is generally very large and substantial amounts of waste material must be removed. Thus, if economic purification is to be achieved, the volume of starting material must be decreased by concentration. Only mild concentration procedures which do not inactivate enzymes can be employed. These procedures include thermal methods, precipitation and to an increasing extent, membrane filtration.

Thermal methods

Only brief heat treatment can be used for concentration because enzymes are thermolabile. Evaporators with rotating components which achieve a thin liquid film (thin-layer evaporator, centrifugal thin-layer evaporator) or circulation evaporators (long-tube evaporator) can be employed.

Precipitation

Enzymes are very complex protein molecules possessing both ionisable and hydrophobic groups which interact with the solvent. Indeed, proteins can be made to agglomerate and, finally, precipitate by changing their environment. Precipitation is actually a simple procedure for concentrating enzymes.

Precipitation with salts: High salt concentrations act on the water molecules surrounding the protein and change the electrostatic forces responsible for solubility. Ammonium sulphate is commonly used for precipitation, hence, it is an effective agent for concentrating enzymes. Enzymes can also be fractionated, to a limited extent, by using different concentrations of ammonium sulphate. Sodium sulphate is another precipitating agent used.

Precipitation with organic solvents: Organic solvents influence the solubility of enzymes by reducing the dielectric constant of the medium. The solvation effect of water molecules surrounding the enzyme is changed, the interaction of protein molecules is increased and therefore, agglomeration and precipitation occur. Commonly used solvents are ethanol and acetone.

Precipitation with polymers: The polymers generally used are polyethyl-enimines and polyethylene glycols of different molecular masses. The mechanism of this precipitation is similar to that of organic solvents and results from a change in the solvation effect of the water molecules surrounding the enzyme. Most enzymes precipitate at polymer concentrations ranging from 15 to 20%.

Precipitation at the isoelectric point: Proteins are ampholytes and carry both acidic and basic groups. The solubility of proteins is markedly influenced by pH and is minimal at the isoelectric point at which the net charge is zero. Because most proteins have isoelectric points in the acidic range, this process is also called acid precipitation.

Ultrafiltration

A semipermeable membrane permits the separation of solvent molecules from larger enzyme molecules, because only the smaller molecules can penetrate the membrane when the osmotic pressure is exceeded. This is the principle of all membrane separation processes including ultrafiltration. In reverse osmosis, used to separate materials with low molecular mass, solubility and diffusion phenomena influence the process, whereas ultrafiltration and cross-flow filtration are based solely on the sieve effect. In processing enzymes, cross-flow filtration is used to harvest cells, whereas ultrafiltration is employed for concentrating and desalting.

Desalting: The desalting of enzyme solutions can be carried out conveniently by diafiltration. The small salt molecules are driven through a membrane with the water molecules. The permeate is continuously replaced by fresh water.

Purification

For many industrial applications, partially purified enzyme preparations will suffice, however, enzymes for analytical purposes and for medical use must be highly purified. Special procedures employed for enzyme purification are crystallisation, electrophoresis and chromatography. However crystallisation and electrophoresis are not relevant for large scale purifications. Chromatography, in contrast, is of fundamental importance to enzyme purification. Molecules are separated according to their physical properties (size, shape, charge, hydrophobic interactions), chemical properties (covalent binding), or biological properties (biospecific affinity). In gel chromatography (also called gel filtration), hydrophilic, cross-linked gels with pores of finite size are used in columns to separate biomolecules. In gel filtration, molecules are separated according to size and shape. Molecules larger than the largest pores in the gel beads, i.e. above the exclusion limit, cannot enter the gel and are eluted first. Smaller molecules, which enter the gel beads to varying extent depending on their size and shape, are retarded in their passage through the column and eluted in order of decreasing molecular mass. Gel filtration is used commercially for both separation and desalting of enzyme solutions.

Ion-exchange chromatography is a separation technique based on the charge of protein molecules. Enzyme molecules possess positive and negative charges. The net charge is influenced by pH and this property is used to separate proteins by chromatography on anion exchangers (positively charged) or cation exchangers (negatively charged). The ability to process large volumes and the elution of dilute sample components in concentrated form make ion exchange very useful.

For hydrophobic chromatography, media derived from the reaction of CNBr-activated Sepharose with aminoalkanes of varying chain length are suitable. This method is based on the interaction of hydrophobic areas of protein molecules with hydrophobic groups on the matrix. Adsorption occurs at high salt concentrations and fractionation of bound substances is achieved by eluting with a negative salt gradient. This method is ideally suited for further purification of enzymes after concentration by precipitation with such salts as ammonium sulphate. In affinity chromatography the enzyme to be purified is specifically and reversibly adsorbed on an effector attached to an insoluble support matrix. Suitable effectors are substrate analogues, enzyme inhibitors, dyes, metal chelates, or antibodies. The insoluble matrix is contained in a column. The biospecific effector, e.g. an enzyme inhibitor, is attached to the matrix. A mixture of different enzymes is applied to the column. The immobilised effector specifically binds the complementary enzyme. Unbound substances are washed out and the enzyme of interest is recovered by changing the experimental conditions, for example by altering pH or ionic strength. Immunoaffinity chromatography occupies an unique place in purification technology. In this procedure, monoclonal antibodies are used as effectors. Hence, the isolation of a specific substance from a complex biological mixture in one step is possible. In this procedure, enzymes can be purified by immobilising antibodies specific for the desired enzyme. In this way, enzymes that usually do not bind to an antibody can be purified by immunoaffinity chromatography. Covalent chromatography differs from other types of chromatography by forming a covalent bond in between the required protein and the stationary phases.

FORMULATION OF THE FINAL ENZYME PRODUCT

Once the enzyme has been purified to the desired extent and concentrated, the manufacturer's main objective is to retain the activity. Enzymes for industrial use are sold on the basis of overall activity. Often a freshly supplied enzyme sample will have a higher activity than that stated by the manufacturer. This is done to ensure that the enzyme preparation has the guaranteed storage life. The manufacturer will usually recommend storage conditions and quote the expected rate of loss of activity under those conditions. It is of primary importance to the enzyme producer and customer that the enzymes retain their activity during storage and use.

Some enzymes retain their activity under operational conditions for weeks or even months. However, most enzymes do not. Most industrial enzyme preparations contain a relatively little amount of active enzyme, the rest being due to inactive protein, stabilisers, preservatives, salts and the diluent which allows standardisation between production batches of different specific activities.

The key to maintaining enzyme activity is maintenance of conformation, so preventing unfolding aggregation of the enzyme molecules and changes in the covalent structure. Three approaches are possible:

1. Use of additives.
2. The controlled use of covalent modification.
3. Enzyme immobilisation.

In general, proteins are stabilised by increasing their concentration and the ionic strength of their environment. Neutral salts compete with proteins for water and bind to charged groups or dipoles. This may result in the interactions between an enzyme's hydrophobic areas being strengthened causing the enzyme molecules to compress and making them more resistant to thermal unfolding reactions. Not all salts are equally effective in stabilising hydrophobic interactions, some are much more effective at their destabilisation by binding to them and disrupting the localised structure of water (the chaotropic effect). From this it can be seen why ammonium sulphate and potassium hydrogen phosphate are powerful enzyme stabilisers whereas sodium thiosulphate and calcium chloride destabilise enzymes. Many enzymes are specifically stabilised by low concentrations of cations which may or may not form part of the active site, for example Ca^{2+} stabilises α-amylases and Co^{2+} stabilises glucose isomerases. At high concentrations (e.g. 20% NaCl) salt discourages microbial growth due to its osmotic effect. In addition, ions can offer some protection against oxidation to groups such as thiols by salting-out the dissolved oxygen from solution.

Low molecular weight polyols (e.g. glycerol, sorbitol and mannitol) are also useful for stabilising enzymes, by repressing microbial growth due to the reduction in the water activity and by the formation of protective shells which prevent unfolding processes. Glycerol may be used to protect enzymes against denaturation due to ice-crystal formation at sub-zero temperatures. Some hydrophilic polymers (e.g. polyvinyl alcohol, polyvinylpyrrolidone and hydroxypropylcelluloses) stabilise enzymes by a process of compartmentalisation, whereby the enzyme-enzyme and the enzyme-water interactions are somewhat replaced by less potentially denaturing enzyme-polymer interactions. They may also act by stabilising the hydrophobic effect within the enzymes.

Many specific chemical modifications of amino acid side chains are possible which may (or, more commonly, may not) result in stabilisation. An useful example of this is the derivatisation of lysine side chains in proteases with N-carboxyamino acid anhydrides.

These derivatissation result in polyaminoacylated enzymes with various degrees of substitution and length of amide-linked side chains. This derivatisation is sufficient to disguise the proteinaceous nature of the protease and prevent autolysis. Enzymes are much more stable in the dry state than in solution. Solid enzyme preparations sometimes consist of freeze-dried protein. More usually they are bulked out with inert materials such as starch, lactose, carboxymethylcellulose and other poly-electrolytes which protect the enzyme during a cheaper spray-drying stage.

Other materials which are added to enzymes before sale may consist of substrates, thiols to create a reducing environment, antibiotics, benzoic acid esters as preservatives for liquid enzyme preparations, inhibitors of contaminating enzyme activities and chelating agents. Additives of these types must, of course, be compatible with the final use of the enzymes product.

In order to ensure safe handling, stability, suitable mixing, functionality, etc. in the various applications, most enzyme preparations are formulated in a variety of liquid and granular forms. Some enzyme preparations are immobilised. Often the precise details of the methods used to stabilise enzyme preparations are kept secret or revealed to customers as a confidential information only.

Nature of Enzyme Products

The nature of enzyme products is influenced by: (i) the enzyme itself and its properties as active compound, (ii) the enzyme source, the fermentation media and conditions and the purification steps (resulting in the enzyme concentrate), (iii) the additives used to formulate the final enzyme preparation. Thus, one has to differentiate between enzyme, enzyme concentrate and enzyme preparation.

Table 21.2 describes the terminology used in enzyme manufacturing. Enzyme preparations vary considerably in terms of purity and formulation (additives) among companies and depending on the particular application. Most enzyme applications do not require high enzyme concentration of the isolate or the preparation. A look on the range of main components of industrial enzyme preparations (Table 21.3) reveals, that the contents of active enzyme protein in the final enzyme preparation is usually very low, typically in the order of 1–5%.

Sugars and inorganic salts are used for stabilising the finished product for storage and distribution. Salts and sometimes carbohydrates such as starch, maltodextrins and sugar alcohols are used to dilute extracted enzymes (the enzyme concentrate) to a standard activity. Preservatives are generally restricted to use in liquid preparations. Thus, by-products from the fermentation process (other proteins, carbohydrates, sugars, salts), additives and preservatives comprise up to 99% of the enzyme preparation. Manufacturing costs are not the only issue, in some cases limited purity and concentration are desirable in terms of formulation of the preparation, improving stability/performance, facilitating homogeneous blending, etc. The increasing use of genetically modified micro-organisms (GMM) and new recovery techniques have generally increased purity and amount of enzyme present in the enzyme concentrate.

Table 21.2: Terminology used in enzyme manufacturing.

Nomenclature	Description
Enzyme	A pure protein with a specified activity, on basis of which the enzyme is identified. The activity has an IUB and a CAS number. Enzyme may also be referred to as active enzyme or active enzyme protein
Enzyme concentrate	An enzyme-containing mixture in its most concentrated form as it occurs during commercial production, thus before formulation. Enzyme concentrates are not commercially available as such, but some may be used directly by the producer as a processing aid for food or to synthesise certain chemicals (i.e. captive use). Enzyme concentrate may also be referred to as enzyme 'isolate'
	The enzyme concentrate is measured as total organic substance (TOS, TOS = 100% - (Water + Ash + Diluents)
Enzyme formulation	An enzyme concentrate, diluted to a standardised activity, stabilised and commercially available on the market
Enzyme blend	A formulation of an intended mixture of enzyme concentrates originating from different sources
Enzyme preparation	Enzyme formulation or enzyme blend
Declared enzyme	The enzyme components in the preparation of significance in terms of application and regulation

Table 21.3: Typical range of main components of industrial enzymes in enzyme preparations.

Component	Range of content (% dry solids)
Proteins and amino acids	10–15
Active enzyme protein	1–5
Complex carbohydrates	5–12
Sugars (and sugar alcohols)	2–20
Inorganic salts	3–40
Preservatives	0–0.3

Enzyme Concentrate

Impurities in the enzyme isolate/concentrate result from the fermentation broth and subsequent purification and consist of proteins, peptides and amino acids, carbohydrates, minerals and other minor components. The relative amounts of these components vary considerably within and between categories of enzyme concentrates. Enzyme content and purity is similar for technical, food and feed enzymes. Enzymes used in personal care products, for therapeutic and analytical or diagnostic application may generally be of higher purity and concentration. Ash constituents comprise small amounts of minerals and diluents are additives for granulation, liquid formulation, stabilisation, preservation, etc.

Enzymes used for food and feed have to comply with purity specifications comprising limits for heavy metals and contaminating micro-organisms and absence of mycotoxins, antibiotics and the production strain. The type and range of impurities, which are present in the enzyme concentrate, is highly varying and depend on the production strain, the media used, the fermentation conditions and the subsequent purification steps. Some of the impurities may also fulfil a technical function during enzyme application. This is particularly true for side activities. An enzyme concentrate will typically contain other enzymes usually referred to as side activities (not to be confused with side activities of a particular enzyme protein). The type and range of these side activities are largely depending on the enzyme manufacturing conditions and the production strain.

These side activities may be more or less important in any given application. An example of this is the use of amylases in baking. Fungal alpha-amylase is added to dough as part of a flour improver. The objective is to hydrolyse starch to provide more fermentable sugars for the yeast. Different enzymes, however, result in different effects on the rheology of the dough and the final product quality. The reason for this is that the enzyme preparation contains a side activity of endo-beta-xylanase which cleaves the hemicellulose in the flour. This enzyme is now manufactured solely for this application.

Side activities can partly be inactivated during the enzyme manufacturing process. Nevertheless, it is quite normal to find a variety of different activities in an industrial enzyme preparation. If a preparation is marketed as a protease, e.g. glycosidases and lipases might be present as well.

Factors Affecting the Purity of Enzyme Concentrates

The purity of the enzyme concentrate is largely influenced by: (i) the fermentation and purification processing applied, (ii) the production organism used and (iii) the media used for fermentation.

Production organisms

The particular production strain is obviously affecting the nature as well as the range of by-products present in the enzyme concentrate. As micro-organisms are known to produce toxins, the production organisms may themselves be sources of hazardous materials and have therefore been a chief focus of attention by the regulatory authorities. Some of these toxins are quite well known. However, there is always the possibility of introducing new toxins. Therefore, production strains which are investigated not to produce toxins and which do have a long history of safe use are preferred.

Media for enzyme production

Media used have a bearing on the cost of the enzyme and media components often find their way into commercial enzyme preparations. Details of components used in industrial scale fermentation broths for enzyme production are not readily obtained. Not surprisingly, as manufacturers do not wish to reveal information that may be of technical or commercial value to their competitors. Also some components of media may be changed from batch to batch as availability and cost of, e.g. carbohydrate feedstock change. Such changes reveal themselves in often quite profound differences in appearance from batch to batch of a single enzyme from a single producer. The effects of changing feedstock must be considered in relation to downstream processing. If such variability is likely to reduce the efficiency of the standard methodology significantly, it might be economical to use a more expensive defined medium of easily reproducible composition.

Clearly defined media are usually out of question for large scale use on cost grounds but may be perfectly acceptable when enzymes are to be produced for high value uses, such as analysis or medical therapy where very pure preparations are essential. Less-defined complex media are composed of ingredients selected on the basis of cost and availability as well as composition. Waste materials and by-products from the food and agricultural industries are often major ingredients. Thus, molasses, corn steep liquor, distillers solubles and wheat bran are important components of fermentation media providing carbohydrate, minerals, nitrogen and some vitamins. Extra carbohydrate is usually supplied as starch, sometimes refined but often simply as ground cereal grains. Soyabean meal and ammonium salts are frequently used sources of additional nitrogen. Most of these materials will vary in quality and composition from batch to batch causing changes in enzyme productivity.

Enzyme Preparations

Enzyme preparations contain a varying amount of additives for the purpose of stabilising the enzyme activity, preservation, granulation, coating or as (de)colouring aids. The choice of formulation ingredients (additives) is not specific for a given application category, but certain substances used for, e.g. technical enzymes may not comply with food, feed and cosmetics regulations, specifications, etc. Some applications may require special substances and technologies, e.g. in the case of immobilised enzymes or enteric coating of digestive aid preparations.

RECENT DEVELOPMENTS IN ENZYME MANUFACTURING

The exploitation of new types of enzymes, improvements of enzyme properties and of the production process are overall goals of innovation in the enzyme manufacturing industry. Although considerable improvements had been made in process engineering, large scale fermentation and downstream purification of enzymes remained to be a significant bottleneck in innovation. Until the mid 1980s, screening for interesting enzymes was mainly confined to groups of organisms which promised a realistic chance of developing a cost efficient production process. Micro-organisms which are adapted to extreme environments and therefore, are difficult to be accessed and cultured, but which are producing enzymes with promising properties, could not be exploited. Moreover, strain improvement in general as well as of enzymes properties in particular were largely restricted to statistical methods such as induced mutagenesis. These methods are both time-consuming and undirected. Only when genetic engineering techniques was introduced into routine research and development of enzyme manufacturing companies a promising tool became available to circumvent these problems. The introduction of genetic engineering is most probably the most important breakthrough in enzyme manufacturing during the last 30 years. Much progress have been made since the first enzyme from genetically modified micro-organisms (GMM) was introduced onto the market in 1987. This section will therefore focus on the state of the art in genetic engineering techniques applied in the improvement of both production strains and enzyme properties. Furthermore, some related technologies such as chemical modification of enzymes will be briefly described. Finally, potential impacts for enzyme regulation of the widespread application of this technology in enzyme manufacturing process will be discussed.

Genetic Engineering

From the perspective of an enzyme-producing company, genetic engineering serves as a core technology which is offering the following advantages:

1. The exploitation of new types of enzymes and new source organisms, even enzymes from organisms which are difficult to handle or non-culturable (e.g. extremophiles).
2. Drastic shortening of development times from screening to marketing.
3. Significant cost reductions in the development and production process of enzymes.
4. Genetic engineering is a prerequisite for the optimisation of enzyme molecular properties by protein engineering.
5. Improved product safety and fewer production risks, due to the production of enzymes from a wide variety of different source organisms in a small number of well-characterised enzyme production organisms.

Increasing enzyme expression

Before the era of genetic engineering, an increase in enzyme yield could only be attempted by traditional strain selection from natural sources or by random mutagenesis induced by mutagenic agents. Using genetic engineering techniques enzyme yield could be further increased dramatically. This increase is made technically feasible either by multiplying the enzyme gene or by constructing an artificial expression system. Both approaches aim at increasing the transcription/translation of the enzyme gene(s) into proteins on the cellular level.

Multiplying enzymes genes could be done just by amplifying the number of copies of the enzyme gene in the source organisms thereby not necessarily ending up in a GMM as defined in by the regulator. Alternatively, the gene could be isolated from the source organism, cloned onto a plasmid which is then introduced in a production strain. Whereas, normally only one set of genomic genes (two sets in the case of diploid organisms) is present in a micro-organism, up to 1000 and more copies of plasmids and consequently of plasmid related enzyme genes could be present in one cell.

The continuous improvement of artificial expression systems is sparked by the fast growing knowledge on regulatory DNA sequences. Strong promoters, enhancing sequences and efficient terminators often derived from different organisms are used in order to enhance the transcription rate of the enzyme gene. Frequently, ribosome binding sites are also optimised in order to increase the translation rate. However, this approach could lead to changes in the amino acid sequence of the enzyme protein. Applying these techniques enzyme yield could be enormously increased, e.g. heterologous expression of a xylanase in *E. coli* lead to an accumulation of the enzymes up to 70% of total cell protein. An increase in enzyme yield of 3 to 50 was reported in several studies. In case of a yeast-glucosidase a factor of 300 was achieved.overview on enzymes from extremophiles and their industrial importance.

Thermophilic enzymes

These enzymes are well adjusted to temperature above 50°C. As perhaps an outstanding example amylase from *Pyrococcus furiosus* is one of the most stable enzymes presently known with significant half-life at 130°C. Whether this already represents the upper limits of thermal stability, remains however a question to be solved. Physical limits seem to be reached around 250°C. At this temperature, peptide bond hydrolysis reactions rapidly occur in (denatured) proteins. For comparison, the temperature optima of normal mesophilic enzymes are in the range of 30° to 40°C with denaturation starting around 50 to 60°C. Interestingly, it could be shown that resistance to high temperature is often accompanied with resistance to proteases, chaotropic agents, low pH, oxidation and high salt concentration. The high stability of enzymes from extreme thermophiles generally resides in the amino acid sequence and three-dimensional structure.

This is shown by stability which is retained upon purification and when genes for stable enzymes are expressed in mesophiles. A few proteins might also require post-translational modification to become fully thermostable. As a consequence, these enzymes are highly attractive for biotechnological applications, such as starch processing, leaching of low grade ores, sugar conversion, detergents, or as a tool for molecular biology (PCR)—as these processes are carried out at elevated temperatures.

Psychrophilic bacteria

These group of organisms optimally grow at or below 15°C, having an upper limit of growth of about 20°C and a lower limit of growth of 0°C or below. Psychrophilic enzymes have a high specific activity at low and moderate temperatures and are inactivated easily by a moderate increase in temperature.

Typically, the specific activity of these cold adopted enzymes is higher than that of their mesophilic counterparts at temperatures of approximately 0–30°C. This increased activity is accompanied by lower thermal stability.

These properties could be of interest for various applications. According to Gerday cold-adapted cellulase could be used in biopolishing and stone-washing of cotton garments. In fabric production, tissues often have cotton fibre ends protruding from the main fibres which reduce smoothness and alter the appearance of the garment. Treatment with cellulases could excise protruding ends. The current treatment, however, is accompanied by a loss of mechanical resistance. A cold-adapted enzyme would enable the decrease of processing temperature and rapid inactivation as a result of thermal liability would be possible. The mechanical resistance of the final product would also be improved as a result of rapid inactivation of the enzyme.

In the beverage industry, pectinases are added in the juice extraction process in order to reduce viscosity and to clarify the final product. In the meat industry, proteases are applied to tenderise meat and in baking processes amylases, proteases and xylanases are used to reduce the dough fermentation time. The use of psychrophilic enzymes in all these processes could be advantageous not only because of their high specific activity, thereby reducing the amount of enzyme needed, but also for their easy inactivation. Psychrophilic enzymes could therefore become interesting alternatives to mesophilic enzymes in brewing and wine industries, cheese manufacturing, animal feed and other applications.

Improving enzyme properties – protein engineering

Basically three approaches exist for generating enzymes with improved properties:

1. First, the enzyme properties can be modulated by an exchange of a single or a few amino acid residues. This is mainly achieved by *oligonucleotide-directed mutagenesis*.
2. Second, by exchanging functional domains between related enzymes, which lead to hybrid enzymes. This could technically be achieved by DNA shuffling.
3. Third, the active site of an enzyme can be introduced into a small protein fragment scaffold of the enzyme that is devoid of its original active site.

These modifications could be made on a rational basis directed to particular changes which impacts on the enzyme properties (rational protein design). Alternatively, modifications are made randomly, however, restricted to the gene of interest and not affecting the whole genome as described above (directed evolution, molecular evolution of enzymes).

As a prerequisite of rational protein design some hypothesis on the functional effects of well defined alterations is needed—usually affecting one or a few amino acids. The latter approach takes into account the limited knowledge of structure—function relations in enzymes. Virtually, no information on how enzyme structure relates to function is needed. This strategy is even more useful as the problems and demand from enzyme application are usually complex and multifactorial. According to industry, the directed evolution approach might therefore be the more promising one.

Major goals for improving enzyme properties are:

1. Increase in catalytic activity (under conditions applied in practice).
2. Enzyme stability.

Improved substrate range: Stability to pH and temperature are important factors, e.g. for detergent enzymes. Starch conversion into sugars needs low pH and high temperature and so the demand for thermostable α-amylase and amyloglucosidase from thermophile organisms is a logical consequence.

Enzymes which are able to do their job at lower temperatures are also of interest. For example, in diary industry, α-galactosidase is used at low temperature to reduce the amount of lactose.

Subtilisin, a bacterial serine protease often used in detergents, might illustrate the manifold goals of protein engineering. Up to present, mutations in well over 50% of the 275 amino acids have been reported in the scientific literature. Although enhanced stability has been the predominant target, these alterations also resulted in changes in catalytic mechanism, substrate specificity, surface activity, folding mechanisms and also in new activities.

In the following section the different approaches and methods in use to modify protein structure will be briefly described.

Oligonucleotide-directed mutagenesis

This method is based on computer modelling of individual amino acid changes, followed by site-directed mutagenesis of the corresponding DNA with pre-designed oligonucleotide primer and expression of the recombinant protein in mutants for testing and evaluation. This method makes it possible to modify one or a few amino acids in the protein.

DNA (gene) shuffling

The technique of DNA shuffling has been derived from imitating natural recombination by allowing *in vitro* homologous recombination of DNA. With this method, a population of related genes is randomly fragmented and subjected to denaturation and hybridisation, followed by extension with PCR reaction. As a result of repeated PCR cycles, the length of fragments is expanded. DNA recombination occurs when a fragment derived from one template primes a template with a different sequence. Combined with well focused selection procedures, this technique makes it possible to rapidly develop a huge variety of enzyme variants.

Improved product safety

In the course of screening for new enzyme producing micro-organisms, it sometimes turns out that growing of technically interesting micro-organisms under industrial conditions will not be possible. Either the micro-organism is pathogenic or toxic or not safe to handle. Enzyme purification could be prohibitively expensive, e.g. because the enzyme was cell associated or contaminated with undesirable compounds. Using production strains without long-term experience will anyhow be accompanied by a higher risk of harmful by-products in the final enzyme isolate. The cloning of enzymes following heterologous expression makes it possible to confine oneself to a small number of production strains which already have been used for enzyme production for decades and even might be considered as GRAS.

Chemical modification of enzymes

Chemical modification of enzymes was in fact the first method available to alter enzyme properties in the sixties. Since the mid eighties, interests on chemical modification have been rekindled. One major goal for chemical modification is to increase enzyme stability and activity, particularly in non-aqueous applications for biocatalysis in organic synthesis.

Examples: Various approaches were reported on covalent modification of enzymes: incorporation of vanadium into phytase for example, which has the *in vivo* role of catalysing phosphate ester hydrolysis is sufficient to convert phytase to a vanadium-dependent peroxidase that could catalyse enantio-selective sulphoxidation.

BIOREACTORS DESIGNS FOR USAGE IN BIOCATALYSIS

A variety of bioreactors are available for immobilising enzymes and cells and these are shown in Fig. 21.6.

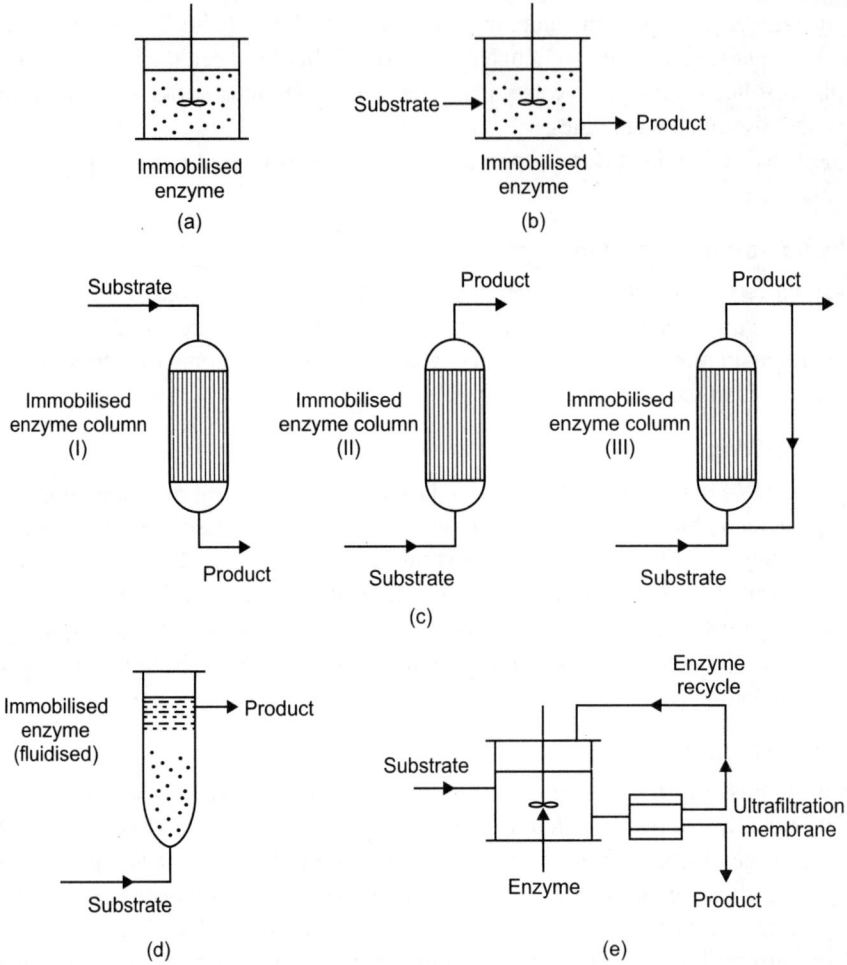

Fig. 21.6: Various designs of bioreactors for use in biocatalysis. (a) batch stirred fermentor, (b) continuous stirred tank, (c) continuous packed-bed (i) downward flow, (ii) upward flow, and (iii) upward flow and recycle, (iv) continuous fluidised-bed, and (v) continuous ultrafiltration.

MANIPULATION OF MICRO-ORGANISMS FOR HIGHER YIELD OF ENZYMES

Until recently higher yields in enzyme production have been achieved, as has been the case with most other industrial microbial products, by empirical means using selection from natural variants, mutants obtained through treatment with various mutagens, and improvement in the environmental conditions of the fermentor or semi-solid medium. With these methods the rate of enzyme production has increased from two to fivehundred times. In recent times more knowledge has accumulated about various aspects of enzyme production including those of molecular and other dimensions, and the manipulation of industrial organisms in general. In this section some of these new developments and their use or possible use in increasing enzyme yield will be discussed. Some of them promise the possibility of producing virtually any enzyme extracelluarly and at will.

Some Aspects of the Biology of Extracellular Enzyme Production

The nature of extra-cellular enzymes secretion: Extracellular enzymes have been defined as those which are secreted into the medium outside the cell *without involving cell lysis*. This distinction is important because most extraccellular enzyme-producing organisms are Gram-positive organisms. Gram-negative organisms, in general produce enzyme in the medium only when the cell is lysed. Most Gram-negative organisms do not therefore, according to this definition, produce 'true' extra-cellular enzymes. However, it has been found in recent times that some Gram-negative organisms do in fact secrete extracellular enzymes. Furthermore many Gram-negative organisms do in fact produce and secrete enzymes across the cytoplasmic membrane. Such enzymes are however held within the periplasmic zone of the Gram-negative cell wall and hence do not find their way to the medium. Thirdly mutants of Gram-negative cells defective in the ability to synthesise cell wall components continue to synthesise and secrete polypeptides into the environment. On account of these observations extra-cellular enzymes have been redefined as those which are secreted across the cell membrane. In terms of industrial microbiology it is an apt definition as methods for deranging the molecular arrangements of cell walls exist, which when successfully applied to Gramnegative bacteria secreting into the periplasmic space convert them to extra-cellular secreters. Such methods include the formation of protoplasts by the prevention of cell wall formation using suitable antibiotics, limited digestion by trypsin, solubilisation of the cell wall with a combination of a detergent (e.g. laurodeoxycholate and a chelating agent).

Some biochemical properties of extra-cellular enzymes: Bacterial extracellular enzymes vary in molecular weight from 12,000 to 500,000 but in the main they range from 20000 to 40000. Secondly, most but not all bacterial exoproteins lack cysteine. It has been suggested, but not entirely accepted, that this absence of cysteine will confer the property of malleability on extra-cellular enzymes thus facilitating their export.

Site of synthesis of extra-cellular proteins: Even in cells actively secreting extracellular proteins, an examination of the cytoplasm shows a complete absence or only a trace amount of the enzymes being excreted. Early report for instance claiming that – amylase of –*amyloliquifaciens* is first produced as a high molecular weight precursor have been shown to be wrong on the basis of radio-isotope (labeling) experiments. Since no evidence exists for the cytoplasmic synthesis of extracellular proteins it has been suggested they are synthesised on ribosomes associated with the cells in much the same way as in eucaryotic cells. In eucaryotic cells, membrance-bound polysomes are engaged in secreting proteins for export whereas polysomes secrete non-exportable proteins. According to the currently accepted model synthesis takes places on the cell membrane and is secreted directly into pores in the cell membrane. Indeed synthesis and secretion are one process, following the system in eucaryotic cells. Some evidence for this are as follows. In many bacteria a considerable but variable fraction of the ribosomes is associated with the cytoplasmic membrane. In exponential phase of *Bacillus licheniformis* for example, 96% of the ribosomes are membrane bound. It is also know that exoenzyme synthesis is more sensitive to antiobiotic inhibition than general protein synthesis. This has been interpreted as being so because of the membrane bound ribosomes are more accessible to the antibiotic.

Control of extra-cellular enzyme secretion by gene cloning: When the terminal portion of the β-*galactosidase* gene of *E. coli* was replaced with a gene that codes for a protein of the outer cell wall membrane of the bacteria, the β-*galactosidase* activity which is normally intracellular was formed extracellularly depending on the size of the latter that attached on the β-*galactosidase* gene. This and other similar experiments show that in due course it may be possible to produce virtually any enzyme extracellularly by gene cloning.

Some methods for increasing enzyme yields: The increased yields which have been observed in enzyme production is based on strain selection, improved environmental factors, regulatory controls and genetic manipulation.

Strain selection: Strain selection from natural variants of the same species or even entirely new species have resulted in the array of enzymes available in some industries, e.g. the starch hydrolysing industries. The natural strains may then be mutagenised for increased variation in the gene pool. Strains have been selected in the above two manners for a wide variety of properties including temperature-tolerant enzymes and resistance to feedback regulation.

Environmental factors: Exoenzymes may be constitutive, but the majority are inducible or partially so. Inducers are therefore important in increasing the yields of many extracellular enzymes. Since many of the substrates are insoluble they cannot enter the cell, and hence their analogues or gratuitious inducers (those that induce the enzyme but are not substrates or breakdown products) have been used. Inducers are usually cheap in order to bring down costs. Thus, corn cobs are hydrolysed to produce xylose which act as inducers for glucose isomerase. Most extracellular enzymes are produced in the idiophase and maximum production is usually in the late log and early stationary phase. This period coincides with the period when the organism is released from catabolite repression. Increased yields may therefore be achieved by feeding low levels of the substrate or feeding them intermittently. Yields may also be increased by increasing cell-wall permeability. Surfactants may be incorporated into the medium for this purpose although how they affect the wall permeability is not fully understood.

Regulatory control: Control of regulation is through induction and catabolite and feedback regulations. Mutants resistant to all three have been produced with consequent boost in production. An example of inducer-resistance is in the case of the glucose isomerase producing antinomycete *Streptomyces phaechromogenes,* the wild strain of which will not germinate on L-lyxose, another form of D-xylose. It will grow on lyxose only if germination is first obtained on xylose. Mutants were selected which would germinate directly on lyxose, thus eliminating the need for xylose. The bypassing of catabolite repression has led to the production of large amounts of enzymes. This has been achieved by using toxic analogues of the substrates. Thus, 2-deoxy-glucose is used as a toxic analogue when seeking for mutants able to over produce glucosidase. Feedback regulation is not very applicable to enzyme synthesis, although some examples are known.

Genetic manipulation: As had been indicated earlier the gene specifying the extracellular secretion may be cloned on those controlling the synthesis of particular enzymes, thus causing the enzyme to be secreted extracellularly.

The number of copies of specific genes may be increased by gene amplification methods thus increasing enzyme yield several times. For example, plasmids specifying particular extra-cellular enzymes may continue to replicate while the parent chromosome is inhibited by, for example, chloramphenicol thereby permitting an amplification of the genes – sometimes up to 2000 copies or up to 40% of the cells total DNA. The result is increased yield of the enzyme.

Chapter 22

Fungal Laccase Enzyme for Biotechnological Application

INTRODUCTION

Laccase belongs to the small group of enzymes called the blue multi copper oxidases. Laccase is widely distributed in higher plants and fungi. In fungi, laccase is present in Ascomycetes, Deuteromycetes, *Basidiomycetes* and is particularly abundant in many white-rot fungi that degrade lignin. Laccases have been subject of intensive research in the last decades due to their broad substrate specificity. In the recent years, their uses span from the textile to the pulp and paper industries and food applications to bioremediation processes. Laccases also have uses in organic synthesis, where typical substrates are phenols and amines and the reaction products are dimers and oligomers derived from the coupling of reactive radical intermediates. More recently, they have found applications in other field such as in the design of biosensors and biofuel cells. In this chapter, the occurrence, mode of action, general properties, production and immobilisation of laccases are discussed.

As discussed above laccase (benzenediol:oxygen oxidoreductase, EC 1.10.3.2) belongs to the small group of enzymes called the blue copper proteins or the blue copper oxidases along with the plant ascorbate oxidase and the mammalian plasma protein ceruloplasmin among others. These proteins are characterised by containing 4 catalytic copper atoms. One copper is placed at the T1 site, where reducing substrate binds and it is responsible in the characteristic blue-greenish colour in the oxidising resting state Cu^{2+}. The other three coppers are clustered in the called T2/T3 site in which molecular oxygen binds. Laccase is widely distributed in higher plants and fungi and has been found also in insects and bacteria. Recently a novel polyphenol oxidase with laccase like activity was mined from a metagenome expression library from bovine rumen microflora.

Fungal laccases have higher redox potential than bacterial or plant laccases (up to +800 mV) and their action seems to be relevant in nature finding also important applications in biotechonology. Thus, fungal laccases are involved in the degradation of lignin or in the removal of potentially toxic phenols arising during lignin degradation. In addition, fungal laccases are hypothesised to take part in the synthesis of dihydroxynaphthalene melanins, darkly pigmented polymers that organisms produce against environmental stress or in fungal morphogenesis by catalysing the formation of extracellular pigments. Concerning their use in the biotechnology area, fungal laccases have widespread applications, ranging from effluent decolouration and detoxification to pulp bleaching, removal of phenolics from wines, organic synthesis, biosensors, synthesis of complex medical compounds and dye transfer blocking functions in detergents and washing powders, many of which have been patented. The biotechnological use of laccase has been expanded by the introduction of laccase-mediator systems, which are able to oxidise non-phenolic compounds that are otherwise hardly or not oxidised by the enzyme alone.

OCCURRENCE OF LACCASES

Laccase is the most widely distributed of all the large blue copper-containing proteins, as it is found in a wide range of higher plants and fungi and recently some bacterial laccases have also been characterised from *Azospirillum lipoferum, Bacillus subtilis, Streptomyces lavendulae, S. cyaneus* and *Marinomonas mediterranea*. The occurrence of laccases in higher plants appears to be far more limited than in fungi. Laccases in plants have been identified in trees, cabbages, turnips, beets, apples, asparagus, potatoes, pears and various other vegetables. The classical demonstration of laccase in *R. vernicifera* is well documented. In addition, the lacquer tree is a member of the Anacardiaceae family, appear to contain laccase in the resin ducts and in the secreted resin. Cell cultures of *Acer pseudoplatanus* have been shown to contain eight laccases, all expressed predominantly in xylem tissue. Other reports are those of Wosilait and others on the presence of a laccase in leaves of *Aesculus parviflora* and in green shoots of tea. Other higher plant species also appear to contain laccases, although their characterisation is less convincing. Laccases have been isolated from Ascomyceteous, Deuteromyceteous and *Basidiomyceteous* fungi. In the fungi, Ascomycetes and Deuteromycetes have not been a clear focus for lignin degradation studies as much as the white-rot *Basidiomycetes*. Laccase from *Monocillium indicum* was the first laccase to be characterised from an Ascomycete showing peroxidative activity.

The white-rot basidiomycetes are the most efficient degraders of lignin and also the most widely studied. The enzymes implicated in lignin degradation are: (i) lignin peroxidase, which catalyses the oxidation of both phenolic and non-phenolic units, (ii) manganese-dependent peroxidase, (iii) laccase, which oxidises phenolic compounds to give phenoxy radicals and quinones, (iv) glucose oxidase and glyoxal oxidase for H_2O_2 production and (v) cellobiose-quinone oxidoreductase for quinone reduction.

The veratryl alcohol oxidase and some esterases may also play roles in the complex process of natural wood decay. The different degrees of lignin degradation with respect to other wood components depend on the environmental conditions and the fungal species involved. It has been demonstrated that there is no unique mechanism to achieve the process of lignin degradation and that the enzymatic machinery of the various micro-organisms differ. *Pleurotus ostreatus,* for instance, belongs to a subclass of lignin–degrading micro-organisms that produce laccase, manganese peroxidase and veratryl alcohol oxidase but no lignin peroxidase. *Pycnoporus cinnabarinus* has been shown to produce laccase as the only ligninolytic enzyme and *Pycnoporus sanguineus* produces laccase as the sole phenol oxidase. In plants, laccase plays a role in lignification, whereas in fungi laccases have been implicated in many cellular processes, including delignification, sporulation, pigment production, fruiting body formation and plant pathogenesis. Only a few of these functions have been experimentally demonstrated. Ligninolytic enzymes have mostly been reported to be extracellular but there is evidence in literature of the occurrence of intracellular laccases in white–rot fungi. Intracellular as well as extracellular laccases were identified for *Neurospora crassa* by Froehner and Eriksson, who suggested that the intracellular laccase functioned as a precursor for extracellular laccase as there were no differences between the two laccases other than their occurrence.

MODE OF ACTION OF THE LACCASE ENZYME

Laccases contain 4 copper atoms termed Cu T1 (where the reducing substrate place) and trinuclear copper cluster T2/T3 (where oxygen binds and is reduced to water). As a one-electron substrate oxidation is coupled to the four-electron reduction of oxygen the reaction mechanism cannot be entirely straight forward. Laccase can be thought to operate as a battery, storing electrons from individual oxidation reactions in order to reduce molecular oxygen. Hence the oxidation of four reducing substrate molecules is necessary for the complete reduction of molecular oxygen to water.

In general terms, substrate oxidation by laccase is a one-electron reaction generating a free radical. The initial product is typically unstable and may undergo a second enzyme-catalysed oxidation or otherwise a non-enzymatic reaction such as hydration, disproportionation or polymerisation. The bonds of the natural substrate, lignin, that are cleaved by laccase include, Cα- oxidation, Cα-Cβ cleavage and aryl-alkyl cleavage (Fig. 22.1a). Laccases are similar to other phenol-oxidising enzymes, which preferably polymerise lignin by coupling of the phenoxy radicals produced from oxidation of lignin phenolic groups. Due to this specificity for phenolic subunits in lignin and its restricted access to lignin in the fibre wall, laccase has a limited effect on pulp bleaching unless redox mediators [e.g. 2,2'-azino-*bis* (3-ethy-benzthizoline-6-sulphonic acid (ABTS)] will be introduced in the reaction (Fig. 22.1b).

LACCASE MEDIATOR SYSTEM

With respect to other ligninolytic enzymes, laccase can oxidise only phenolic fragments of lignin due to the random polymer nature of lignin and to the laccase lower redox potential. Small natural low molecular weight compounds with high redox potential than laccase itself (> 900 mV) called mediators may be used to oxidise the non-phenolic part of lignin (Fig. 22.1b). Recently the discovery of new and efficient synthetic mediators extended the laccase catalysis towards xenobiotic substrates.

A mediator is a small molecule that acts as a sort of 'electron shuttle': once it is oxidised by the enzyme generating a strongly oxidising intermediate, the co-mediator (oxidised mediator), it diffuses away from the enzymatic pocket and in turn oxidises any substrate that, due to its size could not directly enter into the active site. Futhermore, the use mediators allows the oxidation of polymers by side-stepping the inherent steric hindrance problems (enzyme and polymer do not have to interact in a direct manner) (Fig. 22.2).

Alternatively, the oxidised mediator could rely on an oxidation mechanism not available to the enzyme, thereby extending the range of substrates accessible to it. It is therefore of primary importance to understand the nature of the reaction mechanism operating in the oxidation of a substrate by the oxidised mediator species derived from the corresponding mediator investigated. In the laccase-dependent oxidation of non-phenolic substrates, previous evidence suggests an electron-transfer (ET) mechanism with mediator ABTS, towards substrates having a low oxidation potential. Alternatively, a radical hydrogen atom transfer (HAT) route may operate with *N*-OH type mediators, if weak C-H bonds are present in the substrate.

More than 100 mediator compounds have been described but the most commonly used are the ABTS and the triazole 1-hydroxybenzotriazole (HBT). It is well known that cation radicals represent an intermediate oxidation step in the redox cycle of azines and upon extended oxidation and abstraction of the second electron, the corresponding dications can be obtained. The redox potentials of ABTS$^+$ and ABTS^{2+} were evaluated as 0.680 V and 1.09 V respectively. HBT belongs to the *N*-heterocyclic coumpounds bearing *N*–OH groups mediators. Consuming oxygen HBT is converted by the enzyme into the active intermediate, which is oxidised to a reactive radical (R–NO·) and HBT redox potential has been estimated

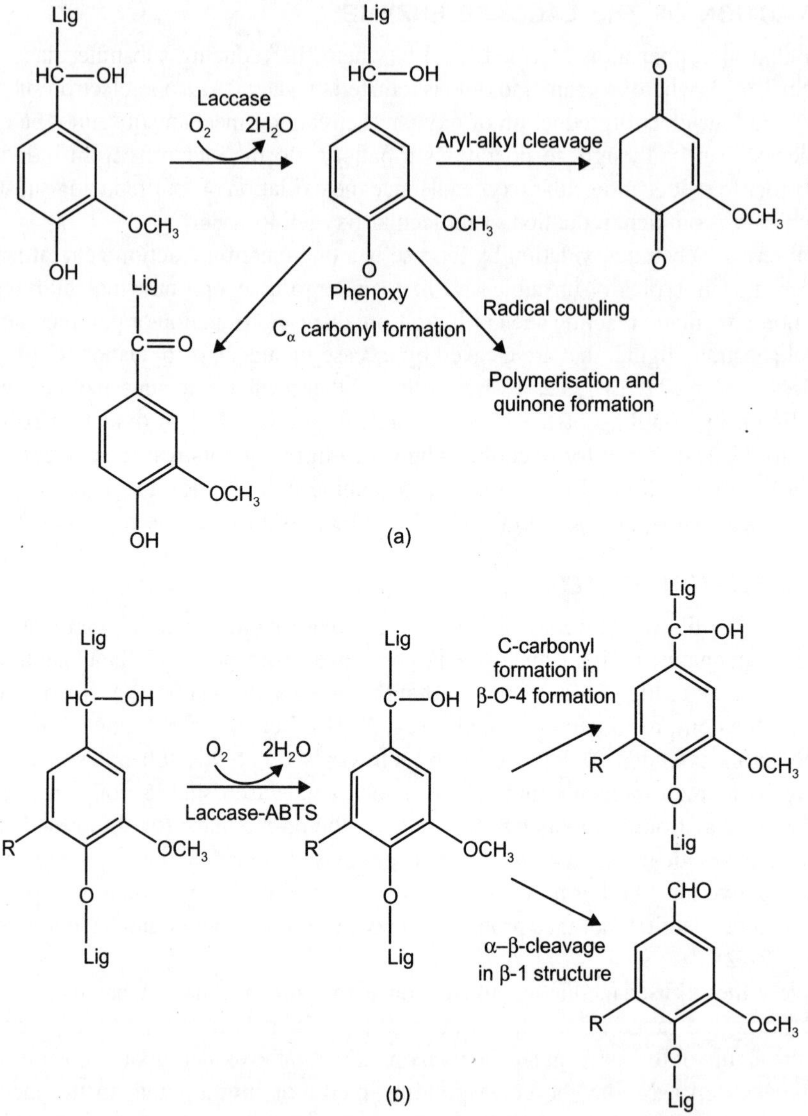

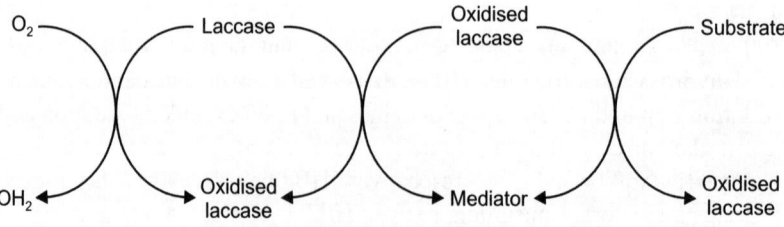

Fig. 22.1: Oxidation of (a) phenolic subunits of lignin by laccase and (b) non-phenolic lignin model compounds by a laccase mediator system.

Fig. 22.2: Catalytic cycle of a laccase-mediator oxidation system.

as 1.1–1.2 V. Mediated laccase catalysis has been used in a wide range of applications, such as pulp delignification, textile dye bleaching, polycyclic aromatic hydrocarbon degradation, pesticide or insecticide degradation and organic synthesis. In pulp and paper industry, novel enzymatic bleaching technologies are attracting increasing attention because of concerns regarding the environmental impact of the chlorine-based oxidants currently being used in delignification or bleaching.

However, synthetic mediators are toxic, expensive and generally at concentrations above 1 mM inactivate the laccase. Novel approaches to over come this hurdles are coming up (from searching for natural mediators such as *p*-coumaric acid, 4-hidroxybenzoic acid, syringaldehyde, etc.), to the directed evolution of laccases.

GENERAL PROPERTIES OF LACCASE ENZYMES

Most fungal laccases are monomeric, dimeric or tetrameric glycoproteins. Glycosylation of fungal laccase is believed to play a role in secretion, susceptibility to proteolytic degradation, copper retention and thermal stability. Upon purification, laccase enzymes demonstrate considerable heterogeneity. Glycosylation content and composition of fungal glycoproteins can vary with growth medium composition. For this reason data can be heterogeneous. The molecular mass of the monomer ranges from about 50 to 100 kDa. An important feature is a covalently-linked carbohydrate moiety (10–45% of total molecular mass), which may contribute to the high stability of the enzyme. The sugar composition has been analysed in several examples, such as *Podospora ansenna* and *Botrytis cinerea*, *Trametes hirsuta*, *Trametes ochracea*, *Cerrena maxima* and *Coriolopsis fulvocinerea* and *Melanocarpus albomyces*.

Isozymes

Many laccase producing fungi secrete isoforms of the same enzyme. These isozymes have been found to originate from the same or different genes encoding for the laccase enzyme. The number of isozymes present differs between species and also within species depending on whether they are induced or non-induced. They can differ markedly in their stability, optimal pH and temperature and affinity for different substrates.

Furthermore, these different isozymes can modulate different roles in the physiology of different species or in the same species under different conditions. *Cerrena unicolour* secreted two laccase isoforms with different characteristics during the growth in a synthetic low-nutrient nitrogen/glucose medium. Various laccase encoding gene sequences have been reported from a range of ligninolytic fungi, these sequences encode for proteins between 515 and 619 amino acid residues and close phylogenetic proximity between them is indicated by sequence comparisons.

Substrate Specificity of Laccase

Laccases are remarkably non-specific as to their reducing substrates and the range of substrates oxidised varies from one laccase to another. These enzymes catalyse the one-electron oxidation of a wide variety of organic and inorganic substrates, including polyphenols, methoxy-substituted phenols, aromatic amines and ascorbate with the concomitant four-electron reduction of oxygen to water. Laccase has broad substrate specificity towards aromatic compounds containing hydroxyl and amine groups and as such, the ability to react with the phenolic hydroxyl groups found in lignin.

Kinetic data of laccases from different sources were reported. K_m values are similar for the cosubstrate dissolved oxygen (about 5–10 M), but V_{max} varies with the source of laccase (50–300 M/s). The turnover is heterogeneous over a broad range depending on the source of enzyme and substrate/type of reaction.

The kinetic constants differ in their dependence on pH. K_m is pH-independent for both substrate and co-substrate, while K_{cat} is pH-dependent.

Influence of pH on Laccase Activity and Stability

The pH optima of laccases are highly dependable on the substrate. When using ABTS as substrate the pH optima are more acidic and are found in the range 3.0–5.0. In general, laccase activity has a bell shaped profile with an optimal pH that varies considerably. This variation may be due to changes in the reaction caused by the substrate, oxygen or the enzyme itself. The difference in redox potential between the phenolic substrate and the T1 copper could increase oxidation of the substrate at high pH values, but the hydroxide anion (OH⁻) binding to the T2/T3 coppers results in an inhibition of the laccase activity due to a disruption of the internal electron transfer between the T1 and T2/T3 centres. These two opposing effects can play an important role in determining the optimal pH of the bi-phasic laccase enzymes. Laccase produced by *Trametes modesta* was fully active at pH 4.0 and very stable at pH 4.5 but its half-life decreased to 125 min at pH 3.0.

Influence of Temperature on Laccase Activity and Stability

The optimal temperature of laccase can differ greatly from one strain to another. The laccases isolated from a strain of *Marasmius quercophilus* were found to be stable for 1 h at 60°C. Farnet and others further found that pre-incubation of enzymes at 40°C and 50°C greatly increased laccase activity. Another technique that can be used to increase the stability of laccase is to immobilise the enzyme on glass powder by means of air-drying. This technique also has potential for the enzyme to be used on the glass powder matrix in specific biotechnology applications where stability is required. The laccase from *P. ostreatus* is almost fully active in the temperature range of 40–60°C, with maximum activity at 50°C. The activity remains unaltered after prolonged incubation at 40°C for more than 4 hr. Nyanhongo and others showed that laccase produced by *T. modesta* was fully active at 50°C and was very stable at 40°C but half-life decreased to 120 min at higher temperature (60°C).

Influence of Inhibitors on Enzyme Activity

In general, laccases responds similarly to several inhibitors of enzyme activity. Many ions such as azide, halides, cyanide, thiocyanide, fluoride and hydroxide bind to the type 2 and type 3 Cu, resulting in the interruption of internal electron transfer and accordingly therefore inhibition of activity. Other inhibitors include metal ions (e.g. Hg^{+2}), fatty acids, sulphydryl reagents, hydroxyglycine, kojic acid, desferal and cationic quaternary ammonium detergents, the reactions with which may involve amino acid residue modifications, confirmational changes or Cu chelation. Confirmational changes are highly dependent on the state of oxidation of the copper atoms. This is one of the reasons for the sensitivity towards chelating agents. The selective removal of Cu by chelating agents (EDTA, dimethyl glyoxime, N,N'-dimethyldithiocarbamate, NTA) leads to a loss of catalytic activity.

PRODUCTION OF FUNGAL LACCASES

Laccase activity was detected in the cultures of a wide range of fungi, from Ascomycetes to *Basidiomycetes* and from wood- and litter-decomposing fungi to ectomycorrhizal fungi. White-rot fungi have been studied extensively for application in biological pulping and bleaching because they are of the only organisms that are able to degrade lignin efficiently. White-rot fungi, such as *Coriolus versicolour* and *P. sanguineus*, are known producers of lignolytic enzymes that are involved in the natural delignification of wood. This group of fungi is the only known micro-organisms that have evolved complex enzymatic systems that enable them to degrade lignin.

In general, laccases occur as extracellular glyco-proteins, which allows for rapid removal from fungal biomass. One of the major limitations for the large scale applications of fungal laccases is the low production rates by both wild type and recombinant fungal strains according to Galhaup and others.

White-rot fungi constitutively produce low concentrations of various laccases when they are cultivated in submerged culture or on wood. Higher concentrations can be induced by the addition of various aromatic compounds such as 2,5-xylidine and ferulic acid. High concentrations of laccase have also been observed in old non-induced cultures. The mechanisms of metabolism in micro-organisms are used and controlled by its environmental conditions and medium composition. There are various response element sites in the promoter regions of laccase genes that can be induced by certain xenobiotic compounds, heavy metals or heatshock treatment.

Induction of Laccase Production

Laccase production has been found to be highly dependent on the conditions for the fungus cultivation and media supporting high biomass did not necessarily support high laccase yields. Ligninolytic systems of white-rot fungi were mainly activated during the secondary metabolic phase and were often triggered by nitrogen concentration or when carbon or sulphur became limiting. Laccases were generally produced in low concentrations by laccase producing fungi, but higher concentrations were obtainable with the addition of various supplements to media. The addition of aromatic compounds such as 2,5-xylidine, lignin and veratryl alcohol is known to increase and induce laccase activity. Many of these compounds resemble lignin molecules or other phenolic chemicals. Veratryl alcohol is an aromatic compound known to play an important role in the synthesis and degradation of lignin. The addition of veratryl alcohol to cultivation media of many white-rot fungi has resulted in an increase in laccase production. Some of these compounds affect the metabolism or growth rate while others, such as ethanol, indirectly trigger laccase production.

Eggert and others found that the addition of 2,5-xylidine as inducer had the most pronounced effect on laccase production. The addition of 10 μM 2,5-xylidine after 24 h of cultivation gave the highest induction of laccase activity and increased laccase activity nine-fold. At higher concentrations the 2,5-xylidine had a reduced effect, probably due to toxicity.

The promoter regions of the genes encoding for laccase contain various recognition sites that are specific for xenobiotics and heavy metals. These can bind to the recognition sites when present in the substrate and induce laccase production. White-rot fungi were very diverse in their responses to tested inducers for laccase. The addition of certain inducers can increase the concentration of a specific laccase or induce the production of new isoforms of the enzyme. Some inducers interact variably with different fungal strains.

Lee and others investigated the inducing effect of alcohols on the laccase production by *Trametes versicolour.* The enhanced laccase activity was comparable to those obtained using 2,5-xylidine or veratryl alcohol. It was postulated that the addition of ethanol to the cultivation medium caused a reduction in melanin formation. The monomers, when not polymerised to melanin, then acted as inducers for laccase production. The addition of ethanol as an indirect inducer of laccase activity offers a very economical way to enhance laccase production.

Lu and others found that there is a strong correlation between hyphal branching and the expression and secretion of laccase. The addition of cellobiose can induce profuse branching in certain *Pycnoporus* species and consequently increase laccase activity. The addition of cellobiose and lignin can increase the activity of extracellular laccases without an increase in total protein concentration. Osma and others showed that soya oil was the best inducer of laccase activities, attaining 4-fold higher than those obtained in the reference cultures.

The addition of low concentrations of Cu^{+2} to the cultivation media of laccase producing fungi stimulates laccase production. Palmieri and others found that the addition of 150 μM copper sulphate to the cultivation media can result in a fifty-fold increase in laccase activity compared to a basal medium. Employing copper sulphate as laccase inducer or supplementing the culture medium with veratryl alcohol, led to maximum values of laccase activity. A new basidiomycete, *Trametes* sp. 420, produced laccase in glucose medium and in cellobiose medium with induction by 0.5 mM Cu^{+2} and 6 mM *o*-toluidine.

Influence of Carbon Sources on Laccase Production

The carbon sources in the medium play an important role in ligninolytic enzyme production. Mansur and others showed that fructose induced 100-fold increase in laccase production of *Basidiomycete* sp. I-62. *T. versicolour* is an excellent producer of laccase in fermentation of mandarin peels. Glucose and cellobiose were efficiently and rapidly utilised by *Trametes pubescens* with high laccase activity. Similarly, the replacement of crystalline cellulose or xylan by cellobiose increased laccase activity of *C. unicolour* by 21- and 70-fold, respectively. Furthermore, in *T. versicolour* lignocellulosic material (barly bran) increased almost 50-fold laccase activity compared to the control culture with glucose. In the medium with the best carbon sources (mandarine peels and grapevine sawdust), both *Pleurotus eryngii* and *P. ostreatus* strain No. 493, showed the highest laccase activity. Glucose showed the highest potential for the production of laccase.

Influence of Nitrogen Sources on Laccase Production

White-rot fungi ligninolytic systems are mainly activated during the secondary metabolic phase of the fungus and are often triggered by nitrogen depletion. Monteiro and De Carvalho reported high laccase activity with semi-continuous production in shake-flasks using a low carbon to nitrogen ratio (7.8 g/g). Buswell and others found that laccases were produced at high nitrogen concentrations, although it is generally accepted that a high carbon to nitrogen ratio is required for laccase production. Laccase was also produced earlier when the fungus was cultivated in a substrate with a high nitrogen concentration and these changes did not reflect differences in biomass. Elisashvili and others observed highest laccase activity in *C. unicolour* IBB 62 in a medium with ammonium sulphate as the nitrogen source. D'Souza-Ticlo and others showed that well defined organic nitrogen sources such as glutamic acid and glycine were better than beef extract and corn steep liquor for laccase production. Heinzkill and others also reported a higher yield of laccase using nitrogen rich media rather than the nitrogen-limited media usually employed for induction of oxidoreductases.

Influence of pH on Laccase Production

There is not much information available on the influence of pH on laccase production, but when fungi are grown in a medium of which the pH is optimal for growth (pH 5.0), the laccase will be produced in excess. Most reports indicated initial pH levels set between pH 4.5 and pH 6.0 prior to inoculation, but the levels are not controlled during most cultivations. Nyanhongo and others reported that an initial pH of 7.0 was the best for optimal growth and laccase production by a newly isolated strain of *T. modesta*.

Influence of Temperature on Laccase Production

It has been found that the optimal temperature for fruiting body formation and laccase production is 25°C in the presence of light, but 30°C for laccase production when the cultures are incubated in the dark. In general the fungi were cultivated at temperatures between 25°C and 30°C for optimal laccase production. When cultivated at temperatures higher than 30°C the activity of ligninolytic enzymes was reduced.

Inhibition of Laccase Production

It seems as if the use of excessive concentrations of glucose as carbon source in cultivation of laccase producing fungal strains has an inhibitory effect on laccase production. An increase in the amount of glucose in the media resulted in a delay of the laccase production. An excess of sucrose or glucose in the cultivation media can reduce the production of laccase, as these components allow constitutive production of the enzyme, but repress its induction when applicable. A simple but effective way to overcome this problem is the use of cellulose as carbon source during cultivation.

Heterologous Expression

Most commercial laccases are produced in *Aspergillus* hosts. The functional expression of the *Myceliophthora thermophila* laccase in *Saccharomyces cerevisiae* by directed molecular evolution was recently reported, becoming the mutant T2 an idoneous scaffold for further improvements towards biotechnological applications. Another efficient expression system was developed for the basidiomycete *P. cinnabarinus* and this was used to transform a laccase-deficient monokaryotic strain with the homologous laccase gene. The yield was above 1.2 g of laccase per litre and represents the best laccase production reported for recombinant fungal strains.

LACCASE IMMOBILISATION

Enzymes exhibit a number of features that make their use advantageous as compared to conventional chemical catalysts. However, a number of practical problems exist that reduce their operational life-time, such as their high cost of isolation and purification, their non-reusability, the instability of their structures and their sensitivity to harsh process conditions. Many of these undesirable limitations may be overcome by the use of immobilised enzymes.

Immobilisation is achieved by fixing enzymes to or within solid supports, as a result of which heterogeneous immobilised enzyme systems are obtained. By mimicking the natural mode of occurrence in living cells, where enzymes for the most cases are attached to cellular membranes, the systems stabilise the structure of enzymes, hence their activities. In the immobilised form enzymes are more robust and more resistant to environmental changes allowing easy recovery and multiple reuse. Compared with the free enzyme, the immobilised enzyme has usually its activity lowered and the Michaelis constant increased. These alterations result from structural changes introduced to the enzyme by the applied immobilisation procedure and from the creation of a microenvironment in which the enzyme works, different from the bulk solution. Enzymes may be immobilised by a variety of methods (adsorption, entrapment, cross-linking and covalent bonding) mainly based on chemical and/or physical mechanisms.

Since the methods for the immobilisation procedures greatly influence the properties of the resulting biocatalyst, immobilisation strategy determines the process specifications for the catalyst.

Laccase immobilisation was extensively studied with a wide range of different methods and substrates. The adsorption of chromophoric-oxidised products on the surface of the immobilisation support often leads to enzyme inactivation phenomena.

APPLICATIONS OF FUNGAL LACCASE IN BIOTECHNOLOGY

A number of industrial applications for fungal laccases have been proposed and they include paper processing, prevention of wine decolouration, detoxification of environmental pollutants, oxidation of dye and their precursors, enzymatic conversion of chemical intermediates and production of chemicals from lignin. Before laccases can be commercially implemented for potential applications, however, an inexpensive enzyme source needs to be made available. Two of the most intensively studied areas in the potential industrial application of laccase are the delignification and pulp bleaching and the bioremediation of contaminating environmental pollutants.

Delignification and Pulp Bleaching

In the industrial preparation of paper the separation and degradation of lignin in wood pulp are conventionally obtained using ClO_2 and O_3. Oxygen delignification process has been industrially introduced in the last years to replace conventional and polluting chlorine-based methods. In spite of this new method, the pre-treatments of wood pulp with laccase can provide milder and cleaner strategies of delignification that also respect the integrity of cellulose. Lignocellulose is a common substrate for laccase and the laccase ability to break down nonphenolic ligno-cellulose is provided by certain phenolic compounds acting as mediators. More recently, the potential of this enzyme for cross-linking and functionalising lignocellulose compounds was discovered. Laccases can be used for binding fibre-, particle- and paper-boards. However, different wood decaying basidiomycetes have shown a highly variable pattern of laccase formation and this subject requires more detailed experiments.

Bioremediation

Laccases have also shown to be useful for the removal of toxic compounds through oxidative enzymatic coupling of the contaminants, leading to insoluble complex structures. The degradation of a variety of persistent environmental pollutants, in particular phenols, was also observed. Phenolic compounds are present in wastes from several industrial processes, as coal conversion, petroleum refining, production of organic chemicals and olive oil production among others. Immobilised laccase was found to be useful to remove phenolic and chlorinated phenolic pollutants. Laccase was found to be responsible for the transformation of 2,4,6-trichlorophenol to 2,6-dichloro-1,4-hydroquinol and 2,6-dichloro-1,4-benzoquinone. Laccases from white rot fungi have been also used to oxidise alkenes, carbazole, N-ethyl-carbazole, fluorene and dibenzothiophene in the presence of HBT and ABTS as mediators. Isoxaflutole is an herbicide activated in soils and plants to its diketonitrile derivative, the active form of the herbicide. Laccases are able to convert the diketonitrile into the acid. The study of the laccase-mediator system in the bioremediation of polycyclic aromatic hydrocarbons (PAHs) has been extensively reported. In particular, the combination of several mediators looking for synergetic effects along with the use of natural mediators open new alternatives in this field.

Alternative Applications

Laccases do not only show potential for biological delignification of pulp but also for other applications. Laccases can be applied for the treatment of and detoxification of soils containing phenolic pollutants as well as other polluted systems due to the broad substrate range of the enzyme. The application of laccase for dyeing of materials with sulphur and reduced vat dyes has been patented. The use of laccase for the treatment of textile and bleach-plant effluents has also been investigated with success. Recently, increasing interest has arises on the application of laccase as a new biocatalyst in organic synthesis. The use of laccase for the production and treatment of beverages and as biosensor for the estimation of phenol or other enzymes in fruit juice has also been proposed. Recently, they have applications in other field such as in the design of biofuel cells.

FUNGAL ENZYMES IN OCCUPATIONAL DISEASE

The industrial utility of fungi has been well known since antiquity. In addition to the role of fungi as saprophytes in the environment, many species have commercial use, for example, mushrooms as food sources, ingredients in food preparation (cheese flavouring *Penicillium roqueforti*), alcoholic fermentation and the conversion of sugars in bread dough to carbon dioxide (*Saccharomyces cerevisiae*). In Asia, *Aspergillus oryzae* is an essential ingredient for the production of soya sauce and the fermented drink, sake. *Rhizopus* spp. secrete a wide variety of enzymes including cellulolytic, proteolytic, lipolytic and pectinolytic enzymes that are used in the production of various foods such as Tempe from Indonesia. *Rhizopus oryzae* has also been identified as a biocatalyst for biodiesel fuel production.

Other fungi such as *Yarrowia lipolytica* have more recent applications in the biodegradation of industrial products. Advances in industrial enzymology following World War II have enabled researchers to identify and utilise various enzymes and proteases that fungi produce to break down carbohydrate and lignin containing plant material in the environment. To date, close to 200 fungal enzymes have been purified from fungal cultures and the biochemical and catalytic properties characterised. These enzymes have great utility in pharmaceutical, agricultural, food, paper, detergent, textile, waste treatment and the petroleum industries. Industrial fungal enzymes are high-molecular-weight proteins that are catalysts. A description of the common enzymes used in various industries is presented in Table 22.1. The most widely used enzymes of occupational importance are derived from the genus *Aspergillus* and include α-amylase, xylanase and cellulase. Other enzymes are also utilised from rhizosphere fungal species belonging to the genera *Rhizopus* and *Humicola*.

These enzymes usually have intracellular or other functional roles associated with apical hyphal growth. It is uncommon for individuals in the general population to be exposed and sensitised to these antigens. In fact, in the general population, the prevalence of sensitisation to fungal enzymes has been reported to be as low as 1% and as high as 15%. However, in the occupational environment, workers that handle purified fungal enzymes are at an increased risk of becoming sensitised to enzymes. This is especially the case for workers whose occupation requires debagging, sieving, weighing, dispensing and mixing enzymes. Eight-hour time-weighted average exposures demonstrate that occupations weighing the enzyme preparations have the lowest average exposure compared to those workers that sieve. These workers are often exposed to levels of dust that exceed 4 mg m^{-3}, the threshold limit value (TLV) for inhalable dust. For other industrial environments that use lipase and cellulose in production, occupational exposure is highest in production areas and laboratories.

Adverse health effects associated with enzyme exposure are well characterised in the baking industry. In some countries, bakery exposures to enzymes are one of the leading causes of occupational allergy. Fungal enzymes are commonly used as baking additives to improve the dough, increase shelf life and decrease production time. Airborne concentrations ranging from 5.3 ng m^{-3} to 200 ng m^{-3} have been reported in occupational environments. Occupational sensitisation to fungal enzymes was first reported

Table 22.1: Fungal enzymes utilised in different industries and associations with occupational sensitisation.

Industry	Fungal enzyme	
	Characterised occupational allergen	*Uncharacterised occupational allergen*
Agriculture	Protease, lipase	
Animal feed	α-amylase, cellulase, lipase, phytase, protease and xylanase	Beta-glucanase, endo-xylanase
Pulp and paper production	Cellulase, hemicellulase, lipase and xylanase	Esterase, laccase, lignin peroxidase, manganese peroxidase, pectinase and mannose
Waste management	Lipase	Esterase, cytochrome P450, laccase, lignin peroxidase, manganese peroxidase and mono-oxygenase
Biotechnology	α-amylase, cellulase, glucoamylase, hemicellulase and protease	Cytochrome P-450 monooxygenase, glucose oxidase, glutathione-transferase, lignin peroxidase and manganese peroxidase
Detergent	α-amylase, cellulase, lipase and protease	
Food processing	α-amylase, cellulase, glucoamylase, lactase, lipase, protease and xylanase	Glucose isomerase, invertase and pectinase
Biofuels	α-amylase, cellulase, glucoamylase, protease and xylanase	
Bakery	α-amylase, cellulase, glucoamylase, hemicellulase, lipase, protease and xylanase	Glucose oxidase, lipoxygenase
Brewing and wine production	Cellulase, glucosidase, protease and xylanase	Alpha-acetolactate decarboxylase, β-glucanase and pectinase
Pharmaceutical	Lactase, lipase and protease	Alpha-galactosidase, catalase, cytochrome P450 oxygenase and glutathione transferase
Textile	α-amylase, cellulase, lipase, protease and xylanase	Catalase
Leather processing	Lipase, protease	
Hygiene products	Glucoamylase, protease	Catalase, glucose oxidase

by Flindt. Later, Baur and others demonstrated IgE sensitisation in workers handling these products. Since the original study, fungal enzymes have been identified as potent allergens in the occupational environment. Prevalence of sensitisation to *Aspergillus* enzymes ranges from 8% for glucoamylase, 11% for xylanase, 13% for cellulase and up to 34% for α-amylase. Sensitisation to α-amylase in bakery workers results in decreased peak expiratory flow and OA. In one report, workers exposed to fungal enzymes induced an immediate bronchospastic reaction. In the United States, the prevalence of work-related wheeze, runny nose, frequent sneesing and specific IgE to fungal enzymes was significantly higher among highly exposed workers. However, other irritantinduced mechanisms associated with high total dust levels have also been reported in a cohort of British bakers. To date, atopy has been hypothesised to be an important risk factor for OA to fungal enzymes.

Occupational exposure to enzymes has been demonstrated in other industries including manufacturing, pharmaceutical, food processing, animal feed and biotechnology. Like in baking environments, workers handling or in direct contact with fungal enzymes and with a history of atopy are at increased risk of becoming sensitised. Sensitisation to proteolytic enzymes has also been demonstrated in the manufacture of detergents. In the future, additional uses for fungal enzymes in industrial environments will be identified. Recent examples include the use of α-amylase and glucoamylase for the production of ethanol in the biofuel industry. If proper methods of exposure prevention are not followed and exposure is not monitored in these industries, it is possible that new groups of workers will suffer adverse health outcomes and become sensitised to enzymes. In the following sections we describe the major fungal enzymes, prevalence of sensitisation and occupational environment that they are most likely to be encountered.

EMERGING OCCUPATIONAL FUNGAL ENZYME EXPOSURES

The utility of fungal enzymes to degrade xenobiotics and organic compounds in the industrial sector continues to be recognised. Fungal enzymes are now being used for a variety of purposes across many different industries. Improved biochemical and molecular technologies have enabled the production of other potentially allergenic proteins. Interestingly, alpha-galactosidase has been associated with delayed anaphylaxis, angioedema, or urticaria in sensitised patients following the ingestion of beef, pork, or lamb. Although the role of alpha-galactosidase and these other enzymes following occupational exposure remains unclear, these studies provide preliminary insight into the possible potency of these allergens in industrial environments.

IMMUNODIAGNOSTIC DETECTION METHODOLOGIES

Occupational allergic sensitisation to fungal enzymes is diagnosed clinically using available *in vivo* SPT reagents, or *in vitro* assays such as Phadia ImmunoCap. However, SPT reagents for most of the fungal enzymes used in industrial settings are not commercially available and have to be either custom ordered or prepared individually by the investigator. Methods for SPT extract preparation that are used by investigators in the field have been previously described by Quirce and others. *In vitro* diagnostic tools that can quantify the amount of specific IgE to an occupational allergen are not readily available except in research laboratories where investigators prepare their own inhibition or radioallergosorbent enzyme-linked immunosorbent assay (ELISA) to quantify specific IgE. To date, α-amylase (k87) is the only fungal enzyme available on the Phadia ImmunoCap testing panel. To confirm OA caused by fungal enzymes, bronchial provocation tests can be undertaken to document immediate or late-phase responses to fungal enzymes. Positive immediate response criteria used in workers exposed to enzymes include a greater than 20% fall in FEV1, whereas a late-phase response has been considered positive when there is a 30% or greater fall in peak expiratory flow rate.

In order to better understand the relationships between occupational fungal enzyme exposure and clinical symptomology, accurate information on the distribution and quantity of the fungal enzyme in the occupational environment will be required. Immunodiagnostic methods that utilise antibodies could provide standardised methods for quantifying fungal enzyme biomarkers in a variety of occupational environments. Following validation and interlaboratory comparison, the assays could be used for exposure assessment to determine the existence of exposure-response relationships. This information is critical for the development of future threshold limit values (TLVs) and other occupational standards.

Several antibodies and immunodiagnostic methods have been produced to detect industrial fungal enzymes, in particular α-amylase. These methods have been employed in field investigations and used to quantify the concentration of the enzyme from collected air samples. Bogdanovic and others used an enzyme immunoassay with a sensitivity of 25 pg/mL to quantify α-amylase in airborne and surface dust samples collected from five bakeries. In the same study, a lateral flow immunoassay for α-amylase was compared to the reference enzyme immunoassay. The sensitivity of the lateral flow assay was 1–10 ng/mL and extracts with >5 ng/mL allergen were positive in the lateral flow assay. In a study of 507 personal air samples, Houba and colleagues used a rabbit IgG capture immunoassay to quantify α-amylase in specific baking job category. Concentrations of α-amylase up to 40 ng m^{-3} were quantified and workers directly involved with dough preparation had the highest exposures.

Using the same rabbit IgG sandwich assay, Nieuwenhuijsen and others identified dispensing and mixing areas to have the highest α-amylase exposure in British bakeries and flour mills.

Two monoclonal antibody- (mAb-) based ELISAs have been developed for the detection of α-amylase in the occupational environment. Assay sensitivities ranged from 0.2 ng/mL to 0.6 ng/mL. A quantitative mAb mediated dot blot assay has also been previously described for cellulase and xylase, the detection limits reported were 20 ng m^{-3} and 2 ng m^{-3}, respectively. mAbs to other fungal enzymes, such as xylanase have been produced and reported in the literature. Similarly, the detergent industry has produced antibodies and immunoassays for several common fungal enzymes and these have been utilised in industrial hygiene safety programmes to mitigate worker exposures. The development of fungal enzyme specific mAbs in combination with immunodiagnostic techniques will further our knowledge of the exposure-response relationships in occupational environments. Using these methods will also help enable the development of standards and focus on the prevention of sensitisation in heavily contaminated work environments.

ALLERGEN AVOIDANCE AND DIRECTIONS FOR THE FUTURE

Exposure to fungal enzymes, in particular α-amylase, is a considerable health risk in a number of industries. Cross-sectional studies have shown that processing workers in high-exposure categories who handled fungal enzymes are up to ten times more likely to be sensitised to fungal enzymes than workers in the low-exposure category. Highest concentrations of enzymes in the inhalable fraction were encountered among workers located in dispensing, mixing, weighing and sieving occupations.

Airborne concentrations as high as 40 ng m^{-3} and in some cases even higher (200 ng m^{-3}) have been reported for sensitised workers located in these handling areas. Concentrations in the low ng m^{-3} range have been associated with an increased frequency of sensitisation. For other fungal enzymes, such as phytase, similar findings have been reported.

The continued utilisation of other previously overlooked enzymes as well as new genetically engineered enzymes in various industries will continue to provide diagnostic challenges, even for the most seasoned occupational medicine professional. It is likely that new cases of occupational allergic disease will emerge following exposure to fungal industrial enzymes during the next decade. In response, identification of exposure-response relationships will be critical for the development of TLVs and occupational exposure levels. However, this will depend on the development of suitable diagnostic antibodies and immunoassays. Currently, subtilisin, a sereine endopeptidase derived from *Bacillus subtilis*, is the only enzyme for which the American Conference of Governmental Industrial Hygienists (ACGIH) has established a TLV value (60 ng m^{-3}). The European Union Directive also classifies the fungal enzymes cellulase and α-amylase with the risk phrase R42 (may cause sensitisation by inhalation). There are currently no consensus standards for other industrially utilised fungal enzymes.

As a precautionary measure, it has been concluded that all enzymes should be regarded as an allergen that can exacerbate respiratory sensitisation in susceptible populations. Baur has further proposed that all enzymes should be classified as R42 according to the European Union Directive criteria. Although intervention in the bakery industry has had little to no effect, installation of engineering controls and implementation of personal protective equipment programmes in animal feed workers exposed to phytase was shown to result in the immediate cessation of hypersensitivity symptoms. Improvements in biotechnology have also included the encapsulation of some enzymes and proteins. These engineering controls have been proposed to reduce occupational exposure to enzymes, however, encapsulation alone may not completely prevent enzyme-induced allergy and OA. To date, the detergent industry has implemented a derived minimal effect level (DMEL) of 60 ng m^{-3} for pure enzyme proteins. Although this DMEL was provided as guidance by the ACGIH, other manufacturers have implemented their own occupational exposure guidelines (OEGs) for fungal enzymes such as α-amylase (5–15 ng m^{-3}), lipase (5–20 ng m^{-3}) and cellulase (8–20 ng m^{-3}). In addition, the detergent industry has developed a medical surveillance programme to identify and correct elevated exposures before occupational illnesses occur. As a result, the incidence of occupational allergy has dropped substantially. Implementation of DMELs and OEGs will further assist in the reduction of occupational exposure. Reducing worker exposure to fungal enzymes in industry by the implementation of engineering controls and other allergen avoidance strategies will continue to mitigate personal exposure and further reduce the occupational health risk.

To sum up, laccases are widespread in nature, being produced by a wide variety of plants, fungi and also bacteria. The functions of the enzyme differ from organism to organism and typify the diversity of laccase in nature. Laccases catalyse the oxidation of phenolic compounds whilst simultaneously reducing molecular oxygen to water. The catalytic ability of laccases has, not surprisingly, led to diverse biotechnological applications of this enzyme.

The introduction of the laccase-mediator system provides a biological alternative to traditional chlorine bleaching processes. The laccase enzyme has a wide field of application including the pulp and paper industries, the treatment of various industrial effluents, enzymatic decolouring of material and bioremediation of soils. One of the limitations to the large-scale application of the enzyme is the lack of capacity to produce large volumes of highly active enzyme. These problems can be solved with the use of recombinant organisms or screening for natural hypersecretory strains. Environmental factors influence the ability of fungi to produce high titres of laccase and different strains react differently to these conditions. One should thus select a strain capable of producing high concentrations of a suitable enzyme and then optimise conditions for laccase production by the selected organism. It is therefore not surprising that this enzyme has been studied intensively since the nineteenth century and yet remains a topic of intense research today.

Chapter 23

Enzymes in Biosynthesis of Nanoparticles

INTRODUCTION

While a large number of microbial species are capable of producing metal nanoparticles (NPs), mechanism of nanoparticle biosynthesis is very important. The metabolic complexity of viable micro-organisms complicates the analysis and identification of active species in the nucleation and growth of metal NPs. Strategies such as enzymatic oxidation or reduction, sorption on the cell wall and in some cases, subsequent chelating with extracellular peptides or poly-saccharides have been developed and used by micro-organisms. Some species can control the membrane transport of heavy metals towards, or their active efflux from, the cell. Metal ion resistance via transport and passive mechanisms leading to extracellular precipitation is more characteristic for prokaryotes.

This chapter discusses several enzymes which play an important role in nanoparticle synthesis by micro-organisms. Variety of techniques for making of metal nanoparticles are chemical recovery using regenerative materials, aerosol technique, electrochemical deposition, photochemical recovery and laser exposure, etc. But as the term of green nanotechnology has emerged that a lot of attention has attracted and includes a wide range of processes that reduce or eliminate toxic substances to restore environment. Green nanotechnology also seek more effective alternatives for energy production (e.g. solar and fuel cells). In green nanotechnology, for the synthesis of nanoparticles micro-organisms are used. It is well known that many micro-organisms, aggregate inorganic material within or outside the cell to form nanoparticles. However, large number of microbial species are capable of producing metal NPs, the mechanism of nanoparticle biosynthesis is very important. Many things about the biochemical and molecular mechanism of these processes remain unknown and should be revealed. In fact, the biochemical mechanisms referred to finding materials like enzymes, which may mediate the biosynthesis mechanism. The studies of the enzyme structure and the genes which code these enzymes may help improve our understanding of how metal nanoparticle synthesis is performed. Improvements in chemical composition, size and shape and dispersity of generated nano-particles could allow the use of nanobiotechnology in a variety of other applications. Cell walls and cell wall proteins are likely to play an important role in the reduction of metal ions. This chapter discusses the importance of enzymes in several micro-organisms in nanoparticle synthesis process.

BIOREDUCTION BY OXIDOREDUCTASE ENZYMES

One mechanism of metal nanoparticles biosynthesis by micro-organisms is bioreduction. In microbial bioreduction processes, myriads of proteins, carbohydrates and biomembranes are involved. Nanoparticles are formed on cell wall surfaces and the first step in bioreduction is the trapping of the metal ions on this surface. This probably occurs due to the electrostatic interaction between the metal ions and positively charged groups in enzymes present at the cell wall. This may be followed by enzymatic reduction of the metal ions, leading to their aggregation and the formation of nanoparticles. The microbial cell reduces metal ions by use of specific reducing enzymes like NADH-dependent reductase or nitrate dependent reductase.

Oxidoreductase in Yeast

In the yeast the membrane bound (as well as cytosolic) oxidoreductases and quinones might have played an important role in the process. The oxidoreductases are pH sensitive and work in alternative manner. At a lower value of pH, oxidase gets activated while a higher pH value activates the reductase. Along with this, a number of simple hydroxy/methoxy derivatives of benzoquinones and toluquinones are elaborated by lower fungi (especially *Penicillium* and *Aspergillus* species). Yeast might be treasuring any other such quinine because it belongs to the same class of fungi thereby facilitating the redox reactions due to its tautomerisation. The transformation seems to be negotiated at two distinct levels, at the cell membrane level immediately after addition of the $TiO(OH)_2$ solution which triggers tautomerisation of quinones and low pH sensitive oxidases and makes molecular oxygen available for the transformation. Such a stress generated response had earlier been suggested in case of *Candida glabarata* cell, challenged with cadmium ion in form of elaboration of an enzyme phytochelatin synthase and a protein HMT-1 which effectively aborted the CdS nanocrystals from Cytosol. Once entered into the cytosol, the $TiO(OH)_2$ might have triggered the family of oxygenases harboured in the endoplasmic reticulum (ER), chiefly meant for cellular level detoxification through the process of oxidation/oxygenation. Taking use of the above-mentioned facts was earlier reported synthesis of metallic selenium, cadmium, silver, titanium, as well as antimony oxide.

NADH-dependent Reductase

Fusarium oxysporum

Protein assays indicate that an NADH-dependent reductase, is the main responsible factor of biosynthesis processes. This reductase gains electrons from NADH and oxidises it to NAD^+. The enzyme is then oxidised by the simultaneous reduction of metal ions. The enzymatic route of *in vitro* synthesis of silver hydrosol of 10–25 nm using α-NADPH-dependent nitrate reductase (44 kDa) from *F. oxysporum* with capping peptide, phytochelatin was demonstrated recently. The mechanistic aspect was explained by Duran and others that apart from enzymes, quinine derivates of napthoquinones and anthraquinones also act as redox centres in the reduction of silver nanoparticles. A similar finding was also reported in the reduction of gold (III) chloride to metallic gold by α-NADPH-dependent sulphite reductase of molecular mass of 35.6 kDa and phytochelatin. A dimeric hydrogenase enzyme (44.5 kDa and 39.4 kDa) of *F. oxysporum* that showed optimum activity at pH 7.5 and 38°C passively reduced H_2PtCl_6 to platinum nanoparticles was also reported. To date, only very few reports have been documented on optimisation in biological process, a 29-kDa 'gold shape-directing protein (GSP)' present in the extract of green algae. Chlorella vulgaris was used in the bioreduction and in the synthesis of shape-and size-controlled distinctive triangular and hexagonal gold nanoparticles. With increase in the concentration of GSP produced gold plates with lateral sizes up to micrometers. Such mechanistic components should be unraveled for efficient biological processes.

Aspergillus flavus

The fungus, *Aspergillus flavus* also resulted in the accumulation of silver nanoparticles on the surface of its cell wall when incubated with silver nitrate solution. Extracellularly produced nanoparticles were stabilised by the proteins and reducing agents secreted by the fungus. A minimum of four high molecular weight proteins released by the fungal biomass have been found in association with nanoparticles. One of these was strain specific NADH-dependent reductase. However, emission band produced by fluorescence spectra indicate the native form of these proteins present in the solution as well as bound to the surfaces of nanoparticles. Further, the reduction of metal ions and surface binding of the proteins to the nanoparticles did not compromise the tertiary structure of the proteins.

Nitrate/Nitrite Reductase

Fusarium oxysporum

One other important enzyme that is responsible for this reduction in some micro-organisms is nitrate dependent reductase. In *Fusarium oxysporum,* this enzyme is conjugated with an electron donor (quinine), reduces the metal ion and changes it to elemental form. In the case of rapid extracellular synthesis, because the reduction happens in very few minutes, complex electron shuttle materials may be involved in the biosynthesis process.

Enterobacteriaceae

The culture supernatants of *Enterobacteriaceae* (*Klebsiella pneumonia*, *E. coli* and *Enterobacter cloacae*) also rapidly synthesised silver nanoparticles by reducing Ag^+ to Ag^0. With the addition of piperitone, silver ion reduction was partially inhibited, which showed the involvement of nitroreductase enzymes in the reduction process.

Sulphate and Sulphite Reductase

Rhodobacter sphaeroides

In the formation of ZnS nanoparticles by *Rhodobacter sphaeroides* (Fig. 23.1) a series of reductase have serious roles: First, soluble sulphate enters into immobilised beads via diffusion and later is carried to the interior membrane of *R. sphaeroides* cell facilitated by sulphate permease. Then, the sulphate is reduced to sulphite by ATP sulphurylase and phosphoadenosine phospho-sulphate reductase and next sulphite is reduced to sulphide by sulphite reductase. The sulphide reacts with *o*-acetyl serine to synthesis cysteine via, *o*-acetyl serine thiolyase and then cysteine produces S^{2-} by a cysteine desulphydrase in presence of zinc. After this process, S^{2-} reacts with the soluble zinc salt and the ZnS nanoparticles are synthesised. Finally, ZnS nanoparticles are discharged from immobilised cells to the solution. In the synthetic process, the particle size is controlled by the culture time of the *R.sphaeroides* and simultaneously the immobilised beads act on separating the ZnS nanoparticles from the rhodobacter sphaeroides.

Cyanobacteria

The mechanisms of gold bioaccumulation by cyanobacteria (Plectonema boryanum UTEX 485) from gold (III) - chloride solutions have documented that interaction of cyanobacteria with aqueous gold (III)

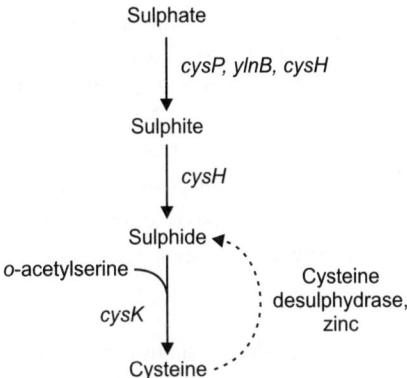

Fig. 23.1: Sulphate assimilation action of *Rhodobacter sphaeroides*. The enzymes present in *Rhodobacter sphaeroides* are indicated by the corresponding genes: *cysP*, sulphate permease, *ylnB*, ATP-sulphurylase, *cysH*, phosphoadenosine phosphosulphate reductase, *cysH*, sulphite reductase, *cysK*, *o*-acetylserine synthase.

- chloride initially promoted the precipitation of nanoparticles of amorphous gold (I)-suphide at the cell walls and finally deposited metallic gold in the form of octahedral (III) platelets near cell surfaces and in solutions. Adding further to the mechanism, a sulphate-reducing bacterial enrichment was used to destabilise gold(I) thiosulphate complex to elemental gold and proposed that this could occur by three possible mechanisms involving iron sulphide, localised reducing conditions and metabolism.

The interaction of *P. boryanum* UTEX485 with $Au(S_2O_3)_2^{-3}$ promoted the accumulation in membrane vesicles with 10–25 nm in size in cubic morphology, clustering inside the cell with ≤10 nm and precipitated in solution with <10–25 nm along with admixed AuS nanoparticles of ≤10 nm. But in the presence of $AuCl_4^-$ precipitation was resulted in octahedral gold platelets of 1–10 μm in solution and ≤10 nm inside the bacterial cells.

HYDROLASES IN FUNGI

Proteins have been implicated in nanoparticle formation in a number of different studies found that during the formation of zirconia nanoparticles, the fungus secreted proteins capable of extracellular hydrolysing compounds with zirconium ions and this was confirmed in subsequent studies with silica and titania. It was found that their fungus was also capable of hydrolysing metal halide precursors under acidic conditions. The studies indicated that the proteins involved in the reduction of metal nanoparticles were cationic proteins with molecular weights of around 21–24 kDa also suggested that a cationic protein of around 55 kDa found in extracellular extracts of *Verticillium* sp., might be responsible for the hydrolysis of $[Fe(CN)_6]_3$ and $[Fe(CN)_6]_4$ found proteins bound to the nanoparticle surface and the presence of sodium (S) atoms around the silver nanoparticles was taken to suggest an association between nanoparticles and fungal proteins.

Cysteine Desulphydrase in *Rhodopseudomonas Palustris*

This enzyme belongs to the family of lyases, specifically the class of carbon-sulphur lyases. The systematic name of this enzyme class is D-cysteine sulphide-lyase. This enzyme participates in cysteine metabolism. A simple route for the synthesis of cadmium sulphide nanoparticles by photosynthetic bacteria *Rhodopseudomonas palustris* has been demonstrated. The purified solution yielded the maximum absorbance peak at 425 nm due to CdS particles in the quantum size regime. Transmission electron microscopic analysis of the samples showed a uniform distribution of nanoparticles, having an average size of 8.01±0.25 nm and its corresponding electro diffraction pattern confirmed the face-centered cubic (fcc) crystalline structure of cadmium sulphide. Furthermore, it was observed that the cysteine desulphydrase producing S^{2-} in the *R. palustris* was located in cytoplasm and the content of cysteine desulphydrase depending on the growth phase of cells was responsible for the formation of CdS nanocrystal, while protein secreted by the *R. palustris* stabilised the cadmium sulphide nanoparticles. In addition, *R. palustris* was able to efficiently transport CdS nanoparticles out of the cell.

Glutathione

The major molecules that contribute to the detoxification mechanisms in yeast cells are glutathione (GSH) and two groups of metal-binding ligands: metallo-thioneins and phytochelatins, GSH with the structure *c*-Glu-Cys-Gly is an important tripeptide involved in various metabolic processes in bacteria, yeasts, plants and animals. The unique redox and nucleophilic properties classify this compound as a detoxificator, actively taking part in the bioreduction and defense against free radicals and xenobiotics. Metallothioneins are low molecular-weight and it classified according to the arrangement of these residues. Although the model has been created on the basis of detoxification of Cu^{2+} ions, it is also accurate for other metals in yeast.

There is evidence that metallothioneins play a similar role in some plants and yeast species, e.g. in *S. cerevisiae* and *C. glabrata*. GSH is also a structural unit in phytochelatin molecules, which is one of its major functions. Although, initially, they were described as cadmium binding peptides, phytochelatin formation is induced by a large number of elements such as Cd^{2+}, Pb^{2+}, Zn^{2+}, Sb^{3+}, Ag^+, Ni^{2+}, Hg^{2+}, $HAsO_4^{2-}$, Cu^{2+}, Sn^{2+}, SeO_3^{2-}, Au^+, Bi^{3+}, Te^{4+} and W^{6+}, when supplemented to the medium. Phytochelatins have the general structure (*c*-Glu-Cys) *n*-Gly, where $n = 2–11$ and a multitude of structural variants has been described in the scientific literature. The enzyme phytochelatin synthase or *c*-Glu-Cys dipeptidyl transpeptidase (EC 2.3.2.15) catalyses the reaction of transpeptidation of *c*-Glu-Cys dipeptide from a GSH molecule to a second molecule of GSH, resulting in phytochelatin PC2, or to a phytochelatin molecule, resulting in an $n + 1$ oligomer. Phytochelatin synthesis begins within minutes after exposing yeast cells to cadmium ions and is regulated by enzyme activation in the presence of metal ions.

The best activators are cadmium ions, followed by ions of Ag, Bi, Pb, Zn, Cu, Hg and Au. In yeasts, Cd-phytochelatin complexes are formed in the cytosol but accumulate in vacuoles. Detailed studies of the fission yeast *S. pombe* revealed that nearly the whole cadmium and phytochelatin amounts are located in vacuoles. Compared to metallothioneins, phytochelatins feature many advantages, derived from their unique structure and especially the repeated *c*-Glu-Cys units. For example, they have better metal-binding capacity. In addition phytochelatins can incorporate large amounts of inorganic sulphur, resulting in increased capacity of these peptides to bind cadmium. For crystal lattice of CdS consists of 85 CdS pairs covered by 30 (*c*-Glu-Cys)*n*-Gly peptides, with $n = 3–5$. In the presence of heavy metal stress, yeast cells increase cellular pools of glutathione and glutathione-like compounds called phyto-chelatins. The resulting metal thiolate complex formation neutralises the toxicity of heavy metal ions and traps them inside the cell. Sulphide anions are readily incorporated into these cadmium- glutathione complexes, resulting in the formation of nanocrystals.

Thus, today, nano metal particles have drawn the attention of scientists because of their extensive application to new technologies in chemistry, electronics, medicine and biotechnology. Beside many physical and chemical methods which have been developed for preparing metal nanoparticles, nanobiotechnology also serves as an important method in the development of clean, nontoxic and environmentally friendly procedures for the synthesis and assembly of metal nanoparticles. This new biotechnological method has important advantages in comparison to conventional methods. For instance, it is an easier and cheaper procedure. To be utilised in different scientific fields, biological synthesis requires an understanding of the biochemical and molecular mechanisms of the reaction for obtaining better chemical composition, shape, size and monodispersity.

As already discussed in nanoparticle biosynthesis, many enzyme such as redutases, syntases and hydrolases, etc. are important and describe specific genes and characteristisation of enzymes involved in the biosynthesis of nano-particles is also required. Therefore, a complete knowledge of the molecular mechanisms involved in the microbial synthesis of nanoparticles is neseccery to control the size, shape and crystallinity of nanoparticles. In regard with nanoparticle biosynthesis considerable advantages, with improvement of these methods they could be the leading large scale production method for nanoparticles in future.

Chapter 24

Nanoparticles in Enzyme Immobilisation

INTRODUCTION

Macro, micro and nanosized chitosan particles are suitable as carriers for enzyme immobilisation. Chitosan nanoparticles due to their highest specific surface area are much proper for immobilisation of higher amount of enzymes. As compared to chitosan macro and microparticles, higher activity values of immobilised enzyme onto chitosan nanoparticles is explained again by better distribution of the enzyme onto the larger surface area of nanoparticles. Higher magnetic property of chitosan nanoparticles reduces self aggregation of immobilised enzymes onto them and increases the stability of immobilised enzymes. Further studies will be needed to explore the kinetic of immobilised enzymes onto chitosan nanoparticles. The immobilisation of enzymes onto nanoparticles and the subsequent attachment of the nanoparticles onto an electrode is also an attractive alternative in biosensor researches and developments, especially in the case of magnetic nanoparticles which can be removed from the electrode by the action of magnetic fields.

Enzymes are proteins that catalyse chemical reactions. Unlike more traditional organic and inorganic catalysts, enzymes are large and fragile molecules. Instability and sensitivity to process conditions are disadvantages of using soluble enzymes. Therefore, application of solid phase biocatalysts has become more and more important during the last decades. Immobilised enzymes have a great importance in industrial bioprocesses especially in food, nutritional and pharmaceutical technologies. There are several reasons for using an enzyme in an immobilised form. In addition to more convenient handling of the enzyme, it provides for its facile separation from the product. It helps to prevent the contamination of the substrate with enzyme/protein or other compounds, which decreases purification costs. Immobilisation also facilitates the efficient recovery and reuse of costly enzymes, with longer half lives and less degradation, in successive batches, or the process can eventually be carried out in a continuously operating reactor. There are several methods used to immobilise enzymes and three of the most common methods are physical adsorption, entrapment (encapsulation) and cross-linking or covalently binding to a support. For practical purposes, carrier beads with size falling into millimeter range are mainly used. However, more and more results are reported on immobilisation of enzymes onto microparticles possessing high specific surface area and numerous active sites available for the enzyme molecules to be fixed. Moreover, because of the smaller size of the support particles, the internal diffusion hindrance diminishes. Hindrance can be described as a function of solute size and geometric properties of the porous network. Diffusion of macro-molecules in porous structures is hindered in comparison to diffusion in free solution. Various supports have been used for enzymes immobilisation such as synthetic organic polymers, biopolymers, hydrogels, smart polymers and inorganic supports. A variety of biopolymers, mainly water insoluble polysaccharides such as cellulose, starch, agarose and chitosan and proteins such as gelatin and albumin have been widely used as supports for immobilising enzymes.

Chitosan with some primary amino groups is derived by deacetylation of chitin. Its pK_a is about 6.3. At lower pH solutions ($<pK_a$), most of the amino groups are protonated, making chitosan a water soluble polyelectrolyte. When the pH is higher than pK_a, the amino groups are deprotonated and chitosan becomes insoluble. Because of its excellent film forming ability, biocompatibility, nontoxicity, high mechanical strength, cheapness and a susceptibility to chemical modifications, chitosan has been extensively used for immobilisation of enzymes. In recent years, nanotechnology has shown a significant attraction to the preparation of immobilised enzymes. Nanomaterials due to their small size and large surface area to volume ratio have special characteristics, which make them favourable for enzyme immobilisation.

IMMOBILISED ENZYME SUPPORTS

Immobilisation is often the key to optimising the operational performance of an enzyme in industrial processes, particularly for use in non aqueous media. The immobilisation of enzymes has been a growing field of research, because it has allowed enzymes to be easily reused multiple times for the same reaction with longer half lives and less degradation and has provided a straight forward method of controlling reaction rate as well as reaction start and stop time. The properties of immobilised enzymes are governed by the properties of both the enzymes and the support materials. The interaction between the two lends an immobilised enzyme specific physico-chemical and kinetic properties that may be decisive for its practical application and thus, a support judiciously chosen can significantly enhance the operational performance of the immobilised system. Suitable materials used as a carrier should have chemical, physical and biological stability during processing, as well as in the reaction conditions, sufficient mechanical strength, specially for its utilisation in reactors and industries, should be nontoxic both for the immobilised enzyme/bioparticle, as well for the product, also should have adequate function groups for binding biocatalyst and high loading capacity. Other criteria, such as structural characteristics (porosity, swelling, compression, material and mean particle behaviour), as well as possibility for microbial growth, biodegradability, high affinity to proteins, solubility, regenerability and ease of preparation in different geometrical configurations that provide the system with permeability and surface area are more application specific. Chemical structure of carrier material determines interaction with enzymes. If the support material is highly porous, pore size and pore size distribution will play an important role in determining the immobilised enzyme properties. A small pore size can cause diffusion limitation resulting in structural rearrangement of the enzymes and subsequent inactivity. However, for very large pore sizes, enzymes can cluster together and thus lose activity. Enzymes immobilisation onto carriers has been extensively studied and applied in many fields, such as biocatalysts medical devices, drug delivery systems and biosensor. Several carriers such as synthetic organic polymers (e.g. Eupergit C and polyurethane), biopolymers (e.g. alginate), hydrogels (e.g. polyvinyl alcohol), smart polymers (e.g. poly-N-isopro-pylacrylamide) and inorganic supports (e.g. alumina, silica and zeolites) have been used in immobilisation of enzymes. Biopolymers are one of the most applied supports for immobilising enzymes, includes two types of mainly water insoluble proteins such as gelatin and albumin and polysaccharides such as starch, alginate and chitosan.

ENZYME IMMOBILISATION TECHNIQUES

Immobilisation often stabilises structure of the enzymes, thereby allowing their applications even under harsh environmental conditions of pH, temperature and organic solvents and thus enable their uses at high temperatures in non aqueous enzymology and in the fabrication of biosensor probes. Different methods for the immobilisation of enzymes have been critically reviewed.

These methods are divided into two main categories namely, reversible and irreversible immobilisation. Adsorption, disulphide bonding and chelation or metal binding are the reversible methods for enzyme immobilisation.

Adsorption: Adsorption is the elementary and probably the simplest method of immobilisation. This method is based on weak forces (e.g. van der Waals), however still enabling an efficient binding process. A wide range of both organic and inorganic materials can be used as a support in this method.

Disulphide: Disulphide binding may be seen as a variation of covalent bonding, as there are stable covalent bonds formed between activated support and free thiol group (e.g. on cysteine) in the biocatalyst.

Chelation: Chelation is based on the ability of the side chains of certain amino acids (e.g. histidine, tryptophan, tyrosine, cysteine and phenylalanine) to substitute weakly bonded ligands in the metal ions that have been immobilised by chelating group covalently bound to a solid support. Schematics of reversible enzyme immobilisation methods are shown in Fig. 24.1.

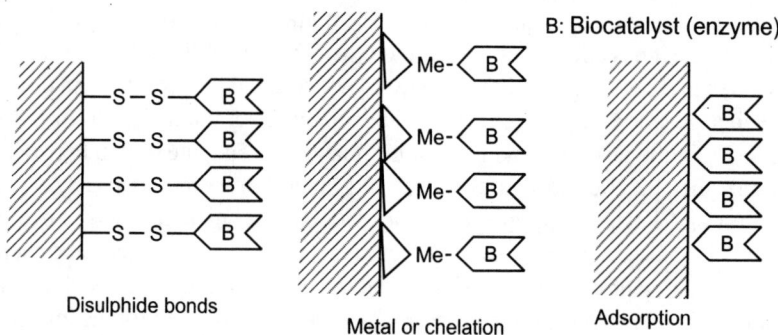

Fig. 24.1: Reversible methods of enzyme immobilisation.

Covalent bonding, entrapment, encapsulation and cross-linking, fall into the irreversible enzyme immobilisation methods. Covalent bonding is one of the most widely used methods for enzymes immobilisation. Popularity of this approach is mainly connected with the stability of the bonds formed between the enzyme and the support, which prevents enzyme release into the environment. Entrapment is based on incorporating enzymes into the lattices of a semi permeable gel or enclosing the enzymes in a semi permeable polymer membrane. Encapsulation is similar to entrapment. In this process, enzymes are restricted by the membrane walls (usually in a form of a capsule). Cross-linking method is widely used to immobilise enzymes and is based on intermolecular cross-linking of enzymes by bi-functional or multi functional reagents. In fact, this method does not require a support and this is the advantage of this method. Immobilisation of an enzyme on a carrier often leads to the loss of more than 50% native activity, specially at high enzyme loadings. The design of carrier bound immobilised enzymes also relies largely on laborious and time consuming trial and error experiments, because of the lack of guidelines that link the nature of a selected carrier to the performance expected for a given application. Schematics of irreversible enzyme immobilisation techniques are shown in Fig. 24.2.

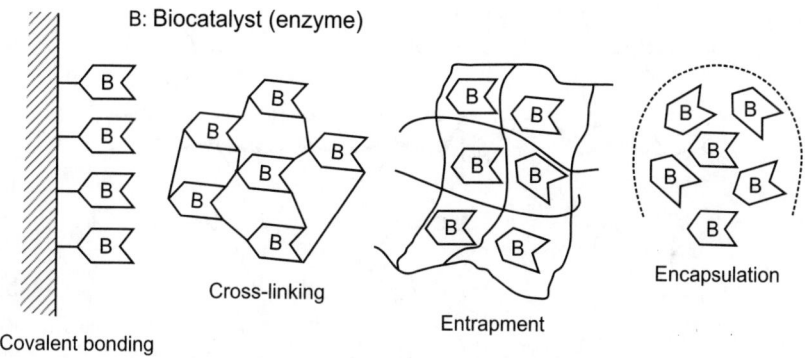

B: Biocatalyst (enzyme)

Covalent bonding

Cross-linking

Entrapment

Encapsulation

Fig. 24.2: Irreversible methods of enzyme immobilisation.

CHITOSAN

Chitin is a copolymer of N-acetyl-β-glucosamine and D-glucosamine units linked with β-(1-4) glycosidic bond, where N-acetyl-d-glucosamine units are predominant in the polymeric chain. The deacetylated form of chitin refers to chitosan. Basically, the process consists of deproteinisation of the raw shell material with a dilute NaOH solution and decalcification with a dilute HCl solution. To result in chitosan, the obtained chitin is subjected to N-deacetylation by treatment with a 40–45% NaOH solution, followed by purification procedures. Thus, production and utilisation of chitosan constitutes an economically attractive means of crustacean shell wastes disposal sought worldwide.

These renewable resources are found in many naturally occurring organisms (e.g. fungi and yeast) and is the principal component in the exoskeleton of sea crustaceans such as crabs, shrimps, lobsters and krills. Chitosan represents long chain polymers having molecular mass up to several million Daltons. Chitosan is relatively reactive natural biopolymer due to presence of reactive amino and hydroxyl functional groups in its structure and can be produced in various forms such as powder, gels and films, sponges, intragastric floating tablets and especially spherical micro and nanoparticles. If degree of deacetylation and molecular weight of chitosan can be controlled, then it would be a material of choice for developing micro and nanoparticles. Chitosan has ability to control the release of active agents and avoid the use of hazardous organic solvents while fabricating particles since it is soluble in aqueous acidic solution. It is a linear polyamine containing a number of free amine groups that are readily available for cross-linking and its cationic nature allows for ionic cross-linking with multivalent anions. Chitosan has muco-adhesive character, which increases residual time at the site of absorption. Chemical structures of chitin and chitosan are shown in Fig. 24.3.

APPLICATIONS OF CHITOSAN

Chitosan due to its unique physico-chemical and biological properties is an attractive material for use in various applications. These properties include: biodegradability, lack of toxicity, antibacterial and antifungal effects, wound healing acceleration and immune system stimulation. Because of excellent biological and physico-chemical properties of chitosan, it has the ability to bind to particular materials including cholesterols, fats, proteins, metal ions and even tumour cells. This allows chitosan to be used as a chelating agent in various applications.

Over the last decade, a number of potential products based on chitosan have been increasingly produced and applied in several areas such as: wastewater treatment, due to its high adhesive and insolubility

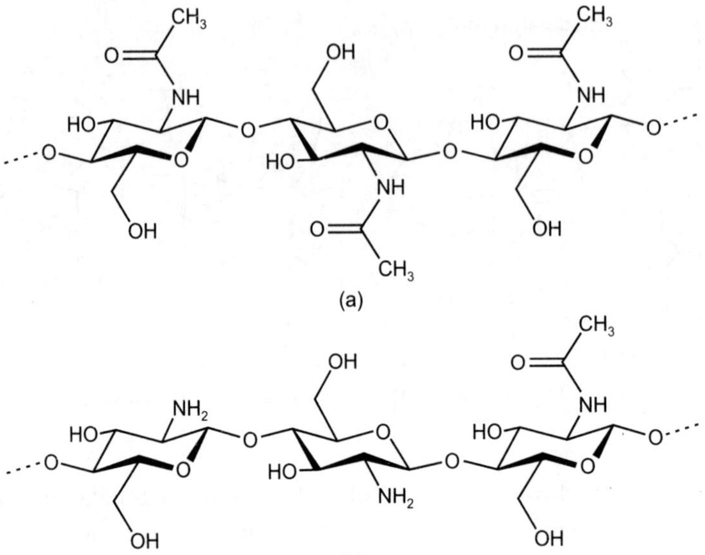

Fig. 24.3: Chemical structure of (a) chitin and (b) chitosan.

properties (e.g. removal of heavy metal ions and membrane purification processes), food industries (e.g. anticholesterol and fat binding, preservative, packaging material and animal feed additive), agriculture (e.g. seed and fertiliser coating, controlled agrochemical release), pulp and paper industries, because of its wet strength to paper (e.g. surface treatment, photographic paper) and cosmetics and toiletries due to its fungicidal and fungistatic properties (e.g. moisturiser, body creams, bath lotion). Fibres made of chitosan are useful as absorbable sutures and wound dressing materials. Chitosan has also important applications in photography due to its resistance to abrasion, optical characteristics and film forming ability. Due to its unique molecular structure, chitosan has an extremely high affinity for many classes of dyes, including disperse, direct, reactive, acid, vat, sulphur and naphthol dyes.

Chitosan ability to development of drug delivery systems is one of the most well known of its characteristics. Different types of chitosan made drug carriers have been conceived for various administration routes, such as oral, nasal, transdermal, parenteral, vaginal, cervical and rectal due to its biocompatibility, biodegradability and ecological safety. Tissue engineering is an important therapeutic strategy for present and future medicine. Recently, functional biomaterial researches have been directed towards the development of improved scaffolds for regenerative medicine. Chitosan and its derivatives are promising candidates as a supporting material for tissue engineering applications owing to their porous structure, gel forming properties, ease of chemical modification, high affinity to *in vivo* macromolecules and so on. Moreover, it is one of the most promising immobilisation matrices with excellent characteristics as a support matrix for enzyme immobilisation.

IMMOBOLISED ENZYMES ONTO CHITOSAN

Chitosan has shown favourable biocompatibility characteristics, as well as the ability to increase membrane permeability. Moreover, it is one of the most promising immobilisation matrices due to an excellent membrane forming ability, good adhesion, low cost, low immunogenicity and nontoxicity, high mechanical strength and hydrophilicity as well as the improvement of stability. These properties have prompted extensive applications of chitosan as a matrix for enzyme immobilisation. In fact, polycationic nature of chitosan and presence of large primary amine groups in its chemical structure (Fig. 24.3b) make it potentially useful in immobilisation of various enzymes. Results of several studies have been shown that after selection of suitable enzyme immobilisation method, chitosan reactive amino and hydroxyl groups offer a good enzyme coupling efficiency.

Chitosan has been extensively used in immobilisation of various enzymes groups because of its many characteristics like improved resistance to chemical degradation, avoiding disturbance of metal ions to enzyme and antibacterial property. Simsek-Ege and Vaillant used chitosan to immobilise carbonic anhydrase and pectin lyase, respectively, which both of them are classified in lyase enzymes group.

Applications of chitosan in immobilisation of oxidoreductase, hydrolase and transferase enzymes are summarised in Tables 24.1 to 24.3, respectively.

Table 24.1: Some of immobilised oxidoreductases onto chitosan.

Enzyme	*Support*	*Type of immobilisation*
Alanine dehydrogenase	Chitosan beads	Covalent binding with G
Alcohol dehydrogenase	Chitosan beads	Covalent binding with G
Alcohol oxidase	Chitosan beads	Covalent binding with G
Catalase	F3GA attached-chitosan beads	Adsorption on support
Galactose oxidase	Chitosan membrane	Covalent binding with G
Glucose oxidase	Chitosan capsules	Micro encapsulation
Glutamate dehydrogenase	Chitosan membrane	Covalent binding with G
Glutamate oxidase	Chitosan membrane	Covalent binding with G
Horseradish peroxidase	Chitosan membrane	Covalent binding with G
Laccase	Magnetic chitosan microspheres	Adsorption followed by cross-linking with G
Lactate oxidase	Chitosan-enzyme beads	Gel inclusion
Leucine dehydrogenase	Chitosan capsules	Micro encapsulation
Octopine dehydrogenase	Chitosan beads	Covalent binding with G
Oxalate oxidase	Chitosan powder	Adsorption followed by cross-linking with G
Putrescine oxidase	Chitosan beads	Covalent binding with G
Sulphite oxidase	Chitosan-PHEMA membrane	Adsorption on support
Uricase	Chitosan membrane	Covalent binding with agents other than G
Xanthine oxidase	Chitosan beads	Covalent binding with G

G: Glutaraldehyde

Table 24.2: Some of immobilised hydrolases onto chitosan.

Enzyme	*Support*	*Type of immobilisation*
Acid phosphatase	Chitosan beads	Adsorption on support
Alkaline phosphatase	Chitosan beads	Covalent binding with G
Alkaline protease	Chitosan powder	Covalent binding with G, adsorption on support
Aminoacylase	Chitosan-coated alginate beads	Gel inclusion
α-amylase	Chitosan beads	Covalent binding with G
β-amylase	Chitosan beads	Adsorption on support

(Cont'd...)

Enzyme	Support	Type of immobilisation
α-l-arabino-furanosidase	Chitosan powder	Adsorption, cross-linking and covalent binding with G
Bromelain	Chitosan beads	Covalent binding with agents other than G
Candida ntarctica lipase	Chitosan-based hydrogels	Covalent binding with G
Cellulase	Chitosan beads	Adsorption
Chitinases	Chitosan beads	Adsorption followed by cross-linking with G
α-Chymotrypsin	Chitosan beads	Adsorption followed by cross-linking with G
Creatininase	Chitosan-g-polyaniline with iron oxide	Covalent binding with G, adsorption on support
Creatinine deaminase	Chitosan membrane	Adsorption on support
Endo-1,4-β-Xylanase	Chitosan powder	Adsorption on support
Ficin	Chitosan beads	Covalent binding with agents other than G
β-galactosidase	Chitosan beads	Covalent binding with G
α-glucosidase	Chitosan beads	Covalent binding with G
β-glucosidase	Chitosan powder	Covalent binding with G
β-glycosidase and precipitate	Chitosan powder	Adsorption followed by cross-linking with G
Invertase	Chitosan powder and solution	Adsorption on support
Lipase	Chitosan flakes and beads	Adsorption on support
Neutral proteinase	Chitosan precipitate	Adsorption followed by cross-linking with G
5-Nucleotidase cross-linking with G	Chitosan beads	Adsorption followed by
Papain	Chitosan beads	Adsorption followed by cross-linking with G
Pectinase	Chitosan beads	Adsorption on support
Pepsin	Chitosan beads	Adsorption followed by cross-linking with G
Phospholipase A2	Chitosan beads	Covalent binding with agents other than G
Proteases	Chitosan beads	Covalent binding with G
Urease	Chitosan-triphosphate beads	Covalent binding with G
β-xylolidase	Chitosan powder and beads	Adsorption on support, covalent binding with G

G: Glutaraldehyde

Table 24.3: Some of immobilised transferases onto chitosan.

Enzyme	Support	Type of immobilisation
Cyclodextrin glycosyltransferase	Chitosan powder	Adsorption on support
Limonoid glucosyltransferase	Chitosan powder	Covalent binding with G
Nucleoside phosphorylase	Chitosan beads	Covalent binding with G
Transglutaminase	Chitosan beads	Covalent binding with G
W-transaminase	Chitosan beads	Covalent binding with G

G: Glutaraldehyde

CHITOSAN NANOPARTICLE

In recent decades, nanostructured materials have attracted much attention because of their unique properties and interesting applications. Nanotechnology is the ability to work on a scale of about 1–100 nm in order to understand, create, characterise and use material structures, devices and systems with new properties derived from their nanostructures. Reducing the particle size of materials is an effective method for improving their properties. Nanoparticles have proportionally larger surface area and consequently more surface atoms than their microscale counterpart, which in turn affects their physico-chemical, optical, catalytic and other reactive properties.

In polymer composites conjugated with nanoparticles, a uniform dispersion of nanoparticles leads to a very large matrix/filler interfacial area, which changes the molecular mobility, the relaxation behaviour and the consequent thermal and mechanical properties of the materials. In contrast to other biopolymers, chitosan is a hydrophilic polymer with positive charge that comes from weak basic groups, which gives it special characteristics from the technological point of view. Recently, chitosan nanoparticles have attracted great attention in several fields due to their unique physico-chemical and biological properties.

CHITOSAN NANOPARTICLES PREPARATION METHODS

Several methods have been used to prepare chitosan nanoparticulate. The selection of any of these methods depends on shape and particle size requirements. Emulsion cross-linking, emulsion-droplet coalescence, coacervation/precipitation, ionotropic gelation, reverse micelles, template polymerisation and molecular self assembly are the main preparation methods of chitosan nanoparticles. In emulsion cross-linking method, the reactive functional amine group of chitosan crosses link with aldehyde groups of the cross-linking agent. In fact, chitosan solution is emulsified in oil phase (water-in-oil emulsion) and the aqueous droplets are stabilised using a suitable surfactant. The stable emulsion is then reacted with an appropriate cross-linking agent such as glutaraldehyde to stabilise the chitosan droplets. The nanoparticles are then washed and dried. Ionotropic gelation method is commonly used to prepare chitosan nanoparticles. This method is based on the electrostatic interactions between the chitosan amine group and a polyanion such as tripolyphosphate. In this method chitosan is dissolved in water or in weak acidic medium. This solution is then added drop wise under constant stirring to the solutions containing other counter ions.

Due to the complexation between oppositely charged species, chitosan undergo ionic gelation and precipitate to form spherical nanoparticles. Much researches have been focused on the preparation of chitosan nanoparticles using inotropic gelation method.

In coacervation/precipitation method, the chitosan solution is spraying into sodium hydroxide, NaOH methanol or ethanediamine alkaline solutions using compressed air, which in turn originates coacervated chitosan droplets in the form of nanoparticles. Emulsion-droplet coalescence method involves both emulsion cross-linking and precipitation. A stable emulsion containing the aqueous chitosan solution in oil and a second emulsion, containing a NaOH solution, is produced. By mixing the both emulsions under high speed stir, droplets of each emulsion collide at random, coalesce and finally precipitate as small size particles. In the reverse micelles method, a surfactant is dissolved in organic solvent to prepare reverse micelles. The aqueous phase containing the chitosan is added to this emulsion with constant vortexing and the nanoparticles forms in the core of the reverse micelles. In template polymerisation, chitosan is first dissolved in an acrylic monomer solution under magnetic stirring. Due to the electrostatic interaction, the negatively charged acrylic monomers align along the chitosan molecules. After complete dissolution of chitosan, the polymerisation is started by adding the initiator ($K_2S_2O_8$) under stirring at 70°C. The complete polymerisation leads to the appearance of an opalescent solution, indicating the nanoparticles formation. Molecular self assembly is based on cationic and hydrophobic properties of chitosan. This method is characterised by diffusion followed by specific association of molecules through non-covalent interactions, including electro-static and/or hydrophobic associations.

APPLICATIONS OF CHITOSAN NANOPARTICLES

Chitosan nanoparticles are natural materials with excellent physico-chemical, antimicrobial and biological properties, which make them a superior environmentally friendly material and they possess bioactivity that does not harm humans. Due to these unique properties, chitosan nanoparticles are being used in a vast array of widely different products and applications, ranging from pharmaceutical, tissue engineering and food packaging to biosensing, enzymes immobolisation and wastewater treatment.

The potential use of chitosan nanoparticles as carriers has led to the development of many different colloidal delivery vehicles. The main advantages of this kind of systems lie in their capacity to cross biological barriers, to protect macro-molecules, such as peptides, proteins, oligonucleotides and genes from degradation in biological media and to deliver drugs or macro-molecules to a target site with following controlled release. Chitosan nanoparticles have frequently used as a controlled release drug carrier for gene transfer in artificial organs and for immune prophylaxis. In addition, chitosan nanoparticles have been used to improve the strength and washability of textiles and to confer antibacterial effects. Several researches have been demonstrated that chitosan nanoparticles are able to improve drug bioavailability, modify pharmaco-kinetics and protect the encapsulated drugs.

In tissue engineering, chitosan nanoparticles improve transmucosal permeability enhancing transport through the paracellular pathway due to good bio and muco-adhesive properties of the nanoparticles and to an induced structural reorganisation of tight junction associated proteins.

During the past decade, there was an increasing interest to develop and use bio-based active films which are characterised by antimicrobial and antifungal activities in order to improve food preservation. Chitosan based films have attracted serious attention in food preservation and packaging technology. This is mainly due to the excellent film forming and gas barrier properties of chitosan and its high antimicrobial activity against pathogenic and spoilage micro-organisms, including fungi and bacteria.

Addition of chitosan nanoparticles into the coating and film formulations, improve film tensile properties and their permeability towards water vapour and simple gases (e.g. oxygen and carbon dioxide). Electrochemical biosensor has been considered as the best choice for the *in situ* monitoring of active compounds (e.g. phenolic) by virtue of its high sensitivity, simple instrumentation, low production cost and promising response speed. Excellent membrane forming ability of chitosan nanoparticles and their small response time and high sensitivity and stability (due to their high surface to volume ratio), low cost and hydrophilicity making them suitable for biosensor applications that are mostly concerned with working of enzymes for detection mechanisms.

Arsenic, molybdenum, lead and copper widely used in industries and are released through industrial wastewater. Sorption processes are found to be capable of adsorbing large number of metal ions from contaminated wastewater. In many studies such as adsorption of metals (e.g. copper and zinc), polymer supported nanoparticles has been prepared and used for selective removal of metal compounds and target metal contaminants. Chitosan due to its high hydrophilicity and presence of a large number of hydroxyl and amino groups with high activity as adsorption sites is found to be more efficient for the removal of metals such as uranium, copper, vanadium and molybdenum. Now-a-days, chitosan nanoparticles have been used effectively in wastewater treatment to remove toxic metal ions such as arsenic. Nano carriers can be effectively controlled by the application of nanotechnology. The catalytic efficiency and stability property of immobilised enzymes can be greatly improved. Furthermore, immobilisation rate of enzyme can be improved using chitosan nanoparticles.

CHITOSAN NANOPARTICLES AS ENZYME IMMOBILISATION SUPPORT

The results of immobilisation, including the performance of immobilised enzymes, strongly depend on the properties of supports, which are usually referred to as material types, compositions and structures. Chitosan is known as an ideal support material for enzyme immobilisation because of its many characteristics like improved resistance to chemical degradation and avoiding disturbance of metal ions to enzyme. Enzymes have been immobilised onto chitosan supports with different particle size. For practical purposes, carrier beads with size falling into millimeter range are mainly used.

Under the scale of nano, nanomaterials have characteristics, such as magnetism and large surface area to volume ratio. These characteristics are in favour of immobilisation of the enzymes. Recent studies indicated that the performance of enzyme immobilisation onto chitosan nanoparticles is higher than that of the biocatalyst immobilised onto chitosan microparticles. The result can be related to better distribution of the enzyme onto the support due to high specific surface area and numerous active functional groups available for fixing the enzyme molecules.

Activity of immobilised enzymes strongly affects by size and size distribution of the support particles. Specific surface available to bind the applied enzyme and the contacting area with the substrate during a reaction mainly depends on particle size. Therefore, a lot of efforts were made to minimise the particle size and maximise the specific surface area. In addition, the physical characteristics of nanoparticles such as enhanced diffusion and particle mobility can impact inherent catalytic activity of attached enzymes.

Biro and others indicated that the activity of β-galactosidase immobilised onto chitosan nanoparticles was higher than that of the biocatalyst immobilised onto chitosan micro and macroparticles. Research studies have also indicated that the stability of immobilised enzymes onto chitosan nanoparticles drastically affects by method of chitosan nanoparticles preparation and type of surfactant. Immobolisation of enzymes onto biopolymers nanoparticles has shown some benefits like improving their stability to pH and temperature, resistance to proteases and other denaturing compounds, as well as an adequate environment for their repeated use or controlled release. In the last decades, immobilised enzymes onto nanoparticles have been also considered for biosensors and food packaging applications. Applications of chitosan nanoparticles in immobilisation of some enzymes are summarised in Table 24.4.

Table 24.4: Some of immobilised enzymes onto chitosan nanoparticles.

Enzyme	*Support*	*Preparation method*
Alcohol dehydrogenase	Fe_3O_4–chitosan nanoparticles	Covalent binding with glutaraldehyde
β-d-galactosidase	Fe_3O_4–chitosan nanoparticles	Covalent binding with glutaraldehyde
β-galactosidase	Chitosan nanoparticles	Precipitation, emulsion cross-linking, ionic gelation
Glucoamylase	Fe_3O_4–chitosan nanoparticles	Ionic adsorption
Glucose oxidase	Chitosan nanoparticles	Covalent binding
Laccase	Chitosan nanoparticles	Reversed phase suspension
l-Laccase	Fe_3O_4–chitosan nanoparticles	Ionic adsorption and covalent binding
Lipase	Fe_3O_4–chitosan nanoparticles	Cross-linking with trypolyphosphate
Neutral proteinase	Chitosan nanoparticles	Ionic gelation
Pullulanase	Fe_3O_4–chitosan nanoparticles	Photochemistry in aqueous suspension
Trypsin	Linolenic acid-modified chitosan nanoparticles	Covalent binding with glutaraldehyde

Liu and others studied trypsin immobilised on linolenic acid-modified chitosan nanoparticles using glutaraldehyde as cross-linker. Their results indicated that the kinetic constant value (K_m) of trypsin immobilised on nano-particles (71.9 mg mL^{-1}) was higher than that of pure trypsin (50.2 mg mL^{-1}). Tang and others used Chitosan nanoparticles as support to immobilise and protect activity of neutral proteinase. Their results indicated that chitosan nanoparticles could improve 13.17% of neutral proteinase activity than that of free neutral proteinase. Nakorn used glucose oxidase immobilised onto chitosan nanoparticles in biosensor for glucose determination.

The coated electrode with immobilised glucose oxidase exhibits a rapid and sensitive current response for the changes of glucose concentration in the prepared solutions and indicates the excellent electrocatalytic behaviour of the electrode. Li and others immobilised *Saccharomyces Cerevisiae* Alcohol Dehydrogenase (SCAD) to magnetic Fe$_3$O$_4$-chitosan nanoparticles. For reduction of phenylglyoxylic acid by immobilised SCAD, the kinetic analysis data indicated that the immobilised SCAD retained 48.77% activity of its original activity. Furthermore, the immobilised SCAD enhanced thermal stability and good durability in the repeated use after recovered by magnetic separations. Fang and others prepared magnetic chitosan nanoparticles and then immobilised laccase onto them. The immobilised laccase exhibited an appreciable catalytic capability (480 units $\cdot$ g^{-1} support) and had good storage stability and operation stability. The K_m of immobilised and free laccase for ABTS [2,2′-azino-*bis* (3-ethyl benzthiazoline-6-sulphunate)] were 140.6 and 31.1 µM in phosphate buffer (0.1 M, pH 3.0) at 37°C, respectively. Pan and others developed a novel and efficient immobilisation of β-d-galactosidase by using magnetic Fe$_3$O$_4$-chitosan nanoparticles. The immobilised β-d-galactosidase showed the same or even higher activity in wider ranges of temperature and pH than that of its free form. In addition, the immobilised enzyme could be stored for a long time with little activity loss. Furthermore, the immobilised enzyme retained 92% of its initial activity after successively utilisation for 15 cycles. Wu and others prepared magnetic Fe$_3$O$_4$-chitosan nanoparticles and immobilised lipase onto them.

The immobilisation of lipase onto the nanoparticles showed good loading ability and little loss of enzyme activity and the stability of the catalyst was very good. In fact, it only lost 12% of enzyme activity after five batches. Zhang and others prepared magnetic chitosan beads (with mean particle size of 50 ± 3 nm) and immobilised pullulanase on them. Results indicated that the kinetic constant value (K_m) of immobilised pullulanase was three times higher than that of free pullulanase. However, the thermal and operational stabilities of immobilised pullulanase were improved greatly. Kalkan and others immobilised laccase onto chitosan-coated magnetic nanoparticles by adsorption and covalent binding after activating the hydroxyl groups of chitosan with carbodiimide or cyanuric chloride.

The results indicated that the immobilised enzyme retained more than 71% of its initial activity at the end of 30 batch uses. Wang and others immobilised glucoamylase onto Fe$_3$O$_4$-chitosan nanoparticles. The results from characterisation and determination remarkably indicated that the immobilised glucoamylase obtained presents excellent storage stability, pH stability, reusability, magnetic response and regeneration of supports.

CHALLENGES IN APPLICATIONS OF NANOPARTICLES TO IMMOBILISE ENZYMES

Application of nanoparticles in enzyme immobilisation has been associated with two main disadvantages. The first is lower storage stability of immobilised enzyme onto nanoparticles as compared to that of the immobilised enzyme onto microparticles. This can be explained by the fact that nanoparticles are aggregated during the storage. In fact, if the nanoparticles are not stabilised, they try minimising their surface energy by clustering together. Biro and others indicated that the stability of β-galactosidase immobilised onto chitosan nanoparticles was lower than that of the biocatalyst immobilised onto chitosan micro and macroparticles. The second disadvantage of using nanoparticles to immobilise enzyme is difficult separation of them from the reaction mixture at the end of the biochemical process, due to their small size. In this case, suitable methods are necessary to apply to facilitate the separation of the catalyst particles. Therefore, the general use of nanosized catalysts in industrial biotechnology needs considerable research efforts yet.

To overcome these issues, nanosized magnetic particles have received increasing attention because of their larger specific surface area for the enzymes immobilisation, their super paramagnetic nature for the reduction of self aggregation and easy separability from the reaction mixture by the application of a magnetic field. Magnetite (Fe_3O_4), silica, Zinc Oxide (ZnO), cellulose and chitosan are some of magnetic materials and their nanoparticles have been used to immobilise of various enzymes.

Chitosan can be used as a material for magnetic carriers, since it has a variety of functional groups which can be tailored to specific application. Additionally, glutaraldehyde cross-linking is shown to be the simple and efficient method to immobilise enzymes. Glutaraldehyde can react with several functional groups of proteins and supports, such as amino groups. Each of glutaraldehydes is expected to form Schiff bases upon nucleophilic attack by the primary amino groups in enzymes and chitosan and due to the linkage formed by the Schiff base reaction, the enzyme stability is improved. Several methods have been developed to synthesise magnetic chitosan nanoparticles, such as micro-emulsion polymerisation, emulsion polymerisation and *in situ* polymerisation.

As clearly observed in Table 24.4, magnetic Fe_3O_4–chitosan nanoparticles have been used as support to immobilise the most of enzymes. In fact, magnetic Fe_3O_4 nanoparticles tend to aggregate in liquid media due to the strong magnetic dipole–dipole attractions between particles. Thus, chitosan with specific functional groups have been used as stabiliser to modify and increase their stability. In addition to, coating magnetic Fe_3O_4 nanoparticles with chitosan can protect the nanoparticles against corrosion an also offers flexibility, favourable functional groups and features (e.g. their ability to fold into a globular state or structure) for various applications.

Magnetic nano and micron sized particles with specific modifications are widely used in biomedical applications such as diagnostics, magnetic separation and purification of biomolecules. In fact, magnetic nanoparticles coated with biopolymers and immobilised with ligand were found to have promising characteristics for the application of these adsorbents in bioseparation processes, particularly in antibody, nucleic acids and enzymes purification.

SECTION VII

Recovery and Purification of Fermentation Products

Chapter 25

Downstream Processing: A Review

INTRODUCTION

Downstream processing refers to the recovery and purification of biosynthetic products, particularly pharmaceuticals, from natural sources such as animal or plant tissue or fermentation broth, including the recycling of salvageable components and the proper treatment and disposal of waste. It is an essential step in the manufacture of pharmaceuticals such as antibiotics, hormones (e.g. insulin and human growth hormone), antibodies (e.g. infliximab and abciximab) and vaccines, antibodies and enzymes used in diagnostics, industrial enzymes, and natural fragrance and flavour compounds. Downstream processing is usually considered a specialised field in biochemical engineering, itself a specialisation within chemical engineering, though many of the key technologies were developed by chemists and biologists for laboratory-scale separation of biological products.

Downstream processing and analytical bioseparation both refer to the separation or purification of biological products, but at different scales of operation and for different purposes. Downstream processing implies manufacture of a purified product fit for a specific use, generally in marketable quantities, while analytical bioseparation refers to purification for the sole purpose of measuring a component or components of a mixture, and may deal with sample sizes as small as a single cell.

STAGES IN DOWNSTREAM PROCESSING

A widely recognised heuristic for categorising downstream processing operations divides them into four groups which are applied in order to bring a product from its natural state as a component of a tissue, cell or fermentation broth through progressive improvements in purity and concentration.

Removal of insolubles: Removal of insolubles is the first step and involves the capture of the product as a solute in a particulate-free liquid, for example the separation of cells, cell debris or other particulate matter from fermentation broth containing an antibiotic. Typical operations to achieve this are filtration, centrifugation, sedimentation, flocculation, electro-precipitation, and gravity settling. Additional operations such as grinding, homogenisation, or leaching, required to recover products from solid sources such as plant and animal tissues, are usually included in this group.

Product isolation: Product isolation is the removal of those components whose properties vary markedly from that of the desired product. For most products, water is the chief impurity and isolation steps are designed to remove most of it, reducing the volume of material to be handled and concentrating the product. Solvent extraction, adsorption, ultrafiltration, and precipitation are some of the unit operations involved.

Product purification: Product purification is done to separate those contaminants that resemble the product very closely in physical and chemical properties. Consequently steps in this stage are expensive to carry out and require sensitive and sophisticated equipment. This stage contributes a significant fraction of the entire downstream processing expenditure. Examples of operations include affinity, size exclusion, reversed phase chromatography, crystallisation and fractional precipitation.

Product polishing: Product polishing describes the final processing steps which end with packaging of the product in a form that is stable, easily transportable and convenient. Crystallisation, desiccation, lyophilisation and spray drying are typical unit operations. Depending on the product and its intended use, polishing may also include operations to sterilise the product and remove or deactivate trace contaminants which might compromise product safety. Such operations might include the removal of viruses or depyrogenation. A few product recovery methods may be considered to combine two or more stages. For example, expanded bed adsorption accomplishes removal of insolubles and product isolation in a single step. Affinity chromatography often isolates and purifies in a single step.

REMOVAL OF INSOLUBLE COMPONENTS

Filtration

Filtration is one of the most common processes used at all scales of operation to separate suspended particles from a liquid or gas, using a porous medium which retains the particles but allows the liquid or gas to pass through. It is possible to carry out filtration under a variety of conditions, but a number of factors will obviously influence the choice of the most suitable type of equipment to meet the specified requirements at minimum overall cost, including:

1. The properties of the filtrate, particularly its viscosity and density.
2. The nature of the solid particles, particularly their size and shape, the size distribution and packing characteristics.
3. The solids: liquid ratio.
4. The need for recovery of the solid or liquid fraction or both.
5. The scale of operation.
6. The need for batch or continuous operation.
7. The need for aseptic conditions.
8. The need for pressure or vacuum suction to ensure an adequate flow rate of the liquid.

A simple filtration apparatus is illustrated in Fig. 25.1, which consists of a support covered with a porous filter cloth. A filter cake gradually builds up as filtrate passes through the filter cloth. As the filter cake increases in thickness, the resistance to flow will gradually increase. Thus, if the pressure applied to the surface of the slurry is kept constant, the rate of flow will gradually diminish. Alternatively, if the flow rate is to be kept constant the pressure will gradually have to be increased. The flow rate may also be reduced by blocking of holes in the filter cloth and closure of voids between particles, if the

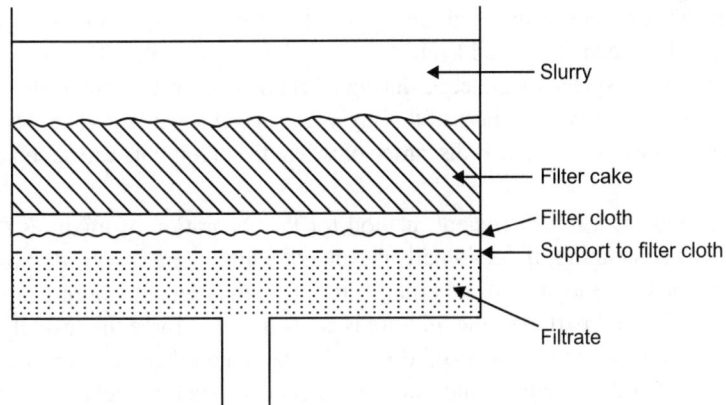

Fig. 25.1: Diagram of a simple filtration apparatus.

particles are soft and compressible. When particles are compressible it may not be feasible to apply increased pressure.

Liquid filtration

There are many different methods of filtration, all aim to attain the separation of substances. Separation is achieved by some form of interaction between the substance or objects to be removed and the filter. The substance that is to pass through the filter must be a fluid, i.e. a liquid or gas. Methods vary depending on the location of the targeted material, i.e. whether it is in the fluid phase or not.

Filter media: Two main types of filter media are employed in the chemical laboratory—surface filter, a solid sieve which traps the solid particles, with or without the aid of filter paper (e.g. Büchner funnel, Belt filter, rotary vacuum-drum filter, crossflow filters), and a depth filter, a bed of granular material which retains the solid particles as it passes (e.g. sand filter). The first type allows the solid particles, i.e. the residue, to be collected intact; the second type does not permit this. However, the second type is less prone to clogging due to the greater surface area where the particles can be trapped. Also, when the solid particles are very fine, it is often cheaper and easier to discard the contaminated granules than to clean the solid sieve. Filter media can be cleaned by rinsing with solvents or detergents. Alternatively, in engineering applications, such as swimming pool water treatment plants, they may be cleaned by backwashing.

Achieving flow through the filter: Fluids flow through a filter due to a difference in pressure-fluid flows from the high pressure side to the low pressure side of the filter, leaving some material behind. The simplest method to achieve this is by gravity and can be seen in the coffee maker. In the laboratory, pressure in the form of compressed air on the feed side (or vacuum on the filtrate side) may be applied to make the filtration process faster, though this may lead to clogging or the passage of fine particles. Alternatively, the liquid may flow through the filter by the force exerted by a pump, a method commonly used in industry when a reduced filtration time is important. In this case, the filter need not be mounted vertically.

Filter aid: Certain filter aids may be used to aid filtration. These are often incompressible diatomaceous earth or kieselguhr, which is composed primarily of silica. Also used are wood cellulose and other inert porous solids. These filter aids can be used in two different ways. They can be used as a precoat before the slurry is filtered. This will prevent gelatinous-type solids from plugging the filter medium and also give a clearer filtrate. They can also be added to the slurry before filtration. This increases the porosity of the cake and reduces resistance of the cake during filtration. In a rotary filter, the filter aid may be applied as a precoat; subsequently, thin slices of this layer are sliced off with the cake. The use of filter aids is usually limited to cases where the cake is discarded or where the precipitate can be separated chemically from the filter.

Alternatives: Filtration is a more efficient method for the separation of mixtures than decantation, but is much more time consuming. If very small amounts of solution are involved, most of the solution may be soaked up by the filter medium. An alternative to filtration is centrifugation—instead of filtering the mixture of solid and liquid particles, the mixture is centrifuged to force the (usually) denser solid to the bottom, where it often forms a firm cake. The liquid above can then be decanted. This method is especially useful for separating solids which do not filter well, such as gelatinous or fine particles. These solids can clog or pass through the filter, respectively.

Batch Filters

Plate and frame filters

A plate and frame filter is a pressure filter in which the simplest form consists of plates and frames arranged alternately. The plates are covered with filter cloths or filter pads. The plates and frames are assembled on a horizontal framework and held together by means of a hand screw of hydraulic ram so that there is no leakage between the plates and frames which form a series of liquid-tight compartments. The slurry is fed to the filter frame through the continuous channel formed by the holes in the corners of the plates and frames. The filtrate passes through the filter cloth or pad, runs down grooves in the filter plates and is then discharged through outlet taps to a channel. Sometimes, if aseptic conditions are required, the outlets may lead directly into a pipe. The solids are retained within the frame and filtration is stopped when the frames are completely filled or when the flow of filtrate becomes uneconomically low. On an industrial scale the plate and frame filter is one of the cheapest filters per unit of filtering space and requires the least floor space, but it is intermittent in operation (a batch process) and there may be considerable wear of filter cloths as a result of frequent dismantling. This type of filter is most suitable for fermentation broths with a low solids content and low resistance to filtration. It is widely used as a 'polishing' device in breweries to filter out residual yeast cells following initial clarification by centrifugation or rotary vacuum filtration. It may also be used for collecting high value solids that would not justify the use of a continuous filter. Because of high labour costs and the time involved in dismantling, cleaning and re-assembly, these filters should not be used when removing large quantities of worthless solids from a broth.

Pressure leaf filters

There are a number of intermittent batch filters usually called by their trade names. These filters incorporate a number of leaves, each consisting of a metal framework of grooved plates which is covered with a fine wire mesh or occasionally a filter cloth and often pre-coated with a layer of cellulose fibres. The process slurry is fed into the filter which is operated under pressure or by suction with a vacuum pump. Because the filters are totally enclosed it is possible to sterilise them with steam. This type of filter is particularly suitable for 'polishing' large volumes of liquids with low solids content or small batch filtrations of valuable solids.

Vertical metal-leaf filter

This filter consists of a number of vertical porous metal leaves mounted on a hollow shaft in a cylindrical pressure vessel. The solids from the slurry gradually build up on the surface of the leaves and the filtrate is removed from the plates via the horizontal hollow shaft. In some designs the hollow shaft can be slowly rotated during filtration. Solids are normally removed at the end of a cycle by blowing air through the shaft and into the filter leaves.

Horizontal metal-leaf filter

In this filter the metal leaves are mounted on a vertical hollow shaft within a pressure vessel. Often, only the upper surfaces of the leaves are porous. Filtration is continued until the cake fills the space between the disc-shaped leaves or when the operational pressure has become excessive. At the end of a process cycle, the solid cake can be discharged by releasing the pressure and spinning the shaft with a drive motor.

Stacked-disc filter

One kind of filter of this type is the metafilter. This is a very robust device and because there is no filter cloth and the bed is easily replaced, labour costs are low. It consists of a number of precision-made rings which are stacked on a fluted rod. This ensures that there will be clearances of 0.025 mm to 0.25 mm when the rings are assembled on the rods. The assembled stacks are placed in a pressure vessel which can be sterilised if necessary. The packs are normally coated with a thin layer of Kieselguhr which is used as a filter aid. During use, the filtrate passes between the discs and is removed through the grooves of the fluted rods, while solids are deposited on the filter coating. Operation is continued until the resistance becomes too high and the solids are removed from the rings by applying back pressure, via., the fluted rods. Metafilters are primarily used for 'polishing' liquids such as beer.

Continuous Filters

Rotary vacuum filters

Large rotary vacuum filters are commonly used by industries which produce large volumes of liquid which need continuous processing. The filter consists of a rotating, hollow, segmented drum covered with a fabric or metal filter which is partially immersed in a trough containing the broth to be filtered. The slurry is fed on to the outside of the revolving drum and vacuum pressure is applied internally so that the filtrate is drawn through the filter, into the drum and finally to a collecting vessel. The interior of the drum is divided into a series of compartments, to which the vacuum pressure is normally applied for most of each revolution as the drum slowly revolves (~1 rpm). However, just before discharge of the filter cake, air pressure may be applied internally to help ease the filter cake off the drum. A number of spray jets may be carefully positioned so that water can be applied to rinse the cake. This washing is carefully controlled so that dilution of the filtrate is minimal.

Cross-flow filtration (tangential filtration)

In the filtration processes previously described, the flow of broth was perpendicular to the filtration membrane. Consequently, blockage of the membrane led to lower rates of productivity and/or the need for filter aids to be added and these were serious disadvantages. In contrast, an alternative which is rapidly gaining prominence both in the processing of whole fermentation broths and cell lysates is cross-flow filtration. Here, the flow of medium to be filtered is tangential to the membrane (Fig. 25.2a) and no filter cake builds up on the membrane. The benefits of cross-flow filtration are:

1. Efficient separation, > 99.9 per cent cell retention.
2. Closed system; for the containment of organisms with no aerosol formation.
3. Separation is independent of cell and media densities, in contrast to centrifugation.
4. No addition of filter aid.

The major components of a cross-flow filtration system are a media storage tank (or the fermenter), a pump and a membrane pack (Fig. 25.2b). The membrane is usually in a cassette pack of hollow fibres or flat sheets in a plate and frame type stack or a spiral cartridge. In this way and by the introduction of a much convoluted surface, large filtration areas can be attained in compact devices. Two types of membrane may be used, microporous membranes with a specific pore size (0.45, 0.22 µm, etc.), or an ultrafiltration membrane with a specified molecular weight cut-off (MWCO). The type of membrane chosen is carefully matched to the product being harvested, with microporous and 1,00,000 MWCO membranes being used in cell separation.

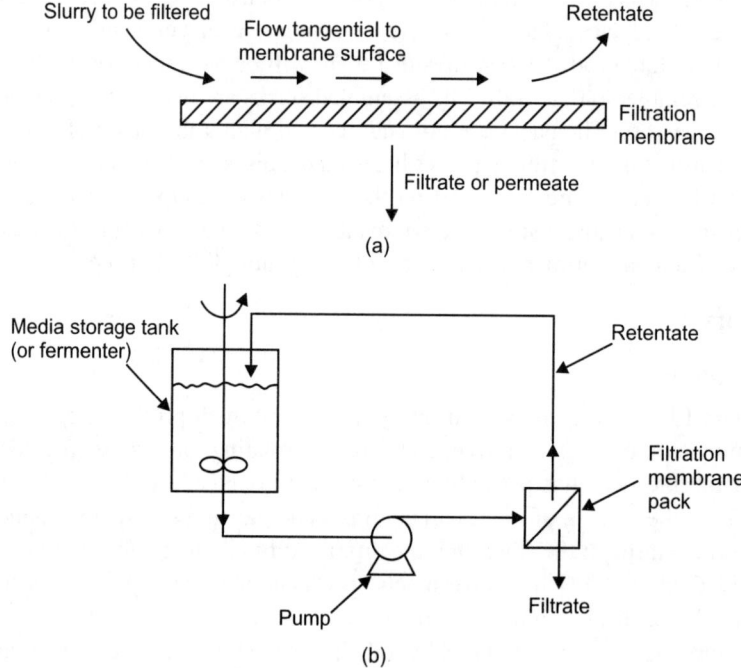

Fig. 25.2: (a) Schematic diagram of cross-flow filtration, and (b) major components of a cross-flow filtration system.

The output from the pump is forced across the membrane surface; most of this flow sweeps the membrane, returning retained species back to the storage tank and generally less than 10 per cent of the flow passes through the membrane (permeate). As this process is continued the cells or other retained species are concentrated to between 5 and 10 per cent of their initial volume. More complex variants of the process can allow *in situ* washing of the retentate and enclosed systems for containment and sterilisation.

Many factors influence filtration rate. Increased pressure drop will, up to a point increase flow across the membrane, but it should be remembered that the system is based on a swept clean membrane. Therefore, if the pressure drop is too great the membrane may become blocked. The filtration rate is therefore, influenced by the rate of tangential flow across the membrane; by increasing the shear forces at the membrane's surface retained species are more effectively removed, thereby increasing filtration rate. Higher temperatures will increase filtration rate by lowering the viscosity of the media, though this is clearly of limited application in biological systems. Filtration rate is inversely proportional to concentration and media constituents can influence filtration rate in three ways. Low molecular weight compounds increase media viscosity and high molecular weight compounds decrease shear at the membrane surface, both leading to a reduction in filtration rate. Finally, broth constituents can 'foul' the membrane, primarily by adsorption onto the membrane's surface, causing a rapid loss in efficiency. This can be controlled by modification of the membrane or media formulation in particular by reducing the use of antifoaming agents. Lee has shown that the pulses of air injected into the flow to a cross-flow filter increase the shear rate at the membrane surface reducing the effects of membrane fouling.

Centrifugation

Centrifugation is a process that involves the use of the centrifugal force for the separation of mixtures, used in industry and in laboratory settings. More-dense components of the mixture migrate away from the axis of the centrifuge, while less-dense components of the mixture migrate towards the axis. Chemists and biologists may increase the effective gravitational force on a test tube so as to more rapidly and completely cause the precipitate ('pellet') to gather on the bottom of the tube. The remaining solution is properly called the 'supernate' or 'supernatant liquid'. The supernatant liquid is then either quickly decanted from the tube without disturbing the precipitate, or withdrawn with a Pasteur pipette.

The rate of centrifugation is specified by the acceleration applied to the sample, typically measured in revolutions per minute (rpm). The particles settling velocity in centrifugation is a function of their size and shape, centrifugal acceleration, the volume fraction of solids present, the density difference between the particle and the liquid, and the viscosity. In the chemical and food industries, special centrifuges can process a continuous stream of particle-laden liquid. It is worth noting that centrifugation is the most common method used for uranium enrichment, relying on the slight mass difference between atoms of U238 and U235 in uranium hexafluoride gas.

Centrifugation in biotechnology

Microcentrifuges and superspeed centrifuges: In microcentrifugation, centrifuges are run in batch to isolate small volumes of biological molecules or cells (prokaryotic and eukaryotic). Nuclei is also often purified via microcentrifugation. Microcentrifuge tubes generally hold 1.5–2 ml of liquid, and are spun at maximum angular speeds of 12,000–13,000 rpms. Microcentrifuges are small and have rotors that can quickly change speeds. Superspeed centrifuges work similarly to microcentrifuges, but are conducted via larger scale processes. Superspeed centrifuges are also used for purifying cells and nuclei, but in larger quantities. These centrifuges are used to purify 25–30 ml of solution within a tube. Additionally, larger centrifuges also reach higher angular velocities (around 30,000 rpm) and also use a larger rotor.

Ultracentrifugation: Ultracentrifugation makes use of high centrifugal force for studying properties of biological particles. While microcentrifugation and superspeed centrifugation are used strictly to purify cells and nuclei, ultracentrifugation can isolate much smaller particles, including ribosomes, proteins, and viruses. Ultracentrifuges can also be used in the study of membrane fractionation.

This occurs because ultracentrifuges can reach maximum angular velocites in excess of 70,000 rpm. Additionally, while microcentrifuges and supercentrifuges separate particles in batch, ultracentrifuges can separate molecules in batch and continuous flow systems. In addition to purification, analytical ultracentrifugation (AUC) can be used for determination of macromolecular properties, including the amino acid composition of a protein, the protein's current conformation, or properties of that conformation. In analytical ultracentrifuges, concentration of solute is measured using optical calibrations. For low concentrations, the Beer-Lambert law can be used to measure the concentration. Analytical ultracentrifuges can be used to simulate physiological conditions (correct pH and temperature).

In analytical ultracentrifuges, molecular properties can be modelled through sedimentation velocity analysis or sedimentation equilibrium analysis. In sedimentation velocity analysis, concentrations and solute properties are modelled continuously over time. Sedimentation velocity analysis can be used to determine the macromolecule's shape, mass, composition, and conformational properties. During sedimentation equilibrium analysis, centrifugation has stopped and particle movement is based on diffusion. This allows for modelling of the mass of the particle as well as the chemical equilibrium properties of interacting solutes.

Sedimentation

Sedimentation is the tendency for particles in suspension or molecules in solution to settle out of the fluid in which they are entrained, and come to rest against a wall. This is due to their motion through the fluid in response to the forces acting on them: these forces can be due to gravity, centrifugal acceleration or electromagnetism. Sedimentation may pertain to objects of various sizes, ranging from large rocks in flowing water to suspensions of dust and pollen particles to cellular suspensions to solutions of single molecules such as proteins and peptides. Even small molecules such as aspirin can be sedimented, although it can be difficult to apply a sufficiently strong force to produce significant sedimentation.

The term is typically used in geology, to describe the deposition of sediment which results in the formation of sedimentary rock, and in various chemical and environmental fields to describe the motions of often-smaller particles and molecules.

Flocculation

Flocculation is, in the field of chemistry, a process where colloids come out of suspension in the form of floc or flakes. The action differs from precipitation in that, prior to flocculation, colloids are merely suspended in a liquid and not actually dissolved in a solution. According to the IUPAC definition, flocculation is 'a process of contact and adhesion whereby the particles of a dispersion form larger-size clusters.' Flocculation is synonymous with agglomeration and coagulation.

For emulsions, flocculation describes clustering of individual dispersed droplets together, whereby the individual droplets do not lose their identity. Flocculation is thus the initial step leading to further ageing of the emulsion (droplet coalescence and the ultimate separation of the phases).

In biology, the process is used to refer to the asexual aggregation of micro-organisms, most commonly brewing yeast at the end of a brew. Flocculation and sedimentation are widely employed in the purification of drinking water as well as sewage treatment, stormwater treatment and treatment of other industrial waste-water streams. Flocculants, or flocculating agents, are chemicals that promote flocculation by causing colloids and other suspended particles in liquids to aggregate, forming a floc. Flocculants are used in water treatment processes to improve the sedimentation or filterability of small particles. For example, a flocculant may be used in swimming pool or drinking water filtration to aid removal of microscopic particles which would otherwise cause the water to be turbid (cloudy) and which would be difficult or impossible to remove by filtration alone.

Cell Aggregation and Flocculation

Following an industrial fermentation it is quite common to add flocculating agents to the broth to aid de-watering. The use of flocculating agents is widely practised in the effluent-treatment industries for the removal of microbial cells and suspended colloidal matter. It is well known that aggregates of microbial cells, although they have the same density as the individual cells, will sediment faster because of the increased diameter of the particles (Stoke's law). The sedimentation process may be achieved naturally with selected strains of brewing yeasts, particularly if the wort is chilled at the end of fermentation and leads to a natural clearing of the beer.

Micro-organisms in solution are usually held as discrete units in three ways. Firstly, their surfaces are negatively charged and therefore repulse each other. Secondly, because of their generally hydrophilic cell walls a shell of bound water is associated with the cell which acts as a thermodynamic barrier to aggregation. Finally, due to the irregular shapes of cell walls (at the macromolecular level) steric hindrance will also play a part.

During flocculation one or more mechanisms besides temperature can induce cell flocculation:

1. Neutralisation of anionic charges, primarily carboxyl and phosphate groups, on the surfaces of the microbial cells, thus allowing the cells to aggregate. These include changes in the pH and the presence of a range of compounds which alter the ionic environment.
2. Reduction in surface hydrophilicity.
3. The high molecular weight polymers, e.g. anionic, non-ionic and cationic polymers can be used, though the former two also require the addition of a multivalent cation.

Flocculation usually involves the mixing of a process fluid with the flocculating agent under conditions of high shear in a stirred tank, although more compact and efficient devices have been proposed. This stage is known as coagulation and is usually followed by a period of gentle agitation when flocs developed initially are allowed to grow in size.

Nakamura described the use of various compounds for flocculating bacteria, yeasts and algae, including alum, calcium salts and ferric salts. Other agents which are now used include tannic acid, titanium tetrachloride and cationic agents such as quaternary ammonium compounds, alkyl amines and alkyl pyridinium salts. Gasner and Wang reported a many hundred-fold increase in the sedimentation rate of *Candida intermedia* when recoveries of over 99 per cent were readily obtained. They found that flocculation was very dependent on the choice of additive, dosage and conditions of floc formation, with the most effective agents being mineral colloids and polyelectrolytes. Nucleic acids, polysaccharides and proteins released from partly lysed cells may also bring about agglomeration. In SCP processes, phosphoric acid has been used as a flocculating agent since it can be used as a nutrient in medium recycle with considerable savings in water usage.

The majority of flocculating agents currently in use are polyelectrolytes, which act by charge neutralisation and hydrophobic interactions to link cells to each other. In processes where the addition of some toxic chemicals is to be avoided, alternative techniques have been adopted. One method is to coagulate microbial protein which has been released from the cells by heating for short periods. Kurane reported the use of bioflocculants obtained from *Rhodococcus erythropolis*. They are suggested as being safer alternatives to conventional flocculants. Warne and Bowden suggest the use of genetic manipulation to alter cell surface properties to aid aggregation. Flocculating agents such as cross-linked cationic polymers may also be used in the processing of cell lysates and extracts prior to further downstream processing. Bentham utilised borax as a flocculating agent for yeast cell debris prior to decanter centrifugation.

Range of Centrifuges

A number of centrifuges will be described which vary in their manner of liquid and solid discharge, their unloading speed and their relative maximum capacities. When choosing a centrifuge for a specific process it is important to ensure that the centrifuge will be able to perform the separation at the planned production rate and operate reliably with minimum manpower. Large-scale tests may therefore be necessary with fermentation broths or other materials to check that the correct centrifuge is chosen.

Basket centrifuge (perforated-bowl basket centrifuge)

Basket centrifuges are useful for separating mould mycelia or crystalline compounds.

Tubular-bowl centrifuge

This is a centrifuge to consider using for particle size ranges of 0.1 to 200 μm and up to 10 per cent solids in the in-going slurry. The centrifuge may be altered to use for:

1. Light-phase/heavy-phase liquid separation.
2. Solids/light-liquid phase/heavy-liquid phase separation.
3. Solids/liquid separation (using a different rotor).

Advantages of this design of centrifuge are: (i) high centrifugal force, (ii) good dewatering, and(iii) ease of cleaning.

Disadvantages are limited solids capacity, difficulties in the recovery of collected solids, gradual loss in efficiency as the bowl fills, solids being dislodged from the walls as the bowl is slowing down and foaming. Plastic liners can be used in the bowls to help improve batch cycle time. Alternatively a spare bowl can be changed over in about 5 minutes.

Solid-bowl scroll centrifuge (secanter centrifuge)

This type of centrifuge is used for continuous handling of fermentation broths, cell lysates and coarse materials such as sewage sludge. The number of variants related to basic design are given below:

1. Cake washing facilities (screen bowl decanters).
2. Vertical bowl decanters.
3. Facility for in-place cleaning.
4. Bio-hazard containment features, steam sterilisation *in situ*, two or three stage mechanical seals, control of aerosols, containment casings and the use of high pressure sterile gas in seals to prevent the release of micro-organisms.

Multichamber centrifuge

Ideally, this is a centrifuge for a slurry of up to 5 per cent solids of particle size 0.1 to 200 μm diameter. In the multichamber centrifuge (Fig. 25.3), a series of concentric chambers are mounted within the rotor chamber. The broth enters via the central spindle and then takes a circuitous route through the chambers. Solids collect on the outer faces of each chamber. The smaller particles collect in the outer chambers

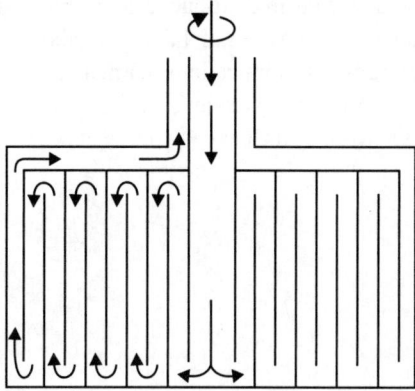

Fig. 25.3: LS of a multichamber centrifuge.

where they are subjected to greater centrifugal forces (the greater the radial position of a particle, the greater the rate of sedimentation). Because of the time needed to dismantle and recover the solids fraction, the size and number of vessels must be of the correct volume for the solids of a batch run.

Disc-bowl centrifuge

This centrifuge relies for its efficiency on the presence of discs in the rotor or bowl (Fig. 25.4). A central inlet pipe is surrounded by a stack of stainless-steel conical discs. Each disc has spacers so that a stack can be built up.

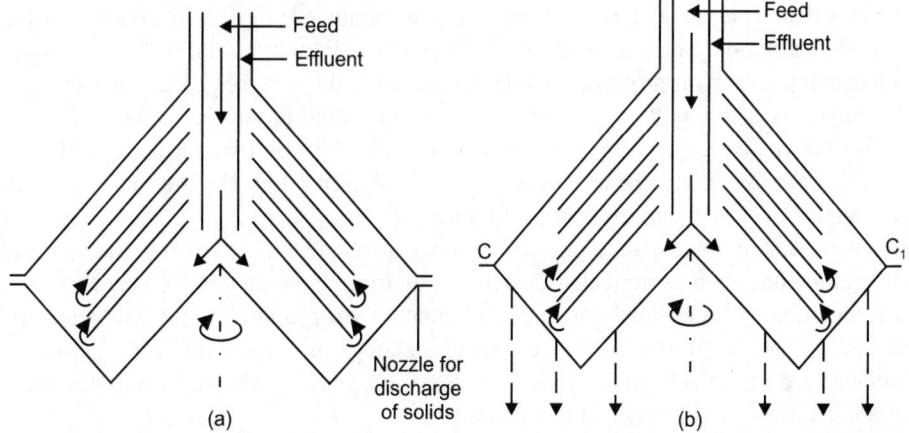

Fig. 25.4: (a) LS of disc-bowl centrifuge with nozzle discharge, and (b) LS of disc-bowl centrifuge with intermittent discharge. (Solids discharged when rotor opens intermittently along the section C–C$_1$).

CELL DISRUPTION

Micro-organisms are protected by extremely tough cell walls. In order to release their cellular contents a number of methods for cell disintegration have been developed. Any potential method of disruption must ensure that labile materials are not denatured by the process or hydrolysed by enzymes present in the cell. Huang report the use of a combination of different techniques to release products from specific locations within yeast cells. In this way the desired product can be obtained with minimum contamination. Although many techniques are available which are satisfactory at laboratory scale, only a limited number have been proved to be suitable for large-scale applications, particularly for intracellular enzyme extraction. Containment of cells can be difficult or costly to achieve in many of the methods described below and thus containment requirements will strongly influence process choice. Methods available fall into two major categories:

Physico-mechanical methods:
1. Liquid shear.
2. Solid shear.
3. Agitation with abrasives.
4. Freeze-thawing.
5. Ultrasonication.

Chemical methods:
1. Detergents.
2. Osmotic shock.
3. Alkali treatment.
4. Enzyme treatment.

Product Isolation

Adsorption

Adsorption is the accumulation of atoms or molecules on the surface of a material. This process creates a film of the adsorbate (the molecules or atoms being accumulated) on the adsorbent's surface. It is different from absorption, in which a substance diffuses into a liquid or solid to form a solution. The term sorption encompasses both processes, while desorption is the reverse process of 'adsorption'.

In simple terms, adsorption is 'the collection of a substance onto the surface of adsorbent solids.' It is a removal process where certain particles are bound to an adsorbent particle surface by either chemical or physical attraction. Adsorption is often confused with absorption, where the substance being collected or removed actually penetrates into the other substance.

Adsorption is present in many natural physical, biological, and chemical systems, and is widely used in industrial applications such as activated charcoal, capturing and using waste heat to provide cold water for air conditioning and other process requirements (adsorption chillers),synthetic resins, and water purification. Adsorption, ion exchange, and chromatography are sorption processes in which certain adsorbates are selectively transferred from the fluid phase to the surface of insoluble, rigid particles suspended in a vessel or packed in a column.

Similar to surface tension, adsorption is a consequence of surface energy. In a bulk material, all the bonding requirements (be they ionic, covalent, or metallic) of the constituent atoms of the material are filled by other atoms in the material. However, atoms on the surface of the adsorbent are not wholly surrounded by other adsorbent atoms and therefore can attract adsorbates. The exact nature of the bonding depends on the details of the species involved, but the adsorption process is generally classified as physisorption (characteristic of weak van der Waals forces) or chemisorption (characteristic of covalent bonding).

Precipitation

Precipitation is the formation of a solid in a solution during a chemical reaction. When the reaction occurs, the solid formed is called the precipitate, and the liquid remaining above the solid is called the supernate. Powders derived from precipitation have also historically been known as flowers.

Natural methods of precipitation include settling or sedimentation, where a solid forms over a period of time due to ambient forces like gravity or centrifugation. During chemical reactions, precipitation may also occur particularly if an insoluble substance is introduced into a solution and the density happens to be greater (otherwise the precipitate would float or form a suspension). With soluble substances, precipitation is accelerated once the solution becomes supersaturated.

An important stage of the precipitation process is the onset of nucleation. The creation of a hypothetical solid particle includes the formation of an interface, which requires some energy based on the relative surface energy of the solid and the solution. If this energy is not available, and no suitable nucleation surface is available, supersaturation occurs.

Chromatography

Chromatography is the collective term for a set of laboratory techniques for the separation of mixtures. It involves passing a mixture dissolved in a 'mobile phase' through a stationary phase, which separates the analyte to be measured from other molecules in the mixture based on differential partitioning between the mobile and stationary phases. Subtle differences in compounds partition coefficient results in differential retention on the stationary phase and thus separation.

Chromatography may be preparative or analytical. The purpose of preparative chromatography is to separate the components of a mixture for further use (and is thus a form of purification). Analytical chromatography is done normally with smaller amounts of material and is for measuring the relative proportions of analytes in a mixture. The two are not mutually exclusive.

Chromatography became developed substantially as a result of the work of Archer John Porter Martin and Richard Laurence Millington Synge during the 1940s and 1950s. They established the principles and basic techniques of partition chromatography, and their work encouraged the rapid development of several types of chromatography method: paper chromatography, gas chromatography, and what would become known as high performance liquid chromatography. Since then, the technology has advanced rapidly. Researchers found that the main principles of Tsvet's chromatography could be applied in many different ways, resulting in the different varieties of chromatography.

Chapter 26

Solid-Liquid Separation

INTRODUCTION

Most of the biological processes in which particulate slurries are handled use some form of solid–liquid separation within their flowsheets. As the name suggests, solid–liquid separation is the separation of two phases, solid and liquid, from a mixture. The technology for carrying out this process is often referred to as 'mechanical separation' because the separation is accomplished by purely physical means. This does not preclude chemical or thermal pretreatment which is increasingly used to enhance the separation that follows. Although some biological slurries separate perfectly well without chemical conditioning, most such suspensions of a widely varying nature can benefit from pretreatment, whether the separation is by sedimentation, filtration or flotation. The problem with biological suspensions is that they often contain very fine, colloidal particles and the particles are always of low density, thus making separation difficult and pretreatment often necessary.

Like other unit operations in biotechnology such as mass or heat transfer, solid–liquid separation is never complete. There may be some solids leaving with the liquid in the overflow stream and there will certainly be some liquid entrained with the solids leaving through the underflow. One of the important concerns in this subject is therefore the question how complete is the separation achieved with a given piece of equipment under a given set of operating conditions. For doing this, there has to be a common, well-defined way of assessing the efficiency of separation.

EFFICIENCY OF SEPARATION

The completeness of separation of the two phases, solid and liquid, may be measured using various criteria which consider both phases equally, in one parameter. In most practical applications, however, the emphasis between the two phases is not equal and it is, therefore, best to assess the separation in two separate criteria: one for the separation of the solids and another for the separation of the liquid.

Starting with the liquid, we can simply define recovery of the liquid to the overflow, but this is only used in mineral processing. In the chemical and process industries, it is more common to measure the separation of the liquid by how much liquid is misplaced to the wrong stream, i.e. how much leaves with the solids in the underflow. Depending on the type of the separator, we measure this either as the concentration of the solids in the underflow (in those processes, where solids leave as a slurry) or as the moisture content of the filtration cake, where the solids leave in a cake. For the solids the situation is more complicated as it is the dispersed phase and many solid–liquid separation processes are strongly size-dependent. We do, of course, define total mass recovery (also called total efficiency) of the solids (to underflow) but this, although essential for mass balance calculations in a process flowsheet, is not useful as a general criterion of separation efficiency because it is strongly affected by the particle size distribution of the solids in the feed.

Grade efficiency $G(x)$ is defined in a similar way as the total efficiency or total recovery except that its value corresponds to one particle size only. As the value of the grade efficiency has the character of a probability, it is sometimes referred to as the partition probability. The concept of grade efficiency is widely used with the dynamic-type separators like the sedimenting centrifuges, hydrocyclones or some settling tanks, because their performance does not change with time under, steady-state conditions. The concept is also useful in filtration, where the performance changes with the amount of solids separated on or within the medium, for the purpose of media selection and rating.

The grade efficiency curve, while it provides the most comprehensive description of separation efficiency, is rather clumsy for correlation with operating variables or for simple equipment comparisons. Such applications call for a single number, independent of the size of the feed solids, as a measure of efficiency. This is available in the form of the 'cut size', which is the particle size corresponding to 50 per cent on the grade efficiency curve. It is, therefore, the size of a particle which has a 50/50 chance of being separated or remaining in the flow.

Processes Related to Solid–Liquid Separation

Various processes related to solid–liquid separation are discussed below.

Washing of solids

Washing is a process designed to replace the mother liquor in the solids stream with a wash liquid. Alternatively, the process may be used to remove one solid from another, such as gluten from starch, and solid–liquid separation is required as part of this process. Washing may often represent a dominant portion of the installation cost because it is usually multistaged and often counter-current.

There are three somewhat different types of washing: washing of filter cakes by displacement, washing by reslurrying of cakes or sludges, and washing by successive dilution.

Cake washing is used to follow cake filtration and it needs quite small amounts of wash liquid, usually only a few times more than the volume of the mother liquor present in the cake to start with. The suitability of different filters to washing varies widely. Cake washing can be cocurrent or counter-current. It often has to be done in several stages, and it is advantageous to enhance it by compression.

Reslurry washing is a good alternative either when the filtration cake is slimy, cracked or generally not very permeable, or when the solids are not in cake form such as deposited in deep bed filters or in the matrix of a high gradient magnetic separator. The solids are simply reslurried (from a cake) or backwashed (from a packed bed) into the wash liquid.

Washing by successive dilution is designed for the cases when the solids are separated into a slurry, such as in gravity thickeners or hydrocyclones. The solids, thickened into a small amount of mother liquor, are diluted into a wash liquid and then separated again, diluted, separated, etc. until clean of mother liquor. The consumption of the wash liquid can be reduced in counter-current washing systems, sometimes referred to as counter-current decantation.

Cake dewatering

Dewatering is another process identified as a separate entity in solid–liquid separation and its aim is simply to reduce the moisture content of filter cakes. It is achieved either by mechanical compression or by air displacement under vacuum, pressure or by drainage in a gravitational or centrifugal system. Dewatering of cakes is enhanced by addition of dewatering aids to the suspensions, in the form of surfactants which reduce surface tension.

In dewatering by mechanical compression, the necessary prerequisite is the compressibility of the cake and this is usually the case with biological cakes. In dewatering by air (or other gas) displacement the important issues are the threshold pressure which has to be exceeded in order that air may enter the filter cake, the irreducible saturation level which gives the least moisture content achievable by air displacement, and finally kinetic dewatering characteristics.

Solid–solid separation

Another special operation closely married to solid–liquid separation is solid–solid separation by particle size, solids density or affinity to water. The purpose here is to use the solid–liquid separation processes to remove only the coarse fraction of the solids or only one material from a mixture.

Taking first the separation by particle size, this is clearly based on the size-dependent nature of some solid–liquid separation principles and equipment such as hydrocyclones or sedimenting centrifuges. Solids are classified by the separation process and the feed is either split into coarse and fine fractions (with the fine fraction usually remaining in suspension) or only a tail removed from either end of the size distribution (de-gritting or de-slimming). Even in cases when solids classification is not required it may still be desirable before the solid–liquid separation stage in order that the material in each size range may be treated by the type of equipment best suited to it.

PRETREATMENT OF SUSPENSIONS

Another important aspect of solid–liquid separation is the role of conditioning or pretreatment of the feed suspension. The purpose of this is to alter some important property of the suspension to improve the performance of a separator that follows. A conditioning effect can be obtained using several processes such as coagulation and flocculation, addition of inert filter aids, crystallisation, freezing, temperature or pH adjustment, thermal treatment and ageing, Only the first two operations are considered in more detail here due to their importance and wide use.

Coagulation and Flocculation

The two words are often used interchangeably because both processes lead to increases of the effective particle size with the accompanying benefits of higher settling or flotation rates, higher permeability of filtration cakes, or better particle retention in deep bed filters. There is, however, a subtle difference between coagulation and flocculation. Coagulation is a process which brings particles into contact to form agglomerates. The suspension is 'destabilised' by addition of chemicals such as hydrolysing coagulants like alum or ferric salts, or lime, and the subsequent agglomeration can produce particles up to 1 mm in size. Some of the coagulants simply neutralise the surface charges on the primary particles, others suppress the double layer (indifferent electrolytes such as $NaCl$, $MgSO_4$) or some even combine with the particles through hydrogen bridging or complex formation.

Flocculation uses flocculation agents, usually in the form of natural or synthetic polyelectrolytes of high molecular weight, which interconnect and enmesh the colloidal particles into giant flocs up to

10 mm in size. Flocculating agents have undergone very fast development in the past decade and this has led to a remarkable improvement in the use and performance of many types of separation equipment. As such agents are relatively expensive, the correct dosage is critical and has to be carefully optimised. Overdosage is not merely uneconomic but may inhibit the flocculation process by coating the particles completely, with the subsequent restabilisation of the suspension, or cause operating problems such as blinding of filter media or mud-balling and underdrain constriction in sand filters. Overdosage may also greatly increase the volume of sludge for disposal. Optimum dosage has been found to correspond to a situation when about one-half of the surface area of the particles is covered with polymer. As surface charges are also affected by pH, the control of it is therefore also essential in pretreatment.

In filtration operations which use gravity settling initially, like the belt presses, large and loosely packing flocs are required. The resulting free-draining sediment can then be subjected to a controlled breakdown over a period of time, ultimately leading to a complete collapse of the cake due to mechanical squeezing between the belts. In gravity thickening, large and relatively fragile flocs are needed, in order to allow high settling rates and fast collapse in the compression zone.

The optimum flocculant type and dosage depend on many factors such as solids concentration, particle size distribution, surface chemistry, electrolyte content and pH value, and the effect of these is very complex. The flocculant selection and optimisation of dosage therefore require extensive experimentation with only some general guidance as to the ionic charge or molecular weight (or chain length) required.

Addition of Inert Filter Aids

Filter aids are rigid, porous and highly permeable powders which are added to the feed suspensions to extend the applicability of surface filtration. Very dilute or very fine and slimy suspensions are normally too difficult to filter by cake filtration due to fast pressure build-up and medium blinding, but addition of filter aids can alleviate such problems. They can be used in either or both of the following modes of operation:

1. To form a precoat (which then acts as a filter medium) on a coarse support material called a septum.
2. To be mixed with the feed suspension as 'body feed' in order to increase the permeability of the resulting cake.

EQUIPMENT AND PRINCIPLES

Figure 26.1 gives the general classification of equipment and principles of solid–liquid separation. There are two separate classes, which differ in the way the particles are collected. In the first group of sedimentation and flotation, the liquid is constrained in a stationary or rotating vessel and the particles move freely within the liquid. The separation is due to mass forces acting on the particles because of an external or internal field of acceleration. That might be the gravity field, centrifugal field or magnetic field. The separation process does not end by the arrival of the particles onto a collecting surface. If the process is to be continuous, the collected particles have to be transported and discharged from the separator vessel. If gravity or centrifugal fields are used (except in the case of flotation), a density difference must exist between the solids and the suspending liquid for the separation to take place. On the whole, the principles in this group can result in continuously operating equipment which often turns out to be cheaper than filtration.

The second group, loosely called filtration, constrains the particles by a medium and the liquid is allowed to flow freely through the medium. Density difference is not necessary in this group but a truly continuous operation is usually difficult or impossible to achieve and the costs might be high. Figure 26.1 lists the conventional separation processes as a further subdivision of the two main classes but it also shows some relatively novel methods developed in recent years.

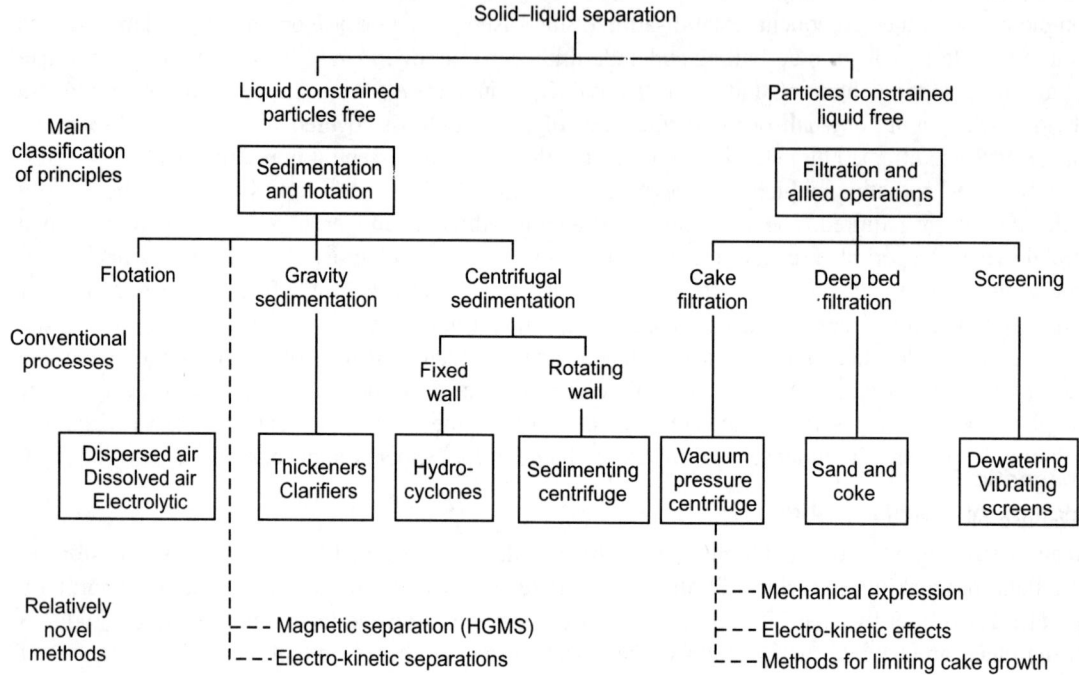

Fig. 26.1: Classification of solid–liquid separation processes.

Flotation

Flotation is a gravity separation process based on the attachment of air or gas bubbles to solid (or liquid) particles, which then are carried to the liquid surface where they accumulate as a 'float' and can be skimmed off. The process consists of two stages: the production of suitably small bubbles and their attachment to the particles. Depending on the method of bubble production, flotation is, classified as dispersed air, dissolved air or electrolytic.

Dispersed air flotation generates the bubbles by injection of air combined with agitation (froth flotation) or by bubbling air through porous media (foam flotation). The size of bubbles is rather large (typically 1 mm), capable of floating large and heavy particles thus making this process unsuitable for most biological separations. Dissolved air (and electrolytic) flotation has made considerable inroads into solid–liquid separation because it can be applied to fine suspensions. It offers a viable alternative to gravity sedimentation because it can operate at much higher overflow rates and use smaller, more compact equipment of lower capital cost (but higher operating cost). Dissolved air flotation is based on the higher solubility of air in water as the pressure increases.

Part, or all, of the feed is saturated with air at high pressure and then the pressure is reduced so that fine air bubbles appear and become available for flotation. The bubbles produced are typically less than 100 μm in size and can therefore float very fine solids. The typical overflow rates (sometimes called 'hydraulic loading') per unit plan area of the flotation cell are from 1.5 to 17 metres per hour. A typical flotation cell is shown in Fig. 26.2. Most industrial applications of dissolved air flotation are in waste water treatment. In electrolytic flotation (electro-flotation), hydrogen and oxygen gas bubbles are generated by electrolysis. The bubbles produced are generally smaller than 30 μm. Instead of the saturator, an expensive rectifier system is required. The flotation cell itself is very similar to the dissolved air unit.

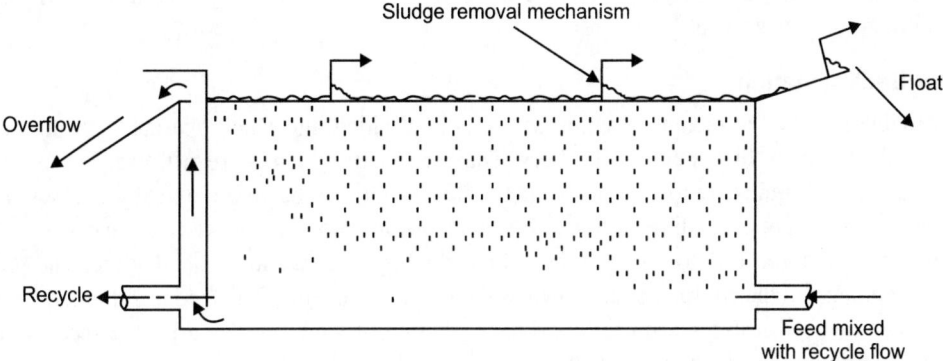

Fig. 26.2: Dissolved air flotation cell.

Magnetic Separation

The use of a magnetic field is an obvious means of removing ferromagnetic particles or agglomerates from a suspension. Particularly strong magnetic fields and gradients can be produced by the high-gradient magnetic separators (HGMS) which have enjoyed considerable attention recently.

The design and operation of the HGMS is in many ways similar to deep bed filters. The 'filtration' takes place through a fine steel wool, loosely packed bed (Fig. 26.3) which is placed in a uniform magnetic field generated by an electromagnet. After the filtration stage, the field is switched off and backwash takes place to remove the particles separated on the steel wool.

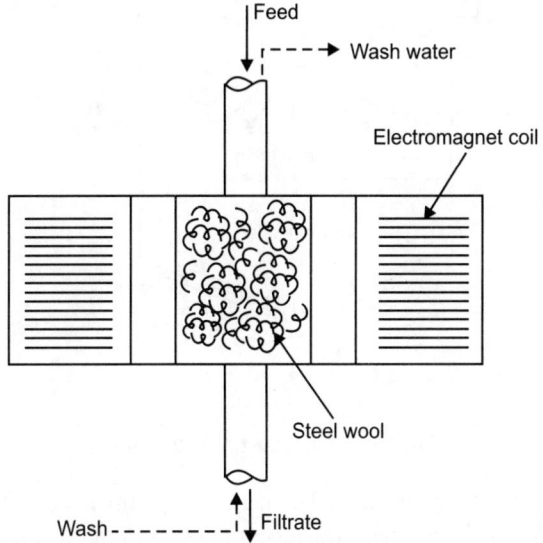

Fig. 26.3: Principle of high-gradient magnetic separation.

The main advantage of HGMS is their high efficiency of separation even at relatively high flow rates and minimum pressure drops across the filter. The capital cost is very high, however, and only large installations are attractive economically. There is a wide range of either actual or potential use of HGMS in industry: in biotechnology they have been used for separation of red blood cells, and also in waste-water treatment, where non-magnetic biological solids can be separated by seeding (and flocculation) with finely ground magnetite.

Gravity Sedimentation

The available gravity sedimentation equipment can be divided into batch operated settling tanks and continuously operated thickeners or clarifiers. The batch settling tanks are still used where relatively small quantities of liquids are to be treated, such as for example in batch biological processes. The bulk of the processing by gravity sedimentation is, however, in continuously operated thickeners and clarifiers.

If the primary purpose is to produce the solids in a highly concentrated slurry then the process is called thickening and the equipment is known as the gravity thickener. The feed to a thickener is usually more concentrated than to a clarifier, the primary purpose of which is to clarify the feed. A correctly designed and operated thickener, however, can accomplish both clarification and thickening in one stage. The most commonly used thickener is the circular basin type. The flocculant-treated feed slurry enters through the central feed well (Fig. 26.4) which disperses the feed gently into the thickener. The overflow is collected in a trough around the periphery of the basin. Raking mechanisms, slowly turning around the centre column, promote solids consolidation in the compression zone and aid solids discharge through the bottom central opening.

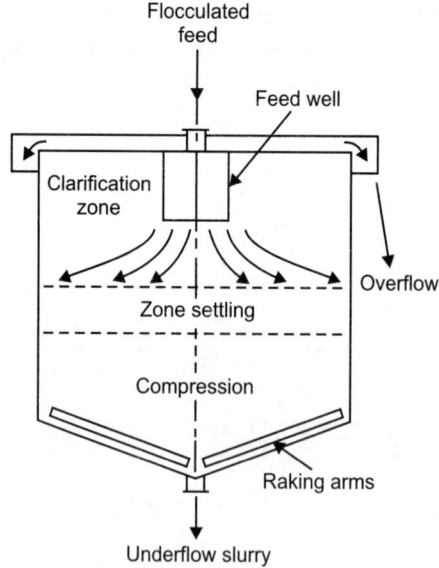

Fig. 26.4: Circular basin thickener.

Thickener types vary with the application. The lamella type thickener affords space saving through the use of flat or corrugated inclined plates (or tubes) in the settling tank to promote solids contacting and settling along and down the plates.

Hydrocyclones

The separation action of hydrocyclones is based on the effect of centrifugal forces. In contrast to sedimenting centrifuges, however, hydrocyclones do not have any rotating parts and the necessary vortex action is produced by pumping the fluid tangentially into a cono-cylindrical body (Fig. 26.5). The cylindrical part is closed on top by a cover, through which the liquid overflow pipe (often called 'vortex finder') protrudes some distance into the cyclone body. The underflow, which carries most of the solids, leaves through the opening in the apex of the cone.

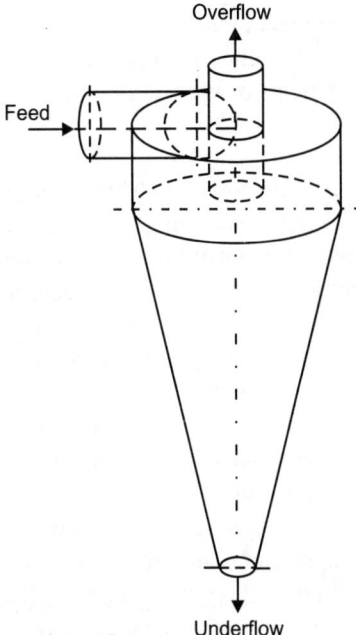

Fig. 26.5: Schematic diagram of a hydrocyclone.

There have been many different designs in use in some specialised fields but, for separation of granular solids, the above-described conventional design is still the best if the particles are rigid enough to survive the shear in flow and at the wall in such designs. A recent research work showed no significant damage to synthetic polysaccharide particles, yeast and blood cells separated in hydrocyclones designed for low shear liquid–liquid separations.

The diameters of individual cyclones range from 10 mm to 2.5 metres, cut sizes for most biological solids range from 6 to 250 μm, flowrates (capacities) of single units range from 0.1 to 7200 cubic metres per hour. The operating pressure drops vary from 0.34 to 6 bar, with smaller units usually operated at higher pressures than the large ones. If hydrocyclones are to be used to produce thick underflows, the total mass recovery of the feed solids has to be sacrificed because throttling the underflow orifice inevitably leads to some loss of the solids to the overflow. A hydrocyclone as a single unit cannot therefore be used for both clarification and thickening at the same time. Typically, the underflow concentrations that can be achieved with hydrocyclones are up to 50 per cent by volume or more.

Another application of hydrocyclones is for solid–solid separation by particle size. As the grade efficiency of a cyclone increases with particle size, it can be used to split the feed solids into fine and coarse fractions. This may be a process requirement, by which coarse and fine solids are separated to follow different routes in the plant. Most frequently in this category, hydrocyclones are used either to remove coarse particles from the product (in a degritting or refining operation) or to remove fine particles from the product (in a deslimming or washing operation).

A good example in the latter category is the, separation of fine granules from wheat starch to produce stilting material in carbonless copy paper production.

Hydrocyclones can also be used for washing of solids by arranging several stages in a counter-current arrangement similar to that used with gravity thickeners. Such systems are found in production of potato or corn starch, or in the chemical industry.

In order to make use of the high efficiency of separation of small diameter hydrocyclones, most manufacturers offer multiple arrangements of varying design.

Segmenting Centrifuges

A sedimenting centrifuge is an imperforate bowl into which a suspension is fed and which is rotated at high speed. The liquid is removed through a skimming tube or over a weir while the solids either remain in the bowl or are intermittently (or continuously) discharged from the bowl. There are five different types of industrial sedimenting centrifuges, classified according to the design of the bowl and of the solids discharge mechanism: tubular, multichamber, imperforate basket, scroll-type and disc centrifuges.

The tubular, multichamber and solids-retaining disc centrifuges are only suitable as liquid clarifiers because the bowls have to be cleaned manually. The imperforate basket type and the solids-ejecting disc type are suitable for moderate feed concentrations and have intermittent solids discharge during which the feed may have to be briefly interrupted. Only the nozzle disc type and the scroll type centrifuges are truly continuous in both the operation and solids discharge, the latter being suitable for particularly high feed solids concentrations of up to 50 per cent by volume.

Most of the centrifuge types are capable of giving cut sizes well into the submicron region. The scroll type gives a cut size in the region of about 2 μm with materials like clay, silica or similar. It is widely used in the chemical industry but its use in biotechnology is rare.

Sedimenting centrifuges, due to the lack of shear in the flow, are suitable for flocculation and suitable agents are often added to the feed upstream of the centrifuge. In biological applications, centrifuges are particularly widely used in the dairy industry.

Cake Filtration

Cake filtration is based on passing a suspension through a permeable, relatively thin medium. The solids are deposited in the form of a cake on the upstream side of the medium. As soon as the first layer of cake is formed, the subsequent filtration takes place on top of this cake and the medium provides only a supporting function. These so-called 'surface filters' are best used for filtration of suspensions of solids concentrations in excess of 1 per cent by volume in order to minimise medium blinding which occurs in the filtration of dilute suspensions. If cake filtration is to be applied to clarification of liquids, which implies low feed concentrations of solids, addition of filter aids is usually necessary. A great majority of the pressure filters are batch operated but there have been some relatively recent developments of continuous pressure filters. These are not known to have been used in biological applications.

This short review of pressure filters would not be complete without a mention of cartridge filters. These use an easily replaceable cartridge made of paper, cloth or various membranes of pore size down to 0.2 μm. Cartridge filtration is nearly always limited to liquid polishing, i.e. removing very low amounts of solids, in order to keep the frequency of cartridge replacement down. Cartridge filters are used to polish beer, juices or pharmaceutical liquids. Testing of cartridges intended for the sterilisation of liquids by filtration is sometimes made by the smallest commercially available bacterium, *Pseudomonas diminuta*, which needs a 0.2 μm pore size to ensure its retention.

Centrifugal filters

The third type of driving force used in cake filtration is the centrifugal force, in the centrifugal filter. A centrifugal filter consists of a rotating perforate basket fitted with a filter cloth or other medium (Fig. 26.6). The action of the centrifugal field leads to two effects: it creates a pressure drop across the filter medium, due to the centrifugal head of the suspension layer rotating in the basket, and it also pulls the liquid out of the cake and medium. The latter effect makes the centrifugal filters ideally suitable for dewatering duties and these indeed represent the bulk of their industrial applications.

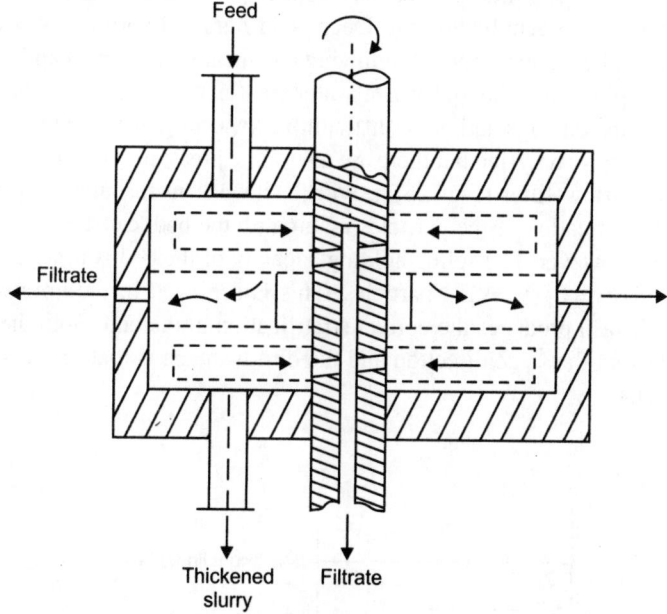

Fig. 26.6: Schematic diagram of a centrifugal filter.

Centrifugal filters can be classified into two fundamentally different classes: discontinuously fed, fixed-bed machines, and continuously fed, moving-bed machines.

Starting with the fixed-bed, batch or cyclic centrifuges, these have a cylindrical screen onto which the feed suspension is fed while the cake is stationary. This feature makes it possible to separate relatively finer particles than in the continuous, moving-bed machines. The perforate basket (or 'three-point') centrifuge and the peeler type centrifuge are examples in this category, used in the sugar industry.

The continuously fed, moving-bed machines use a metal screen instead of filter cloth and this makes them suitable for dewatering of coarse solids only. The baskets are either conical, with or without a scroll (or a vibration device) to aid the movement of the solids across the screen, or cylindrical, in the pusher centrifuges. The pusher centrifuges use a reciprocating piston to push the solids along the basket. They require high feed concentrations of the solids to enable the formation of a sufficiently rigid cake to transmit the thrust of the piston. By the very nature of their operation, the pusher centrifuges are suitable for washing, particularly in their multistage screen versions.

Deep Bed Filtration

Deep bed filtration is fundamentally different from cake filtration both in its principle and its application. The filter medium is a deep bed of pore size much greater than the particles it is meant to remove. No cake should form on the face of the medium but particles penetrate into the medium where they separate due to gravity settling, diffusion and inertial forces. Attachment to the medium is due to molecular and electrostatic forces. Sand is the most common medium and multimedia filters also use garnet and anthracite. The filtration process is cyclic because when the bed is full of solids and the pressure drop across the bed is excessive, the flow is interrupted and solids are backwashed from the bed (this is sometimes aided by air scouring or wash jets). In order to keep the frequency of backwash and the wash water demand down, deep bed filtration is usually applied only to very dilute suspensions of solids concentrations less than 0.1 per cent by volume. Deep bed filters were originally developed for potable water treatment as the final polishing process following chemical pretreatment and sedimentation. They are also increasingly applied in industrial waste-water treatment under somewhat harsher operating conditions of higher solids loadings and more difficult backwashing of the media.

Most deep bed filters are gravity-fed but there are also some pressure types in use. The most common arrangement is the downflow filter (Fig. 26.7), with backwash in the upward direction in order to fluidise the bed. The stratification of the particles making up the bed, caused by the fluidisation (fines on top), is not desirable however. The solids-holding capacity of the bed is best utilised if the filtration flow encounters progressively finer sand particles. This is achieved in the upflow filters, where the fluidisation due to backwash produces the correct stratification in the bed. Both the filtration flow and the backwash take place in the same direction and the disadvantage is that the wash water goes to the 'clean side' of the filter.

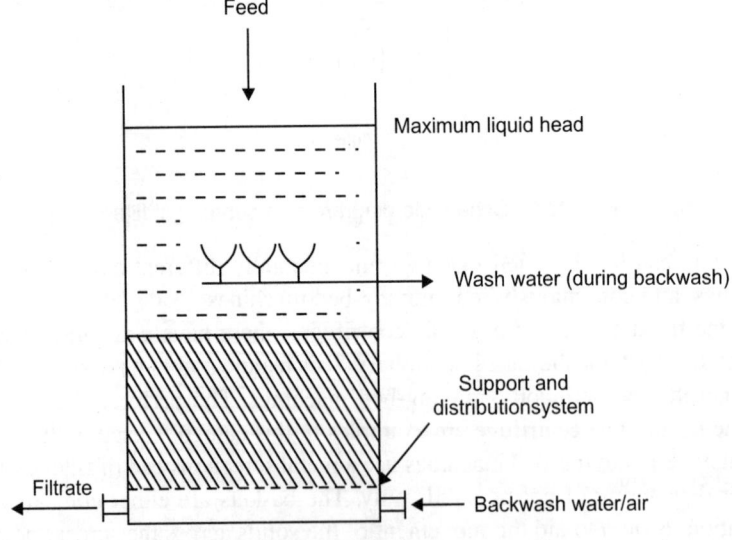

Fig. 26.7: Downflow gravity deep bed filter.

The trend in the use of deep bed filters in water treatment is to eliminate conventional flocculators and sedimentation tanks, and to employ the filter as a 'flocculation reactor' for direct filtration of low turbidity waters. The constraints of batch operation can be removed by using one of the available continuous filters which provide continuous backwashing of a portion of the medium. Such systems include moving bed filters, radial flow filters or travelling backwash filters. Further development of continuous deep bed filters is most likely.

Screening and Straining

Screening is an operation by which particles are introduced to a screen of a given aperture size and thus have an opportunity of either passing through if they are smaller than the apertures, or being retained on the screen if they are larger. The bulk of the screens in industry are used for size grading but they are also used in a dewatering function, often combined with washing.

Virtually any type of screen can be used for dewatering but some screens have been designed specifically for this purpose. The DSM sieve bend is a good example. It consists of a curved, wedge wire deck with the wires or bars at right angles to the feed. The feed suspension is distributed on the top, across the full width of the screen. Coarse screens are gravity fed, while finer ones are pressure fed through a set of nozzles. Polyurethane surface or woven wire cloth have also been used in place of the wedge wire. The applications include thickening organic or waste slurries, starch fibre washing, gluten recovery, paper pulp fibre screening and many other similar operations. The cut size varies from 40 to 2000 μm.

Electrokinetic Effects

There are several effects due to the existence of the double layer on the surface of most particles suspended in liquids, and they can be used to enhance solid–liquid separation processes.

The best known effect is electrophoresis. This term is given to the migration of small particles suspended in a polar liquid, in an electric field, towards an electrode. The second effect, which often accompanies electrophoresis, is electro-osmosis. This is the transport of the liquid past a surface or through a porous solid, which is electrically charged but immovable, towards the electrode with the same sign of charge as that of the surface.

It can be said that electrophoresis reverts to electro-osmotic flow when the charged particles are made immovable, and, if the electro-osmotic flow is forcibly prevented, pressure builds up and is called electro-osmotic pressure.

Electrophoretic settling and electro-osmotic dewatering

This is a straight forward use of electrophoresis when settling velocities are enhanced by placing one electrode (usually negative and perforated) floating on top of the slurry and another at the bottom of the tank. Furthermore, electro-osmosis thickens the settled sludge at the same time. The settling consists in fact of three successive processes. In the first instance, suspended particles settle with the resultant velocity of the gravitational and the electrophoretic components. In the second process, the sludge is thickened by compression and electro-osmosis. In the third process, when compression due to gravity is completed, the sludge is thickened only by electro-osmosis.

Enhancement of gravity settling by electrophoresis is also used with liquid–liquid dispersions in solvent extraction and for the separation of oil–water emulsions.

Electrokinetic filtration

Electrophoresis and electro-osmosis can also be used to enhance conventional cake filtration. Electrodes of suitable polarity are placed on either side of the filter medium (as most particles carry negative charge, the electrode upstream of the medium is usually positive) so that the incoming particles move towards the upstream electrode, away from the medium. The electric field can cause the suspended particles to form a more open cake or, in the extreme, to prevent cake formation altogether by keeping all particles away from the medium.

There is an additional pressure drop across the cake developed by electro-osmosis which leads to increased flowrates through the cake (and further dewatering at the end of the filtration cycle). The filtration theory proposed for electro-filtration assumes the simple superposition of electro-osmotic pressure on the hydraulic pressure drop.

Chapter 27

Aqueous Two-phase Extraction Systems

INTRODUCTION

One of the principal features of the separation of biological molecules which discriminate them from many conventional products of the process industries is the sensitivity of particular species to the environment in which they are processed. This is especially true of proteins and in particular of enzymic products whose activity and functionality are especially sensitive to environmental criteria such as pH, temperature and ionic strength. The small-scale preparation, recovery and purification of enzymes has therefore typically involved a complex series of selective precipitation stages using chemical separation techniques and mechanical solid–liquid separations such as centrifugation and filtration. The use of solid–liquid separation techniques is controlled by the size of the particles to be separated and in the case of biological products the particle sizes can be very small (of the order of microns), thus requiring elegant and expensive processes which may involve consumption of precoats and other filter aids. The primary recovery of intracellular enzymes after cell disruption involves processing in the presence of cell fragments which further exacerbate the problem of separation. In order to scale-up on the basis of this philosophy, complex flowsheets may be required, incurring high capital equipment costs.

There are two major factors which influence the development and choice of downstream separation process. Firstly, as the throughput requirements of plants producing both bulk and speciality enzyme products increase, there is nearly always an economic incentive to consider a range of separation processes which yield advantage in respect of operating costs and overall recovery values. A second major factor is the need for a biocompatible environment which allows application of rigorous separation criteria without prejudice to final product performance. In the case of enzyme recovery by aqueous two-phase extraction both of these factors are very relevant. Alongside a number of other major applications of solvent extraction as a downstream process, separation by this method can offer a straightforward sequential process for the separation of complex mixtures component by component in essentially a liquid phase environment. This approach has been successfully adopted in the nuclear industry for the separation of uranium and plutonium from chemically complex mixtures, in the pharmaceutical industry in the recovery of low molecular weight antibiotics for example, and also within the recent past in the precious metals industry for the recovery of platinum group metals. In each of these cases very significant simplifications in process flowsheets have been achieved by the adoption of liquid–liquid extraction as the principal separation process. Aqueous two-phase extraction also offers a suitable environment for protein separation because of the significant presence of water at all stages of the process, an important requirement for maintenance of enzymic activity. The basic feature of these liquid–liquid systems is that there is a high concentration of water in each of the liquid phases. The achievement of partial or almost complete immiscibility of two liquid phases in the presence of water in both phases is achieved using hydrophilic polymers, for example polyethylene glycol (PEG) which is perhaps the compound most commonly encountered in this context.

Extraction in two-phase aqueous systems has therefore seen a significant growth both in application and research over the past ten years. It is now available as an industrial unit operation for the downstream processing not only of industrial-enzymes but also of other biological active species such as other biologically active proteins. The technology has been extended to the production of biocatalysts, analytical enzymes and pharmaceutical proteins. In addition to the separation of enzymes, aqueous two-phase systems have been used for the separation of DNA and virus particles.

Very recent work is reported on the potential of aqueous two-phase extraction in the recovery of high-value protein products from waste streams arising in the food and beverage industry. There has been an exponential growth in the literature on the subject of aqueous two-phase extraction systems over this period, which perhaps indicates the level of potential importance of this as a biological product separation technique.

For commercial-scale operations it is possible that aqueous two-phase extraction can be more economic than the conventional filtration → ultrafiltration → precipitation → filtration route for protein recovery. A major question is the comparative economics of aqueous two-phase extraction with other novel downstream processes including for example membrane processes based upon cross-flow filtration → microfiltration → ultrafiltration → adsorption. One of the major advantages which aqueous two-phase extraction offers is its compatibility with continuous processing with the attendant advantages of small equipment size, low inventory, process stability and product uniformity.

In common with many established solvent extraction processes, aqueous two-phase extraction is a flexible separation technique which allows a high degree of control of extraction and recovery conditions by control of the chemical environment. Important parameters include the molecular weight of the solvent, the pH, the salt concentration and temperature. There is a degree of choice of the molecular weight of the extracting solvent in many aqueous two-phase extraction systems since the base component is frequently polymeric in nature. The ability to vary the molecular weight is also important in relation to the control of phase density characteristics which is very important in liquid–liquid contacting processes for efficient contacting and phase disengagement.

In a recent review of aqueous two-phase partitioning, Walter reported that the most recent advances in the field include the discovery of inexpensive polymers with favourable partition properties. This is particularly important since the use of high-cost polyethylene glycols and pure dextran would prejudice the economics of recovering many enzymes by this method. Development of biospecific separation techniques for both enzymes and for whole cells is another area where significant advances have been made. The discovery of aqueous two-phase systems which perform at higher and lower temperatures, together with systems whose partitioning properties can be modified by the addition of small quantities of water-soluble organic solvents thereby facilitating partition of materials with low water solubilities, is another significant step forward.

AQUEOUS TWO-PHASE SYSTEMS

The formation of two immiscible aqueous phases in mixed polymer–salt or polymer–polymer systems has been known for approximately a century. The earliest recorded occurrence was in the case of mixtures of agar with soluble starch and gelatine. The application of the systems to the partition of cells and macromolecules within laboratory-scale analytical techniques is also long-standing.

Consider for example the classical case of polyethylene glycol (PEG) and dextran in aqueous solution. At low concentrations of each of the solutes, a homogeneous single-phase liquid is obtained. However, as the solute concentrations are increased the solution becomes turbid and under stagnant conditions two liquid layers are formed, which are in fact two immiscible liquid phases at equilibrium. In this system the upper phase is rich with respect to polyethylene glycol and the lower phase rich with respect to dextran. The mutual insolubility of two essentially hydrophilic components is explained by differences in molecular form each of which result in mutual repulsion. Dextran is essentially a globular molecule with little tendency to dipole formation. Polyethylene glycol is a linear chain polymer with a high density of lone electron pairs. Each polymer molecule has a tendency to surround molecules of similar shape, size and polarity and thus the repulsive forces between the molecules of different type exceed the forces of mutual attraction associated with their hydrophilicity. The polymers therefore segregate, forming two distinct phases.

Some water-soluble polymers exhibit a quite different behaviour when the two polymers exhibit strong mutually attractive forces when the water component is largely excluded upon mixing. In these cases the polymers collect in a single phase leaving the second liquid phase as essentially pure solvent. This process is known as complex coacervation. In the absence of strongly, attractive or repulsive forces complete miscibility is obtained. For the majority of polymeric systems this would be an exception.

Primary extraction and purification of proteins and other macromolecules derived biologically have been successfully demonstrated using mixtures of PEG and salts and mixtures of PEG with other polymers. The effectiveness of the extractant can be determined by a number of factors which include concentration of polymer, structure and concentration of added salt, pH, and temperature.

The properties of aqueous biphasic systems are strongly dependent upon several other factors which include:

1. Polymer molecular weight.
2. System viscosity.
3. Temperature.
4. Density.
5. Viscosity.
6. Interfacial tension.

Use of high molecular weight polymers in the two-phase system favours good phase separation. Thus if higher molecular weight polymers are used in the extraction system relatively low concentrations of polymer will be required for good physical performance relative to polymers of lower molecular weight. In systems involving two polymers, large differences in molecular weight affect the symmetry of the binodal solubility curve and when the difference in molecular weight is large the curve becomes distinctly asymmetrical. The viscosities of the two liquid phases formed are very important in relation to the mechanics of contact and phase separation and in relation to mass transfer coefficients. The strong relationship between viscosity and polymer molecular weight provides the process design with a considerable degree of choice in ensuring that the viscosity of the system does not unduly inhibit contacting and mass transfer processes. This is especially important in the context of primary recovery of proteins from untreated broth in which many viscoelastic contaminants may already be present and could seriously influence the kinetics of contacting processes.

Temperature also exerts a significant influence upon the physical aspects of contacting in two-phase aqueous systems, not only in relation to viscosity but also in relation to phase equilibria. For example in the system polyethylene glycol–water–potassium phosphate, increase of temperature leads to decrease in miscibility thus enhancing phase separation. By contrast, the system polyethylene glycol–water–dextran exhibits better miscibility at higher temperatures. The density ratio of the phases in aqueous two-phase systems tends to be close to unity, but as a broad observation phase inversion is unusual; for example the heavier phase will remain so throughout an entire process. However, there are exceptions to this. Significant density difference is preferred for successful contacting processes.

Although low interfacial tension is beneficial with respect to dispersion and creation of high interfacial area, some aqueous two-phase systems do exhibit relatively poor phase separation behaviour in certain cases and this would have implications for scaled processing operations particularly in relation to column contactors.

Partition

The distribution of proteins and other biopolymers between two aqueous-rich liquid phases is a complex function of hydrogen bonding, the molecular weight of the phase-forming components, pH, type and concentration of salting in compounds, charge interaction, van der Waals interactions, the amphipathic nature of the protein and steric effects. One of the major benefits which would accrue from the use of aqueous two-phase systems focuses upon the possibility of direct extraction of the desired enzyme from whole broth. Indeed, this approach is essential if application to many intracellular enzymes is to be economic. Therefore the distribution of the cells and cell debris is of vital significance alongside the distribution of the soluble components. There are, therefore, a number of criteria by which partition in aqueous two-phase systems may be measured:

1. The distribution of the desired product protein relative to other proteins and biopolymers in the system, for example polysaccharides and nucleic acids. In general terms protein affinity for the extracting phase is according to the magnitude either of hydrogen bonding mechanisms or due to van der Waals interactions with the polymer molecules comprising the extract phase. The distribution of the associated salts is also important but in most cases the distribution of fully ionised salts will be closed to unity and therefore will not significantly influence phase equilibrium during extraction.

2. The selectivity of each phase for the solid material present. The selective separation of this material during the liquid–liquid contact is very important. Successful transfer or retention of such material in the reject phase reduces the need for solid–liquid separation processes during final purification and recovery operations.

3. The activity of the product protein must be maintained throughout the sequence of contacts and therefore the conditions at each stage of partition must be compatible with this criterion.

Cell debris partition

The task of removing cell debris from homogenates is often difficult and cumbersome and many of the protein recovery techniques described in the literature refer to separation from clarified liquor. Aqueous two-phase systems are particularly well suited to direct recovery from unclarified homogenates. The preferred general scheme is that the cell debris should partition into the bottom phase, which should also form the dispersed phase. These requirements are directly linked to the process engineering requirements for efficient operation of centrifugal separators which would be used in a commercial-scale process. Generally speaking, centrifugal separators do not perform well when dispersed particulates are present in the top phase. Another factor, which is discussed later on, is the viscous nature of the phase containing cell debris: better separation is achieved in the separation equipment if the phase of higher viscosity is dispersed.

The ability to partition from cell homogenates in addition to whole-broths is an important feature of aqueous two-phase systems. The process of homogenisation, apart from releasing intracellular enzyme and improving the available contact between extractant and protein, results in the formation and release of subcellular species such as organelles and cell wall fragments. Particulates of this type display a wide size distribution typically in the range $0.3 \times 10^{-6} - 5 \times 10^{-6}$ m and some particulates can be colloidal in nature. At this level such components can be removed by mechanical separation techniques, but only with difficulty. In the case of colloidal particulates, membrane separation techniques may be necessary.

Appropriate choice of solvent should permit substitution of a very difficult mechanical separation process with liquid–liquid partition step which is relatively easy to conduct and to control. Kula discussed the possibility of preferential collection of cell debris and fragments arising from the homogenisation process in the reject phase. In order to be effective the liquid–liquid system should, if possible, be designed to ensure that the protein and the cell material move into different phases. The method of cell disintegration is not considered to have an important influence upon the ease of separation of enzyme from cell material by this method.

Use of direct extraction techniques can improve overall yield and can lead to fewer processing stages. In order to be economic the ratio of cell homogenate to extractant should be as high as possible in order to achieve a high concentration of enzyme in the extract phase. Extraction under conditions of high concentration means that the polymer concentration in the extractant is very significant leading to changes in rheology and in the phase partition behaviour. The existence of higher viscosities is likely to reduce rates of mass transfer but the presence of additional polymer in the extract phase will tend to shift the binodal curve of the carrier system and can improve phase separation. Golden and Hatton demonstrated this benefit during the partition of a 20 per cell homogenate into an extractant of 18 per cent PEG 1550/7 per cent potassium phosphate. Cell debris was partitioned into the heavier reject phase with a high degree of efficiency. High cell debris concentrations in the reject phase will of course result in significant increases in viscosity.

Product purification and solvent recovery

The protein extracted into the polymer-rich phase can be recovered either by single-stage extractions or by use of chromatographic techniques. For enzyme production further purification is essential for most applications since the presence of impurities and other enzymes may impair performance. The specifications required for analytical or pharmaceutical applications are very much higher compared with those appropriate for bulk industrial enzymes. In these cases it is unlikely that single-stage extraction methods would be sufficiently good and preparative chromatography may be required. In this context the use of two-phase extraction is a means of removing thickening compounds and other impurities which may interfere with the chromatographic column performance. The use of the latter for operations at commercial scale is yet to become a mature technology.

Kula cites the example of fumarase and aspartase. Aspartase present in preparations of fumarase does not affect the activity or performance of the fumarase providing that ammonium ion is absent. By contrast, if fumarase is present as a contaminant in preparations of aspartase the fumarase will react with fumaric acid to produce L-malate by water addition. This is undesirable and, therefore, requires selective removal from the binary protein mixture.

Recovery of the solvent phase after the partition and purifications steps of the process is necessary for both commercial and environmental reasons. The cost of the polyethylene-glycol polymers and pure dextrans is very high and there is a strong incentive to successfully recover and recycle these compounds. In many cases it is possible to recover the final protein product by transfer at the last stage of contact into the salt-rich phase. This can be achieved by adjustment of pH at the appropriate point in the process and the protein-lean PEG-rich phase is available for recycle. Separation of the enzyme from the salt-rich phase may be achieved by other means such as membrane ultrafiltration, precipitation or ion-exchange.

EQUIPMENT AND SCALE-UP

The contacting process methods for aqueous two-phase extraction systems are chosen on the basis of criteria relevant to conventional liquid–liquid extraction system.

Centrifugal Devices

For continuous and large-scale operations the combined use of in-line or static mixers with centrifugal separators has been favoured. The low interfacial tension of aqueous two-phase systems means that only low levels of mixing energy are required since excessive shear results in very fine dispersions with poor settling properties. Another factor associated with high levels of shear during mixing is the detrimental effect on protein activity due to chain breakage and resulting loss of active product yield. Generally, interfacial mass transfer processes are rapid in these systems and short residence time contacts are sufficient. In many aqueous two-phase extractions a single stage of contact is sufficient for high recovery providing that the partition conditions are optimised according to solvent composition, pH and salt concentration.

Column contactors and others

There are many designs of countercurrent column contactors and the achievement of stable operation for long periods with low maintenance resources associated with columns is well known for many industrial scale extractions. The main limitation of column contactors is in relation to hydraulic capacity at small drop sizes where entrainment and backmixing can prejudice the overall efficiency of operation. The low interfacial tension of many aqueous two-phase systems means that very small drop sizes can be easily achieved, indeed in some circumstances too easily, and thus the hydraulic limits of a particular column design may be necessarily tight and make column contact less economic compared with centrifugal contact.

In aqueous two-phase extraction systems the density difference between the phases can also be very small and therefore the gravitational effects upon drop motion are low, further prejudicing hydraulic capacity in column contactors. Another limitation which can result from small density differences and high viscosity is poor phase disengagement rate and this has been reported in some aqueous two-phase, extraction systems.

TWO-PHASE EXTRACTION PROCESSES

The scale-up characteristics of aqueous two-phase extraction processes appear to be good with little loss of extraction efficiency in the progression from laboratory-scale trial to pilot-scale operations which may involve processing of upto 500 kg biomass per day. The process concept is based on a sequential stepwise removal of products and impurities by adjustment of chemical and physical conditions at each stage, maintaining operations wherever possible in the liquid phase. Separations involving a significant number of solid–liquid filtration steps in which the filtrate contains the desired product tend to be less efficient overall.

The first step in the process is removal of cell debris, which is achieved by partition of the cell debris into the heavier salt-rich phase. The aqueous homogenate feed is contacted with the solvent phase, in this case shown as a PEG–salt–water mixture. The proteinaceous product is recovered in the PEG-rich lighter phase to which is added a further solution of phosphate at fixed pH and fixed ionic strength. The latter is controlled by adjustment of the overall salt concentration.

Some examples of enzymes which have been recovered commercially or at demonstration scale using continuous aqueous two-phase extraction are shown in Table 27.1. In each case direct recovery from cell homogenate has been demonstrated and the recovery values compare very favourably with those obtained by conventional precipitation methods.

Table 27.1: Some examples of enzymes which have been recovered commercially or at demonstration scale using continuous aqueous two-phase extraction.

Enzyme	Organism	No. of steps	Overall yield (%)
Aspartase	E. coli	3	82
Formate dehydrogenase	Candida boidini	3	78
Fumarase	Brevibacterium ammoniagenes	2	75
Glucose dehydrogenase	Bacillus	2	97
D-2-hydroxy isocapoate-dehydrogenase	Lactobacillus casei	2	85
Beta-interferon	Human fibroblasts	1	100
Penicillin acylase	E. coli	2	78
Pullulanase	Klebsiella pneumoniae	4	60

The majority of conventional protein recovery and purification processes have been based on batch operations. One of the fundamental advantages of aqueous two-phase extraction is that the principle of the process comprises the two steps of mixing and phase separation. Therefore, operation on a continuous basis is relatively straight forward.

Potential Developments

The partition of a wide range of enzymes into aqueous two-phase systems has been evaluated and reported in the literature. Many of these would be for small-scale specialist applications such as analytical tools and for biomedical applications. The recovery of bulk proteins and the extraction of useful proteins from low-grade process streams are areas of application worthy of brief mention in relation to future developments. Recent research has shown the possibility of successfully recovering and purifying alkaline proteases from whole broths of *Bacillus lichenformis*. The study showed the high pH dependency of the partition in two aqueous two-phase systems. Both systems were based upon mixtures of two polymers, the first being a mixture of PEG and Reppal 200, this latter polymer being a partly hydrolysed hydroxypropyl derivative of starch.

The second mixture comprised ethylhydroxycellulose and Reppal 200. Both of these systems showed excellent partition characteristics, comparable to PEG/dextran-based extractants and with much reduced solvent cost in each case. Another very recent study examined the recovery of bulk intracellular proteins from waste yeast arising from brewery operations. This is an area where there is great potential for the recovery of valuable products from effluent streams.

Currently the large yields of waste yeast associated with beer production are converted to low-value animal feed supplements. The results of this study revealed that PEG/phosphate aqueous two-phase extractants offered the best prospect for bulk partition of the intracellular proteins from these yeast reject streams.

The final example mentioned is the recovery of beta-galactosidase or lactase from homogenised yeast by aqueous two-phase extraction. This is an enzyme of considerable commercial importance, particularly in the production of added value products in the dairy industry. Availability of commercial quantities of lactase is a vital requirement for such large-scale applications and the use of this enzyme has been further expanded with the onset of immobilisation techniques.

Gonzalez studied the recovery of lactase from homogenised *Kluyveromyces fragilis* at laboratory scale and showed that aqueous two-phase extraction gave purification factors and recoveries comparable to other methods including solvent extraction with organic solvents and affinity and ion-exchange chromatography.

Chapter 28

Chromatography

INTRODUCTION

Chromatography lies at the core of all biotechnology purification processes. Almost all the published literature on chromatography has been written for analytical applications. It is important to appreciate that analytical chromatography has a completely different set of aims compared to preparative chromatography. Usually when the literature speaks of optimising a separation, it means achieving the maximum number of resolved peaks. In preparative chromatography there is only one peak or real value, i.e. that of the target molecule. If all other molecules were eluted in only two peaks, one before the product and one after, this would represent an optimised manufacturing process. The emphasis in process chromatography is merely to ensure that the target peak is sufficiently separated from the closest peaks on either side. This should be the central aim of any changes made in elution conditions. One potential opportunity from this approach is the use of ion-exchange media in isocratic mode. If the conductivity and pH are carefully selected it is possible to load an ion-exchange media with feedstock and have the target protein retarded, while nonbinding proteins pass through the column and more tightly binding proteins are retained. Although requiring more development effort, this approach removes the need for gradient formation and can allow much higher loads of target protein to be separated in each purification cycle.

CHROMATOGRAPHY

In many fermentation processes, chromatographic techniques are used to isolate and purify relatively low concentrations of metabolic products. In this context, chromatography will be concerned with the passage and separation of different solutes as liquid is passed through a column, i.e. liquid chromatography.

Depending on the mechanism by which the solutes may be differentially held in a column, the techniques can be grouped as follows:

1. Adsorption chromatography.
2. Ion-exchange chromatography.
3. Gel permeation chromatography.

4. Affinity chromatography.

5. Reverse phase chromatography.

6. High performance liquid chromatography.

Chromatographic techniques are also used in the final stages of purification of a number of products. The scale-up of chromatographic processes can prove difficult and there is much current interest in the use of mathematical models and computer programmes to translate data obtained from small-scale processes into operating conditions for larger scale applications.

Adsorption Chromatography

Adsorption chromatography involves binding of the solute to the solid phase primarily by weak van der Waals forces. The materials used for this purpose to pack columns include inorganic adsorbants (active carbon, aluminium oxide, aluminium hydroxide, magnesium oxide, silica gel) and organic macro-porous resins. Adsorption and affinity chromatography are mechanistically identical, but are strategically different. In affinity systems selectivity is designed rationally whilst in adsorption selectivity must be determined empirically.

Dihydro-streptomycin can be extracted from filtrates using activated charcoal columns. It is then eluted with methanolic hydrochloric acid and purified in further stages. Active carbon may be used to remove pigments to clarify broths. Penicillin-containing solvents may be treated with 0.25 to 0.5 per cent active carbon to remove pigments and other impurities.

Macro-porous adsorbants have also been tested. The first synthetic organic macro-porous adsorbants, the Amberlite XAD resins, were produced by Rohm and Haas. These resins have surface polarities which vary from non-polar to highly polar and do not possess any ionic functional groups. Voser considers his most interesting application to be in the isolation of hydrophilic fermentation products. He stated that these resins would be used at Ciba-Geigy in recovery of cephalosporin C (acidic amino acid), cefotiam (basic amino acid), desferrioxamine B (basic hydroxamic acid) and paramethasone (neutral steroid).

Ion Exchange

Ion exchange can be defined as the reversible exchange of ions between a liquid phase and a solid phase (ion-exchange resin) which is not accompanied by any radical change in the solid structure. Cationic ion-exchange resins normally contain a sulphonic acid, carboxylic acid or phosphonic acid active group. Carboxy-methyl cellulose is a common cation exchange resin. Positively charged solutes (e.g. certain proteins) will bind to the resin, the strength of attachment depending on the net charge of the solute at the pH of the column feed. After deposition solutes are sequentially washed off by the passage of buffers of increasing ionic strength or pH. Anionic ion-exchange resins normally contain a secondary amine, quaternary amine or quaternary ammonium active group. A common anion exchange resin, DEAE (diethylaminoethyl) cellulose is used in a similar manner to that described above for the separation of negatively charged solutes. Other functional groups may also be attached to the resin skeleton to provide more selective behaviour similar to that of affinity chromatography. The appropriate resin for a particular purpose will depend on various factors such as bead size, pore size, diffusion rate, resin capacity, range of reactive groups and the life of the resin before replacement is necessary. Weak-acid cation ion-exchange resins can be used in the isolation and purification of streptomycin, neomycin and similar antibiotics.

There is currently an increasing need for improved production processes for biological drugs. In the past, biotherapeutics like erythropoietin and growth hormones were generally only administered in small doses; the situation with therapeutic recombinant proteins like monoclonal antibodies is that they are required in far greater quantities. Among the biomolecules in use or in testing are nucleic acids and viral vectors for gene therapy and proteins as well as peptides. The yearly demand for such drugs can be as high as several hundred kilograms. This high demand has to be considered when establishing new purification schemes for commercial use of biologically derived biotherapeutics. However, producing high-quality, efficacious and safe proteins in sufficient amounts for the clinic is not trivial. Recovery and purification operations are the most time-consuming steps in manufacturing. Whether the products are derived from microbial or mammalian cells, the downstream purification scheme is expensive to develop. The development of a downstream purification process includes the responsibility of implementing appropriate and state-of-the-art chromatography media, techniques and equipment. All purification issues should consider technology transfer ability of the process to current good manufacturing practice (cGMP) conditions. In any case, the aim of the downstream process is to use the shortest route from the biological source to the specified final product. The yield has to be maximised while at the same time the biological activity must be maintained.

The downstream process can be divided into different parts: the first step after the harvest is the capture of the target protein, followed by intermediate purification step(s) and a final polishing step before the formulation. The capture step, that is in many cases an ion-exchange column, has the objective to recover as much product as possible and to provide a product pool that is suitable for subsequent chromatography. The reduction of the volume (mainly by eliminating water) and the removal of other impurities has to be achieved during this step. The captured material is then purified by column chromatography using a process that removes all host contaminants and product related impurities up to a certain level. The final polishing that is applied to separate degraded or modified forms of the product is often the most challenging step, because of the biochemical similarities that typically exist between the final product and the modified forms. Finally, the end of the purification process is bulk formulation of the drug.

Principles of ion-exchange chromatography

Preparative ion-exchange chromatography is a technique for separating biomolecules based primarily on their electrostatic interactions with charged stationaryphase materials. It is simply an adsorption/desorption process utilising the principle of ion-exchange and is therefore reliant on the chemical properties of the solute molecules to be separated as well as on the stationary phases used. This technique is the most widely employed chromatographic separation technique in the biotechnology industry today. It is used for both capture and high-resolution separations. Different generalisations of the simplest coulombic model describing the attraction of groups of opposite charge on the protein surface and the stationary phase, respectively, have been proposed. They can be divided into stoichiometric and non-stoichiometric models. Stoichiometric models describe the multifaceted binding of the protein molecules to the stationary phase as a stoichiometric exchange of mobile-phase protein and bound counterions. To enable efficient process development rational strategies for screening and selection of chromatographic media and process conditions must be developed where one combines the experimental results with theoretical considerations and model calculations.

A chromatographic separation is developed through a number of steps including screening of different media and techniques to select appropriate candidates to investigate the retention behaviour in dependence of the possible process variables that usually comprise particle size, pH, type and concentration of salt, solvents and additives, and temperature. When more material becomes available capacities must be estimated. Process optimisation includes column size and column aspect ratio, flow rate, buffer composition, and gradient length. All this can of course be performed by the trial-and-error method, but ultimately model-assisted development should be beneficial.

Ion-exchange chromatography can be used as a capture, intermediate purification as well as a final polishing step. For product recovery (capture), the capacity and production rate are paramount. In addition, it is often necessary to process large volumes of relatively dilute feedstock, which often contain particles.

Scale-up of Ion-exchange Chromatography

During early drug research, a high throughput of many samples is required with respect to their potency. However, after a specific protein has been identified to become a potential drug, more protein material is needed for further clinical testing. In contrast to the first material that is made in bioreactors containing a few hundreds of litres, the fermentation will be increased to several thousands of litres and huge volumes have to be processed. Scale-up is not a simple increase of the size and volume of the laboratory equipment. Usually, production-scale equipment, larger piping dimensions, larger column diameters, larger dead volumes and different types of pumps cannot be compared directly with bench-top workstations even if an optimal system configuration has been chosen. In particular, columns containing several hundreds of litres of resin are often customised, and column packing and qualification becomes a challenging part of the work. Due to decreased wall support, large columns behave completely different compared to pilot- or laboratory-scale columns.

The transfer of a method starts from the milligram level and ends up at the kilogram scale. The easiest way to scale-up the procedure is to change the column dimensions by keeping the height of the gel bed constant and increasing the column diameter. In many production suites large ion-exchange columns containing several hundreds of litres of media are installed. Column diameters usually are in the range between 60 and 160 cm or even more. The scale-up is typically, done in two steps: the first step from laboratory to pilot plant is on the order of 50 to 100 fold and the final scale-up from the pilot plant to full-scale manufacturing is 10 to 50 folds.

Application Areas of Ion-exchange Chromatography

Ion-exchange chromatography is easy to scale-up to production scale and the results are very reproducible. The robustness of ion-exchange chromatography combined with the high protein-binding capacity makes this method very useful for most of the purification strategies, either for capture or intermediate steps, or even polishing. Many protein purification protocols contain one or more ion-exchange chromatography steps. When using small beaded ion-exchange chromatography media, a relatively good resolution can be achieved and, thus, occasionally ion-exchange chromatography is performed late in the procedure for polishing. The main task of this particular step is to remove possible impurities, such as structurally very similar or closely related forms of the product like aggregates, deamidated or oxidised product, or isoforms. Principal applications of ion-exchange chromatography are purifications of cytosolic proteins, plant proteins plasma-derived or recombinant blood coagulation factors and other proteins from different sources including recombinant proteins. Membrane proteins and lipo-proteins can also be isolated with this technique.

Ion-exchange chromatography is the method of choice for the intermediate purification of monoclonal and polyclonal antibodies, but can also be used for capture. In contrast to purification strategies for plasma-derived antibodies, where weak ion-exchangers are widely used, strong ion-exchangers are preferred if monoclonal antibodies have to be manufactured. Antibodies have been reported to have isoelectric points between 5 and 8 and, therefore, can be purified at $pH < pI$ on cation exchangers. High purities and high recoveries can be obtained using weak cation exchangers carrying a carboxy group changing the degree of ionisation depending on pH. However, the main ligand for cation-exchange chromatography is in many cases the sulpho group. Cation exchangers are able to remove unfolded forms of the monoclonal antibody as well as aggregates of the monoclonal antibody and small molecular weight impurities. Both modes of ion-exchange chromatography are able to separate protein A leakage products and host cell proteins. Strong anion-exchange chromatography can efficiently remove DNA and, in addition, was described to contribute to the virus clearance.

DISPLACEMENT CHROMATOGRAPHY OF BIOMACROMOLECULES

The displacement chromatography is an exclusively nonlinear form of chromatography. This makes it more difficult to apply and optimise. However, much progress has been made over the last two decades both in the development of computer tools and mathematical algorithms for the simulation of nonlinear chromatography. In addition, the question of what constitutes a suitable displacer for typical applications in protein or peptide separation and how to prepare them has been addressed on both a theoretical and practical basis. Applying the displacement approach today may still be challenging, but has become considerably easier as a result of these efforts. Concomitantly, the many advantages of displacement chromatography should be considered, most prominently the ability to use the stationary-phase capacity to an extreme extent, often allowing semi-preparative purifications on analytical columns, the ability to work isocratically with just a single displacer step even in the case of protein chromatography, but also the possibility to co-enrich and separate a large number of substances that differ considerably in concentration, in any valuation of this chromatographic mode.

Whereas counter-current chromatography is regularly employed, in both academia and industry, for the separation of natural product extracts, synthetic mixtures and intermediate-polarity compounds, the purification of biomolecules by counter-current chromatography is a technique still in its infancy and thus its full potential has yet to be realised. A number of different approaches exist for the purification of compounds such as peptides, proteins, hormones, lipids, sugar derivatives and enzymes, but currently none have been completely investigated and the full benefits of the technique in this field of research are still to be determined.

However, some impressive examples of biomolecule purifications exist in the scientific literature, many where labile or otherwise difficult compounds were purified by counter-current chromatography when other techniques failed. With its inherent advantages over solid-phase chromatography, such as high loading, total sample recovery and the ability to accept particulates, counter-current chromatography looks set to become an important process in the field of bioseparations. Furthermore, the ever-increasing understanding of the processes that affect the separation and the development of easy-to-follow protocols for the selection of solvent systems combine to make counter-current chromatography a technique that has a definite future in the bioseparations business.

The difficulty of a given downstream process depends mainly on the complexity of the feed and the target molecule concentration therein together with the required final purity and composition. Feed product concentrations may vary between several grams per litre (antibiotics) and some milligrams per litre (recombinant blood factors). In order to gain approval for a pharmaceutical, for example, it must be guaranteed that a sufficient amount can be provided to satisfy the fore seeable medical need. Even if this may mean only a few kilograms per year, the sheer size of the product stream can pose a problem given the typical product concentrations. Moreover, substances which typically are produced at rather low product titers in the bioprocess tend to require the very highest levels of final purity, e.g. parenteralia, intended for use in human beings. Concentration plus isolation are therefore the somewhat contradictory goals of many downstream processes in biotechnology. Intellectually, the downstream process is often divided into three distinct stages that differ somewhat in goal. These stages are capture, intermediate purification and final polishing. These three stages are typically preceded by a preparatory stage in which the product stream is rendered suitable for (standard) downstream processing. In particular, intermediate and final purification steps tend do depend heavily on chromatographic principles to achieve separation. This has largely to do with the high selectivity and 'biocompatibility' of chromatographic operations. Concomitantly, however, chromatographic steps today contribute drastically to the difficulties and cost of the corresponding isolation process.

In displacement chromatography, this competition for the binding sites is used to drive the separation. The first step of a displacement separation is the adsorption of the substance mixture on the column under conditions favourable to binding. During loading some degree of separation is already achieved due to a frontal chromatographic effect. A considerable portion of the stationary phase capacity may be exploited during that phase.

The growing number of successful applications of displacement chromatography in bioseparation shows, however, that displacement chromatography is less limited by these circumstances than one would assume from the theoretical considerations.

The development of a separation by displacement chromatography will usually involve the following steps: (i) choosing the stationary phase/chromatographic mode, (ii) choosing the displacer, (iii) choosing/optimising the carrier composition, (iv) adjusting the column length/sample size, (v) adjusting the flow rate and perhaps, and (vi) adjusting the temperature, as an increase in temperature may result in beneficial effects such as a decrease in viscosity or improved mass transfer and reaction kinetics.

Composition of the mobile phase/flow rate

In displacement chromatography it is often advised to use a mobile phase that supports strong binding of the substances and the displacer to the stationary phase. However, it should be kept in mind that the separation factor rather than the individual retention factors determine the separation of any two components and that even the displacer retention factor cannot be considered individually, but must be optimised in relation to the displacer concentration. While a higher displacer concentration means a faster separation and high product concentration factor, it also diminishes the length of the zones and thus increases the amount of substance found in the shock layers. In practical terms, however, the solubility of the displacer will usually be the limiting factor in that regard.

Displacers for Biopolymer Displacement Chromatography

The displacer is a unique feature of displacement chromatography. The importance of the displacer for a successful displacement separation can hardly be over estimated. In general, the ideal displacer should have the following features. It should be nontoxic and stable. It should combine high solubility in the carrier with a high binding tendency towards the stationary phase. Column regeneration must nevertheless be possible. In addition, the displacer should be highly uniform, since displacer impurities/heterogeneity may make column regeneration difficult or pollute the substance zones depending on their relative affinity. Detectability, costs and the possibility to recycle the displacer are other considerations. Ideally the displacer should also be easy to remove from the recovered substance fractions. For pharmaceutical applications it might even be necessary to sterilise the material.

Special Variants of Displacement Chromatography

Most applications of displacement chromatography are more or less straightforward variants of the experimental setup as outlined above. However, apart from the standard scheme and application area, a number of derivations and highly specialised forms of displacement chromatography exist, which may also be useful to the biotechnologist and applicants in related fields.

Complex displacement chromatography

Complex displacement chromatography is related to ordinary displacement chromatography less through its mechanism and more through the chemicals used. A typical application of complex displacement chromatography was the isolation of an antibody (cationic protein) using a cation-exchange column and substances such as the CM-Ds ordinarily used as displacers in anion-exchange displacement chromatography.

Selective displacement chromatography

The efficacy of low-molecular-weight displacers depends much more than that of high-molecular-weight displacers on both the mobile phase salt and the displacer concentration. This method enhances the resolving power of the displacement approach by establishing conditions under which mainly the product is displaced, while impurities of lower binding strength are eluted ahead of the displacement train in the induced salt gradient and impurities of higher binding strength are either retained on the column or desorbed in the displacement zone.

In this approach, the sample mixture is applied as a spot to the standard planar stationary phases. The use of spacers, i.e. the spacer displacement chromatography approach, is necessary to keep the substance zones apart and available for visualising. TLDC has, for example, been used to isolate the metabolites of radio labelled Deprenyl and its metabolites in rats urine, and for the screening of ecdysteroids (a class of important steroid hormones) from plants.

Analytical aspects of displacement chromatography

Displacement chromatography is correctly considered a predominantly preparative technique. As such it may, however, become a useful tool in analytical biochemistry. Many analytical procedures in biochemistry, molecular biology and related fields involve the separation of a complex mixture, e.g. a peptide digest, prior to a closer analysis of the individual components of the mixture or at least the resulting less-complex subsets of the mixture. The displacement process, which focuses even trace components into highly concentrated zones while enriching all mixture components to a high extent, is a prime choice for such sample pretreatment steps. At the same time no previous knowledge of the exact physical nature of the 'impurity' is necessary. Trace components, which may be difficult to isolate by conventional chromatographic methods, can be obtained in sufficient amounts to allow chemical characterisation by displacement chromatography. Displacement chromatography also played a key role in the discovery of two previously unknown amino acids, amarine and feline.

Displacement chromatography has also repeatedly been used instead of conventional elution chromatography for concentration/separation in hyphenated techniques.

The displacement approach could also be a much more suitable 'first-dimension' proteome analysis scheme than either electrophoresis or elution chromatography, due to the fact that the concentration of a given component in the original sample is of little consequence. Substances varying by several orders of magnitude in concentration can be co-enriched by displacement chromatography without interference.

No matter how long the zone of a major component, it will always be followed or preceded by the individual zone of a minor or trace component.

Miscellaneous aspect of displacement chromatography

The concept of displacement chromatography has been adapted to centrifugal partitioning chromatography and continuous annular chromatography. In case of centrifugal partition chromatography, an ionic liquid, i.e. benzalkonium chloride, was used as strong anion exchanger together with iodine as displacer.

Applications of Displacement Chromatography for Separations in Biotechnology

Although displacement chromatography is far from being an accepted standard operation in industrial bioseparation, results accumulated over the last decade and a half have demonstrated that the technique can be a powerful tool for the purification of antibiotics, peptides and proteins. Certain features of displacement chromatography make the method even ideally suited to application in proteomics or the efficient isolation of a few grams of pure protein using analytical equipment in the biopharmaceutical industry. While the number of applications of displacement chromatography for bioseparation is still small compared to the elution mode, it is, however, already much too large to discuss each case in detail.

Separation and isolation of amino acids, peptides and antibiotics

Reversed-phase chromatography dominates this particular area of displacement separations and the separation of the product of a (solid-phase) peptide synthesis from its closely related by-products remains one of the typical applications. Although less common, the separation of amino acids by displacement chromatography is possible. The separation of antibiotics is another common application of reversed-phase displacement chromatography.

Protein separation

Among the first applications of displacement chromatography to protein isolation was the purification of crude β-galactosidase from *Aspergillus oryzae* on a weak anion exchanger with chondroitin sulphate as displacer. A comparison with the elution mode demonstrated the superiority of displacement chromatography in terms of the utilisation of the stationary- and mobile-phase capacity, throughput, and waste production.

Separation of isomers

While the large-scale separation of isomer mixtures in the pharmaceutical industry today is a typical application of the SMB technology, displacement chromatography has also been recognised as a method for the efficient separation of optical and structural isomers at high throughputs and concentrations, perhaps at the intermediate rather than large scale.

Applications of Displacement Chromatography

Displacement chromatography has also been used for the isolation of many other substances. Huang and Jin reported on the use of displacement chromatography to purify methyl esters of polyunsaturated fatty acids. Oleic acid was used as displacer. As already mentioned, displacement chromatography has also been suggested as an economically sound way for the large-scale purification of oligonucleotides.

Reversible chemical reactions are another area of application for displacement chromatography. Such reactions require the removal of the product in order to reach high conversions. An interesting application in this context is the combination of displacement chromatography with packed-bed enzyme reactors for integrated reactor/separator schemes.

Reverse Phase Chromatography (RPC)

This chromatographic method utilises a solid phase (e.g. silica) which is modified so as to replace hydrophilic groups with hydrophobic alkyl chains. This allows the separation of proteins according to their hydrophobicity. More-hydrophobic proteins bind most strongly to the stationary phase and are therefore eluted later than less-hydrophobic proteins. The alkyl groupings are normally eight or eighteen carbons in length (C_8 and C_{18}). RPC can also be combined with affinity techniques in the separation of, for example, proteins and peptides.

High Performance Liquid Chromatography (HPLC)

HPLC is a high resolution column chromatographic technique. Improvements in the nature of column packing materials for a range of chromatographic techniques (e.g. gel permeation and ion-exchange) yield smaller, more rigid and more uniform beads. This allows packing in columns with minimum spaces between the beads, thus minimising peak broadening of eluted species. It was originally known as high pressure liquid chromatography because of the high pressures required to drive solvent through silica based packed beds. Improvements in performance led to the name change and its widespread use in the separation and purification of a wide range of solute species, including bio-molecules. HPLC is distinguished from liquid chromatography by the use of improved media (in terms of their selectivity and physical properties) for the solid (stationary) phase through which the mobile (fluid) phase passes.

The stationary phase must have high surface area/unit volume, even size and shape and be resistant to mechanical and chemical damage. However, this will lead to high pressure requirements and cost. This may be acceptable for analytical work, but not for preparative separations. Thus, in preparative HPLC some resolution is often sacrificed (by the use of larger stationary-phase particles) to reduce operating and capital costs. For very high value products large-scale HPLC columns containing analytical media have been used. Affinity techniques can be merged with HPLC to combine the selectivity of the former with the speed and resolving power of the latter.

Continuous Chromatography

Although the concept of continuous enzyme isolation is well established the stage of least development is continuous chromatography. Fox developed a continuous-fed column for this purpose. It consisted of two concentric cylindrical sections clamped to a base plate. The space (1 cm wide) between the two sections was packed with the appropriate resin or gel giving a total column capacity of 2.58 dm^3. A series of orifices in the circumference of the base plate below the column space led to collecting vessels. The column assembly was rotated in a slow-moving turntable (0.4–2.0 rpm). The mixture for separation was fed to the apparatus by an applicator rotating at the same speed as the column, thus allowing application at a fixed point, while the eluent was fed evenly to the whole circumference of the column. The components of a mixture separated as a series of helical pathways, which varied with the retention properties of the constituent components.

This method gave a satisfactory separation and recovery but the consumption of eluent and the unreliable throughput rate were not considered to be satisfactory for a large-scale method. However, the development of such continuous separation equipment suitable for large scale extraction would considerably simplify the use of chromatographic separation.

PURIFICATION OF BIOMOLECULES BY COUNTER-CURRENT CHROMATOGRAPHY

Counter-current chromatography is a form of liquid–liquid separation technology. There is no solid stationary phase, as in the more 'traditional' forms of chromatography such as high-performance liquid chromatography (HPLC) or flash chromatography. In counter-current chromatography instruments, tubing is wound on a drum (called a bobbin) which is centrifugally rotated in planetary motion, i.e. revolving around the central axis of the sun gear while simultaneously rotating about its own axis at the same angular velocity. A two-phase solvent system is used for the separation. For most counter-current chromatography separations, this two-phase system comes from an organic/aqueous solvent mix. A simple example would be heptane/water, but a more likely system might consist of heptane, ethyl acetate, methanol and water. This gives a two-phase system with an upper organic layer consisting mainly of heptane/ethyl acetate and a lower aqueous layer consisting mainly of methanol/water. The tubing is initially filled with the solvent phase intended to be stationary and, with the instrument spinning, the mobile phase is pumped through it.

A small initial displacement of stationary phase occurs before an equilibrium is set up, with the planetary motion retaining up to 85 per cent of the stationary phase in the coil and the mobile phase passing through, eluting at the far end without displacing any more stationary phase. Counter-current chromatography can be used in normal or reverse-phase mode, depending on which phase (upper or lower) is selected to be stationary. Furthermore, material that elutes very slowly can be recovered without any compound losses by pumping out the stationary phase while maintaining resolution.

Counter-current Chromatography Compared to Solid-phase Chromatography

Both counter-current chromatography and solid-phase chromatography have the same ultimate aim – to separate a mixture of solutes into their component compounds in order to isolate one or more of them. The manner in which separations are achieved is similar. Both techniques keep one phase stationary as a second phase passes through it. Furthermore, apart from the actual separation columns, much of the equipment is similar and shares the same technology, e.g. the pump, injector and detector apparatus. However, there the similarities end.

The most fundamental difference between solid-phase chromatography and counter-current chromatography is that the former has a solid stationary phase and thus adsorption, ion-exchange or hydro, phobic interaction is the general mechanism of sample retention, whereas the latter has a liquid stationary phase and so liquid partitioning is the sample retention process.

Solvent System Selection Process

The whole heart of a counter-current chromatography separation is the solvent system. For the majority of counter-current chromatography purifications, the two-phase system is created from a mixture of organic and aqueous solvents. In the past, developing a solvent system for counter-current chromatography required lots of time, operator skill and experience. With biphasic systems coming from two-, three-, four- or even five- or more-component solvents, this made the range of options almost limitless. On the plus side, this gave the technique the ability to separate compounds ranging from extremely polar to extremely nonpolar.

The down side was that it was difficult to know where to begin. Various systematic, step-by-step protocols have now been developed that take an inexperienced user logically through the process. The solvent selection part has even been fully automated using a liquid-handling robot and HPLC.

Counter-current Chromatography of Polar Biomolecules

For many biomolecules, a simple organic/aqueous solvent system is unsuitable. Biomolecules such as peptides and sugar derivatives are extremely water soluble, while compounds such as proteins, enzymes and nucleic acids can be readily denatured by contact with organic solvents.

Considering first the problem of extremely water-soluble molecules, several options are available which are given below:

1. pH adjustment of the aqueous layer.
2. Addition of salts to the aqueous layer.
3. pH zone-refining counter-current chromatography.
4. Addition of affinity ligands to the stationary phase.
5. Room temperature ionic liquids.

If, however, the biomolecules to be purified are sensitive to organic solvents, e.g. most proteins, then an aqueous–aqueous polymer system must be developed.

CONTINUOUS CHROMATOGRAPHY IN THE DOWNSTREAM PROCESSING OF PRODUCTS OF BIOTECHNOLOGICAL AND NATURAL ORIGIN

Continuous chromatography has found an attractive niche in the production of small-molecule pharmaceutical compounds with several production systems running and a multitude of systems in research and development for the preparation of first-kilogram amounts. With the lessons learned in small-molecule production, continuous chromatographic processes are now ready to make the next step into biopharmaceutical manufacturing.

The continuous operation itself might be improved enough as it fits well into continuous fermentation systems, where it helps to avoid the intermediate storage of labile products. The counter-current operation, e.g. in simulated moving bed (SMB) mode, will show its benefits in difficult separations, e.g. in size-exclusion chromatography, where otherwise long columns have to be packed to achieve the necessary column efficiency. SMB chromatography in size-exclusion chromatography mode will show much higher productivities and will help size-exclusion chromatography to gain more impact in biopharmaceutical production. New modes of counter-current operation will be developed and fitted to the needs of bio-chromatography. These modes will combine the advantages of continuous and counter-current operations with the flexibility in mobile phases, which are typically used in biochromatography.

Some obstacles still have to be overcome to show the full potential of continuous counter-current chromatography in biopharmaceutical manufacturing. The availability of suitable stationary phases will be no problem as today rigid and pressure stable phases for all operation modes (e.g. ion-exchange, size exclusion chromatography, hydrophobic interaction) have been developed and shown to be easily packed into large-scale production columns. Major concerns regarding SMB, ISEP and production-scale continuous annular chromatography are related to the contact of the sometimes labile product with a multitude of stainless steel columns, tubes, pumps and valves. The residence time of the product inside the systems is also an issue as is the cleanability of the system. Appropriate CIP and sterilisation-in-place protocols have to be developed under production conditions to show the equivalence of the new chromatographic methods. Modelling and simulation of the new processes is still a major task as the separation problems are becoming more and more complex, with a multitude of very different impurities and changing mobile-phase compositions during the elution. Together with the better understanding of the chromatographic processes it is absolutely necessary to teach the new concepts and complex operation modes so that the operators in biopharmaceutical production feel safe in the daily use of the new systems.

Nevertheless, continuous and, in particular, counter-current chromatographic operations are ready to play their role in modern biopharmaceutical production, helping to fulfill the production demands of modern biopharmaceutical drug compounds. Two promising areas of application can be seen in the field of industrial biotechnology, where productive and economic downstream processes are needed to isolate complex value compounds, e.g. from oligo saccharide feed streams. The other potential application arises with the advent of proteins produced in the milk of transgenic animal. In-line downstream processing of proteins from animal milk with continuous biochromatographic systems is both a challenge and a dream. This dream might come true some day, with farmers herding transgenic cattle and being as familiar with SMB chromatography for product isolation as they are today with continuous centrifugation for defatting milk. Intensive work has been done to improve single chromatographic units for the separation of small molecules. In particular, for difficult separations with low selectivities and larger production amounts in the multi-ton range, continuous chromatographic concepts like the simulated moving bed (SMB) have proven their ability to improve the process performance and thus are accepted techniques applied to an industrial scale.

The time has come to bring the SMB technology in the pharmaceutical industry to the next level, approaching more complex separations under difficult constraints. When designing a complete downstream process, which often consists of three or more chromatographic steps, it is not only the choice of single chromatographic principles (e.g. affinity, ion-exchange or size-exclusion) and their interconnection that secures the success of a project. The mode of operation might also contribute and improve the overall process performance. Apart from the classical batch operation in a single column, continuous-process modes have been available for a long time, but their potential has not yet been realised in macromolecule production. In this section we will show where and how the continuous chromatographic purification of macromolecular pharmaceutical compounds will deliver a valuable contribution in the race towards new and economically manufactured drug products.

CONTINUOUS ANNULAR CHROMATOGRAPHY

Continuous feed introduction has been a goal in preparative chromatography for a long time. In 1949, Martin suggested a system for continuous feed introduction that operates in an annular mode. The idea was to pack the selective sorbent into an annulus, which slowly rotates around a central axis, while fresh eluent is introduced over the whole cross-section at the top of the bed. This mode of operation results in a cross-current movement of solid and fluid phase. Depending of the affinity of the different substances towards the solid adsorbent, they can be collected and purified at different positions at the column outlet.

Liquid chromatography is often the first and sometimes the only choice for the isolation of biological macromolecules such as proteins from complex multicomponent mixtures, including typical feeds from the biotechnology industry such as culture supernatants or cell lysates. This continued popularity is due to a number of factors—chromatography is universally applicable, versatile, has high-resolution capabilities and can be operated under 'physiological' conditions. Biomolecules (proteins, antibodies, nucleic acids, etc.) can be isolated using a number of interaction ranging from electrostatic and hydrophobic interactions (ion-exchanges, respectively, reversed-phase and hydrophobic-interaction chromatography) to specific biological interaction (affinity chromatography). If necessary, several orthogonal separation principles can be used in series. In addition, the relative simple separation by size (size-exclusion chromatography, gel filtration) is often applied. The development of a chromatographic separation is also aided by the ever-increasing selection of dedicated chromatographic stationary phases that become commercially available. At the preparative scale, however, the capacity/loadability of the column and thus the throughput of the chromatographic separation frequently presents a limiting factor for such operations. Scale-up of the typical batch column is simply done by increasing the cross-sectional area (diameter) of the column, while keeping the column length constant. Under such circumstances the resolution should stay comparable. However, this approach is limited by the maximum diameter, which is still compatible with packing a uniform and stable bed. Even with radial and axial compression, such columns will rarely be much more than a 1 m wide. The column length (bed height), on the other hand, is limited by the applicable pressure. In most production environments, high pressure is not an option.

The idea of combining the adaptation of the chromatographic separation process to large-scale with the introduction of continuous feed injection and product withdrawal has therefore been discussed for more than 50 years as a means to make chromatography more competitive in the industrial sector in terms of reduced residence time for the feed and simplified process control if the continuous system reaches the steady-state, for example. One approach to continuous chromatography, i.e. the simulated moving bed (SMB) approach is eminently suited to the separation of two-component/two-fraction mixtures. However, many separations, especially in recombinant biotechnology, require multicomponent separations at several stages, while the scale of the operation (kilogram amounts) is often below that most suited to the SMB. In such cases another concept for continuous chromatography, i.e. continuous annular chromatography, may be considered.

Continuous annular chromatography may be performed on a continuous (annular) bed or be approximated using a carrousel of columns. Both approaches have been realised in the past, and both have their distinct advantages and disadvantages. The homogeneous packing of the involved chromatographic beds is of primary importance in both cases, as differences in flow resistance will obviously bias the separation. In the case of a column carrousel, the number of columns limits the separation. Introduction of feed and eluent is technically involved, and becomes more so as the number of columns increases. Fraction collection at distinct points, on the other hand, can be done by a simple fraction collector. In the case of an annular column, the chromatographic bed is continuous. However, the number of distinct collectable fractions will be limited by the number or outlet points. The application of pressure to such a column requires the design of a dedicated 'head', while the seal between the rotating column and the fixed outlet points may present a weak point in column design.

Applications of Continuous Annular Chromatography Separation in Biotechnology

From the viewpoint of bioprocess development, one of the most attractive possible applications of the continuous annular chromatography approach seems to be for the processing of large volumes of complex, low-titer product streams, e.g. in molecular or cellular biotechnology. Especially when interfaced to a continuous production process, continuous annular chromatography may considerably contribute to a reduction of the scale of the production facility. A continuous fermentation process is an efficient method for biomolecule production, the cell density of which can be significantly increased by running the process in the perfusion mode, resulting in very high space-time yields. It is, therefore, a logical consequence also to operate the downstream side as much as possible continuously in order to keep the residence time of the culture supernatant at a minimum and productivity at high level. Here, the integration of continuous annular chromatography into a continuous production process is certainly an option.

Chapter 29

Membrane Separation Processes

INTRODUCTION

A membrane is a layer of material which serves as a selective barrier between two phases and remains impermeable to specific particles, molecules or substances when exposed to the action of a driving force. Some components are allowed passage by the membrane into a permeate stream, whereas others are retained by it and accumulate in the retentate stream. Membranes can be of various thickness, with homogeneous or heterogeneous structure. Membrane can also be classified according to their pore diameter. Membranes can be generally classified into three groups: inorganic, polymeric or biological membranes. These three types of membranes differ significantly in their structure and functionality.

This chapter discusses the major membrane processes employing synthetic organic and inorganic membranes and to indicate where they have found use in the biotechnology industry.

CRITERIA FOR CONSIDERING THE USE OF MEMBRANES

Before examining how membrane processes operate, it is worthwhile giving some thought to the characteristics of the process material and the separation desired in order to assess the suitability of membranes for that separation. Although the final decision to use a membrane separation step will be based on a far more detailed study of the process, the following are some useful rules of thumb. The list is not exhaustive but forms a preliminary guide.

Consider membranes if:

1. Concentration of a dilute feed is being considered—membranes may offer a lower energy alternative to evaporation.
2. The difference in size of the molecules to be separated is at least a factor of ten.
3. A heat-sensitive material is being concentrated.
4. Sterilisation of heat sensitive materials is required.
5. A very clear liquid is desired, e.g. drinks production.
6. Separation is possible on the basis of charge—use electrodialysis.
7. Trace organics are to be removed from aqueous streams—consider pervapouration.

Do not give membranes serious consideration if there is:

1. A high concentration of low molecular material—because of its high osmotic pressure. This mainly applies to reverse osmosis.
2. A high solids content (say >25 per cent w/w)—because of problems with pumping this material past the membrane.

Membrane separations cover a wide range of fundamentally different processes, from microfiltration to electrodialysis. The common factor linking this diverse range of separation operations is the physical arrangement of the process. In all, the separation occurs between two fluids that are separated by a thin physical barrier or interphase. This interphase constitutes the membrane and permits the selective transport of components of the two phases thus allowing some materials to pass through the membrane while others are retained. Typically in many modern membrane systems the thickness of this membrane is in, the region of 100 μm although the active layer is usually less than a micron. Membrane processes operate on a variety of physical principles, and therefore it can be misleading to group them together, casually, under the single title of 'membrane processes'. As will be shown later the mechanism, or driving force, by which the separation is affected is not the same for all membrane operations.

In common with many other separation processes membrane separations are rarely absolute, and in some the fluid under separation can influence the performance of the membrane. However, the potential advantages of membrane separations are significant. Membrane separations usually have a low energy requirement: most do not involve a phase change. This contrasts with established processes such as distillation, crystallisation or evaporation, which involve significant energy inputs. The conditions under which the separations are conducted are normally mild, which is an important consideration when complex heat-sensitive materials are involved. The mild operating conditions and low energy requirements are among the principal reasons why membrane processes have been found to be suitable for a range of biotechnological separations.

MEMBRANE MATERIALS AND APPLICATIONS

Membrane Materials

The first synthetic materials to be formed as membranes were developed in the first part of the twentieth century. Early materials exhibited a selectivity for species at the molecular level but the flux of these species was low. The first membranes were isotropic (also called symmetrical); that is, the membranes had a broadly similar pore structure and chemical composition throughout. Various techniques were developed to produce synthetic membranes with defined pores. Leachable pore formers were added to the polymers to increase the permeability of the membrane but the membrane remained essentially a homogeneous structure with continuous pores traversing the membrane. Many of the first membranes were based on cellulose. These were later superseded by modified celluloses which combined good material properties and flux characteristics for a wide range of separations. Indeed, many commercial membranes are still manufactured in cellulose-based materials. The flux of species through semipermeable materials is approximately inversely proportional to the thickness of the material, thus the search continued for thinner membranes. In the early 1960s a method of casting membranes was developed which produced membranes in which the active layer was very thin. The technique produces membranes by a process of phase inversion. This yields membranes with an anisotropic (or asymmetric) structure in which one side of the membrane possesses a thin layer of dense polymer with either no pores or small ones, supported on an open very porous substructure of the same polymer.

These membranes have the advantage of improved separation combined with high strength and higher fluxes. Phase inversion can be brought about by a number of different mechanisms, but most commercial membranes are formed by immersion precipitation. A solution of a polymer in a solvent is formed into a thin film or layer and then brought in contact with a non-solvent (for the polymer). The solvent and nonsolvent must be miscible. The exchange, by diffusion, of the solvent and nonsolvent, at the interface, produces a thermodynamically unstable condition, causing the polymer solution to separate, or demix, into two stable phases: a polymer-rich phase and a polymer-lean phase. This process generally takes place very rapidly. The final structure of the membrane is determined by the rate of this process and other factors such as the polymer concentration and the choice of non-solvent.

The thin dense layer can either be porous or non-porous. This layer forms the membrane. The lower highly porous substructure is of the same polymer but plays no significant part in the separation process. Instead it acts a structural support for the thin upper surface.

A large number of polymers can be formed into membranes; however, only a few form membranes which are of any practical use. Cellulose or modified cellulose used in the first isotropic membranes can also be cast as an asymmetric membrane. Consequently many of the early phase-inversion membranes were made from modified cellulose, but currently the most popular polymer is probably polysulphone due to its chemically inert nature and high maximum operating temperature. Another common membrane polymer is polyamide (nylon). Although membranes have been developed which demonstrate a resistance to a wide range of chemical conditions, a drawback of most organic membranes is the low maximum operating temperature. Recent research has focused, therefore, on the development of inorganic membranes, e.g. sintered ceramics and stainless steel. These new materials demonstrate a high chemical and temperature resistance but difficulties remain in the production of large, defect-free sections. In the context of bioseparations operating temperatures are likely to be low, but for sterile operation membranes may need to withstand several steam or chemical sterilisation cycles.

While the membranes produced by phase inversion are adequate for microfiltration and ultrafiltration, the selectivity of tighter membranes for lower molecular weight separations is often less than expected. This is due in part to defects or pinholes in the membrane. To overcome imperfections in the membrane and thus improve its selectivity, techniques have been developed that coat or deposit a tight continuous skin of a highly permeable polymer on the surface. In the mid-1970s thin film composite membranes were developed. In these an additional layer of a selective polymer is deposited on top of the dense skin of an asymmetric membrane in order to improve selectivity and chemical resistance. These second generation membranes find use in reverse osmosis, gas separation and pervapouration.

Applications of Membranes

The mild operating conditions of many membrane separations make them favourable for a number of biotechnological separation steps. The list below sets out some of the uses for which membranes are now used routinely. Because kidney dialysis units are single-use items, dialysis represents the most significant market for membranes in economic terms. In the biotechnology sector microfiltration, mainly for sterilisation, represents the largest application of membranes followed by ultrafiltration in the food and dairy industries. The market for gas separation and pervapouration is still small in economic terms but it is anticipated that these markets are likely to grow more rapidly over the next decade.

1. Microfiltration, MF (also called membrane filtration):
   - Sterilisation using, for example a 0.2 μm cut-off membrane.
   - Separation of cells from fermentation broth.

2. Ultrafiltration, UF:
   - Protein concentration, e.g. enzymes.
   - Antibiotic concentration.
   - Antibody concentration.
   - Milk concentration.
   - Plant cell recovery.
   - Desalting of blood.
3. Reverse osmosis, RO (also called hyperfiltration):
   - Desalination.
   - Ultra-pure water production.
   - Effluent treatment.
   - Concentration of:
   (a) Antibiotics
   (b) Amino acids
   (c) Vitamins
   (d) Flavours
4. Gas separation:
   - Production of enriched nitrogen or oxygen from air.
   - Landfill methane purification.
5. Dialysis:
   - Artificial kidneys—haemodialysis.
   - Production of low alcohol beer.
6. Pervapouration:
   - Alcohol dehydration
   - Removal of organics from aqueous solutions.

One possible classification of membrane separation processes is by mean pore size within the membrane, the driving force across the membrane and the transport mechanism. Table 29.1 sets out the normally accepted classification. Alternatively, membrane processes can be classified on the basis of the size of the molecule being separated. Bear in mind that these classifications overlap; they are not rigidly defined.

Table 29.1: Normally accepted classification of membrane separation processes.

Process	Pore size	Driving force	Mechanism
Microfiltration	0.02–10 μm	Pressure 0–1 bar	Sieving
Ultrafiltration	0.001–0.02 μm mol. wt. 10^3–10^6	Pressure 0–10 bar	Sieving
Reverse osmosis	Non-porous mol. wt. < 1000	Pressure 0–100 bar	Solution-diffusion
Gas separation	Non-porous	Pressure 0–100 bar	Solution-diffusion
Dialysis	10–30 Å	Concentration difference	Sieving plus diffusivity differences
Electrodialysis	mol. wt. < 200	Electrical potential	Ion migration
Pervapouration	Non-porous	Partial pressure difference	Solution-diffusion

MEMBRANE PROCESSES

Ultrafiltration and Reverse Osmosis

Both processes utilise semi-permeable membranes to separate molecules of different sizes and therefore act in a similar manner to conventional filters.

Ultrafiltration

Ultrafiltration (Fig. 29.1) can be described as a process in which solutes of high molecular weight are retained when the solvent and low molecular weight solutes are forced under hydraulic pressure (around 7 atmospheres) through a membrane of a very fine pore size. It is therefore used for product concentration and purification. A range of membranes made from a variety of polymeric materials, with different molecular weight cut-offs (500 to 500,000), are available which makes possible the separation of macro-molecules such as proteins, enzymes, hormones and viruses. It is practical only to separate molecules whose molecular weights are a factor of ten different due to variability in pore size. Because the flux through such a membrane is inversely proportional to its thickness, asymmetric membranes are used where the membrane (~0.3 μm thick) is supported by a mesh around 0.3 mm thick.

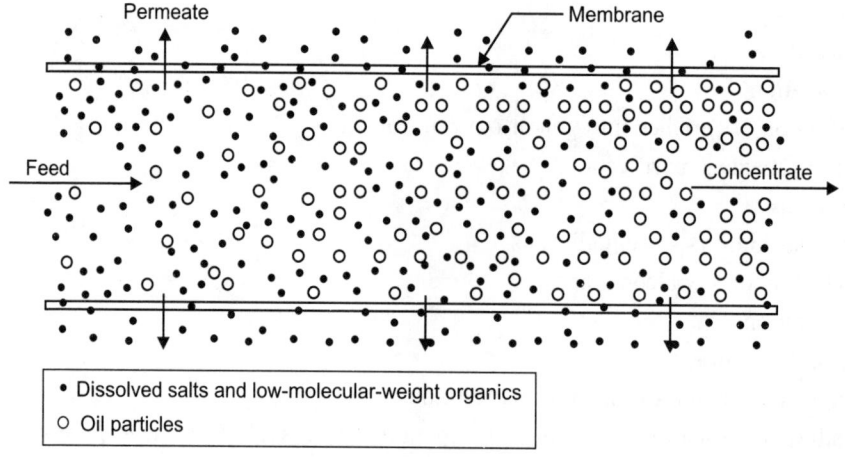

Fig. 29.1: Ultrafiltration basics.

When considering the feasibility of ultrafiltration it is important to remember that factors other than the molecular weight of the solute affect the passage of molecules through the membranes. There may be concentration polarisation caused by accumulation of solute at the membrane surface which can be reduced by increasing the shear forces at the membrane surface either by conventional agitation or by the use of a cross-flow system. Secondly, a slurry of protein may accumulate on the membrane surface forming a gel layer which is not easily removed by agitation. Formation of the gel layer may be partially controlled by careful choice of conditions such as pH. Finally, equipment and energy costs may be considerable because of the high pressures necessary; this also limits the life of ultrafiltration membranes.

There are numerous examples of the use of ultrafiltration for the recovery of bio-molecules: viruses, antibiotics. Affinity ultrafiltration is a novel separation process developed to circumvent difficulties in affinity chromatography. It offers high selectivity, yield and concentration, but it is an expensive batch process and scale-up is difficult.

Reverse Osmosis

Reverse osmosis (RO) is a pressure-driven process. It separates low molecular weight solutes such as dissolved salts and amino acids from a solvent. The transport mechanism is primarily solution-diffusion rather than the sieving action of microfiltration and ultraflltration. Reverse osmosis was used initially in the desalination of brackish water but is now employed extensively in the food and dairy industries to concentrate fruit juices, vegetable juices, and milk, all of which contain a high level of dissolved minerals and salts which need to be retained. One application which is of increasing importance is the production of ultra-pure water for the pharmaceutical and electronics industries. Principle of reverse osmosis is shown in Fig. 29.2.

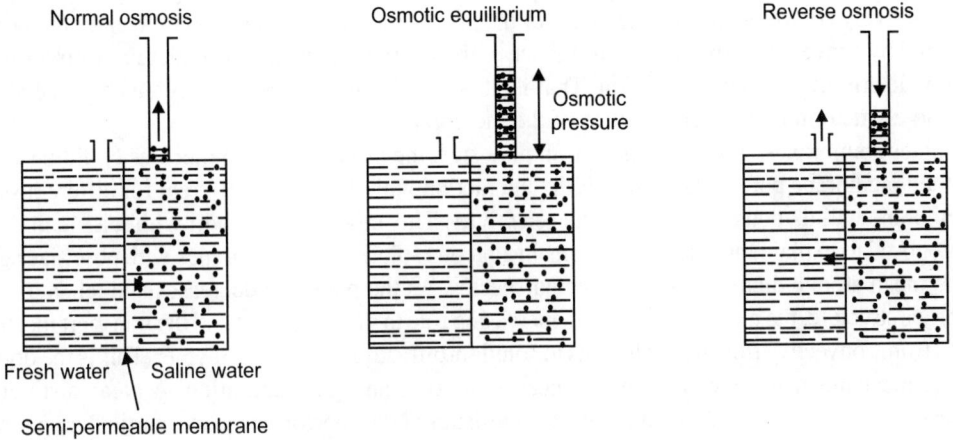

Fig. 29.2: Principle of reverse osmosis.

In RO, a solvent (usually water) is removed from a solution in which the solute is a small molecule and so a non-porous membrane is required. The membranes are anisotropic with a skin of active membrane 0.2–0.5 μm thick on a porous support film 50–100 μm thick. Nanofiltration is a comparatively new process which can be regarded as a variant on RO. The membranes are slightly more porous than those for RO and can be used to separate molecules up to molecular weight 500. Hence it is possible to remove salts from sugars or to separate some salts, e.g. monovalent from divalent salts. Thus, reverse osmosis is a separation process where the solvent molecules are forced by an applied pressure to flow through a semi-permeable membrane in the opposite direction to that dictated by osmotic forces and hence is termed reverse osmosis. It is used for the concentration of smaller molecules than is possible by ultrafiltration. Concentration polarisation is again a problem and must be controlled by increased turbulence at the membrane surface.

Liquid Membranes

Liquid membranes are insoluble liquids (e.g., an organic solvent) which are selective for a given solute and separate two other liquid phases. Extraction takes place by the transport of solute from one liquid to the other. They are of great interest in the extraction and purification of biologicals for the following reasons:

1. Large area for extraction.
2. Separation and concentration are achieved in one step.
3. Scale-up is relatively easy.

Their use has been reported in the extraction of lactic acid and citric acid using a supported liquid membrane. The utilisation of selective carriers to transport specific components across the liquid membrane at relatively high rates has increased interest in recent years. Liquid membranes may also be used in cell and enzyme immobilisation and thus provide the opportunity for combined production and isolation/ extraction in a single unit. The potential use of liquid membranes has also been described for the production of alcohol reduced beer as having little effect on flavour or the physico-chemical properties of the product.

Membrane Selectivity

In porous membranes, especially microfiltration membranes and to a lesser degree ultrafiltration membranes, the membrane acts principally as a sieve. Possessing relatively large pores the membrane does not actively contribute to the selectivity of the process. Species small enough to enter the pores are transported by convective flow. The performance of the membrane separation under these circumstances is largely determined by the application. That is, it is the characteristics of the retained process stream which largely determine the performance of the membrane.

In tighter membranes, with a smaller mean pore size, the probability of molecules colliding with the wall of the membrane pore is increased. If the interaction with the wall is different for different species, the membrane will demonstrate an intrinsic selectivity. This model of transport within a membrane has been proposed to explain the rejection of salts in aqueous solution by membranes with pore sizes greater than the mean atomic diameter of the rejected species. As the pore size decreases and tends towards a non-porous structure, the membrane plays an increasing role in the separation and the transport mechanism changes from convective flow through pores to solution-diffusion within the polymer. This is the dominant transport mechanism in reverse osmosis, pervapouration and gas separation. A clear distinction is necessary between the intrinsic membrane characteristics and its performance in a particular application. For example, if an ultrafiltration membrane were to be used to sterilise a water supply or to concentrate a fruit purée, under the same operating conditions of flow and pressure the permeate flux would be significantly different for each application. If we are to consider membranes for a separation step we have to take into account the relative influence of the process streams. In microfiltration these effects are likely to dominate, whereas in gas separation although the composition of the gas stream will affect the driving force it will not usually have a marked effect on the membrane selectivity.

External Factors Affecting Membrane Performance

There are two major external influences which alter the selectivity and adversely affect the flux of membrane, especially microfilters and ultrafilters: concentration polarisation is short-term and recoverable; fouling is long-term and frequently permanent.

Concentration polarisation

In most membrane processes the concentration of non-permeating species is higher at the membrane surface than in the bulk liquid since it is carried towards the membrane by the convective flux but remains on the retentate side of the membrane. This effect is called concentration polarisation and it generally reduces the flux through the membrane. The membrane separation characteristics may also be affected. The phenomenon is especially important in ultrafiltration.

Fouling: In virtually all membrane processes the flux through the membrane falls slowly with time. The reasons for this decline in flux are numerous and include: slime accumulation due to bacterial growth, deposition of organic macromolecules, colloid deposition and physical compaction of the membrane material. Long-term flux decline is often permanent or not easily irreversible and thus differs from concentration polarisation.

Cross-flow vs. Dead-end Filtration

In conventional filtration, dead-end filtration is usually used. The fluid to be filtered is fed perpendicularly to the filter: all the fluid passes through and the solids build up on the filter. The cake of accumulated solids, which is usually reasonably porous, can be continuously removed, e.g. from a rotary filter using a fixed blade or from a bag filter using a backflush arrangement. Alternatively, the filter can be discarded periodically when it is more or less blocked. In many situations an inert material termed a filter aid can be added to the medium during filtration. The filter aid assists in the creation of a porous filter cake and it is the cake which forms the filter. The filter cloth primarily acts as a support for the filter cake.

MEMBRANE CONFIGURATIONS

The growing number of membrane, applications has been accompanied by an increase in the range of techniques and arrangements to house and secure membranes in physical devices. The large range of alternative constructions has arisen for a combination of economic and operational reasons. Despite rapid progress in the development of thin membrane materials and significant improvements in their separation characteristics, large membrane areas are still required for industrial applications. As a consequence considerable care is required in the design and specification of a membrane module to ensure that the most advantageous hydrodynamic conditions exist within the module, thus minimising the size and cost of the unit. Tubular membranes require either internal or external support to sustain them in operation, whereas the smaller diameter capillary and hollow fibre membranes are self-supporting.

Plate-and-Frame

The membranes in these modules are cut from flat sheets which are mounted in a plate-and-frame system similar in basic layout to that of a filter press or plate heat exchanger. Flat sheet rigs are useful for laboratory-scale work because the technology for making sheet membranes is more straightforward and the units to hold them are simpler in construction than other membrane forms. In many applications, this configuration is favoured because cleaning of the membranes is easier.

Spiral-wound

A large length of sheet membrane in a double layer is wound into a spiral. The feed is passed axially or spirally over the outside of this double layer. The permeate moves spirally and is removed by a pipe at the centre of the spiral. Like hollow fibres, spiral-wound membranes have a high surface-to-volume ratio and require clean feeds.

Tubular

This configuration is useful for feeds of high viscosity, or containing particulate material. For example, tubular membranes would be preferred for concentrating materials like tomato purée.

Hollow Fibre

Hollow fibres and capillary fibres are installed in a shell-and-tube arrangement and provide a very high surface area for a given volume of unit. Because of the small bore of the fibres, the feed must be free of particulate matter.

Dialysis

Dialysis is the term given to the transport of a solute across a membrane by diffusion resulting from a concentration difference. If two solutions of different concentration employing the same solvent are separated by a suitable membrane, the solute will diffuse from the more to the less concentrated solution. A pressure differential across the membrane is deliberately avoided so that concentration provides the sole driving force. A complication is that even in the absence of pressure-driven flows through the membrane, solvent will be transported across the membrane if there is a difference in osmotic pressure between the two solutions. The first dialysis membranes were manufactured in regenerated cellulose but recently other materials such as polysulphone have been used.

Although there are as yet few industrial biotechnological applications for dialysis it is included because a very significant proportion of the total membrane market is attributed to membranes used for the haemodialysis of patients suffering renal failure. The development of membranes for this application and the associated techniques of producing sterile modules suggests their use for certain pharmaceutical and animal cell culture applications where it is necessary to add and simultaneously remove low molecular weight solutes while retaining high value high molecular weight products.

Since concentration difference provides the driving force, this difference should be large and the membrane should be thin to reduce the diffusion path. The process is slow compared with UF or RO in which pressure is applied. It is worth noting that in dialysis it is the solute which passes through the membrane; in UF and RO it is the solvent which passes through the membrane, although a proportion of the solute may be carried through also.

To prevent the convective transport of the solution arising as a result of small pressure differences across the membrane, the pores must be very small. In fact, one can choose to think of dialysis membranes as a swollen non-porous polymeric film. Co-current flow is usually used to avoid pressure drops across the membrane. For example, if both streams enter a unit at 0.1 bar and the pressure drop on each side of the membrane is 0.1 bar, then there is no pressure difference across the membrane at either end of the unit. The separation capabilities of dialysis arise from the small pores and the diffusion process. Large molecules cannot pass through the membrane at all and small molecules diffuse faster then large ones — at a first approximation, diffusion coefficients are inversely proportional to the square root of molecular weight. As dialysis is a diffusion process and pressure driving forces can be considered to be absent, transport across the membrane can be described by a simple relationship between the diffusion coefficient of the solute, membrane thickness and the concentration difference.

Osmotic pressure

Consider two compartments, one containing water and the other an aqueous solution, separated by a membrane. The presence of the solute lowers the chemical potential of the water in the solution giving a driving force for water transport into the solution. The pressure which must be applied to the solution to avoid this water transport is the osmotic pressure of the solution. If a pressure greater than the osmotic pressure is applied, the water diffuses from the solution into the water compartment.

Pervapouration

Pervapouration is one of the second generation membrane processes—using coated membranes—for which the number of potential applications is growing. It differs from the other processes so far described in that the separation is accompanied by a phase change. A consequence of this is that heat of vapourisation of the transported species must be supplied to sustain the process.

There are two methods of operation. In one the liquid to be separated is circulated on one side of the membrane and a vacuum is applied to the other. In the second arrangement, the vacuum is replaced by an inert carrier gas which transports the permeant away from the membrane, thus maintaining a low permeant partial pressure. As in reverse osmosis, the transport mechanism is solution-diffusion; species are preferentially sorbed into the membrane, diffuse through it and leave the permeate side as a vapour. As the preferred species is sorbed membrane swelling usually occurs and at high concentrations of the preferentially sorbed component selectivity can decrease dramatically. Because pervapouration relies on differences in solubility and diffusion of species and not on differences in volatility as in distillation, the process has potential applications in separating azeotropic mixtures such as ethanol and water.

Gas Separation

In gas separation, a gas mixture is pressurised on one side of a membrane and a low pressure maintained on the permeate side. As with reverse osmosis and pervapouration, dense and often coated membranes are used and transport can be considered to be by solution-diffusion.

Chapter 30

Affinity Precipitation

INTRODUCTION

Affinity precipitation is a relatively simple, convenient and reproducible technique that results in high target molecule recovery at high specificity. Using afinity macroligands (AMLs) based on homogeneous oligomers, the ligand efficiency is very high and the (expensive) affinity ligands are effectively used during the purification procedure. The method is robust and shows flexibility towards variation in the process parameters. Since only mixing of AMLs into the target solution is involved, the scale-up potential is high. At preparative scale the most appropriate separation operation for the precipitate is filtration. Future work should be directed to further optimisation of the AML (precursor), the establishment of more applications and towards the development of a robust, integrated purification process.

Precipitation is a commonly used unit operation in the downstream processing of biologicals. Moreover, precipitations are used for this purpose at various scales from the very small [microlitres, e.g. plasmid DNA (pDNA) preparation] to the extremely large (e.g. technical enzymes). Perhaps the most impressive application of the principle of precipitation in the general area of bioseparation is that of plasma fractionation, i.e. the process used to produce certain therapeutic plasma proteins from human blood donations. Plasma fractionation today is carried out at a scale that surpasses any other established biopharmaceutical process. The worldwide manufacturing scale for human albumin, immunoglobulins (IgGs) and other plasma-derived therapeutic proteins is of the order of hundreds of tons annually, i.e. much larger than even that of recombinant human insulin.

The plasma fractionation process was originally developed by Cohn and colleagues in 1946 and is still used in this form today in the USA, while the modified version introduced by Kistler and Nitschmann in 1962 is preferred in Europe. The process fractionates the human plasma proteins in a series of steps deploying the unique physico-chemical properties of each plasma protein, i.e. its solubility (hydrophobicity) and its isoelectric point (pI). Following a cryoprecipitation (depletion) step to capture certain sensitive blood factors, successive fractions of the plasma proteins are precipitated via adjustment of the pH, temperature, ionic strength and ethanol concentration in the environment. The precipitating proteins first form colloidal particles, which then grow and aggregate. The particle size distribution of the resulting precipitate as well as its mechanical strength and stability are controlled by a dedicated ageing treatment. In the case of the most common batch process, the ageing treatment consist of mixing at defined shear force and time. This way the efficient separation of the precipitate by the subsequent centrifugation or body-feed filtration is ensured. The supernatants obtained in the various steps after separation of the precipitate are subjected to additional extraction/precipitation steps carried out in the same manner, but with modified parameters.

Other commonly found types of precipitation-based separations in the area of biopolymer and in particular protein separation are the addition of nonionic polymers such as poly(ethylene glycol) (PEG) or poly(vinyl pyrolidone) (PVP), but also the so-called salting-out methods. The latter approach is a commonly used first-capture and concentration step for many fairly high-titer protein products. The method is based on the fact that protein aggregation and subsequent precipitation call be enforced by the addition of salt. The effect is generally interpreted as the result of a hydrophobic interaction and the salting-out potential of a given salt, thus depends on its position in the Hofmeister series. Ammonium sulphate is a very strong salting-out agent and is widely used as such in preparative protein separation. Salting-out (and precipitation in general) is typically applied during the early stages of the bioseparation process, as it is quite tolerant to impurities/feed composition and convenient to use even at large scale. That the product fraction is concentrated and that some partial purification usually takes place are additional benefits of this method. In sequential combination with caprylic acid it was even possible to achieve a crude antibody purification by precipitation alone.

The limits of the outlined precipitation methods are generally the stability of the target molecule and the lack of specificity of the method. In the case of low-titer products, the yields are also often quite low. In such cases precipitation is typically not even considered, instead an approach combining enrichment with specific capture is chosen, most commonly affinity chromatography. Affinity chromatography has supplied the bioseparation community for more than 50 years with an efficient one-step technique for the specific isolation-cum-concentration of proteins. The method is based on immobilising so-called affinity ligands, i.e. molecules capable of a (bio)specific interaction (molecular recognition) to the chromatographic support. These ligands then selectively retain and separate the target molecules even from a very diluted and complex feed. Elution can be achieved by a buffer, which no longer supports the noncovalent affinity interaction. Often a pH shift or a chaotropic, respectively, competing agent is used.

Affinity chromatography is very popular in research laboratories as well as in bioproduction plants; however, compared to a precipitation, the use of a chromatographic column for separation during early stages of the downstream process is also known to be beset with certain difficulties. Scale-up, column fouling and flowrate limitations frequently cause problems. A particular problem with affinity chromatography is the sometimes inconveniently slow association rate of the target protein molecule with the immobilised biospecific ligand as a result of pronounced mass transfer limitations, but also the restriction of the available stationary phase capacity in the case of the commonly used porous beads because of steric hindrances (access to the pores). Affinity chromatography is also awkward to use with many raw process streams. Consequently, some effort is still spent on the development of alternative affinity techniques, which retain the principle of biospecific interaction up to the use of the same affinity ligands, but which might overcome some of the known disadvantages of affinity chromatography yet retain its unparalleled selectivity.

In the late 1970s, these efforts led to the development of two concepts for 'affinity precipitation' — one by the group of Klaus Mosbach and the other by Michel Schneider and coworkers. The two principles have little in common, but the same nomenclature was used. Precipitation is an attractive concept in this context, since the required solid–liquid separation is extremely well understood and can be handled at various scales in the production as well as in the research environment. The two concepts developed for affinity precipitation were later distinguished as 'primary-effect' and 'secondary-effect' affinity precipitation.

Affinity chromatography is a separation technique with many applications since it is possible to use it for separation and purification of most biological molecules on the basis of their function or chemical structure. This technique depends on the highly specific interactions between pairs of biological materials such as enzyme-substrate, enzyme-inhibitor, antigen-antibody, etc. The molecule to be purified is specifically adsorbed from, for example, a cell lysate applied to the affinity column by a binding substance (ligand) which is immobilised on an insoluble support (matrix). Eluent is then passed through the column to release the highly purified and concentrated molecule. The ligand is attached to the matrix by physical absorption or chemically by a covalent bond. The pore size and ligand location must be carefully matched to the size of the product for effective separation. The latter method is preferred whenever possible.

Coupling procedures have been developed using cyanogen bromide, bisoxiranes, disaziridines and periodates, for matrixes of gels and beads. Four polymers which are often used for matrix materials are agarose, cellulose, dextrose and polyacrylamide. Agarose activated with cyanogen bromide is one of the most commonly used supports for the coupling of amino ligands. Silica based solid phases have been shown to be an effective alternative to gel supports in affinity chromatography.

Purification may be several thousand-fold with good recovery of active material. The method can, however, be quite costly and time consuming and alternative affinity methods such as affinity cross-flow filtration, affinity precipitation and affinity partitioning may offer some advantages. Affinity chromatography was used initially in protein isolation and purification, particularly enzymes. Since then many other large-scale applications have been developed for enzyme inhibitors, antibodies, interferon and recombinant proteins and on a smaller scale for nucleic acids, cell organelles and whole cells. In the scale-up of affinity chromatographic processes bed height limits the superficial velocity of the liquid, thus scale-up requires an increase in bed diameter or adsorption capacity.

PRIMARY-EFFECT AFFINITY PRECIPITATION

For plasma protein precipitation, size increase will usually entail precipitation. A protein molecule, which becomes insoluble due to changed environmental conditions (forming thus a small colloidal particle), acts as a nucleation centre for further diffusion controlled growth via the attachment of others (aggregation). Larger particle can then be formed by collision-induced agglomeration. Finally, particles will reach a sufficient size to form a macroscopically observable 'precipitate', which spontaneously separates from the liquid phase. In analogy, the concept of primary-effect affinity precipitation is based on the controlled growth of protein molecules to a network, which becomes insoluble once a given size limit is surpassed. Such a controlled network formation is possible if there are at least two points of specific interaction between the involved molecules.

AFFINITY PRECIPITATION BY STIMULI-RESPONSIVE MATERIALS

Affinity interaction and precipitation are directly linked in primary-effect affinity precipitation. While this makes the process very straight forward, it is also the cause of many of the above-mentioned disadvantages of this bioseparation method. In secondary-effect or indirect affinity precipitation, these two aspects are no longer linked; hence, they can be performed and controlled independently. Secondary-effect affinity precipitation, from here onward simply called 'affinity precipitation' as it has become the more ubiquitous form of affinity precipitation, makes instead use of stimuli-responsive materials to bring about precipitation. Stimuli-responsive or intelligent materials react by pronounced property changes to a small change in an environmental parameter. The effect is usually fully reversible, i.e. the material reverts to its original state once the stimulus has been removed. For application in the context of protein isolation, materials that undergo pronounced changes in their water compatibility (hydrophobicity) are very interesting. Materials responding to a wide variety of stimuli have been described, and many of them have been already used for bioseparation purposes and will be discussed below.

The affinity complex forms in solution, i.e. with very little steric hindrance and a minimum of mass transfer resistance. Then the stimulus is applied, the affinity complex precipitates and can be removed, while the impurities stay in solution. Subsequently, the target molecule can be eluted either by redissolution of the complex in dissociation buffer under conditions that promote redissolution of the material followed by stimuli-induced separation of only the AML or via direct elution from the separated precipitate under conditions that promote continued precipitation of the AML.

Stimulus: pH

The first description of a stimuli-induced affinity precipitation was published in 1981 by Schneider and coworkers. The target molecule was trypsin, i.e. again an enzyme, but in this case an enzyme that contained only a single binding site for the ligand (monovalent enzyme) and hence could not have been purified by primary-effect affinity precipitation, where multivalency is required. The AML in this case was based on a ter-polymer composed of acrylamide, N-acryloyl-p-aminobenzoic acid and N-acryloyl-m-aminobenzamidine. The benzamidine units in this molecule served as affinity ligands, as depending on the pH they represent strong and specific inhibitors of the protease trypsin.

The entire AML was water-soluble above a pH of 4. Below a pH of 4, the acidic residues on the polymer backbone became neutralised (protonation) and hydrophobic interactions between the then uncharged polymer backbones caused aggregation. As a result, the AML precipitated whenever the pH was reduced below a value of 4. The process was reversible and the AML-affinity complex did resolubilise when the pH was again raised. The related bioseparation process for the purification of the protease trypsin from a crude pancreatic extract at pH 8 can be considered typical for pH-induced affinity precipitation and is therefore given in some detail here. In particular, the AML was added to the extract at a load of 0.5 wt%. The biospecific interaction between the trypsin and its inhibitor (AML) took place in homogeneous solution. The AML-trypsin complex was then:

1. Precipitated by lowering the pH to 4.
2. Separated by centrifugation from the supernatant, in which putative impurities remained dissolved.
3. Washed once with water.
4. Resuspended at pH 2 for dissociation of the affinity complex.

Stimulus: Salt Addition

Another common means to bring about the precipitation of certain water-soluble biopolymers is the addition of salts. Alginate, for example, is a soluble polysaccharide that precipitates reversibly from solution upon the addition of bivalent ions such as calcium (cross-linking agent). This natural polymer is composed of linear, unbranched blocks of guluronic acid and mannuronic acid. The size and sequence of these blocks determine the chemical and physical properties of the alginate. Alginate possesses an inherent biological affinity for some enzymes such as amylase, pectinase and lipase. As alginate is of natural resource, lot-to-lot variability and contamination, e.g. by endotoxins, are unavoidable. The separation of alginate and endotoxin, which are both negatively charged, is laborious and finally makes alginate for bioseparation purposes an expensive material.

Stimulus: Temperature

Another class of stimuli-responsive materials with potential for affinity precipitation are the thermoresponsive ones, i.e. polymers that are soluble in (cold) water, but precipitate once a certain critical solution temperature (CST) is surpassed. Thermoresponsiveness is often observed in amphiphilic polymer molecules, where dissolution is characterised by a negative dissolution entropy together with a negative dissolution enthalpy. Such molecules dissolve well in cold water. As the temperature increases, the unfavourable dissolution entropy starts to dominate the behaviour and at a certain temperature (the CST)—the formerly soluble macromolecules become insoluble; macroscopically, this manifests itself as precipitation. The fact that the H-bridges, which typically aid the dissolution of such polymers in water, become weaker as the temperature increases also contributes to this effect. Since thermoresponsive polymers bear little to no charges, while they only show moderate hydrophobicity, their tendency for nonspecific interaction with biological molecules is low—another advantage for their application in affinity precipitation.

Stimulus: Photo

Affinity precipitation is an extremely versatile purification method, because a broad range of stimuli-responsive materials can be used, where each material has its specific chemical and physical properties. In addition to pH- and temperature-responsive materials and polymers, which precipitate upon an increase of ionic strength, a fourth option in affinity precipitation has been reported by Desponds and Freitag, particularly the use of photo-responsive polymers as AML precursors. Strictly speaking, the proposed photo-responsive molecules are variants of the thermoresponsive ones. Chain-transfer copolymerisation of NIP AM and *N*-acryloxysuccinimide using a biotinylated chain-transfer agent and conjugation of (3-aminopropyloxy)azobenzene chromophores to the activated polymer backbone produced a prototype of a photo-responsive biotin–AML. This AML shows a critical solution temperature in pure water of 16°C when the azo groups in the side-chains are predominately in the (stable) *trans*-state. Irradiation with ultraviolet light (330 nm) switches the azo group into the more hydrophilic *cis*-state and the critical solution temperature rises to 18°C. Irradiation with visible light (above 440 nm) switches the group back to the *trans*-state. Adjusting the temperature to an intermediate level, the photo-AML was used to demonstrate the concept of photo-affinity precipitation, i.e. the specific capture and recovery by light-induced precipitation of a target molecule (avidin) from a serum-containing cell culture supernatant. The avidin was obtained in highly purified form without nonspecific coenrichment of protein impurities.

APPLICATION OF AFFINITY PRECIPITATION IN BIOSEPARATION

The attractiveness of affinity precipitation stems from the fact that the well-understood and manageable unit operation 'precipitation' is combined with the specificity and efficiency of the 'affinity' approach in homogeneous solution. In their perennial review, Labrou and Clonis placed (secondary-effect) affinity precipitation highest among the major affinity purification technologies in terms of purification power combined with large-scale potential. Affinity precipitation also shows broad flexibility towards variation in the process parameters. Despite these advantages, however, affinity precipitation has to date hardly made a major impact on biotechnical downstream processing.

Scale-up and Technical Realisation

The theoretical scale-up potential of affinity precipitation is indisputable. At present, however, the technical realisation of such large-scale processes is difficult due to the lack of a suitable large-scale separation technology for the precipitated polymers and the affinity complex. High-speed centrifugation is an efficient method except for oligomeric materials. The recovered precipitate, however, tends to present a compact gel, the resolubilisation of which can be very time consuming. Entrapment of impurities in the wet gel is almost unavoidable, making three precipitation/resolubilisation washing/cleaning steps necessary as a rule of thumb. Senstad and Mattiasson were the first to address the necessity of developing alternative modes of precipitate recovery at large scale in order to avoid the centrifugation steps. They proposed flotation as a relatively mild operation for this purpose. In particular, chitosan-based AMLs were precipitated by raising the pH of the solution above 8. Afterwards the mixture (liquid containing the precipitate) was put under pressure (3 bar, 15 minutes) and then transferred into a flotation chamber at atmospheric pressure. Air bubbles, which formed as result of the pressure expansion, adsorbed to the surface of the precipitated material causing the flotation of the particles and thereby their accumulation at the surface, where they could easily be recovered.

Protein Purification

Most previously reported affinity precipitations were designed for protein recovery. Although affinity precipitation has not yet been established for industrial protein purification, a number of processes for the isolation of (recombinant) proteins from real matrices have been described among these applications. Table 30.1 shows application of affinity precipitation for the recovery of biomacromolecules from complex raw solutions.

Table 30.1: Application of affinity precipitation for the recovery of biomacromolecules from complex raw solutions.

Target molecule/feed	AML/stimulus	Comments
Enzymes		
Trypsin/crude pancreatic extract	*ter*-Polymer (acrylamide/ N-acryloyl-*p*-aminobenzoic acid/ N-acryloyl-*m*-aminobenzamidine *stimulus*: pH	Elution from the precipitate
Trypsin/crude extract	Eudragit S-100-STI *stimulus*: pH	
Alcohol dehydrogenase/crude yeast extract	κ-Carrageenan-Cibacron Blue 3GA *stimulus*: K^+ ions	
Alcohol dehydrogenase/yeast extract	Eudragit S-100–Cibacron Blue 3GA *stimulus*: $CaCl_2$/temperature	
(Recombinant) proteins other than enzymes		
Avidin/cell culture supernatant containing 5% FCS	polyNIPAM-iminobiotin *stimulus*: temperature	
His-tagged proteins/cell lysates	polyNIPAM–Ni-NTA *stimulus*: temperature	Generic protein separation process
Concanavalin A/jack bean extract	Eudragit S-100–*p*-aminophenyl- α-D-glucopyranoside *stimulus*: pH	
Protein A/*Staphylococcus aureus* lysate	Hydroxypropyl methylcellulose-IgG *stimulus*: pH	
His-tagged single-chain	Cu^{2+}/Ni^{2+}-loaded copolymer of	

(Cont'd...)

Target molecule/feed	AML/stimulus	Comments
Fv-antibody fragments/cell-free *E. coil* culture supernatants	NIPAM and vinylimidazole *stimulus*: salt/temperature	
α-Amylase inhibitor/wheat meal	Cu^{2+}-loaded copolymer of NIPAM and vinylimidazole *stimulus*: salt/temperature	Reuse of AML possible
Antibody/hybridoma culture supernatant	Eudragit-antigen *stimulus*: pH	
Rabbit C-reactive protein/rabbit acutephase serum	polyNIPAM-*p*-aminophenylphosphorylcholine *stimulus*: temperature	
Oligonucleotides		
mRNA/cell lysates	PolyNIPAM-T_8 *stimulus*: temperature	Comparison to magnetic beads
Plasmid DNA/bacterial lysates	PolyNIPAM–$(CTT)_7$ *stimulus*: temperature	THAP process
Miscellaneous		
Taxol/homogeneous immunoassay	Elastin-like polypeptide-protein A *stimulus*: temperature	Proof-of-feasibility of a novel immunoassay format
Peanut lectin/peanuts	Guar gum-linked alginate *stimulus*: addition of $CaCl_2$	Cheap and efficient single-step procedure

Nucleic Acid Purification

Compared to the state-of-the-art in protein purification, the isolation and purification of nucleic acids, particularly pDNA at larger scale, is still a bottleneck in downstream processing. Given the typical sizes of pDNA molecules, mass transfer limitations and steric hindrances of chromatographic materials are even more likely for this biomacromolecule class than for the proteins. While efficient affinity (hybridisation) methods exists for single-stranded oligonucleotides, the concept of biospecific affinity interactions is less often used in the case of double-stranded polynucleotides such as pDNA.

While some effort for the integration of a specifically interactive sequence into the molecule is necessary in the case of DNA purification by affinity interaction, single-stranded oligonucleotides can be specifically captured via simple hybridisation. One example for such a purification, which is routinely used in many molecular biology laboratories, is the preparation of mRNA from eukaryotic cells and tissues. This process is an important step in many genetic engineering protocols, e.g. for the creation of recombinant production organisms, but also in the analysis of gene structure and regulation. In the case of eukaryotic mRNA this isolation typically relies on the poly(A) tail, which is present on most mature eukaryotic mRNAs, whereas the other RNA (transfer RNA, ribosomal RNA) species normally do not carry such a tag. In a typical separation process, biotinylated oligo(dT) probes are first mixed into the crude mRNA preparations. Once the probes and the poly(A) mRNA have annealed, the complexes can be specifically captured by any (strept)avidin-based separation technique. Streptavidin-coated magnetic particles are especially popular in this context. After a series of high-stringency washing steps, water is used to release the poly(A) mRNA into solution.

Recently, the efficient capture of mature eukaryotic poly(A) mRNA by affinity precipitation using an oligomeric avidin–AML has been described, and the advantages in terms of cost, handling and scalability for affinity precipitation compared to more conventional approaches were discussed. Putative problems with RNases could be prevented by a suitable treatment of the AML preparation. The RNA yield and quality were at least equal to that of a standard approach using streptavidin-coated paramagnetic beads. The produced poly(A) mRNA proved to be an excellent target for reverse transcription-polymerase chain reaction amplification. While magnetic beads require a two-step (hybridisation followed by capture) protocol, affinity precipitation could be set up as a one-step protocol using the oligo(dT)-activated AML directly in solution followed by *in situ* precipitation.

In earlier studies for RNA purification, a copolymer of NIPAM and vinylderivatised (dT)s was used for the separation of dA oligonucleotides from a mixture of its one-point-mismatched oligonucleotides at high NaCl concentration. Longer oligo(dA)s (more than five) tended to be precipitated more efficiently than shorter ones, and the AML could be used repeatedly (5 times) without loss in precipitation efficiency. A study of separation of RNA from pDNA by metal chelate affinity precipitation showed that a copper-loaded NIPAM-vinyl imidazole copolymer interacts with exposed purine residues. RNA could be eluted in that case by addition of imidazole. The presence of imidazole after RNA elution, however, hindered copolymer reprecipitation, requiring the addition of high salt concentration for that particular step.

Chapter 31

Solvent Extraction

INTRODUCTION

A major item of equipment in an extraction process is the solvent-recovery plant which is usually a distillation unit. It is not normally essential to remove all the raffinate from the solvent as this will be recycled through the system. In some processes the more difficult problem will be to remove all the solvent from the reaffinate because of the value of the solvent and problems which might arise from contamination of the product. Distillation may be achieved in three stages:

1. Evaporation, the removal of solvent as a vapour from a solution.
2. Vapour-liquid separation in a column, to separate the lower boiling more volatile component from other less volatile components.
3. Condensation of the vapour, to recover the more volatile solvent fraction.

Evaporation is the removal of solvent from a solution by the application of heat to the solution. A wide range of evaporators are available. Some are operated on a batch basis and others continuously. Most industrial evaporators employ tubular heating surfaces. Circulation of the liquid past the heating surfaces may be induced by boiling or by mechanical agitation. In batch distillation (Fig. 31.1) the vapour from the boiler passes up to column and is condensed. Part of the condensate will be returned as the reflux for counter-current contact with the rising vapour in the column. The distillation is continued until a satisfactory recovery of the lower-boiling (more volatile) component(s) has been accomplished. The ratio of condensate returned to the column as reflux to that withdrawn as product is, along with the number of plates or stages in the column, the major method of controlling the product purity.

A continuous distillation is initially begun in a similar way as with a batch distillation, but no condensate is withdrawn initially. There is total reflux of the condensate until ideal operating conditions have been established throughout the column. At this stage the liquid feed is fed into the column at an intermediate level. The more volatile components move upwards as vapour and are condensed, followed by partial reflux of the condensate. Meanwhile, the less volatile fractions move down the column to the evaporator (re-boiler). At this stage part of the bottoms fraction is continuously withdrawn and part is reboiled and returned to the column.

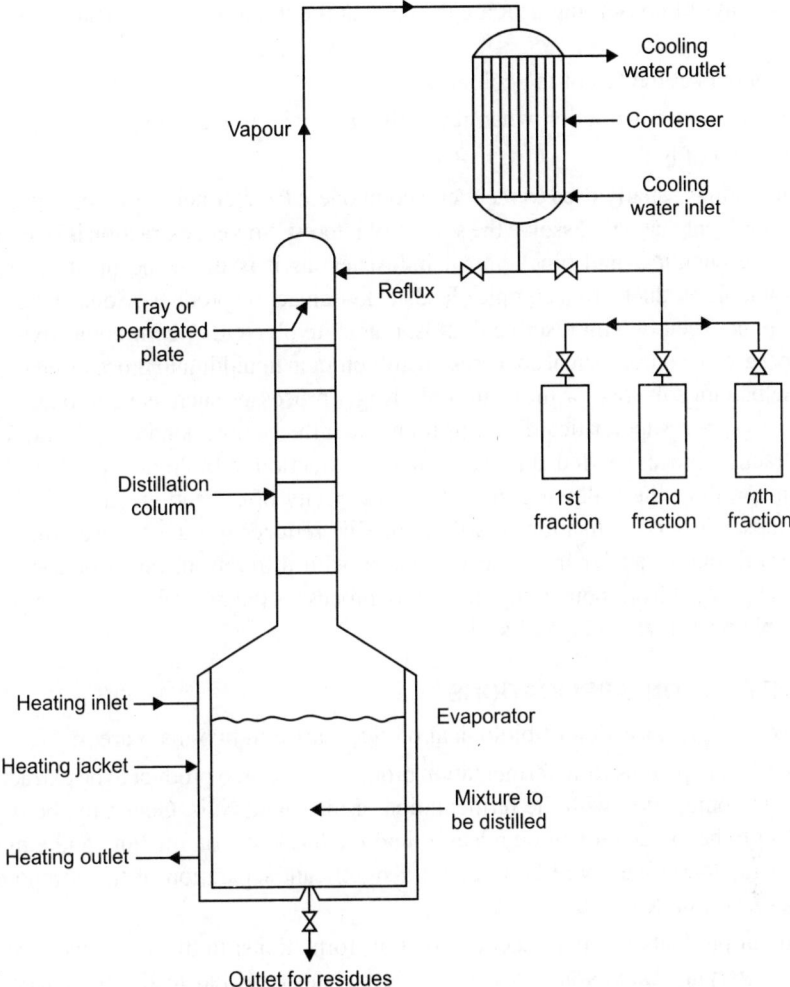

Fig. 31.1: Diagram of a batch distillation plant with a tray or perforated-plate column.

Counter-current contacting of the vapour and liquid streams is achieved by causing:

1. Vapour to be dispersed in the liquid phase (plate or tray column).

2. Liquid to be dispersed in a continuous vapour phase (packed column).

The plate or tray column consists of a number of distinct chambers separated by perforated plates or trays. The rising vapour bubbles through the liquid which is flowing across each plate and is dispersed into the liquid from perforations (sieve plates) or bubble caps. The liquid flows across the plates and reaches the re-boiler by a series of overflow wiers and down pipes.

A packed tower is filled with a randomly packed material such as rings, saddles, helices, spheres or beads. Their dimensions are approximately one-tenth to one-fiftieth of the diameter of the column and are designed to provide a large surface area for liquid-vapour contacting and high voidage to allow high throughput of liquid and vapour. The heat input to a distillation column can be considerable.

The simplest ways of conserving heat are to pre-heat the initial feed by a heat exchanger using heat from:

1. The hot vapours at the top of the column.
2. Heat from the bottoms fraction when it is being removed in a continuous process.
3. A combination of both.

Solvent extraction is usually used to recover a component from either a solid or liquid. The sample is contacted with a solvent that will dissolve the solutes of interest. Solvent extraction is of major commercial importance to the chemical and biochemical industries, as it is often the most efficient method of separation of valuable products from complex feedstocks or reaction products. Some extraction techniques involve partition between two immiscible liquids, others involve either continuous extractions or batch extractions. Because of environmental concerns, many common liquid/liquid processes have been modified to either utilise benign solvents, or move to more frugal processes such as solid phase extraction. The solvent can be a vapour, supercritical fluid, or liquid, and the sample can be a gas, liquid or solid.

Solvent extraction processes find application in the extraction of biological products for reasons that can include one or all of the following factors: (i) selectivity of extraction, (ii) fit with other required purification steps, e.g. crystallisation, distillation, (iii) reduced product losses due to degradation (e.g. hydrolytic) through transfer into a second phase with different physical or chemical properties, (iv) isolating the product from potentially degrading processes (e.g. metabolic or microbial processes), and (v) applicable over a wide range of scales.

SOLVENT EXTRACTION APPLICATIONS

Typical uses for solvent extraction of biological products arise in two main areas:

1. Extraction of compounds from fermentation broths: The desired product to be extracted is produced during or in conjunction with the fermentation of microbial cells. Generally, but not exclusively, the product to be extracted will be released into the fermentation medium and a major aim of the solvent extraction process will be to effect a significant separation of the compound from other similar molecules released by the cells.

2. Extraction of products from bioreactions/biotransformations: In this case whole cells or enzymes purified to varying extents are used to convert a substrate fed to the bioreaction to a desired product. In contrast to the extraction from fermentation broth case, the major task for the solvent extraction process may be to achieve separation of the reaction product from unreacted substrate.

Such applications for solvent extraction with biological systems can be further divided into two main groups. There are fundamental differences in the solvent extraction processes used between the groups which depends on the size of molecule being extracted. The two groups can be defined as:

1. Small molecules: Approximately <1000 daltons molecular weight. These include compounds such as antibiotics, organic acids, etc. which can be liquid extracted by methods similar to that used with non-biologically derived compounds, namely extraction into an organic solvent.

2. Macromolecules: Approximately >1000 daltons molecular weight. Compounds in this group include enzymes, antibodies, etc. For this second group of compounds the conventional techniques of liquid extraction are not feasible without substantial modification.

Proteins are generally not soluble in organic solvents and will be denatured (loss of activity and structural features) by such contact, through effects on the interactions between charged groups on the protein surface which stabilise the nature protein configuration. Although simple extraction into organic solvents is not feasible, extraction systems have been designed to allow large molecules such as proteins to be extracted while preserving the aqueous environment.

A two-phase aqueous extraction system can be set up using hydrophilic solutes (polymer/polymer or polymer/salt) which display incompatibility when dissolved in aqueous solution above critical concentrations. Two phases then form with each preferentially enriched in one component. Proteins will partition between the two phases based on the surface properties of the protein.

Organic solvents can be used to extract biologically active proteins provided the system is shielded from denaturation or insolubility in the organic phase within the polar core of surfactant aggregates or 'reversed micelles' in apolar solvents. Surfactants which exhibit high solubility in organic solvents can form reversed micelles, aggregates in which the surfactant head group forms a polar core and the hydrocarbon tails extend themselves outwards. These aggregates are able to solubilise substantial quantities of aqueous solution in their polar cores, forming an aqueous droplet which is shielded from the organic environment by the surfactant shell.

COMPARISON WITH CONVENTIONAL EXTRACTION SYSTEMS

At this point it should be noted that the following descriptions of the important variables in the application of solvent extraction to biological products will also hold for a separate but related area, two-phase bioreactions, where a solvent phase is present as a means of influencing bioreaction productivity.

Solvent extraction is a process which has been widely used in the chemical and hydrometallurgical industries for a considerable length of time. Typical processes have included benzene, toluene, xylene extraction (sulpholane process) in the petroleum industry and processes for extraction of copper. Because of this there is a perceived belief that solvent extraction is a mature separation process in terms of its application and theoretical understanding. However, it needs to be appreciated that solvent extraction applications are far ahead of the underlying design data. Techniques for reliable scale-up and design of extraction equipment are in a relatively primitive state compared, for example, with distillation.

These areas where fundamental process understanding is lacking are highlighted by the areas which require specific attention in applying solvent extraction to biological product recovery: (i) solvent selection, (ii) mass transfer behaviour, (iii) phase separation behaviour, and (v) equipment design and selection. In applying solvent extraction processes to biological products problems can arise in the following areas:

1. Complexity and multicomponent nature of the biological system: This includes both compositional complexity where due to the number of species potentially possible, complete (or even partial) characterisation will be difficult to achieve, and also phase complexity. Extractions involving biological products will generally involve solids (cells, media components, etc.) being present to variable extents. The influence of solids is a feature which has not been considered conventionally in solvent extraction processes.

2. Mass transfer rates: Mass transfer will be influenced by the presence of surface active soluble species and insoluble species. Generally these are considered to be detrimental to mass transfer processes.

3. Phase separation behaviour: The presence of insoluble solids and soluble surface active components in the extraction process has considerable impact, mostly detrimental, on the rate of phase separation in extraction processes and also the extent of phase recovery possible. This may be the most problematic area in all the aspects of applying solvent extraction to biological systems.

4. Product instability: The desired product may be unstable due to metabolic or microbial action or it may be chemically unstable under the conditions necessary to achieve efficient extraction. Penicillin extraction is a well-documented example of this latter problem.

5. Time-dependent processing behaviour: Conventional extraction processes exhibit little evidence of time-dependent behaviour. In biological systems key parameters which will affect solvent extraction processes such as rheology may exhibit time-dependent behaviour during processing.

MASS TRANSFER FUNDAMENTALS

Mass transfer between two liquid phases has been formulated into a wide range of models. The most widely known of these is two-film theory, due to Lewis and Whitman. The theory presumes that turbulence in the two phases disappears near the phase interface and the resistance to mass transfer is considered to be due to two films, one on each side of the interface.

Mass transfer is assumed to take place through films, with equilibrium at the interface.

There are cases where, in addition to a diffusional resistance, there is an additional resistance due to a slow heterogeneous reaction at the interface. Then the total resistance to solute transfer is made up of both diffusional and kinetic components. In practice solvent extraction processes are carried out by dispersing one phase in the other as a dispersion of droplets so that mass transfer occurs to or from the dispersed droplets. Both phases, aqueous or solvent, can be arranged to be the dispersed phase although at extremes of phase ratio dispersions can become unstable and may invert (change continuous phase) depending on a complex function of agitation rate, drop size, and phase properties.

Mass transfer rates are sensitive to the rigidity or mobility of the droplet interface being promoted by circulation within the droplets. Circulation within droplets depends greatly on the system purity so that mass transfer rates in systems containing traces of surfactant are reduced. The mass transfer process itself can enhance or hinder droplet–droplet coalescence depending on the direction of mass transfer. This effect is due to surface tension gradients acting in the region of close approach of the droplets.

SOLVENT EXTRACTION PROCESSES

A number of ways exist to achieve the basic features of a solvent extraction process–contact of an aqueous phase with a solvent phase giving selective extraction of a solute. Solvent extraction methods fall into a number of categories which are discussed below.

Physical Extraction

Physical extraction involves preferential solubilisation of the desired solute by an organic solvent. A major restriction in the application of physical extraction to a separation process is the need to identify a solvent which gives a sufficiently high partition coefficient for the desired solute between the organic and aqueous phases.

Ion Pair/Reactive Extraction

Selective extraction involves uptake into an organic solvent/extractant system by formation of a solvent soluble complex with the desired solute.

Two main types of extractant have been studied for such applications:

1. Phosphorus bonded oxygen donor extractants: Weak organic acids are extracted by organophosphorus compounds with a significantly higher distribution ratio than by carbon bonded oxygen donor extractants under similar conditions.

2. Amine extractants: The extraction of proton-bearing organic compounds from aqueous media by long chain aliphatic amines dissolved in water-immiscible solvents is a feasible process and is used in large-scale recovery of citric acid from fermentation liquors. The extractability of the organic acids depends on the composition of the organic phase, the amine extractant and the diluent. Typical alkylamine extractants include: (i) tri-octylamine, (ii) di-octylamine.

In both cases the extractant would be dissolved in a diluent. The diluent has to meet several important parameters and may influence the association of the extractant with the solute.

The following factors are important in diluent selection:

1. Distribution coefficient (for extraction and stripping): The distribution coefficient should be above 1.0 in the extraction step and below 0.1 in the stripping step to obtain a high concentration in the stripping raffinate. The diluent can affect the distribution coefficient by specifically solvating the extractant/solute complex.

2. Selectivity: As few impurities as possible should be extracted by non-specific extraction. This requirement favours non-polar diluents.

3. Toxicity: For food grade or pharmaceutical compounds a low toxicity solvent is necessary. Long chain paraffins, due to their low toxicity and low water solubility, would thus be preferred to chlorinated solvents.

4. Water solubility: Low water solubility to minimise solvent recovery.

5. Viscosity and density: Low viscosity and low density diluents will make phase separation easier.

6. Stability: Hydrocarbons are more resistant to degradation than alcohols, esters and halogenated hydrocarbons.

Examples of the use of ion pair/reactive extraction processes to recover biological products are:

1. Citric acid recovery: Citric acid is a bulk commodity chemically produced by fermentation. In the industrial process molasses is used as carbohydrate source by *Aspergillus niger* in submerged culture, citric acid concentrations of the order of 15 per cent w/v being achieved. Existing separation technology developed around precipitation and subsequent dissolution using calcium hydroxide and sulphuric acid. Thus a substantial volume of waste was generated by the process.

An alternative citric acid extraction process based on a tertiary amine in a paraffin diluent has been developed. Since it is necessary to recover citric acid as the free acid the option of removing the acid from the solvent by washing with a solution of high pH was not available. However, a satisfactory separation of citric acid from the organic phase to the aqueous phase can be achieved by increasing the temperature of the extraction due to a reduction in the distribution coefficient.

2. Penicillin recovery by ion pair extraction: The problem of penicillin instability at the conditions necessary to give efficient physical extraction has been discussed above.

While ion pair extraction systems can provide the advantages of a highly efficient, selective extraction system, there is the potential drawback compared with physical extraction that the solute cannot be recovered directly from the solvent phase. The solute must be back-extracted from the solvent phase into an aqueous phase to recover the solute and allow the solvent (extractant/diluent) to be reused. An additional drawback of ion pair extraction is the potential toxic hazard associated with the ion pair extractant (in particular phosphate esters). Hence the extraction process needs to ensure the achievement of sufficiently low levels of the ion pair extractant in the final product.

RELATED PROCESSES BASED ON SOLVENT EXTRACTION

The factors which relate to the application of solvent extraction of biological products also apply to two further general areas in biotechnology. These are: extractive fermentation and two liquid phase biocatalysis.

Extractive Fermentation

Extractive fermentation is the name given to a method of intensifying fermentation processes by removing the product before concentrations reach a level where the production of the product is inhibited or the cells producing the product fail to continue metabolism. A major potential use of this technique is in the production of bulk chemicals such as ethanol or acetone, butanol by fermentation, as capital costs can be reduced. In the case of ethanol production by yeast grown on a carbohydrate feedstock, the maximum ethanol concentration present at the end of the fermentation is of the order of 14–17 per cent v/v. This level of ethanol inhibits yeast metabolism and results in the decline in ethanol production and the death of the micro-organism. By extracting the ethanol from the aqueous phase into an organic solvent, the ethanol concentration standing in the aqueous phase is kept low. Thus by continual removal of ethanol from the aqueous phase, ethanol production will continue.

The major problem associated with this process is solvent selection. It is necessary to have a solvent which will extract ethanol itself but which will not inhibit microbial growth or microbial processes. The potential advantage is in the increase in volumetric productivity of the fermentation. For the ethanol case an improvement in productivity of more than six times has been considered feasible. This would be a major cost saving in the production of bulk chemicals, reducing the size of plant and hence capital investment for a given throughput.

A suitable solvent for extractive fermentation must meet the following requirements:

1. Nontoxic towards the micro-organism.
2. Good distribution coefficient for the solute.
3. Good phase separation behaviour.

Two Liquid Phase Biocatalysis

The techniques of liquid extraction find application in the use of enzymes as biocatalysts. Although enzymes are typically employed in aqueous media there are several important advantages for enzyme mediated catalysis taking place in the presence of organic solvents. Many compounds which would be of interest for use in enzyme catalysed reactions are poorly soluble in water but soluble in organic media. In such cases an organic medium makes it possible to obtain higher reactant and/or product concentrations than with a solely aqueous reaction system.

Organic reaction media can also reduce other problems of aqueous enzymic reactions such as product inhibition which is directly related to the concentration of products in the environment of the enzyme.

Often in a synthetic reaction the desired product is preferentially organic soluble. If such a reaction is carried out in an aqueous phase in contact with an immiscible organic phase, the product partitions away from the enzyme into the organic phase. This can effectively alleviate product inhibition and can drive equilibrium controlled reactions towards completion.

Water can react with substrates, products or other species in the reaction medium producing by-products. In such cases reduction or near elimination of water from the system by using organic solvents can significantly increase yields.

Reactions studied with two liquid phase bioreactions have been of four general types each with potential commercial application:

1. Oligopeptide synthesis.
2. Esterification or *trans* esterification.
3. Oxidation or reduction.
4. Hydrolysis.

MASS TRANSFER WITH BIOLOGICAL SYSTEMS

The presence of surfactants is known to influence the mass transfer performance of liquid extraction processes. Mechanisms for this alteration in performance, a decrease in mass transfer rate, are ascribed to effects such as increasing film rigidity and, hence, reducing the level of interfacial phenomena such as the Marangoni effect known to promote mass transfer. Other mechanisms involved are:

1. Simple blockage of the interfacial area available for mass transfer.
2. The alteration of interfacial tension preventing/damping oscillations and circulations associated with swarms of droplets.

Thus, it would be anticipated that, for biological extractions with their high concentrations of interfacially active material, the mass transfer rates would be affected both by this and by the presence of solids such as cells and cell debris in the aqueous phase.

Organic/Aqueous Systems

Available data supports a simple mechanism for solids present in the solution acting to block the available area for mass transfer and create a diffusion barrier. From studies that have been made of ethanol extraction from yeast fermentations it can be seen that the influence of interfacial solids on the mass transfer rate can be substantial. Measurements of the ethanol mass transfer rates were made with a Lewis-type cell with and without yeast cells being present. A reduction in mass transfer rates of almost ten times was observed for the case with cells present compared with the case without cells. Furthermore, it was found that there was a dramatic drop in mass transfer over a narrow range of yeast cell concentrations as yeast concentration in the aqueous phase was increased. Microscopic examination of the droplets indicated that the reduction in mass transfer was due to the absorption of the cells at the, liquid–liquid interface. The degree of reduction of mass transfer could be correlated to a layer of yeast cells packed at the interface with an average packing density between cubic and hexagonal close packing. The large drop in mass transfer was, calculated to arise when a layer of one to five yeast cells thick built up at the interface. Droplet size studies on whole broth systems using both yeast and mycelial systems have shown that the presence of interfacially active species and whole broth solids result in an increase in droplet and interfacial rigidity over a range of droplet sizes. Studies of protein absorption at liquid–liquid interfaces and their effect on mass transfer have also correlated this reduction to an effect on the film diffusivity and to damping of interfacial turbulence.

The performance of extraction equipment with clarified and whole broth has been reported. With identical phase ratios and residence time for the phases and identical operating conditions, the extraction efficiency was found to drop by 10–15 per cent for a system with whole broth compared with clarified broth. Even with clarified fermentation broth though, mass transfer performance was significantly reduced over that reported for similar extractor types operating with non-biochemically derived systems. In extraction equipment, the degree of mixing and turbulence in the dispersion may be sufficient to free the liquid–liquid interface of bulk solids, at least on a time averaged basis. Hence mass transfer performance with whole or clarified broth could be similar. Mass transfer performance will be reduced compared with a pure system, however, due to absorption of interfacially active components of the fermentation broth such as proteins, lipids, etc. as well as any solids precipitated by contact of the aqueous phase and solvent phase.

EQUIPMENT OPTIONS AND SELECTION

Careful selection of equipment for successfully carrying out biological extractions is necessary due to system characteristics, as discussed above. These include high viscosities, low density differences, potentially high solids content and the presence of compounds which would stabilise dispersions and emulsions. Of all the various types of liquid–liquid contactor possible, the following types of contactor have been reported as suitable for feeds containing emulsifying systems:

1. Non-mechanically agitated:
    (a) Spray column.
    (b) Baffle plate column.
    (c) Packed column.
2. Mechanically agitated:
    (a) Raining bucket contactor.
    (b) Rotary film contactor.
    (c) Centrifugal contactor.

The use of other mechanically agitated systems such as mixer settlers or rotating disc columns have been reported with biological systems and were successful when agitation rates were minimised to avoid too high an energy input to the system. A lower energy input reduced the degree of emulsification. The contactors were also operated at low throughputs to allow long coalescence times. In practice, though, most experience has been obtained in carrying out biological extractions using centrifugal contactors either being used for phase contact and phase separation, or with an external mixer system feeding emulsion to a centrifugal separator for phase separation. The use of centrifugal contactors to simultaneously disperse and separate phases has been essential in systems such as penicillin extraction where the extraction must take place in as short a time as possible to reduce degradation of the product.

Whole Broth and Clarified Broth Extraction

Prior to extraction the fermentation broth can contain in the region of 10–30 per cent v/v suspended solids. The standard method of extracting materials such as penicillin from such systems has involved removal of the broth solids by filtration, usually rotary vacuum filtration. The resultant filtered broth is then adjusted to the optimum pH for extraction and is contacted with the selected organic solvent. The loaded solvent phase containing the solute is separated off. To overcome phase separation problems a centrifugal extractor is used. This type of extraction procedure has a number of drawbacks: (i) solids removal, (ii) solute losses.

The first of these problems, solids removal, arises from difficulties in filtering typical fermentation broths. The size of the cells to be filtered and their morphology together with other variables such as the degree of lysis make removal of the solids by filtration extremely difficult and can require large consumption of filter aid to reduce filter blockage. The second factor is solute loss. When removing the broth solids by filtration the losses of solute can be up to 15 per cent due to the hold-up of filtrate in the filter cake itself and adsorption of the solute onto the solids themselves. Addition of filter aid will increase the amount of solids present and lead to an increased loss of solute. The extent of loss of solute by hold-up on the filter will depend on the effectiveness of washing of the filter cake. On a rotary vacuum filter which has been commonly used for this duty the effectiveness of washing is low and hence losses can be high.

It is also necessary to reduce the washings to a minimum in order to reduce the volume of aqueous phase to be extracted and in order to reduce dilution of the solute in the filtrate. Disposal of the filter cake will also be necessary, which can present difficulties. In some cases, incineration is the only recognised disposal route. However, by extracting the whole broth directly these problems can be overcome. The physical loss of material onto the filter cake is eliminated and material is recovered from the broth solids by direct contact with the extracting solvent. Savings are also made in terms of capital expenditure since a process unit (a filter plus holding tank) is not required. Material costs for filter aid and prefilter are eliminated and labour and maintenance costs reduced. The total cost saving for extracting from whole broth compared with extracting from filtered broth has been estimated to be of the order of 30 per cent.

Centrifugal Contactors

The main types of centrifugal contactors used for biological extractions fall into the following two categories:

1. Single stage:
   (a) Disc stack separators.
2. Differential contactors:
   (a) Podbielniak contactor.
   (b) Alfa Laval contactor.
   (c) Decanter extractor.

Disc Stack Separators

Each centrifuge represents a single extraction stage with the broth and solvent mixed either within the machine or externally in a separate mixer. To provide multistage extraction a number of machines are linked together to form a counter-current cascade. The separation is achieved by a collection of conical separating discs stacked above each other with a narrow spacing between the discs. The incoming broth and solvent mixture is pumped to the base of the disc stack; phase separation takes place while passing through the disc stack.

Podbielniak Contactor

These devices have been used for many years for antibiotic extraction. The contactor consists of a rotor containing a series of concentric perforated plates. Feed and product streams are added and removed along the rotation axis of the rotor. A series of channels and drillings convey the heavy liquid to the rotation axis while the light liquid is fed to the rim of the rotor. As the rotor rotates the heavy liquid is forced down through the holes in the plates displacing the light liquid. Hence a dispersion of the two phases results and counter-current flow set up giving multistage extraction.

Alfa Laval Contactor

The Alfa Laval multistage contactor has been widely used in the past like the Polbielniak contactor for extraction with biochemical systems. The feature of this contactor is the rotor which contains a long spiral channel fitted with baffles and screens to aid mixing of the phases. The light phase is fed to the rim of the rotor and the heavy liquid phases pass each other in counter-current flow in the channels in the rotor, mixing being promoted by the perforated baffles and the shearing action of the flow.

Decanter Extractor

Although originally developed for dewatering solids from suspensions, decanters have been modified to operate as counter-current liquid extractors with a solids handling capacity. They have the following advantages over the previously discussed contactor types:

1. Tolerant to high solids content in the feed.

2. No closely spaced separating discs or narrow flow channels prone to solids blockage.

3. Unlike the disc separator, counter-current flow takes place within the equipment itself.

The heavy phase plus solids slurry leaves over the beach of the decanter bowl while the light phase is pumped from the opposite end of the bowl. Although-decanters are able to handle a higher solids concentration in the feed than other centrifugal contactors, they run at lower radial accelerations and hence they are less effective separators. The decanter extractor is a more recent development than the other centrifugal contactor types. Because of its ability to handle a high solids content in the feed and the tolerance of the contactor to solids blockage it has become the preferred contactor for whole broth systems, successful use being reported for a number of different systems.

Chapter 32

Drying and Crystallisation

INTRODUCTION

Drying involves the removal of moisture (either water or other volatile compounds) from solids, solutions, slurries, and pastes to give solid products, which often, after drying, are final products ready to be packaged. In the feed to a dryer, the moisture may be a liquid, a solute in a solution, or a solid. In the first two cases, the moisture is evaporated, in the latter case, the moisture is sublimed. The term drying is also applied to a gas mixture in which a condensable vapour is removed from a non-condensable gas by cooling, and to the removal of moisture from a liquid or gas by sorption.

Drying is widely used in industrial processes. Applications include the removal of moisture from: (i) crystalline particles of inorganic salts and organic compounds to cause them to be free-flowing, (ii) biological materials, including foods, to prevent spoilage and decay from micro-organisms that cannot live without water, (iii) pharmaceuticals, (iv) detergents, (v) lumber, paper, and fibre products, (vi) dyestuffs, (vii) solid catalysts, (viii) milk, and (ix) films and coatings.

Drying can be expensive, especially when large amounts of water, with its high heat of vapourisation, must be evaporated. Therefore, it is important, before drying, to remove as much moisture as possible by mechanical means such as expression, gravity, vacuum, or pressure filtration, settling, and by centrifugal means.

Because drying involves vapourisation or sublimation of the moisture, heat must be transferred to the material being dried. The most commonly employed modes of heat transfer for drying are: (i) convection from a hot gas in contact with the material, (ii) conduction from a hot, solid surface in contact with the material, (iii) radiation from a hot gas or hot surface in view of the material, and (iv) heat generation within the material by dielectric, or microwave heating. These different modes can sometimes be used to advantage, depending on whether the moisture to he removed is on the surface of the solid and/or inside the solid.

Of importance in the drying of solids is the temperature at which the moisture evaporates. When the first mode is employed and the moisture is a continuous liquid film or is rapidly supplied to the surface from the interior of the solid, the rate of evaporation is independent of the properties of the solid and can be determined by the rate of convective heat transfer from the gas to the surface. Then, the temperature of the evaporating surface is the wet-bulb temperature of the gas provided that the dryer operates adiabatically. If the convective heat transfer is supplemented by radiation, the temperature of the evaporating surface will be higher than the wet-bulb temperature of the gas. In the absence of contact with a convective-heating gas, as in the latter three modes, and when a sweep gas is not present, such that the dryer operates non-adiabatically, the temperature of the evaporating moisture is its boiling-point temperature at the pressure in the dryer. In evaporators, if the moisture contains dissolved, nonvolatile substances, the boiling-point temperature will be elevated.

Thus, the drying of any product (including biological products) is often the last stage of a manufacturing process. It involves the final removal of water from a heat-sensitive material ensuring that there is minimum loss in viability, activity or nutritional value. Drying is undertaken because:

1. The cost of transport can be reduced.
2. The material is easier to handle and package.
3. The material can be stored more conveniently in the dry state.

It is important that as much water as possible is removed initially by centrifugation or in a filter press to minimise heating costs in the drying process. Driers can be classified by the method of heat transfer to the product and the degree of agitation of the product. In contact driers the product is contacted with a heated surface. An example of this type is the drum drier, which may be used for more temperature stable bio-products. A slurry is run onto a slowly rotating steam heated drum, evaporation takes place and the dry product is removed by a scraper blade in a similar manner as for rotary vacuum filtration. The solid is in contact with the heating surface for 6–15 seconds and heat transfer coefficients are generally between 1 and 2 kW m^{-2} K^{-1}. Vacuum drum driers can be used to lower the temperature of drying.

A spray drier is most used for drying of biological materials when the starting material is in the form of a liquid or paste. The material to be dried does not come into contact with the heating surfaces, instead, it is atomised into small droplets through for example a nozzle or by contact with a rotating disc. The droplets then fall into a spiral stream of hot gas at 150°F to 250°F. The high surface area: volume ratio of the droplets results in a rapid rate of evaporation and complete drying in a few seconds, with drying rate and product size being directly related to droplet size produced by the atomiser. The evaporative cooling effect prevents the material from becoming overheated and damaged. The gas-flow rate must be carefully regulated so that the gas has the capacity to contain the required moisture content at the cool-air exhaust temperature (75°F to 100°F). In most processes the recovery of very small particles from the exit gas must be conducted using cyclones or filters. This is especially important for containment of biologically active compounds.

The jet spray drier is particularly suited to handling heat sensitive materials. Operating at a temperature of around 350°F, residence times are approximately 0.01 second because of the very fine droplets produced in the atomising nozzle. Spray driers are the most economical available for handling large volumes and it is only at feed rates below 6 kg min^{-1} that drum driers become more economic.

Freeze drying is an important operation in the production of many biologicals and pharmaceuticals. The material is first frozen and then dried by sublimation in a high vacuum. The great benefit of this technique is that it does not harm heat sensitive materials. The process is often termed lyophilisation when the solvent being evaporated is water. Fluidised bed driers are used increasingly in the pharmaceutical industry. Heated air is fed into a chamber of fluidised solids, to which wet materials is continuously added and dry material continuously removed. Very high mass-transfer rates are achieved, giving rapid evaporation and allowing the whole bed to be maintained in a dry condition.

DRYING EQUIPMENT

Many different forms of materials are sent to drying equipment, including granular solids, pastes, slabs, films, slurries, and liquids. No one device can handle efficiently such a wide variety of materials. Accordingly, a large number of different types of commercial dryers have been developed. These dryers can be classified in a number of ways. Perhaps most important is the mode of operation with respect to the material being dried. Batch operation is generally indicated when the production rate is less than 500 lb/hr of dried solid, while continuous operation is preferred for a production rate of more than 2000 lb/hr. In the example above, the production rate is 2900 lb/hr and a continuous drying operation was selected. A second method of classification is the mode used to supply heat to evaporate the moisture. As mentioned above direct-heat (also called adiabatic or convection) dryer contact material with a hot gas, which not only provides the required energy to heat the material and evaporate the moisture, but also sweeps away the moisture. When the continuous mode of operation is used the hot gas can flow counter-currently, co-currently or in cross-flow to the material being dried. Counter-current now is the most efficient configuration, but co-current now may be required if the material being dried is temperature-sensitive. Indirect-heat (also called nonadiabatic) dryers provide the heat to the material indirectly by conduction and/or radiation from a hot surface. The energy may also be generated within the material by dielectric, radio frequency, or microwave heating. Indirect-heat dryers may also be operated under vacuum to reduce the temperature at which the moisture is evaporated. A sweep gas is not necessary, but can be provided to help remove the moisture. In general, indirect-heat dryers are more expensive than direct-heat dryers. Therefore, the former type is generally used only when the material is either temperature-sensitive or subject to breakage of crystals, with dust or fines formation.

A third method for classifying dryers is the degree to which the material to be dried is agitated. In some dryers, the material is stationary. At the opposite extreme is the fluidised-bed dryer. Agitation increases the rate of heat transfer to the material but, if too severe, can cause crystal breakage and dust formation. Agitation in a continuous dryer may be necessary if the material is sticky.

Batch Operation

Equipment for drying batches includes: (i) tray (also called cabinet, compartment or shelf) dryers, and (ii) agitated dryers. Together, these two types cover many of the modes of heat transfer and agitation discussed above and can handle a wide variety of wet-solid feeds, such as slurries, filter cake, and particulate materials.

Tray Dryers

The oldest and simplest batch dryer is the tray dryer is particularly useful when low production rates of multiple products are involved and where drying times vary from hours to days. The material to be dried is loaded to a depth of typically 0.5–4 inches in removable trays that may measure 30 × 30 × 3 inches and are stacked on shelves about 3 inches apart in a cabinet or on a truck that is wheeled into a chamber. If the wet solids are granular or shaped into briquettes, noodles, or pellets, with appreciable voids, the tray bottom can be perforated or can be a screen so that the heating gas can be passed down through the material (called through-circulation).

Agitated Dryers

As discussed by Smith, indirect heat with agitation and, perhaps, under vacuum, may be desirable for batch drying then any of the following conditions exist: (i) material oxidises, becomes explosive, or becomes dusty during drying, (ii) moisture is valuable, toxic, flammable, or explosive, (iii) material tends to agglomerate or set up if not agitated, and (iv) maximum product temperature is less than about 30°C. In most applications, the rate of heat transfer is controlled by contact resistance at the inner wall of the jacketed vessel and conduction into the material being dried. A wide variety of heating fluids can be used, including hot water, steam, dowtherm, hot oil, and molten salt.

Tunnel Dryers

The simplest, most widely applicable, and perhaps oldest continuous dryers are the tunnel dryers, which are suitable for any material that can be placed into trays and is not subject to dust formation. The trays are stacked onto wheeled trucks, which are conveyed progressively in series through a tunnel where the material in the trays is contacted by cross-circulation of hot gases.

Belt or Band Dryers

A truly continuous operation can be achieved by carrying the solids as a layer on a belt conveyor, with hot gases passing over the material. The endless belt is constructed of hinged, slotted-metal plates or, preferably a thin metal band which is ideal for slurries, pastes, and sticky materials. The bands are up to 1.5 m wide × 1 mm thick. Particle sizes typically range from 30 mesh to 2 inches. Hot-gas superficial velocities through the bed typically range from 0.5 to 1.5 m/s, with maximum bed pressure drops of 50-mm head of water. Heating gases are usually provided by heat transfer from condensing steam in finned-tube heat exchangers to temperatures in the range of 50°–180°C, but temperatures up to 325°C are feasible by other means. Continuous, through-circulation conveyor dryers have been used to remove moisture from a wide variety of materials, some of which are listed in Table 32.1, which includes, in parenthesis, the method of preforming, if necessary.

Table 32.1: Materials dried in through-circulation conveyor dryers.

Aluminium hydrate (scored on filter)

Aluminium stearate (extruded)

Asbestos fibre

Breakfast food

Calcium carbonate (extruded)

Cellulose acetate (granulated)

Charcoal (briquetted)

Cornstarch

Cotton linters

Cryolite (granulated)

Dye intermediates (granulated)

Fluorspar

Gelatine (extruded)

Kaolin (granulated)

Lead arsenate (granulated)

Lithopone (extruded)

Magnesium carbonate (extruded)

Mercuric oxide (extruded)

(Cont'd...)

Nickel hydroxide (extruded)

Polyacrylic nitrile (extruded)

Rayon staple and waste

Sawdust

Scoured wool

Silica gel

Soap flakes

Soda ash

Starch (scored on filter)

Sulphur (extruded)

Synthetic rubber (briquetted)

Tapioca

Titanium dioxide (extruded)

Zinc stearate (extruded)

Turbo-Tray Tower Dryers

When floor space is limited, but head-room is available, the turbo-tray or rotating-shelf dryer, may be a good choice for rapid drying of free-flowing, nondusting, granular solids.

Materials successfully handled in turbo-tray dryers include calcium hypochlorite, urea, calcium chloride, sodium chloride, antibiotics, antioxidants, and water-soluble polymers. The unit is particularly useful when product contamination must be avoided and the wet solids contain volatiles besides water. Capacities of up to 24,000 lb/hr of dried product are quoted.

Direct-Heat Rotary Dryers

A widely used dryer for the evaporation of water from free-flowing granular, crystalline, and flaked solids of relatively small size, especially when breakage of the solids can be tolerated, is the direct-heat rotary dryer. Wet solids enter the cylinder through a chute at the high end and dry solids discharge from the low end. Hot gases (heated air, flue gas, or superheated steam) generally flow counter-current to the solids, but co-current flow can also be employed for temperature-sensitive solids. With co-current flow, the cylinder may not need to be inclined because the gas will help move the solids to the discharge end.

Roto-Louvre Dryer

A further improvement in the rate of heat transfer from hot gas to solids in a rotating cylinder is the through-circulation action achieved in the Roto-Louvre dryer. A double wall provides an annular passage for hot gas, which passes through louvres and then through the rotating bed of solids. Because the gas pressure drop through the bed may be significant, both inlet and outlet gas blowers are often provided to maintain an internal pressure close to atmospheric. These dryers range in size from 3 to 12 ft in diameter and 9–36 ft long, with water-evaporation rates reported as high as 12,300 lb/hr. Roto-Louvre dryers are useful for processing coarse, free-flowing, dust-free granular solids.

Indirect-Heat, Steam-Tube, Rotary Dryers

When materials are: (i) free-flowing and granular, crystalline, or flaked, (ii) wet with water or organic solvents, and (iii) subject to undesirable breakage, dust formation, or contamination by air or flue gases, an indirect-heat, steam-tube, rotary dryer is often selected. Wet solids are fed into one end of the cylinder through a chute or by a screw conveyor. A gentle solids-lifting action is provided by the tubes. The dried product discharges from the other end after suitable contact with the hot-tube surfaces. The moisture (water or solvent) evaporates at about the boiling temperature, but can be swept out by a small purge of inert gas. Steam enters the tubes through a central revolving inlet manifold.

Screw-Conveyor Dryers

Used less often than rotary dryers is the screw-conveyor dryer, as shown in Fig. 32.1, consisting of a trough or cylinder that carries a hollow screw inside of which steam condenses to provide heat for drying the material being conveyed. Additional heat transfer can be provided by jacketing the trough or cylindrical shell. A wide range of materials can be dried including slurries, solutions, and solvent-laden solids. The boiling moisture can be purged with a small amount of inert gas. Standard conveyor dryers are as large as 3 ft in diameter by 20 ft long. More drying time can be provided by arranging a number of units in series, with one unit above another to save floor space. The last unit can be a cooler. Overall heat-transfer coefficients are comparable to, but somewhat less than, those for indirect-heat steam-tube rotary dryers. Major applications include removal of solvents from solids and drying of fine and sticky materials.

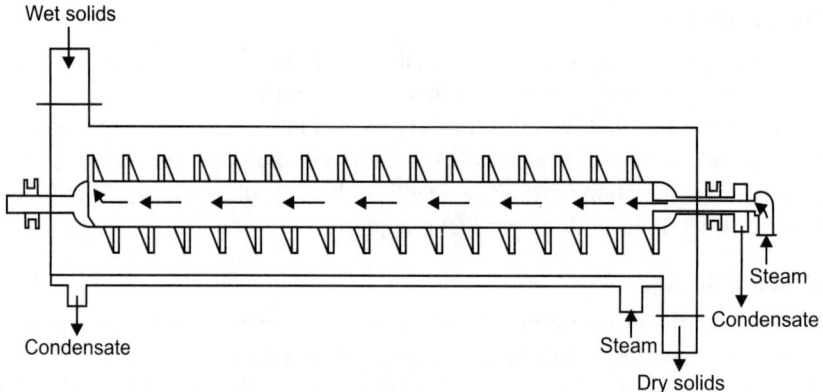

Fig. 32.1: Screw-conveyor dryer.

Fluidised-Bed Dryers

Free-flowing, moist particles can be dried continuously with a residence time of a few minutes by contact with hot gases in a fluidised-bed dryer. This dryer consists of a cylindrical or rectangular fluidising chamber to which wet particles are fed from a bin through a star valve or by a screw conveyor and fluidised by hot gases that are blown through a heater and into a plenum chamber below the bed, from where they pass into the fluidising chamber through a perforated distributor plate. The hot gases pass up through the bed, transferring heat for evaporation of the moisture, and pass out the top of the fluidising chamber and into a cyclone for dust removal. The solids are circulated by the action of the hot gases in the bed, but eventually pass out of the chamber through an overflow duct, which also serves to establish the height of the fluidised bed.

At low gas velocities, solids are not fluidised, but form a fixed bed through which the gas flows upward with a decrease in pressure due to friction and drag of the particles. As the gas velocity is increased, the gas pressure drop across the bed increases until the minimum fluidisation velocity is reached where the pressure drop is equal to the weight of the solids per unit cross-sectional area of bed normal to gas flow.

At this point, the pressure drop is sufficient to support the weight of the bed. Further increases in gas velocity cause the bed to expand with little or no increase in gas pressure drop. Typically, fluidised-bed dryers are designed for gas velocities of about twice the minimum required for fluidisation. That value depends mainly on the particle size and density, and gas density and viscosity. Superficial gas velocities in fluidised-bed dryers generally are in the range of 0.5–5.0 ft/s, which provides stable, bubbling fluidisation. Higher velocities can lead to undesirable slugging of large gas bubbles through the bed. The capital and operating cost of a blower to provide sufficient gas pressure for the pressure drops across the distributor plate and the bed can be substantial. Therefore, the required solids residence time for drying is achieved by a shallow bed height and a large chamber cross-sectional area. Because of intense mixing, the temperatures of the gas and solids in a fluidised bed are equal and uniform at the temperature of the discharged gas and solids. However, there is a substantial residence-time distribution for the particles in the bed.

CRYSTALLISATION

Crystallisation is the (natural or artificial) process of formation of solid crystals precipitating from a solution, melt or more rarely deposited directly from a gas. Crystallisation is also a chemical solid-liquid separation technique, in which mass transfer of a solute from the liquid solution to a pure solid crystalline phase occurs. The crystallisation process consists of two major events, nucleation and crystal growth. Nucleation is the step where the solute molecules dispersed in the solvent start to gather into clusters, on the nanometer scale (elevating solute concentration in a small region), that becomes stable under the current operating conditions. These stable clusters constitute the nuclei. However, when the clusters are not stable, they redissolve. Therefore, the clusters need to reach a critical size in order to become stable nuclei. Such critical size is dictated by the operating conditions (temperature, supersaturation, etc.). It is at the stage of nucleation that the atoms arrange in a defined and periodic manner that defines the crystal structure — note that 'crystal structure' is a special term that refers to the relative arrangement of the atoms, not the macroscopic properties of the crystal (size and shape), although those are a result of the internal crystal structure.

The crystal growth is the subsequent growth of the nuclei that succeed in achieving the critical cluster size. Nucleation and growth continue to occur simultaneously while the supersaturation exists. Supersaturation is the driving force of the crystallisation, hence the rate of nucleation and growth is driven by the existing supersaturation in the solution. Depending upon the conditions, either nucleation or growth may be predominant over the other, and as a result, crystals with different sizes and shapes are obtained (control of crystal size and shape constitutes one of the main challenges in industrial manufacturing, such as for pharmaceuticals). Once the supersaturation is exhausted, the solid-liquid system reaches equilibrium and the crystallisation is complete, unless the operating conditions are modified from equilibrium so as to supersaturate the solution again.

Many compounds have the ability to crystallise with different crystal structures, a phenomenon called polymorphism. Each polymorph is in fact a different thermodynamic solid state and crystal polymorphs of the same compound exhibit different physical properties, such as dissolution rate, shape (angles between facets and facet growth rates), melting point, etc. For this reason, polymorphism is of major importance in industrial manufacture of crystalline products. For crystallisation to occur from a solution it must be supersaturated. This means that the solution has to contain more solute entities (molecules or ions) dissolved than it would contain under the equilibrium (saturated solution).

This can be achieved by various methods, with: (i) solution cooling, (ii) addition of a second solvent to reduce the solubility of the solute (technique known as antisolvent or drown-out), (iii) chemical reaction, and (iv) change in pH being the most common methods used in industrial practice. Other methods, such as solvent evaporation, can also be used. The spherical crystallisation has some advantages (flowability, bioavailability) for the formulation of pharmaceutical drugs.

Used to improve (obtaing very pure substance) and/or verify their purity. Crystallisation separates a product from a liquid feedstream, often in extremely pure form, by cooling the feedstream or adding precipitants which lower the solubility of the desired product so that it forms crystals.

Well formed crystals are expected to be pure because each molecule or ion must fit perfectly into the lattice as it leaves the solution. Impurities would normally not fit as well in the lattice, and thus remain in solution preferentially. Hence, molecular recognition is the principle of purification in crystallisation. However, there are instances when impurities incorporate into the lattice, hence, decreasing the level of purity of the final crystal product. Also, in some cases, the solvent may incorporate into the lattice forming a solvate. In addition, the solvent may be 'trapped' (in liquid state) within the crystal formed, and this phenomenon is known as inclusion.

Crystallisation is an established method used in the initial recovery of organic acids and amino acids and more widely used for final purification of a diverse range of compounds. In citric acid production, the filtered broth is treated with $Ca(OH)_2$ so that the relatively insoluble calcium citrate crystals will be precipitated from solution. Checks are made to ensure that the $Ca(OH)_2$ has a low magnesium content, since magnesium citrate is more soluble and would remain in solution. The calcium citrate is filtered off and treated with sulphuric acid to precipitate the calcium as the insoluble sulphate and release the citric acid. After clarification with active carbon, the aqueous citric acid is evaporated to the point of crystallisation. Crystallisation is also used in the recovery of amino acids. Samejima has reviewed methods for glutamic acid, lysine and other amino acids. The recovery of cephalosporin C as its sodium or potassium salt by crystallisation has been described by Wildfeuer.

Chapter 33

Electrokinetic Separation Processes for Biochemical Products

INTRODUCTION

The general factors governing separation processes for biological products include the complexity of the feed mixtures, the sensitivity of the product molecule to process conditions and the low concentration of the system. The major recent developments in using electrokinetic processes for separation in biological systems appear to have centred on the recovery and separation of high molecular weight, low-volume high-value products such as therapeutic and diagnostic enzymes and proteins, by some form of electrophoretic process. In many of these cases' the commercial product throughput is so small that scale-up from laboratory scale, is relatively straight forward and carries few process engineering implications. The anticipated growth in the volume of many biologically produced agents in the high molecular weight category has added impetus to the scale-up and development of novel electrophoretic processes. The priority for medium-term application of electrokinetic processes for high volume, high molecular weight product separations may lie in the food industry area, for example in dairy product processing for protein recovery.

The use of membrane-based electrokinetic processes such as electrodialysis is also likely to be important in this area for protein treatment and desalting. Processes such as electrodialysis are already well developed in the context of water treatment and therefore further engineering development is likely to focus on the development of novel, fouling-resistant membranes, sterilisation of membrane systems, and integration of the separation process with fermentation and other upstream operations. This latter aspect is very relevant to the development of low-cost fermentation routes to medium-volume products such as lactic acid and citric acid where product dilution and product inhibition are important limiting factors. The use of electrokinetic separations at commercial scale for solid–liquid separation in biotechnological systems is probably more distant except for small-scale applications.

This is also the case of electrically enhanced solvent extraction where scale-up development is required at a basic level for this technique to be exploited, for example in commercial-scale pharmaceutical recovery. The successful demonstration of the ability to electrostatically spray untreated viscous fermentation solutions is important not only in the context of solvent extraction and recovery but also has the potential to enhance the kinetics of biotransformation processes in liquid–liquid systems. This technique typically involves contact of a cell-rich aqueous solution with an organic phase containing substrate for conversion. The inherent problem of slow kinetics may potentially be overcome by electrostatic spraying technique.

There are three broad categories of process which can be defined involving the use of electrical fields for separation:

1. Separation processes which are based upon differences in electrical mobility of the species comprising the mixture. Such differences are used to achieve selective separation of species which are difficult to separate by other means.

2. Processes which use electrostatic forces to enhance the kinetics of a process. This type of process involves the induction of an electrostatic charge upon a macro species, for example a particle or a droplet, which is present in the system. In this category, processes usually would work in the absence of an electric field but at a rate which is unacceptably slow or which is limited in terms of total degree of separation achievable. Also in this category would fall processes which are augmented by other means, for example by operation at high temperature and pressure, or by injection of high-intensity mechanical energy.

3. Electrochemical conversion processes which rely upon electrode processes to achieve a redox type transformation. This application of electric fields to biochemical transformation is particularly relevant to enzyme catalysed reaction systems involving cofactors and their regeneration.

Within the above framework of applications the use of electrokinetic phenomena is relevant to a wide range of species, which could range in size from ionic species through macromolecule, colloidal species, up to and including particles of visual size. Documented applications and potential applications include:

1. Macromolecule—colloid separations.
2. Solid–liquid separations.
3. Separation and removal of ionic species.
4. Enhancement of membrane and other filtration processes.
5. Enhancement of rate processes in liquid/liquid systems.
6. Cofactor regeneration in continuous biotransformation processes.

ELECTROSTATICALLY ENHANCED DROPLET/PARTICLE FORMATION

The essential principle involved in enhancing rates of mass transfer in liquid–liquid separation systems is the electrokinetic acceleration of liquid droplets through a second immiscible liquid phase. Figure 33.1 shows in outline a liquid–liquid extraction process in a spray column in which a dispersion of droplets of one phase is created in a second continuous phase. Enhancement of the process is achieved by inducing an electrostatic charge on the droplets as they are formed. Providing the charge is maintained for a significant time after formation the droplet (or particle) may be accelerated through the second liquid by imposition of an electric field across the continuum formed by the second phase. If droplet formation at a single nozzle is considered, the additional accelerative force acting upon the droplet due to electrostatic forces depends upon the magnitude of the charge on the droplet (i.e. the charge density), and the local field strength experienced by the droplet both during formation and subsequently.

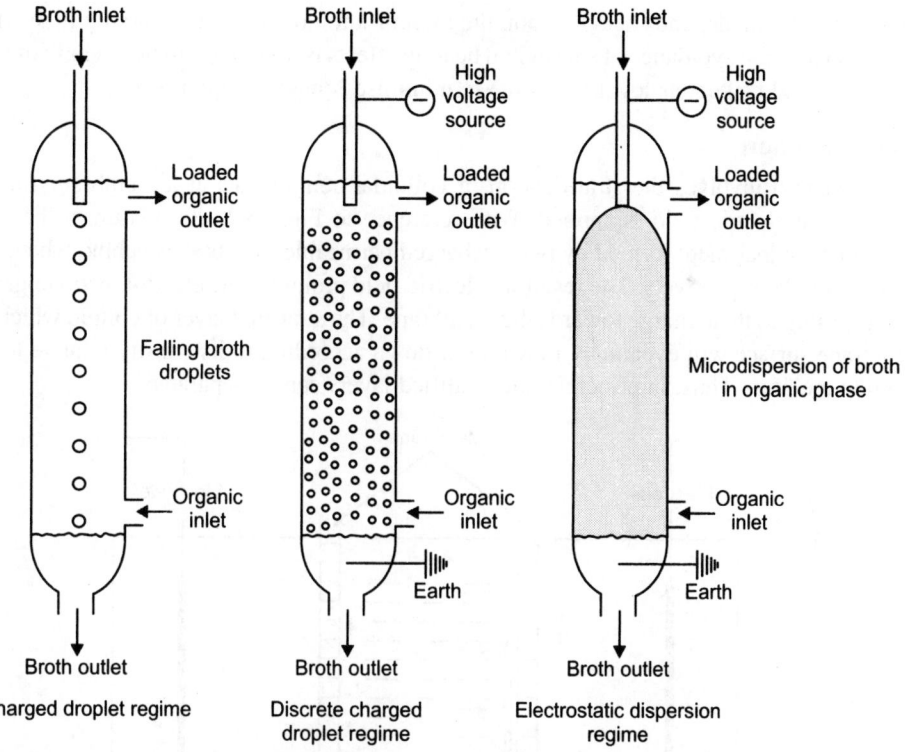

Fig. 33.3: Liquid–liquid extraction process.

APPLICATIONS OF ELECTROPHORETIC PROCESSES

Electrophoresis offers a powerful and versatile method for the separation of biochemicals. Traditionally many types of electrophoresis are used for routine biochemical analyses but in the past ten years there have been a number of important developments in preparative electrophoresis as part of the manufacture of low-volume high-value compounds and reagents. Exploitation of differences in electrophoretic mobility under the influence of an electric field can be used for separation of a number of species which could include ions, colloids, cellular materials, organelles and whole cells. The two fundamental advantages of the process are high resolution, and the ability to retain the bioactivity of the products.

Disadvantages of the process generally accrue during scale-up when the problems of Joule heating, electro-osmosis, and dispersive mixing of the separated products can reduce separation efficiency. The basic principles of several important types of electrophoresis are outlined as follows.

Zone Electrophoresis

This is a variation of moving boundary electrophoresis. The sample for separation is surrounded by buffer solution and separation into a number of discrete zones is achieved. The resolution effect, based upon differences in electrophoretic mobility, can be used in a continuous zone electrophoresis process. In this process the mixture is continuously injected as a narrow source into a body of carrier fluid flowing between two electrodes. The solute is resolved according to electrophoretic mobility and a number of fractions may be taken off at some suitable point downstream of the inlet.

Efficient resolution depends heavily upon the minimisation of convective mixing in the direction of migration and thus the avoidance of substantial heating effects is important. Ionic strength of background electrolytes should be kept as low as possible to minimise Joule heating effects.

Electrodecantation

Electrodecantation involves the stratification of colloidal components at a membrane/fluid interface across which an electric field is applied. With reference to Fig. 33.2, the mixture to be separated is contained in a compartment formed by two uncharged permeable membranes behind which are located anode and cathode respectively. The resulting electric field promotes the electrokinetic migration of the species according to their charge towards the membrane. The stratified layer of colloid which collects at the membrane surface will eventually move up or down according to its density relative to that of the surrounding medium. Thus, in principle, the stratified layers can be separated.

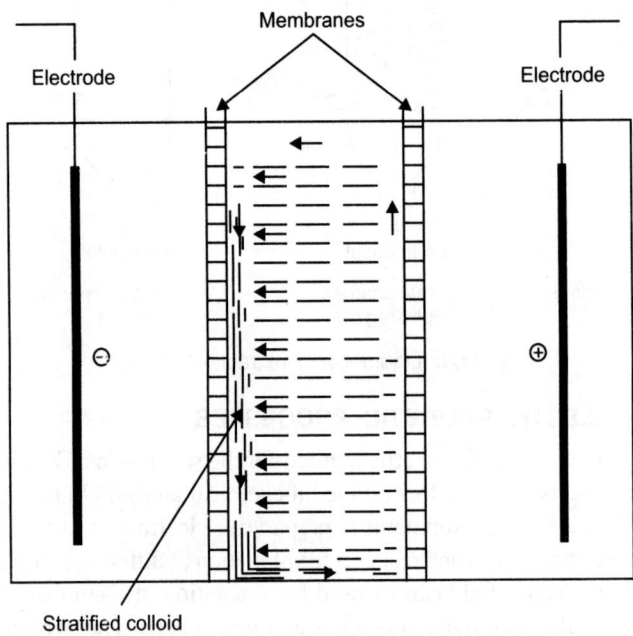

Fig. 33.2: Colloid separation by electrodecantation.

Further refinement of the technique involves the use of electro-osmosis to augment particle coagulation at the membrane/fluid interface producing a compacted layer which can be periodically released from the surface by reversal of electrode polarity. At this point the frictional resistance between the deposit and the membrane is greatly reduced leading to immediate stratification of the deposit.

Examples of applications of the technique include:

1. Separation of proteins and biological sera from non-migrating globulins.
2. Separation of tetanus and diphtheria toxins from protein fractions.
3. Isolation of the active immunoglobulin hepatitis-associated antigen.
4. Sedimentation of albumin by electrodecantation of serum proteins.

5. Enzyme purification from pig kidney extract and from sheep testicular extract.
6. Latex rubber concentration.

Dielectrophoresis

Uncharged particles can be selectively separated using a non-uniform electric field. With reference to Fig. 33.3, uncharged particles placed in an electric field will be subjected to induced charge on each side according to orientation. If the field is non-uniform such as is present in the case of a pin-and-plate electrode pair shown, the species will migrate to the point of maximum field strength, i.e. to where the lines of electrostatic force are of the maximum intensity. Under these conditions the field polarity is inconsequential and the species will migrate in the same direction regardless. Thus an alternating field can be employed to achieve the separation effect.

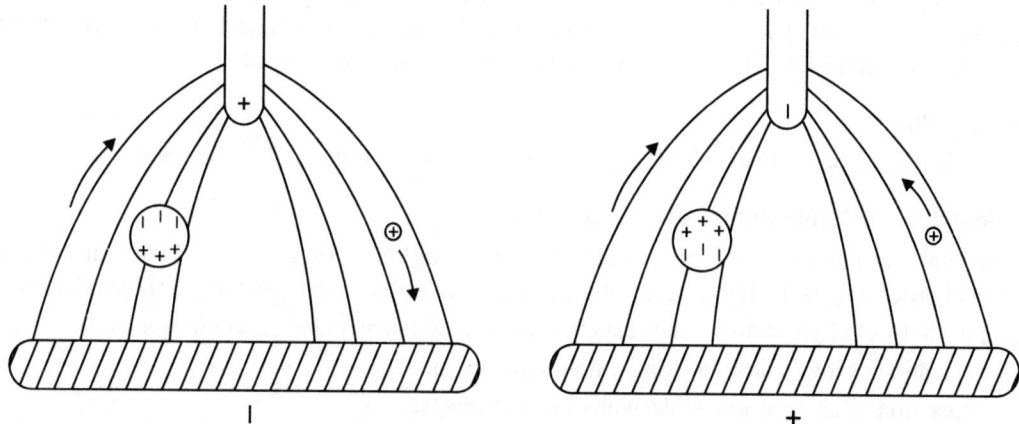

Fig. 33.3: Principle of dielectrophoretic separation.

OTHER ELECTROKINETIC PROCESSES

Electrodialysis

Electrodialysis has been developed widely for the treatment of brackish and saline waters as a route to potable water. The principle of the technique relies upon the ion selectivity of anionic and cationic ion-exchange membranes as a basis for selective separation. The driving force for the process is a current-driven ionic flux across a stack of compartments formed by alternate cationic and anionic ion-exchange membranes. Application of an electric field across the compartments results in electrokinetic transport of positively charged species through the cationic membranes and negatively charged species through the anionic membranes. Thus for a continuous flow system, at steady state, the effluents, from alternate compartments will comprise concentrated and depleted streams with respect to ionic species. Selective separation of mixtures of ionic species is possible and relies upon the differing ionic mobilities within the ion-exchange membranes.

The maximum ion removal rate is mass transfer limited by a phenomenon analogous to electrode polarisation which occurs in electrolysis. At high current densities, concentration gradients form adjacent to each membrane until the ionic concentration at the interface falls to zero.

Under these conditions water splitting may occur resulting in pH changes in both concentrated and dilute effluent streams which may cause membrane fouling. Other variations of conventional electrodialysis are also worthy of mention.

Ion replacement

Here membranes of only one charge are used.

Transport depletion

In this mode of operation the anionic membranes in the conventional stack are replaced by neutral membranes which allow transport of both cations and anions. Application of the electric field results in transport of cations towards the cathode through both sets of membranes. The anions, by contrast, only move through the neutral membranes and their flux is due to charge compensation in addition to electrokinetic transport. The main advantage of this technique for biochemical applications is that the use of anionic membranes which are particularly prone to fouling is avoided.

Ion substitution

Ion substitution involves three different types of compartment, configured in the stack.

Applications of Conventional Electrodialysis

The general advantage of conventional electrodialysis and the variations described with respect to biological processing is the high degree of separation achievable under relatively mild conditions.

Applications of electrodialysis and related techniques to biochemical separations include:

1. Separation of low molecular weight non-ionic products.
2. Separation of charged low molecular weight products.
3. Recovery and purification of reagents.

Recovery of non-ionic low molecular weight products frequently involves removal of ionic species. For example, desalting is important in processes such as recovery of proteins from whey where the removal of minerals is essential without despoilage of the product. The conversion of cheese whey involves the separation of milk proteins and uncharged lactose from salts.

The salts are separated by electrodialysis into a concentrated waste stream through membranes with low water transport properties. Whey is essentially the serum remaining from the cheese production process and contains lactose, lactalbumin, lactoglobulin and minerals. The protein is a potentially high-value product if separated from the salts. Compared with ion-exchange which may also be used for whey desalting, electrodialysis offers a number of advantages:

1. Lower protein denaturation.
2. Continuous processing.
3. Reduced losses of biodegradable material and lower effluent discharge.

Separation of charged low molecular weight species includes conventional deionisation processes around which the majority of commercial-scale electrodialysis processes have developed. Conventional deionisation processes include desalting, adjustment of pH, electrolytic conversion and waste treatment, and pretreatment processes (prior to electrophoresis for protein separation it is desirable to reduce the conductivity of the solution by removal of ionic salt species).

Electrodialysis can be used to concentrate salts and small charged species (of molecular weight less than 300) to very significant concentrations and thus achieve efficient separation of larger species and uncharged species. Ion-exchange membranes will normally reject products having molecular weights in excess of 500. A number of applications of interest to the biotechnologist are listed as follows:

1. pH adjustment of process streams can be achieved without increase in overall salt concentration. This can be done by introduction of an acid stream into the process stream followed by removal of other cations and anions by electrodialysis.

2. Ionic modification of biological solutions without change of pH can be achieved. This would have an application in the formulation of, for example, sodium-free diets, where a sodium-free stream is required. Sodium ions may be replaced by potassium or calcium without product denaturation using electrodialysis.

3. There are a variety of food processing applications of electrodialysis including deionisation of fruit juices, wine, milk, sugar molasses and whey. An example is in the demineralisation of skimmed milk. However, application to this process can be inhibited due to membrane fouling (particularly the anionic membranes) by proteins, although this can be minimised by using acetyl cellulose membranes in conjunction with the ion exchange membranes. The molecular weight of the acetyl cellulose is lower than that of the fouling proteins.

4. There is increasing interest in the use of electrodialysis as part of extractive fermentation processes. Here the overall conversion of substrate to product is enhanced, by continuous removal of the product, thus offsetting the effects of product inhibition on the micro-organisms. The mild processing conditions and the absence of any extractive reagents in the electrodialysis process are attractive in the context of fermentation systems. Examples of processes which could benefit from this approach include lactic acid production by glucose fermentation, and acetic acid production.

The desalting capability of electrodialysis can be used for the fractionation and separation of higher molecular weight species. The sensitivity of protein solubility to salt concentration is one major effect which may be exploited. For example in the case of plasma proteins, reduction in the salt concentration results in rapid precipitation of the non-albumin proteins.

Another application exploits the differing isoelectric points of amino acids to effect separation. The principle of the process is pH adjustment of a mixture of amino acids to ensure a significant difference in the charge held by the desired species compared with others present. Selective migration of the desired species towards the electrode of opposite polarity is achieved in the electrodialysis stack.

ELECTRICALLY ENHANCED LIQUID–LIQUID EXTRACTION

Liquid–liquid extraction offers a highly selective means of product separation for biochemical products. Operation at ambient temperature and a high degree of selectivity control offer considerable advantages when compared with other methods of separation, particularly in the context of high-volume products occurring in dilute solutions from fermentation. Conventional liquid–liquid extraction of untreated fermentation broth can be relatively inefficient due to general reductions in interfacial mass transfer rates which occur as a result of the accumulation of cell debris and other biological surfactant material at the liquid/liquid interface. A second difficulty is the high propensity for the liquid–liquid mixtures involved in the extraction to form poorly separating emulsions. This can result in high entrainment levels and large losses in overall process efficiency.

The operation of liquid–liquid extraction processes in the presence of electric fields results in several fundamental advantages which have the potential to ameliorate some of the worst problems of whole broth processing. Production of a dispersion of aqueous broth at a charged nozzle in contact with a continuum of low conductivity extractant can result in the production of charged droplets. An electric field may be readily imposed across the continuum of solvent through which the charged droplets are then accelerated by the resulting electrostatic force acting upon them.

The mass transfer rate is enhanced due to the following effects:

1. Increased droplet slip velocity results in greater interfacial shear and reduction in the continuous phase mass transfer resistance.

2. Increased droplet oscillation results in reduction in the dispersed phase mass transfer resistance.

3. Increased interfacial area for transfer results due to the smaller droplet sizes achieved at the inlet nozzle.

References

Adams Martin, R., *Fermentation and Food Safety*, Springer, US.

Arceivala, K.J., *Media for Industrial Fermentations*, Marcel Dekker Inc., New York.

Benaim Pinto, C., *Microbial Biomass and Economic Microbiology*, Prentice-Hall, London.

Budyko, M.I., *Bioactive Microbial Products,* Van Nostrand Reinhold, New York.

Cambell, K.E. and Lemer, H.A., *Industrial Fermentation*, Academic Press, London.

Chang, J.C., *Biochemical Engineering and Biotechnology Handbook*, John Wiley & Sons, New York.

Charles, W.B, *Food Fermentation and Micro-organism*, Wiley-Blackwell, New York.

Downe, S.A., *Industrial Microbiology and Biotechnology*, John Wiley & Sons, New York.

Dugan, P.R., *Animal Cell Biotechnology*, Plenum Publishing Corporation, London.

Goldman, M., *Economic Analysis of Fermentation Process*, Gordon and Breach, Science Publishers, New York.

Gould, G.W., *Textbook of Microbiology*, Van Nostrand Reinhold, New York.

Halady, M.K., *Innovation Technologies for Fermented Food Industries*, Springer, US.

Harding, G.L., *Fundamentals of Biochemistry*, Prentice-Hall, London.

Jarvis, B.N., *Biological Methods for Waste Treatment*, John Wiley & Sons, New York.

Jencks, W.P., *Biological Treatment of Aqueous Wastes*, Academic Press, London.

Karp, M.C., *Bioseparation Techniques*, Marcel Dekker, New York.

Lewis, R.L., *Industrial Processing with Membranes*, Chilton Book Co., USA.

McCaull, J. and Crossland, J., *Molecular Biology and Biotechnology*, Harcourt Brace Jovanovich, New York.

Montet Didier, *Fermented Foods*, CRC Press, New York

Nyer, E.K., *Cell Biology*, Van Nostrand Reinhold, New York.

Odum, S.K., *Down Stream Processing in Biotechnology*, Van Nostrand Reinhold, New York.

Phillips, D.J.H., *Effluent Treatments in Biochemical Industries*, Applied Science Publishers, London.

Reid, G.K., *Modelling and Control of Fermentation Processes*, Academic Press, London.

Samson, R.A., *Instruments for Monitoring and Controlling Reactors*, Pergamon Press, Oxford, London.

Soccol, R.C., *Fermentation biotechnology*, CRC Press, New York.

Teal, J.M., *Computers in Fermentation Technology*, Pergamon Press, New York.

Vikas Mishra, *Fermentation Microbiology and Biotechnology*, Academic Press, London.

Vollenweider, R.A., *Scale-up Methodology for Chemical Principles*, Blackwell Scientific Publications, New York.

Wyatt, G.M., *Fermentation and Enzyme Technology*, Reston Publishing Co., Reston, Virginia.

Yu, P.L., *Fermentation Technologies: Industrial Application*, Springer, Netherland.

Index